**Hydrogen as a
Future Energy Carrier**

*Edited by
Andreas Züttel
Andreas Borgschulte
Louis Schlapbach*

Further Reading

B. Elvers (Ed.)

Handbook of Fuels

Energy Sources for Transportation

2007
ISBN 978-3-527-30740-1

G. A. Olah, A. Goeppert, G. K. S. Prakash

Beyond Oil and Gas: The Methanol Economy

2006
ISBN 978-3-527-31275-7

K. Sundmacher, A. Kienle, H. J. Pesch,
J. F. Berndt, G. Huppmann (Eds.)

Molten Carbonate Fuel Cells

Modeling, Analysis, Simulation, and Control

2007
ISBN 978-3-527-31474-4

Hydrogen as a Future Energy Carrier

Edited by
Andreas Züttel, Andreas Borgschulte,
and Louis Schlapbach

WILEY-VCH Verlag GmbH & Co. KGaA

The Editors

Prof. Dr. Andreas Züttel
Empa
(Swiss Federal Lab for Materials
Research and Testing,
an ETH Domain Institute)
CH-8600 Dübendorf
Switzerland

Dr. Andreas Borgschulte
Empa
(Swiss Federal Lab for Materials
Research and Testing,
an ETH Domain Institute)
CH-8600 Dübendorf
Switzerland

Prof. Dr. Louis Schlapbach
Empa
(Swiss Federal Lab for Materials
Research and Testing,
an ETH Domain Institute)
CH-8600 Dübendorf
Switzerland

All books published by **Wiley-VCH** are carefully produced. Nevertheless, authors, editors, and publisher do not warrant the information contained in these books, including this book, to be free of errors. Readers are advised to keep in mind that statements, data, illustrations, procedural details or other items may inadvertently be inaccurate.

Library of Congress Card No.: applied for

British Library Cataloguing-in-Publication Data
A catalogue record for this book is available from the British Library.

Bibliographic information published by the Deutsche Nationalbibliothek
Die Deutsche Nationalbibliothek lists this publication in the Deutsche Nationalbibliografie; detailed bibliographic data are available in the Internet at http://dnb.d-nb.de

© 2008 WILEY-VCH Verlag GmbH & Co. KGaA, Weinheim

All rights reserved (including those of translation into other languages). No part of this book may be reproduced in any form – by photoprinting, microfilm, or any other means – nor transmitted or translated into a machine language without written permission from the publishers. Registered names, trademarks, etc. used in this book, even when not specifically marked as such, are not to be considered unprotected by law.

Cover Design   Grafik-Design Schulz, Fußgönheim
Typesetting   Aptara Inc., New Delhi, India
Printing   betz-druck GmbH, Darmstadt
Bookbinding   Litges & Dopf GmbH, Heppenheim

Printed in the Federal Republic of Germany
Printed on acid-free paper

ISBN: 978-3-527-30817-0

Contents

Preface IX

List of Contributors XI

1 Introduction 1
Andreas Züttel
References 6

2 History of Hydrogen 7
Andreas Züttel, Louis Schlapbach, and Andreas Borgschulte
2.1 Timeline of the History of Hydrogen 7
2.2 The Hindenburg and Challenger Disasters 14
References 20

3 Hydrogen as a Fuel 23
3.1 Fossil Fuels 23
Sonja Studer
3.2 The Carbon Cycle and Biomass Energy 37
Samuel Stucki
3.3 The Hydrogen Cycle 43
John D. Speight
References 68

4 Properties of Hydrogen 71
4.1 Hydrogen Gas 71
Andreas Züttel
4.2 Interaction of Hydrogen with Solid Surfaces 94
Andreas Borgschulte and Louis Schlapbach
4.3 Catalysis of Hydrogen Dissociation and Recombination 108
Ib Chorkendorff
4.4 The Four States of Hydrogen and Their Characteristics and Properties 125
Seijirau Suda

4.5	Surface Engineering of Hydrides 132

Seijirau Suda
References   139

5 Hydrogen Production   149

5.1	Hydrogen Production from Coal and Hydrocarbons 149

Andreas Borgschulte and Andreas Züttel

5.2	Electrolysis: Hydrogen Production Using Electricity 155

Ursula Wittstadt
References   163

6 Hydrogen Storage   165

6.1	Hydrogen Storage in Molecular Form 165

Andreas Züttel

6.2	Hydrogen Adsorption (Carbon, Zeolites, Nanocubes) 173

Michael Hirscher and Barbara Panella

6.3	Metal Hydrides 188

Andreas Züttel

6.4	Complex Transition Metal Hydrides 195

Klaus Yvon

6.5	Tetrahydroborates as a Non-transition Metal Hydrides 203

Shin-ichi Orimo and Andreas Züttel

6.6	Complex Hydrides 211

Borislav Bogdanović, Michael Felderhoff, and Ferdi Schüth,

6.7	Storage in Organic Hydrides 237

Andreas Borgschulte, Sandra Goetze, and Andreas Züttel

6.8	Indirect Hydrogen Storage via Metals and Complexes Using Exhaust Water 244

Seijirau Suda and Michael T. Kelly
References   256

7 Hydrogen Functionalized Materials   265

7.1	Magnetic Heterostructures – A Playground for Hydrogen 265

Arndt Remhof and Björgvin Hjörvarsson

7.2	Optical Properties of Metal Hydrides: Switchable Mirrors 275

Ronald Griessen, Ingrid Anna Maria Elisabeth Giebels, and Bernard Dam
References   327

8 Applications   335

8.1	Fuel Cells Using Hydrogen 335

K. Andreas Friedrich and Felix N. Büchi

8.2	Borohydride Fuel Cells 364

Zhou Peng Li

8.3	Internal Combustion Engines *371*	
	Gerrit Kiesgen, Dirk Christian Leinhos, and Hermann Sebastian Rottengruber	
8.4	Hydrogen in Space Applications *381*	
	Robert C. Bowman Jr. and Bugga V. Ratnakumar	
	References *410*	

Index *415*

Preface

Thermodynamically speaking a human being, like any animal, is an engine combusting an energy carrier like food into work. In other words, the difference between living matter and dead matter is the ability to convert energy. Therefore, energy is essential for our existence and development. Natural processes led to a spontaneous accumulation of carbon- and hydrogen-based energy carriers – the so-called fossil fuels. A large amount of fossil fuels was given by nature, which allowed us to start the industrial age and to develop enormously over the last century. However, we are now starting to realize that these reserves of stored solar energy are finite and that the combustion of fossil fuels leads to climate change. No matter how much fossil fuel is left and who consumes most of it we have to develop a world sustainable energy economy for the future. Today most of our energy demand is covered by energy carriers; however, all renewable energy sources deliver energy fluxes. Therefore, a synthetic energy carrier is needed. At present, the only energy carrier that can be synthesized efficiently without materials limitations and in real-time is hydrogen. Furthermore, the combustion of hydrogen leads to the release of water to the atmosphere and the cycle is closed naturally.

Thus, this book starts with a short description of the history of the energy cycle from biomass to the fossil fuels and finally the outlook to the future hydrogen-based energy cycle. The provision of energy by depletion of natural carbon sources should be abandoned, because the burning of carbon results in a massive release of the greenhouse gas CO_2 into the atmosphere. Therefore, carbon has to be avoided in future energy carriers. Hydrogen is a renewable energy carrier, produced from water and a renewable energy source, e.g. solar energy. Furthermore, if a future society is able to operate fusion reactors and produce heat and electricity from this nearly unlimited energy source an energy carrier like hydrogen will be needed in order to store and transport the energy.

This book is a comprehensive overview of the basic knowledge about hydrogen and the opinions about the potential and future development of hydrogen science and technology. The primary focus of the book is the description of the technology and science of the hydrogen cycle, including hydrogen production, hydrogen storage and hydrogen conversion. In addition, it contains chapters on new applications using hydrogen, e.g. in functionalized materials. The oldest hydrogen-based application is nearly 200 years old and thus the book contains an interesting survey of the history of hydrogen.

The publication of the assessment report of the intergovernmental panel on climate change in 2006 triggered awareness of the need for a change in the current energy policy. Accordingly, the recent progress in research is enormous, and can hardly be summarized in one book. Still, this book can serve as a basis for future research, and is a short introduction to the most important challenge of the 21st century. We thank the authors for their valuable contributions which have allowed the realization of this book.

Dübendorf, January 2008

Andreas Züttel
Andreas Borgschulte
Louis Schlapbach

List of Contributors

Boris Bogdanović
Max-Planck Institut für
Kohlenforschung
Kaiser-Wilhelm-Platz 1
45470 Mülheim an der Ruhr
Germany

Andreas Borgschulte
Empa
(Swiss Federal Lab for
Materials Research and Testing,
an ETH Domain Institute)
Laboratory for Hydrogen & Energy
8600 Dübendorf
Switzerland

Robert C. Bowman Jr
Jet Propulsion Laboratory
California Institute of Technology
Pasadena, CA 91109
USA

Felix N. Büchi
Electrochemistry Laboratory
Paul Scherrer Institut
5232 Villigen PSI
Switzerland

Ib Chorkendorff
Technical University of Denmark
Danish National Research
Foundation's
Center for Individual Nanoparticle
Functionality (CINF)
Department of Physics
Building 312
2800 Kongens Lyngby
Denmark

Bernard Dam
VU University Amsterdam
Faculty of Sciences
Department of Physics
De Boelelaan 1081
1081 HV Amsterdam
The Netherlands

Michael Felderhoff
Max-Planck Institut für
Kohlenforschung
Kaiser-Wilhelm-Platz 1
45470 Mülheim an der Ruhr
Germany

K. Andreas Friedrich
Deutsches Zentrum für Luft- und
Raumfahrt (DLR)
Institut für Technische
Thermodynamik
Elektrochemische Energietechmik
Pfaffenwaldring 38–40
70569 Stuttgart
Germany

Ingrid A.M.E. Giebels
VU University Amsterdam
Faculty of Sciences
Department of Physics
De Boelelaan 1081
1081 HV Amsterdam
The Netherlands

Sandra Götze
University of Amsterdam
Swammerdam Institute for
Life Sciences
BioCentrum Amsterdam
Kruislaan 318
1098SM Amsterdam
The Netherlands

Ronald Griessen
VU University Amsterdam
Faculty of Sciences
Department of Physics
De Boelelaan 1081
1081 HV Amsterdam
The Netherlands

Michael Hirscher
Max-Planck-Institut für
Metallforschung
Heisenbergstrasse 3
70569 Stuttgart
Germany

Björgvin Hjörvarsson
Uppsala University
Department of Physics
Box 530
751 21 Uppsala
Sweden

Michael T. Kelly
Millennium Cell, Inc.
One Industrial Way West
Eatontown, NJ 07724
USA

Gerrit Kiesgen
BMW AG
Development Powertrain
80788 München
Germany

Dirk Christian Leinhos
BMW AG
Development Powertrain
80788 München
Germany

Zhou Peng Li
Zhejiang University
College of Materials Science &
Chemical Engineering
Department of Chemical and
Biochemical Engineering
Zheda Road 38
Hangzhou 310027
P.R. China

Shin-Ichi Orimo
Tohoku University
Institute for Materials Research
Sendai 908-8577
Japan

List of Contributors

Bugga V. Ratnakumar
Jet Propulsion Laboratory
California Institute of Technology
Pasadena, CA 91109
USA

Arndt Remhof
Empa
Materials Science & Technology
Division Hydrogen & Energy
Überlandstrasse 129
8600 Dübendorf
Switzerland

Hermann Sebastian Rottengruber
BMW AG
Development Powertrain
80788 München
Germany

Louis Schlapbach
Empa
(Swiss Federal Lab for
Materials Research and Testing,
an ETH Domain Institute)
Laboratory for Hydrogen & Energy
8600 Dübendorf
Switzerland

Ferdi Schüth
Max-Planck Institut für
Kohlenforschung
Kaiser-Wilhelm-Platz 1
45470 Mülheim an der Ruhr
Germany

John D. Speight
University of Birmingham
Department of Metallurgy and
Materials
Pritchatts Road
Birmingham B15 2TT
United Kingdom

Samuel Stucki
Paul Scherrer Institut
Labor für Energie und Stoffkreisläufe
5232 Villigen PSI
Switzerland

Sonja Studer
Erdöl-Vereinigung
(Swiss Petroleum Association)
Head Transport Fuels and
Environment
Löwenstrasse 25
8001 Zürich
Switzerland

Seijirau Suda
Kogakuin University
Department of Environmental and
Chemical Engineering
2665-1, Nakano-Machi
Hachioji-shi
Tokyo 192-0015
Japan

Ursula Wittstadt
Fraunhofer Institut für Solare
Energiesysteme ISE
Department Thermal Systems and
Buildings
Heidenhofstrasse 2
79110 Freiburg
Germany

Klaus Yvon
University of Geneva
Laboratoire de Cristallographie
24 Quai Ernest Ansermet
1211 Genève 4
Switzerland

Andreas Züttel
Empa (Swiss Federal Laboratories for
Materials Research and Testing, an
ETH Domain Institute)
Laboratory for Hydrogen & Energy
8600 Dübendorf
Switzerland

and

University of Fribourg
Physics Department
Condensed Matter Physics
Pérolles
1700 Fribourg
Switzerland

and

Vrije Universiteit
Faculty of Science
Division of Physics and Astronomy
De Boelelaan 1081
1081 HV Amsterdam
The Netherlands

1
Introduction
Andreas Züttel

1.1
Introduction

Human beings developed on Earth on the basis of plants, that is biomass, as the only energy carrier. The average power consumed by a human body at rest is 0.1 kW and approximately 0.4 kW for a hard working body, delivering about 0.1 kW of work. The consumption of plants by humans and animals did not change the atmosphere because the carbon dioxide liberated by humans and animals was reabsorbed by the plants in the photosynthesis process. The only mechanical work available from nonliving systems were the windmills and the waterwheels, where solar energy was converted into mechanical power.

With the discovery of the steam engine in 1712 by Thomas Newcomen [1] humanity had for the first time a nonliving machine available, consuming carbon or hydrocarbons and delivering mechanical power on demand. This initialized the industrialization process and thereby changed society completely, in particular the demand for more and more energy. The energy for the steam engine was found in the form of mineral coal, solar energy stored in the Earth's crust over millions of years.

Coal as a solid energy carrier was later complemented by liquid crude oil and natural gas. Not only did the state of the energy carrier change with time, from solid to liquid and finally gas, but also the amount of hydrogen in the fuel increased from zero to four hydrogen atoms per carbon atom.

The world energy consumption increased from 5×10^{12} kWh/year in 1860 to 1.2×10^{14} kWh/year today. More than 80% is based on fossil fuels, such as coal, oil and gas [2].

The population of human beings has increased in the last century by a factor of 4 and the energy demand by a factor of 24. The world wide average continuous power demand is 2 kW/capita. In the USA the power consumption is on average 10 kW/capita and in Europe about 5 kW/capita [3], while two billion people on Earth do not yet consume any fossil fuels at all. However, the reserves of, for example crude oil on Earth are limited and predictions based on extrapolation of the energy consumption show that the demand will soon exceed the supply [4]. The reserves

Hydrogen as a Future Energy Carrier. Edited by A. Züttel, A. Borgschulte, and L. Schlapbach
Copyright © 2008 WILEY-VCH Verlag GmbH & Co. KGaA, Weinheim
ISBN: 978-3-527-30817-0

of other fossil fuels, for example coal, are larger [3]. Still, no matter how long the fossil fuels will last their amount is finite. The demand of fossil fuels has also a strong impact on social, political and economic interactions between the various countries. For example two thirds of the crude oil reserves are located in the Near East region, but most of the petrol is consumed in the USA, Europe and Japan. The transportation over long distances by, for example pipelines or tank ships, causes signifianct damage to the environment [4]. The imbalance of the distribution of resources is one of the driving forces of political instability leading to war and terrorism. This is the first important fact forcing the world to search for an energy solution not depending on fossil fuels but on unlimited renewable energy.

The consumption of fossil fuels together with deforestation leads to the liberation of 7×10^{12} kgC/year in the form of CO_2. The plants are able to absorb 2×10^{12} kg/year by means of the photosynthesis process and the same amount is dissolved in the ocean(see Fig. 1.1). Therefore the net increase in CO_2 in the atmosphere due to human activities is approximately 3×10^{12} kg/year [5]. This corresponds to an annual increase of 0.4% of the CO_2 concentration in the atmosphere. The CO_2 concentration in the atmosphere is continuously measured at the Mauna Loa observatory in Hawaii. Due to the seasonal variation of the solar intensity the CO_2 concentration shows oscillations. The plants absorb more CO_2 in the growth period, the summer time, than during the winter. However, plants are not able to absorb all the additionally liberated CO_2 in real time. Carbon dioxide is a greenhouse gas and causes an increase in the average temperature on Earth. A careful investigation of the climate and atmospheric history of the past 420 000 years from the Vostok ice core in Antarctica has shown that the variations in the temperature on Earth correlates with the variations in the concentration of greenhouse gases in the atmosphere. This correlation, together with the uniquely elevated

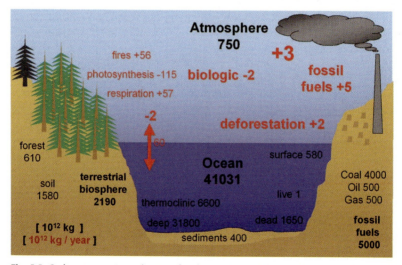

Fig. 1.1 Carbon reservoirs and rates of exchange. From (Climate Change (2001), Ref. [5]).

concentrations of CO_2 today, is of great relevance with respect to the continuing debate on the future of the Earth's climate.

Currently we are in a period of decreasing solar activity and therefore the average surface temperature on Earth should decrease. However, for 100 years the temperature on Earth has been increasing. Furthermore, for the last 10 years an unusually steep increase in the average temperature versus time has been observed. This is the second important reason for the necessity to develop a new energy carrier which is free of carbon.

Fossil energy has a very strong influence on the development of a society and the standard of living. A dependence of the gross national product on the average amount of energy consumed per person can be found. Countries with low energy consumption per person have a low gross national product and vice versa. This led to the conclusion, that today the economic gain of the industrialized world is to a significant part due to the energy from fossil fuels.

Fossil fuels are energy carriers given to human beings by nature. Basically we know two types of energy carriers. The first is a reversible system which can be charged with energy in the form of mechanical or electrical work and delivers the energy again on demand. Examples of such a system are capacitors, batteries and flywheels. We can define a charge and discharge efficiency and the overall efficiency of the system is the product of all partial efficiencies. The energy density in such systems is given by the physical properties of the active material and is, in general, limited to 1 to 2 eV per atom. The second system stores energy by means of the reduction of a compound, liberating oxygen to the atmosphere and producing a semi-stable product, for example photosynthesis ($6CO_2 + 6H_2O \rightarrow C_6H_{12}O_6 + 6O_2$) or ($2ZnO \rightarrow Zn + O_2$). The product can be reacted with oxygen again, liberating energy and leading to volatile or nonvolatile compounds. If the compound is volatile a natural process in the atmosphere should transport it. The advantage of the second system over the first is the natural transport and the storage of the oxygen and the volatile products in the atmosphere. Therefore, rather high-energy storage densities are possible and can, in many cases, exceed the storage densities of the first system. But more processes are involved and this limits the overall energy efficiency of the system.

The natural cycles in the atmosphere are able to transport only a few compounds namely oxygen, nitrogen, water and carbon dioxide. Many more gases could of course be transported; however, they would significantly affect life on Earth.

Considering the historical development of energy carriers towards more hydrogen rich fuels and the necessity to avoid the carbon dioxide emission that causes the greenhouse effect, one concludes that hydrogen should be the energy carrier of the future (see Fig. 1.2)

$$C \text{ (coal)} \rightarrow -CH_2- \text{ (oil)} \rightarrow CH_4 \text{ (natural gas)} \rightarrow H_2 \text{ (hydrogen)}$$

This series also shows a development from a solid to a liquid and then finally to a gaseous state energy carrier. Hydrogen occurs on Earth chemically bound as H_2O in water and some is bound to liquid or gaseous hydrocarbons. The production of hydrogen by means of electrolysis consumes electricity, which is physical

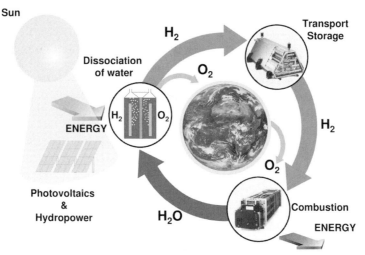

Fig. 1.2 In the hydrogen cycle, solar energy and process of water electrolysis are used to generate, store and combust hydrogen with oxygen, thus returning it to water.

work. Therefore, a primary energy source is necessary to produce hydrogen. In contrast to the biomass produced by all plants using sunlight, there are only a few natural processes leading directly to hydrogen gas. The fundamental question to be answered is: what are the possible sources of energy for a hydrogen-based society.

The answer today is nuclear fission or nuclear fusion. Human controlled nuclear fission is limited by the mining of the materials used in nuclear reactors and the handling of the products of the fission reaction. Furthermore, the reserves of highly concentrated fission materials in the Earth's crust are finite and would not allow production of the world energy demand for more than a century. Natural fission in the Earth leads to a reasonable amount of heat. This heat is the source of geothermal power used in suitable regions, like Iceland, as a primary energy source. However, estimates of the potential amount of energy from geothermal sources show that they cover only a small percentage of the actual worldwide energy demand [6].

On a geological timescale the only source of primary energy is the nuclear fusion of hydrogen. The two options are terrestrial fusion or extraterrestrial fusion on the sun. The terrestrial fusion is human controlled and could deliver the energy in a centralized and concentrated form at a high temperature. While the principle of the release of an enormous amount of energy is realized in nuclear H-bombs, the peaceful use of fusion as a primary energy source has not been achieved yet due to massive technical problems. The Sun on the other hand delivers fusion energy over the whole surface of the Earth in a continuously oscillating way with a frequency of 24 h on a base frequency of one year. The energy arrives on Earth with a rather low average intensity of 165 W m^{-2}. The challenge is to convert and concentrate the energy by means of hydropower, windpower, solar-thermal, photovoltaics or even biomass. The conversion via photovoltaics, for example can be estimated as follows: The solar constant is $1.369\,\text{kW}\,\text{m}^{-2}$ and approximately 50 % of the solar radiation reaches the surface of the Earth. Photovoltaic systems have an efficiency of approximately 10 %. In the best case, half of the time is night and, therefore, under ideal conditions about $473\,000\,\text{km}^2$ ($80\,\text{m}^2$/capita) covered with photovoltaic cells are necessary to produce the world energy demand of today. This number can be compared to the settlement area per capita in Switzerland, which is $397\,\text{m}^2$/capita. The total area corresponds to a square with a side of length 700 km located in northern Africa, for example.

Energy from sunlight is converted into electricity, for example by means of photovoltaic cells. Electricity from a renewable energy source is used for the electrolysis of water. The oxygen is released into the atmosphere and hydrogen is stored, transported and distributed. Finally, hydrogen, together with the oxygen, is combusted and the energy is released as heat and work leaving water or steam for release into the atmosphere. Therewith the hydrogen cycle is closed.

Electrolysis of water is an established technology and is used today to produce high purity hydrogen. Electrolysis at ambient temperature and pressure requires a minimum voltage of 1.481 V and therefore, a minimum energy of $39.4\,\text{kWh}\,\text{kg}^{-1}$ hydrogen. Today electrolyser systems consume approximately $47\,\text{kWh}\,\text{kg}^{-1}$ hydrogen, that is the efficiency is approximately 82 % [7]. The only impurities in the hydrogen gas from electrolysis are water and oxygen. The chemical energy per mass of hydrogen ($39.4\,\text{kWh}\,\text{kg}^{-1}$) is three times larger than that of other chemical fuels, for example liquid hydrocarbons ($13.1\,\text{kWh}\,\text{kg}^{-1}$). In other words, the energy content of 0.33 kg of hydrogen corresponds to the energy content of 1 kg of oil. There is, however, a technical and an economic challenge to overcome before the hydrogen energy economy becomes reality.

The technical challenge is real time production, the safe and convenient storage and the efficient combustion of hydrogen. In order to satisfy the world demand for fossil fuels, more than 3×10^{12} kg hydrogen will have to be produced per year. This is roughly 100 times the current hydrogen production. The number of single processes (hydrogen production, conditioning, distribution, transportation, conversion) reduces the overall energy efficiency; and the installation of the corresponding devices requires an enormous amount of energy to be spent in advance, coupled to a large financial effort.

The economic challenge is the cost of the hydrogen production. The world economy today is based on free energy naturally stored over millions of years. The price we are used to paying for fossil fuels is only the mining costs. In order to adapt the world to a synthetic fuel like hydrogen the world economy has to be convinced to pay also for the conversion of solar energy into a fuel.

This book is a comprehensive review of the fundamentals of such a hydrogen economy. Chapter 2 is devoted to an outline of the history of hydrogen and relevant events which coined the overstated concerns about the safety of hydrogen. In order to underline the need for a new energy strategy, Chapter 3 is a precise analysis of the world's energy status and the currently used fossil fuels. The discussions reflect the various opinions on this subject, in particular how a sustainable future should be developed. Hydrogen is, apart from its use as an energy carrier, one of the most versatile elements and is frequently taken as a physical model system. Therefore, Chapter 4 addresses the properties of hydrogen, ranging from the gas over its various chemical states to interactions of hydrogen with matter. The previous discussion highlighted the technical challenges of a hydrogen economy. These are discussed in depth in Chapter 5 (hydrogen production) and Chapter 6 (hydrogen storage). The field of potential applications of hydrogen has developed enormously over the last ten years. Some of these applications are described in Chapters 7 and 8. Chapter 7 emphasizes thin film applications, while Chapter 8 concentrates on technical realizations in mobile applications and even in space.

References

1 Encyclopaedia Britannica (2001) *Incorporated*, 15th edn; Wilson, S.S. (1981) *Spektrum der Wissenschaft*, pp. 99–109 (Okt. 1981).
2 Martin-Amouroux, J.-M. (2003) IEPE, Grenoble, France, personal communication; BP Statistical Review of World Energy (2004) United Kingdom.
3 International Energy Agency (IEA, AIE) (2002) *Key World Energy Statistics*, 2002 edn.
4 The subject of oil depletion has always been discussed very controversially. Colin J. Campbell forecasted the peak in conventional oil production to occur around 2010. Campbell, C.J. (1997) *The Coming oil Crisis*, Multi-Science Publishing Co. Ltd., Brentwood. See also   Hall, C., et al. (2003) Hydrocarbons and the evolution of human culture. *Nature*, **426**, 318–22; http://www.lifeaftertheoilcrash.net/; and Ref. 3. A contrary opinion is given in Chapter 4.4.
5 Watson, R.T. (ed.) (2001) *Climate Change 2001: Synthesis Report*, Cambridge University Press, Cambridge, UK. Published for the Intergovernmental Panel on Climate Change (IPCC).
6 Clotworthy, A. (2000) *Proceedings World Geothermal Congress*, http://en.wikipedia.org/wiki/Geothermal_energy#_note-heat.
7 Häussinger, P., Lohmüller, R., Watsin, A.M. *Ullmann's Encyclopedia of Industrial Chemistry, Chap.: Hydrogen*, 5th, Completely Revised Edition, vol. A13, p. 333.

2
History of Hydrogen
Andreas Züttel, Louis Schlapbach, and Andreas Borgschulte

The history of hydrogen is linked to great scientists and engineers, fascinating discoveries and technical breakthroughs but also to throwbacks and tragedies. Hydrogen technology has become a keystone for daily life. For example, hydrogen in metals is used in batteries, sensors, ferromagnets, switchable mirrors, and heat pumps; and it is responsible for embrittlement causing corrosion in metallic materials. Hydrogen appears mainly in the form of hydrocarbons and is widely used in chemical engineering, in particular in the petrochemical industry. Hydrogen is involved in the processes of life: for example in photosynthesis and energy conversion in living cells [1]. Still, hydrogen has a bad reputation. A questionnaire in Munich in 1998 brought to the fore that 12.9 % of those interrogated associated "hydrogen" with "hydrogen bomb", followed by nonspecified dangers (risk of explosion, etc.) [2]. While some of the connotations originate from missing scientific knowledge, predominantly tragic events coined the image of hydrogen in the population. This chapter consists of two sections: the first gives a comprehensive overview of the scientific and technological developments connected to hydrogen, whereas the second focuses on two tragic catastrophies and their precise analysis.

2.1
Timeline of the History of Hydrogen

Jan Baptista van Helmont (1577 to 1644) was one of the first to reject the basic elements of *Aristoteles*. He discovered that air is not an element and that another "air" with different properties exists. He called it, based on the Greek word "chaos" which means "empty space", according to the Dutch spelling "gas".

In the Middle Ages *Paracelsus* (1493 to 1541) is reported to have noted that a gas is yielded when iron is dissolved in "spirit of vitriole". *Turquet De Mayerne* (1573–1655) noted that this gas was inflammable. However, hydrogen was first separated and identified in the second half of the eighteenth century; *Robert Boyle* (1627 to 1691) produced "facticious air" from diluted sulfuric acid and iron. He showed that facticious air only burns when air is present and that a part of the air disappears during the reaction. He also noticed that the combustion products

Hydrogen as a Future Energy Carrier. Edited by A. Züttel, A. Borgschulte, and L. Schlapbach
Copyright © 2008 WILEY-VCH Verlag GmbH & Co. KGaA, Weinheim
ISBN: 978-3-527-30817-0

are heavier than the starting material. Therefore, he also rejected the elements of Aristotoles but his findings were not adequately acknowledged. The German chemist *Georg Ernst Stahl* (1659 to 1734) was the doctor of King Friedrich Wilhelm of Preussen and published in 1697 the phlogiston theory. According to this theory all inflammable materials contain phlogiston, a hypothetical substance, which is liberated during combustion.

According to the phlogiston theory (Greek: phlogistos, burned) a material is more flammable and burns more violently the larger the content of phlogiston. For the first time it was possible to describe several chemical reactions by means of the phlogiston theory. Lead, for example is composed of lead oxide and phlogiston and the phlogiston is liberated during combustion leaving behind lead oxide.

The phlogiston theory was very useful and was therefore still defended when an obvious weakness of the theory appeared and an arbitrary assumption was necessary. After Boyle's observation that the mass of a substance increases during combustion, for example, the mass of lead oxide is greater than the mass of lead, a negative mass was attributed to phlogiston.

Water (H_2O) was considered in all theories to be a basic element in the sense of Aristotoles. Still, in 1756, the Scot *Joseph Black* saw gases as a form of the element air, therefore he called the gas, known today as carbon dioxide, quick air because it reacts with magnesium to give magnesium carbonate.

Henry Cavendish (1731–1810) proved that there were different types of air, one of which was "inflammable air" and that a number of metals, when dissolved in acid, produced various amounts of this gas. He assumed that the metal was the source of the inflamable air. This was wrong, however, in accordance with the phlogiston theory. He published [3] in 1766 precise values for the specific weight and density. During the late 1770s, he performed experiments with electrical discharges in a hydrogen–oxygen mixture, thereby producing water. Cavendish's discovery stimulated the search for new gases and the Swedish scientist *Carl Wilhelm Scheele* and the English scientist *Joseph Priestley* independently found a gas which is a component of the air, this was called fireair. 1781 Cavendish burnt his inflammable air with the fireair and obtained nothing else but water.

On 5th June 1783 the Montgolfier Brothers gave the first public demonstration of a model hot-air balloon and in September – in the presence of King Louis XVI and Marie Antoinette – they flew a balloon carrying a sheep, a duck and a cockerel to demonstrate that it was possible to survive in the sky. Some weeks later Pilatre de Rozier, a science teacher, and the Marquis d'Arlandes, an infantry officer, became the first human air travellers when, in a hot-air balloon, they flew for 9 km (5.5 miles) over Paris. *Jacques Alexandre César Charles* (1746–1823) realized that hydrogen was lighter than air, he built the first balloon made of paper and filled with 25 m^3 hydrogen gas and, on 27th August 1783, the balloon ascended to a height of nearly 914 m. The hydrogen was produced by the reaction of iron with sulfuric acid. Upon landing outside Paris, it was destroyed by terrified peasants. On December 1st, 1783, he, along with Ainé Roberts, ascended to a height of 549 m in the newly constructed balloon "La Charlière". Charles is best known for his formulation in 1787 of one of the basic gas laws, known as Charles's law, which

Fig. 2.1 Antoine Laurent Lavoisier (1743–1794) with his wife and secretary Marie-Anne Paulze (1758–1836), Painting by Louis David [4].

states that, at constant pressure, the volume occupied by a fixed weight of gas is directly proportional to the absolute temperature.

Soon after, the French scientist *Antoine Lavoisier* (1743–1794) (Fig. 2.1) confirmed Cavendish's experiments and wrote: "It seams that the phenomena can be explained without the use of phlogiston [5]." According to Lavoisier's antiphlogistic theory it was not the properties of matter that were seen as elements but rather chemical elements were substanses with specific properties. From todays view phlogiston was the energy (heat) which is liberated during combustion.

At the same time, a group of French chemists started to introduce a new nomenclature in chemistry. Especially misleading names were replaced by Greek words, which described the most obvious property of a substance.

In 1787 Lavoisier presented the following proposal, formulated by this group of chemists, to the Academy of Science in Paris [6]. The fire air should be called "oxygene" from the Greek words "oxys" for acidic and "genes" for genesis, because Lavoisier assumed that oxygen was the reason for the acidity. For the inflammable air the word *"hydrogene"* was proposed, based on the Greek word "hydor" meaning water former.

For use as a fuel, a property of hydrogen, that is of even greater importance than flammability is the large amount of energy released during combustion. Lavoisier and Pierre Laplace measured the heat of combustion of hydrogen in 1783–1784 using an ice calorimeter. The experiment took 11.5 h and the amount of ice melted was equivalent to about 9.7×10^7 J per kg of hydrogen. This was much higher than

Fig. 2.2 The Döbereiner Platinum lighter, from Ref. [7].

values obtained for other substances and, whether for this reason or because of other uncertainties, the results were not published until 1793. The Lavoisier–Laplace value was not too far off the correct value of 1.20×10^8 J per kg hydrogen.

Nicholson and *Carlisle* in 1800 split water into oxygen and hydrogen by passing an electric current through it. Water was the first substance to be "electrolyzed".

From the year 1800 on the use of hydrogen was limited to water gas (a mixture of hydrogen and carbon monoxide) for illumination and as an additive to town gas (a mixture of methane, carbon monoxide and hydrogen) for heating. These gases were displaced by natural gas as a gaseous energy carrier in the middle of the twentieth century.

In 1823, *Johann Wolfgang Döbereiner* invented the first "pocket" lighter to light cigars [7]. The "Döbereiner Platinum lighter" (Fig. 2.2) was the first mass produced hydrogen device and over 20 000 were sold. Hydrogen, produced by a chemical reaction of zinc with sulfuric acid, streams over a platinum sponge and reacts spontaneously with oxygen to form water at the metal surface. Subsequently, the released heat ignites also the hydrogen in the gas phase and a flame emerges. The so-called catalytic combustion on catalytic metal surfaces is one of the most important physical effects of hydrogen–metal interactions and is also the basis of the following device.

Sir *William Grove* (1811–1896) constructed in 1839 a "gas voltaic battery" which was the forerunner of modern fuel cells (Fig. 2.3). He based his experiment on the fact that sending an electric current through water splits the water into its component parts hydrogen and oxygen. Grove allowed the reaction to reverse – combining hydrogen and oxygen to produce electricity and water. The term "fuel cell" was coined later, in 1889, by *Ludwig Mond* and *Charles Langer*, who attempted to build the first practical device using air and industrial coal gas. However the development of the fuel cell to a powerful current source was very difficult and in 1866 the first dynamoelectric generator was demonstrated which efficiently converts every type of mechanical energy into electricity. Therefore, fuel cells lost

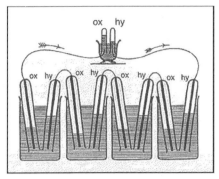

Fig. 2.3 Sir William Robert Grove demonstrated in 1839 the first fuel cell with four galvanic elements in series. Diluted sufuric acid was used as the electrolyte and platinum wires as the electrodes. In the upper part of the glass cylinders the electrode was in contact with hydrogen (hy) and oxygen (ox), respectively. The generated current was again used to electrolyze water (Ostwald, 1896 [8]).

their importance as electricity generators and were not further developed until the middle of the twentieth century.

Gustav Kirchhoff and *Robert Bunsen* analyzed 1861 the emitted spectrum of the sun and found hydrogen to be the major constituent of the sun [9].

Soon after the discovery of the metal palladium in 1803 by *W.H. Wollaston* [10], *T. Graham* [11] reported that this metal could absorb large amounts of hydrogen by forming a metal hydride. While, in those times, the discovery was more a scientific curiosity, it is now the basis for the promising technology of hydrogen storage in metal hydrides.

On 10th May 1898, *James Dewar* [12] used regenerative cooling to become the first to statically liquefy hydrogen. Using liquid nitrogen he precooled gaseous hydrogen, under 180 atmospheres, then expanded it through a valve in an insulated vessel, also cooled by liquid nitrogen. The expanding hydrogen produced about 20 cm^3 of liquid hydrogen, about 1 % of the hydrogen used.

In 1909 the German Chemist *Fritz Haber* discovered a catalyzed process [13], which allowed the synthesis of ammonia (NH_3) from the elements hydrogen and nitrogen. He received the Nobel prize in chemistry for his discovery. The Nobel prize for Fritz Haber was a subject of controversy because Haber is also the inventor of war gas (phosgene $COCl_2$), which killed hundreds of thousands of soldiers in World War I. Conscience-stricken, Haber's wife committed suicide. *Carl Bosch* succeeded to scale up Haber's synthesis from the laboratory scale to industrial production. After World War I other industrialized countries also introduced ammonia synthesis and therefore the consumption of hydrogen increased rapidly.

In 1929 Bonhoeffer and P. Harteck [14] successfully prepared the first pure para-hydrogen sample. In 1931 Urey, Brickwedde and Murphy [15] investigated the visible atomic Balmer series spectra of hydrogen samples and discovered the hydrogen isotope H^2, deuterium. In 1935 Oliphant, Harteck and Lord Rutherford [16] synthesized "superheavy hydrogen", H^3, tritium, through neutron bombardment of deuterated phosphoric acid.

On 1st March 1954 the USA ignited the first hydrogen bomb on the tiny Bikini Atoll in the Marshall Islands, contaminating a passing Japanese fishing boat and showering nearby villagers with radioactive ash. The bomb was 1000 times more powerful than the one dropped on Hiroshima. Three weeks later it emerged that a Japanese fishing boat, called Lucky Dragon, was within 80 miles (129 km) of the test zone at the time. Its 23 crew were severely affected by radiation sickness. They were among 264 people accidentally exposed to radiation because the explosion and fall-out had been far greater than expected.

Hydrogen is frequently used as a process gas in the petrochemical industry. Historically, it started with the production of water gas (a mixture of CO and H_2). Its strongest impact was on the Lurgi process, developed in Germany in the late 1930s. LURGI was the cable address of Metallurgische Gesellschaft founded on 5th February 1897 [17]. The Lurgi process permits the conversion of coal into methane, from which can be synthesized other hydrocarbon fuels such as gasoline. Such processes were operated on a large scale by Germany during World War II but are not now economically competitive with hydrocarbon fuels obtained from oil feedstocks. However, for modern conditioning of crude oil to obtain the desired end products (gasoline, etc.) similar process technologies are used.

Steam reforming, hydrogen reforming or catalytic oxidation, is a method of producing hydrogen from hydrocarbons. On an industrial scale, it is the dominant method for producing hydrogen. In 2000, the total annual hydrogen production worldwide amounted to 500×10^9 m^3 (STP) or aproximately 45×10^6 tons. The major sources are natural gas and coal, accounting for 78 % of the total production, 48 and 30 % respectively. 50% of the total hydrogen production is used for the synthesis of ammonia and 25 % for the cracking and purification of crude oil.

In 1955 Justi [18] (Fig. 2.4) described the utilization of hydrogen as an energy carrier medium. In 1969 an overall hydrogen energy concept using the favorable properties of hydrogen and including nonconventional energy systems was developed by Bockris, Gregory, Marchetti, Veziroglu and others [19].

The rapid development of solid state physics in the twentieth century also promoted the development of new solid state devices connected to hydrogen. Examples are sensors (e.g. hydrogen sensitive Pd-MOS structures [20]), nickel metal hydride batteries (E. Justi 1968 [21]), hydrogen switchable mirrors (R. Griessen 1995 [22]), and amorphization of bulk samples by hydrogen [23]. The metal hydride battery was the first realization of hydrogen-based energy storage, because a large amount of hydrogen has to be stored in one of its electrodes. The breakthrough was the discovery of a cheap metal alloy with superior hydrogen storage capacities, LaNi$_5$ by Philips, Eindhoven [24]. However, hydrogen storage materials for application in automobiles have to meet higher hydrogen storage capacities. It is too early to designate the ultimate hydrogen storage material but certainly, a breakthrough on the road towards high hydrogen content is the discovery by Bogdanovic in 1996 of a catalyst catalyzing light-weight alanates [25].

Fig. 2.4 Eduard Justi, copyright Institut für Angewandte Physik, Ta Brannschweig.

2.1.1
Hydrogen in Transportation

Cavendish, who called hydrogen "inflammable air", was also the first to measure its density. He reported in his 1766 paper that hydrogen is 7 to 11 times lighter than air (the correct value is 14.4). Cavendish's results not only opened up a new chapter in the history of gases, but also attracted attention to hydrogen as an alternative to hot air as a buoyant gas. *Jacques Alexandre César Charles* was the first to take advantage of this, soon after the first public demonstration by the Mongolfier brothers on 15th June 1783 of a hot-air balloon . After four days of struggling with his iron–acid hydrogen generation equipment, Charles launched his 4-m balloon on 27th August 1783. Just over three months later, he and one of his balloon builders, Aine Robert, became the first men to ascend in a hydrogen balloon. With all the enthusiasm for ballooning that began with the Mongolfier brothers and Charles in 1783, it was inevitable that the good and bad properties of hydrogen would meet. The worst happened on 15th June 1785 when Pliatre de Rozier and an assistant, P. A. Ronaon, attempted to cross the English Channel in a hydrogen balloon carrying a small hot-air balloon for altitude control. Thirty minutes into the flight the hydrogen ignited and the two men perished. Hydrogen's flammability was the underlying cause of the first air tragedy. Nevertheless, the attractiveness of hydrogen as a readily available buoyant gas was to outweigh the danger of flammability for 150 years of lighter-than-air flight. The Hindenburg disaster on 6th May 1937 in Lakehurst ended the air applications of hydrogen after 150 years.

Fig. 2.5 Hydrogen car by Karl Kordesch (1970) Copyright Karl Kordesch.

The air application of hydrogen is the reverse of hydrogen's later role as a fuel, where its flammability is a major advantage and its low density the disadvantage that inhibits its use for flight in the atmosphere. Experiments using hydrogen as an engine fuel had already been carried out in 1820 by W. Cecil. These experiments were followed by investigations in car engines up until World War II and partly afterwards (Ricardo, Burstal, Erren). Various transport systems using hydrogen were tested, for example in the USA by. R. Billings or in Germany (Daimler Benz, BMW) in the last decade of the twentieth century. The potential of hydrogen as an aviation fuel was established in1957 with the lift-off of a hydrogen-powered B-57 Canberra twin-engine jet bomber. From 1963 onwards liquid hydrogen–liquid oxygen propelled rockets were launched. For the Apollo moon flights $12 \times 10^3 \, m^3$ of liquid hydrogen was necessary to tank the Saturn carrier rockets. In 1970 Dr. Karl Kordesch built an alkaline fuel cell [26] (6 kW, Union Carbide)/battery hybrid electric car based on an A-40 Austin sedan [27] (Fig. 2.5) and drove it for his own personal transportation needs for over three years.

The fuel cell was installed in the trunk of the car with compressed hydrogen tanks on the roof, leaving room for four passengers in the four door car. It had a driving range of 180 miles (300 km). Thus, Kordesch was the first person in the world to have produced and driven a practical fuel cell/battery electric car. In 1988 a triple-jet hydrogen powered Tupolew Tu-154 flew in the former Soviet Union. In 1994 Daimler-Benz demonstrated Necar 1 with a 50kW PEM fuel cell from Ballard and compressed hydrogen gas cylinders. In 1996 Toyota demonstrated RAV4 with a 10 kW PEM fuel cell and hydrogen stored in a metal hydride. Toyota discontinued the production of its RAV4 Electric Vehicle in 2003.

2.2
The Hindenburg and Challenger Disasters

The LZ-129 *Hindenburg* and her sister-ship LZ-130 "Graf Zeppelin II" were the two largest aircrafts ever built. The *Hindenburg* was named after the President of Germany, Paul von Hindenburg. It was a brand new all-aluminum design: 245 m long, 41 m in diameter, containing $211\,890 \, m^3$ of gas in 16 bags or cells, with a

Fig. 2.6 On May 6, 1937, at 19:25 the German zeppelin **Hindenburg** caught fire and was utterly destroyed within a minute while attempting to dock with its mooring mast at Lakehurst Naval Air Station in New Jersey. Of the 97 people on board, 13 passengers and 22 crew members were killed. One member of the ground crew also died, bringing the death toll to 36.

useful lift of 112 000 kg, powered by four 820 kW engines giving it a maximum speed of 135 km/h. It could carry 72 passengers (50 transatlantic) and had a crew of 61. For aerodynamic reasons the passenger quarters were contained within the body rather than in gondolas. It was skinned in cotton, doped with iron oxide and cellulose acetate butyrate impregnated with aluminum powder. Constructed by Luftschiffbau Zeppelin in 1935 at a cost of £500 000, it made its first flight in March 1936 and completed a record double crossing in five days, 19 h, 51 min in July.

The *Hindenburg* was intended to be filled with helium but a United States military embargo on helium forced the Germans to use highly flammable hydrogen as the lift gas. Knowing of the risks with the hydrogen gas, the engineers used various safety measures to keep the hydrogen from causing any fire when it leaked, and they also treated the airship's coating to prevent electric sparks that could cause fires.

The disaster [28] is remembered because of extraordinary newsreel coverage (Fig. 2.6), photographs, and Herbert Morrison's recorded radio eyewitness report from the landing field. Morrison's words were not broadcast until the next day. Parts of his report were later dubbed onto the newsreel footage (giving an incorrect impression to some modern eyes accustomed to live television that the words and film had always been together). Morrison's broadcast remains one of the most famous in history – his plaintive words "oh the humanity" resonate with the memory of the disaster.

Herbert Morrison's famous words should be understood in the context of the broadcast, in which he had repeatedly referred to the large team of people on the field, engaged in landing the airship, as a "mass of humanity". He used the phrase when it became clear that the burning wreckage was going to fall onto the ground, and that the people underneath would probably not have time to escape it. It is not clear from the recording whether his actual words were "Oh, the humanity" or "all the humanity".

There had been a series of other airship accidents (none of them Zeppelins) prior to the *Hindenburg* fire, most due to bad weather. However, Zeppelins had accumulated an impressive safety record. For instance, the Graf Zeppelin had flown safely for more than 1.6 million km (1 million miles) including making the first complete circumnavigation of the globe. The Zeppelin company was very proud of the fact that no passenger had ever been injured on one of their airships.

But the *Hindenburg* accident changed all that. Public faith in airships was completely shattered by the spectacular movie footage and impassioned live voice recording from the scene. Because of this vivid publicity, Zeppelin transport came to an end, marking the end of the giant, passenger-carrying rigid airships.

Questions and controversy surround the accident to this day. There are two major points of contention: (i) How the fire started and (ii) Why the fire spread so quickly.

At the time, sabotage was commonly put forward as the cause of the fire, in particular by Hugo Eckener, former head of the Zeppelin company and the "old man" of the German airships. The Zeppelin airships were widely seen as symbols of German and Nazi power and, as such, they would have made tempting targets for opponents of the Nazis. However, no firm evidence supporting this theory was produced at the formal hearings on the matter.

Although the evidence is by no means conclusive, a reasonably strong case can be made for an alternative theory that the fire was started by a spark caused by static buildup. Proponents of the "static spark" theory point to the following: The airship's skin was not constructed in a way that allowed its charge to be evenly distributed and the skin was separated from the aluminum frame by nonconductive ramie cords. The ship passed through a moist weather front. The mooring lines were wet and therefore conductive. As the ship moved through the moist air the skin became charged. When the wet mooring lines connected to the aluminum frame touched the ground they grounded the aluminum frame. The grounding of the frame caused an electrical discharge to jump from the skin to the grounded frame. Witnesses reported seeing a glow consistent with a St. Elmo's fire.

The controversy around the rapid spread of the flames centers around whether blame lies primarily with the use of hydrogen gas for lift or the flammable coating used on the outside of the envelope fabric.

Proponents of the "flammable fabric" theory [29] contend that the extremely flammable iron oxide and aluminum impregnated cellulose acetate butyrate coating could have caught fire from atmospheric static, resulting in a leak through which flammable hydrogen gas could escape. After the disaster the Zeppelin company's engineers determined that this skin material, used only on the Hindenburg, was indeed more flammable than the skin used on previous crafts. Cellulose acetate butyrate is of course flammable but iron oxide increases the flammability of aluminum powder. In fact iron oxide and aluminum can be used as components of solid rocket fuel or thermite.

Hydrogen burns invisibly (emitting light in the UV range) so the visible flames (see Fig. 2.6) of the fire could not have been caused by the hydrogen gas. Moreover, motion picture films show downward burning whereas hydrogen, being less dense than air, burns upward. Some speculate that the German government placed the blame on flammable hydrogen in order to cast the US helium embargo in a bad light.

Also, the naturally odorless hydrogen gas in the *Hindenburg* was "odorized" with garlic so that any leaks could be detected, but nobody reported any smell of garlic during the flight or at the landing prior to the disaster.

It is also pointed out that none of those who died were burned by hydrogen. Of the 36 victims, 33 died because they jumped or fell out of the airship, two died of burns from the fabric and diesel fuel, and Allen Hagaman of the ground crew was killed when one of the motors fell on him.

Opponents [30] of the "flammable fabric" theory contend that it is a recently developed analysis focused primarily on deflecting public concern about the safety of hydrogen. These opponents contend that the "flammable fabric" theory fails to account for many important facts of the case.

The space shuttle was developed by NASA between 1972 and 1979. The first launch was April 12th 1981. The total weight of the space shuttle at launch is 2055 ton and the thrust at launch is 32 600 kN allowing the space shuttle to reach orbits of 185 to 965 km. The empty orbiter has a weight of 68 ton and is manufactured by Rockwell. The orbiter itself contains 15 ton of fuel and is equipped with 3 fuel cells of 2–12 kW electric power. The total weight at launch of the orbiter is 126 ton.

The three main engines, from Rocketdyne, of the space shuttle operate with liquid hydrogen and liquid oxygen as fuel and develop a thrust of 1750 kN at launch for about 10 s. The fuel is stored in the external tank from Martin Marietta. The external tank contains 616 ton liquid oxygen (1991 l) and 102 ton liquid hydrogen (14 500 l) and has a total weight of 756 tons and an empty weight of 35 ton. The solid rocket boosters (SRB) from Thiokol each contain 503 ton of fuel (16 % atomized aluminum powder as fuel, 69.83 % ammonium perchlorate as oxidizer, 0.17 % iron-powder as catalyst, 12 % polybutadiene acrylic acid acrylonite as binder, 2 % epoxy curing agent) and the total weight (tank + fuel) is 590 ton. Each SRB develops a thrust of 13 800 kN and operates for about 120 s.

January 28, 1986, 11:38:00 a.m. EST the first shuttle lift-off was scheduled from Pad B. Launch set for 3:43 p.m. EST, Jan. 22, slipped to Jan. 23, then Jan. 24, due to delays in mission 61-C. Launch reset for Jan. 25 because of bad weather at transoceanic abort landing (TAL) site in Dakar, Senegal. To utilize Casablanca (not equipped for night landings) as an alternate TAL site, T-zero moved to morning lift-off time. Launch postponed a day when launch processing unable to meet new morning lift-off time. Prediction of unacceptable weather at KSC led to launch rescheduled for 9:37 a.m. EST, Jan. 27. Launch delayed again by 24 h when the ground servicing equipment hatch closing fixture could not be removed from the orbiter hatch. Fixture sawed off and attaching bolt drilled out before close-out completed. During delay, cross winds exceeded return-to-launch-site limits at KSC's Shuttle Landing Facility. Launch Jan. 28 delayed 2 h when hardware interface

Fig. 2.7 On January 28, 1986, at 11:38 AM Eastern Standard Time, Challenger space shuttle left Pad 39B at Kennedy space center in Florida for Mission 51-L, the tenth flight of Orbiter Challenger. Seventy three seconds later the space shuttle was completely destroyed due to an explosion of the hydrogen tank and all 7 crew members (6 astronauts, 1 civilian) were killed. The solid rocket boosters can be seen speeding away from the gulf of smoke caused by the exploding Challenger.

module in launch processing system, which monitors fire detection system, failed during liquid hydrogen tanking procedures.

The solid rocket boosters (SRB) were ignited, and the thundering noise started. Figure 2.7 shows lift-off and the resulting disaster. Just after lift-off at 0.678 s into the flight, photographic data show a strong puff of gray smoke was spurting from the vicinity of the aft field joint on the right SRB. Computer graphic analysis of film from pad cameras indicated the initial smoke came from the 270 to 310-degree sector of the circumference of the aft field joint of the right SRB. This area of the solid booster faces the external tank. The vaporized material streaming from the joint indicated there was not complete sealing action within the joint.

Eight more distinctive puffs of increasingly blacker smoke were recorded between 0.836 and 2.500 s. The smoke appeared to puff upwards from the joint. While each smoke puff was being left behind by the upward flight of the Shuttle, the next fresh puff could be seen near the level of the joint. The multiple smoke puffs in this sequence occurred at about four times per second, approximating the frequency of the structural load dynamics and resultant joint flexing. As the Shuttle increased its upward velocity, it flew past the emerging and expanding smoke puffs. The last smoke was seen above the field joint at 2.733 s.

The black color and dense composition of the smoke puffs suggest that the grease, joint insulation and rubber O-rings in the joint seal were being burned and eroded by the hot propellant gases.

At approximately 37 s, Challenger encountered the first of several high-altitude wind shear conditions, which lasted until about 64 s. The wind shear created forces on the vehicle with relatively large fluctuations. These were immediately sensed and countered by the guidance, navigation and control system. The steering system (thrust vector control) of the SRB responded to all commands and wind shear effects. The wind shear caused the steering system to be more active than on any previous flight.

Both the Shuttle main engines and the solid rockets operated at reduced thrust approaching and passing through the area of maximum dynamic pressure of 720 pounds per square foot. Main engines had been throttled up to 104 % thrust and the SRBs were increasing their thrust when the first flickering flame appeared on the right SRB in the area of the aft field joint. This first very small flame was detected on image enhanced film at 58.788 s into the flight. It appeared to originate at about 305 degrees around the booster circumference at or near the aft field joint.

One film frame later from the same camera, the flame was visible without image enhancement. It grew into a continuous, well-defined plume at 59.262 s. At about the same time (60 s), telemetry showed a pressure differential between the chamber pressures in the right and left boosters. The right booster chamber pressure was lower, confirming the growing leak in the area of the field joint.

As the flame plume increased in size, it was deflected rearward by the aerodynamic slipstream and circumferentially by the protruding structure of the upper ring attaching the booster to the external tank. These deflections directed the flame plume onto the surface of the external tank. This sequence of flame spreading is confirmed by analysis of the recovered wreckage. The growing flame also impinged on the strut attaching the SRB to the external tank.

The first visual indication that the swirling flame from the right SRB breached the external tank was at 64.660 s when there was an abrupt change in the shape and color of the plume. This indicated that it was mixing with leaking hydrogen from the external tank. Telemetered changes in the hydrogen tank pressurization confirmed the leak. Within 45 ms of the breach of the external tank, a bright sustained glow developed on the black-tiled underside of the Challenger between it and the external tank.

Beginning at about 72 s, a series of events occurred extremely rapidly that terminated the flight. Telemetered data indicate a wide variety of flight system actions that support the visual evidence of the photos as the Shuttle struggled futilely against the forces that were destroying it.

At about 72.20 s the lower strut linking the SRB and the external tank was severed or pulled away from the weakened hydrogen tank permitting the right SRB to rotate around the upper attachment strut. This rotation is indicated by divergent yaw and pitch rates between the left and right SRBs.

At 73.124 s, a circumferential white vapor pattern was observed blooming from the side of the external tank bottom dome. This was the beginning of the structural failure of the hydrogen tank that culminated in the entire aft dome dropping away. This released massive amounts of liquid hydrogen from the tank and created a sudden forward thrust of about 2.8 million pounds, pushing the hydrogen tank upward into the intertank structure. At about the same time, the rotating right SRB impacted the intertank structure and the lower part of the liquid oxygen tank. These structures failed at 73.137 s, as evidenced by the white vapors appearing in the intertank region.

Within milliseconds there was massive, almost explosive, burning of the hydrogen streaming from the failed tank bottom and liquid oxygen breach in the area of the intertank.

At this point in its trajectory, while traveling at a Mach number of 1.92 at an altitude of 46 000 feet, the Challenger was totally enveloped in the explosive burn. The Challenger's reaction control system ruptured and a hypergolic burn of its propellants occurred as it exited the oxygen–hydrogen flames. The reddish brown colors of the hypergolic fuel burn are visible on the edge of the main fireball. The Orbiter, under severe aerodynamic loads, broke into several large sections which emerged from the fireball. Separate sections that can be identified on film include the main engine/tail section with the engines still burning, one wing of the Orbiter, and the forward fuselage trailing a mass of umbilical lines pulled loose from the payload bay.

The explosion 73 s after lift-off claimed crew and vehicle. The cause of the explosion was determined to be an O-ring failure in the right SRB. Cold weather was a contributing factor. The temperature at ground level at Pad 39B was 36 °F, that was 15 °F colder than any other previous launch by NASA. The last recorded transmission from Challenger was at 73.62 s after launch, when it truly fell apart.

References

1 In photosynthesis, water is split, releasing oxygen as a by-product. The hydrogen is used to produce ATP and NADPH, the energy carriers of the living cell. See Purves, W.K., Savada, D., Orians, G.H., Heller, H.C. (2003) *Life: The Science of Biology*, Sinauer Associates, Inc. Publishers, Sunderland, USA.

2 Ludwig-Bölkow-Systemtechnik GmbH (LBST) together with the Ludwig-Maximilians-Universität München; see http://www.hydrogen.org/akzepth2/summ.html.

3 Partington, J.R. (1961) *A History of Chemistry*, volumes 2, 4, Macmillan, London, p. 526; *Philosophical Transactions* (1766), p. 141; Cavendish's papers are in *The Scientific Papers of the Honourable Henry Cavendish, F.R.S.*, volume 2 (ed. Sir Edward Thorpe), University Press, Cambridge (1921). *Chemical and Dynamical*, volume 11, was the source used. See also Partington, *History of Chemistry*, volume 3, pp. 302–62.

4 David, J.-L., Painter, F. (1748–1825) *Neoclassicism: Portrait of Monsieur Lavoisier and His Wife*, Oil on Canvas, 8'8" × 7'4" 1788, The Metropolitan Museum of Art, New York.

5 Weber, R. (1988) *Der sauberste Brennstoff*, Band 3 der OLYNTHUS-Reihe, Oberbözberg, ISBN 3-907175-10-7.

6 Beretta, M. (2001) *Ann. Sci.*, 58 (4), 327–56.

7 Krüger, J. (1887) *Grundzüge der Physik*, G. F. Amelangs Verlag, Leipzig.

8 Ostwald, W. (1896) *Elektrochemie*, Von Veit & Comp., Leipzig.

9 Kirchhoff, G., Bunsen, R. (1860) *Ann. Phys. Chem. (Poggendorff)*, 110, 161–89.

10 http://www.public.iastate.edu/~pcharles/scihistory/Wollaston.html.

11 Graham, T. (1866) *Philos. Trans. R. Soc. (London)*, **156**, 399.
12 Dewar, J. (1927) *Collected Papers of Sir James Dewar* (ed. L. Dewar), University Press, Cambridge, pp. 678–91.
13 Haber, F., Klemensiewicz, Z. (1909) *Z. Phys. Chem.*, **67**, 385.
14 Bonhoeffer, K.F., Harteck, P. (1929) *Z. Elektrochem.*, **35**, 621.
15 Urey, H.C., Brickwedde, F.G., Murphy, G.M. (1932) *Phys. Rev.*, **40** (1), 1–15.
16 Oliphant, M.L.E., Harteck, P., Rutherford, L. (1934) *Proc. R. Soc. London Ser.A*, **144**, 692.
17 http://de.wikipedia.org/wiki/Lurgi_AG; http://www.lurgi.com.
18 Bockris, P.D., Justi, E.W. (1988) *Wasserstoff – die Energie für alle Zeiten. Konzept einer Sonnen-Wasserstoff-Wirtschaft*, Gebundene Ausgabe, Bauverlag, Gütersloh.
19 Bockries, J.O., Veziroglu, T.N. (1983) *Int. J. Hydrogen Energy*, **8** (5), 323–40.
20 Dannetun, H.M., Petersson, L.G., Soderberg, D., Lundtrom, I. (1984) *Appl. Surf. Science*, **17**, 259; Greber, T., Schlapbach, L. (1989) *Z. Phys. Chem. NF*, **164**, 1213.
21 As typical for modern R&D, several inventors participated in the NiMH-battery. The principal idea was patented by Justi, E. (1968) Braunschweig, Patent file 1803122, Germany, the materials were optimized by Philips, Eindhoven [24], and CNRS, Paris.
22 Huiberts, J.N., Griessen, R., Rector, J.H., Wijngaarden, R.J., Dekker, J.P., de Groot, D.G., Koeman, N.J. (1996) *Nature*, **380**, 231.
23 Samwer, K. (1988) *Phys. Rep.*, **161** (1); *J. Less-Common Met.*, **140**, 25 (1988).
24 van Deutekom, H.J.H., Eindhoven, P. (1977) Patent file 7702259, The Netherlands.
25 Bogdanovic, B., Schwickardi, M. (1997) *J. Alloys Compd.*, **253**, 1.
26 Kordesch, K. (1960) *Ind. Eng. Chem.*, **52**, 4, 296–8.
27 Kordesch, K. (1981) *Chimia*, **35** (9), 348–51.
28 Duggan, J. (2002). *LS 129 "Hindenburg" – The Complete Story*, Zeppelin Study Group, Ickenham, UK, ISBN 0-9514114-8-9.
29 Bain, A., van Vorst, W.D. (1999) *Int. J. Hydrogen Energy*, **24**, 399–403.
30 Dürr, E., Bock, B., Hoffmann, D. (1937) Bericht des deutschen Untersuchungsausschusses über das Unglück des Luftschiffes "Hindenburg" am 6.5.1937 in Lakehurst, USA, vom 2.11.1937 (http://spot.colorado.edu/~dziadeck/zf/LZ129fire.htm).

3
Hydrogen as a Fuel
Sonja Studer, Samuel Stucki, and John D. Speight

3.1
Fossil Fuels
Sonja Studer

Few commodities have had such a fundamental impact on modern society as fossil fuels. Before human beings learned to exploit the energy stored in the hydrocarbon bonds of fossil fuels, they relied solely on muscular effort, direct solar, wind and water energy and the energy stored in biomass. In the last two hundred years, the remarkable rise of the coal, oil and more recently the natural gas industries has led to radical changes in almost every aspect of life.

3.1.1
Two Centuries of Growth

The oldest evidence of the use of fossil fuels dates back to prehistoric times. Accumulations of oil and tar were occasionally found at the surface of the earth and used for lighting, heating and building purposes. However, systematic exploitation of fossil fuels did not occur until modern times.

The invention of the coal-powered steam engine at the end of the eighteenth century was a landmark for the industrial revolution. A few decades later, high-pressure mobile steam engines were introduced on a broad scale and rapidly changed transportation habits. The next energy transition took place in the middle of the nineteenth century and was of no less significance. Its beginning was marked by the emergence of the oil industry in Pennsylvania on the east coast of the United States. In 1859, oil was first drilled by self-proclaimed Colonel Edwin Drake near the small town of Titusville. This event is generally regarded as the start of the modern petroleum industry. In the 1860s, a young bookkeeper named John D. Rockefeller bought control of a refinery business in Cleveland. Twenty years later, he was one of the most powerful men in the United States. His Standard Oil Trust controlled almost the entire American oil industry. Rockefeller's corporate monopoly was not broken until 30 years later.

Hydrogen as a Future Energy Carrier. Edited by A. Züttel, A. Borgschulte, and L. Schlapbach
Copyright © 2008 WILEY-VCH Verlag GmbH & Co. KGaA, Weinheim
ISBN: 978-3-527-30817-0

In the early years of the industry, the most important oil product was lamp petroleum. The development of the internal combustion engine at the end of the nineteenth century shifted the focus of the industry to the lighter gasoline fractions and led to an increasing demand for automotive fuels. The oil industry soon expanded from the United States to all parts of the world. After the break-up of the Standard Oil Trust in 1911, a large number of independent companies arose. Seven of these soon became major powers. These "Seven Sisters" – Exxon, Shell, Mobil, Texaco, Chevron, Gulf and British Petroleum – were to dominate the world oil business in the following decades. The first large Middle Eastern oilfields were discovered during the 1930s, and the industry's period of most rapid expansion began about a decade later. Throughout the world, production and refining facilities were developed with astonishing speed [1, 2].

In 1960, Saudi Arabia, Iran, Iraq, Kuwait and Venezuela founded the Organisation of Petroleum Exporting Countries (OPEC) to counter the dominance of the major oil companies. The power struggle between the multinational majors, a large number of independent smaller oil companies and the OPEC and non-OPEC oil-producing countries has shaped the eventful history of the oil industry up to present times [1]. Public awareness of the political, economical and social implications of the ever-growing dependence on oil was first raised by the oil crisis in 1973 and 1979. This did not diminish the importance of oil as a major global energy source, but it facilitated the growth of alternative energy industries such as the natural gas industry.

The natural gas industry originated much like the oil industry in the nineteenth century, but took much longer to become a significant player in the global energy market. The earliest gas markets emerged in the USA in the middle of the nineteenth century. In those early days, gas was only used in small quantities for lighting and cooking. For a long time, it was mainly seen as a useless by-product of crude oil production and simply flared at the production sites. Advances in pipeline technology in the 1940s and 1950s facilitated transportation of natural gas over longer distances and led to a rapid growth of the industry. At the same time, the USSR started to develop their immense gas resources. After the fall of the Soviet Union, the Russian Federation continues to be the most important player in the natural gas industry both in terms of production and of proven reserves [3]. Overall, the natural gas industry has grown to become the third largest source of primary energy worldwide.

In 1850, wood still supplied 70 % of the world's commercial energy. Fossil fuels began to dominate the global energy market sometime at the end of the nineteenth century. In the course of the twentieth century, their use increased between 12- and 16-fold [2, 4]. Figure 3.1 and Table 3.1 show global primary energy consumption since 1965. Presently, fossil fuels cover about 85 % of the world's energy needs. Only two nonfossil energy sources, hydroelectricity and nuclear energy, make significant contributions to commercial energy use on a global scale. Together, they account for less than 13 % of global energy consumption. Oil alone, on the other hand, provides more than 37 % of primary energy and is thus the single most important commercial energy source worldwide. Nevertheless, the importance of coal should not be underestimated. In 2002, coal accounted for 25 % of global primary energy

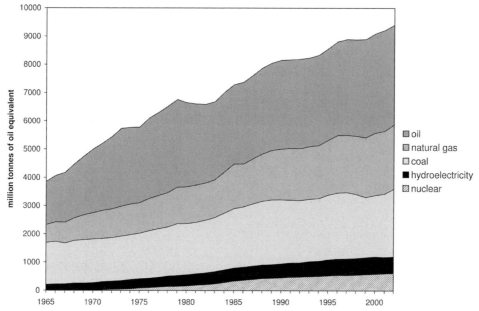

Fig. 3.1 Global primary energy consumption since 1965 in millions of tonnes of oil equivalent. Data from [3].

consumption. In the western world, coal is nowadays almost exclusively used to generate electricity and to produce metallurgical coke. In less developed economies, however, it still supplies the major part of the energy for domestic heating. The economic growth in the Far East, especially China, is expected to give rise to a future increase in global coal consumption [5]. The consumption of natural gas has increased rapidly in the last thirty years to reach 24 % in 2002 [3].

3.1.2
Advantages and Uses of Fossil Fuels

What enabled the fossil fuel industries to attain such a dominant position in the global energy market in such a short time? The main reason is the fact that fossil fuels constitute excellent energy sources with a very high specific energy. One tonne of oil equivalent equals 42.7 gigajoules, one thousand cubic meters of natural gas approximately 37 gigajoules and one tonne of coal 25 gigajoules. One million tonnes of oil produces about 4.5 terawatt-hours of electricity in a modern power plant. The high energy density of fossil fuels is paired with good combustion characteristics which make oil and natural gas products ideal fuels for specialized combustion systems, such as the spark ignition (Otto) engine and the ignition compression (Diesel) engine. Moreover, many fossil fuels are in the liquid state at ambient temperatures and atmospheric pressure, which allows convenient handling, storage and transportation. Their relatively low cost and good availability are further advantages.

Table 3.1 World primary energy consumption 1965–2003 (in million tonnes of oil equivalent, Mtoe).

Year	Nuclear	Hydroelectricity	Coal	Gas	Oil
1965	0	209.8	1485.8	632.1	1528.3
1966	0	224.5	1511.6	686.4	1644.9
1967	0	230.5	1448.8	733.6	1762.2
1968	11.8	241.6	1513.8	797.9	1911.3
1969	0	256.6	1540.2	875.9	2075
1970	0	268.6	1553.3	923.9	2253.1
1971	24.8	280.4	1538	988.1	2375.4
1972	34.1	293.8	1540.7	1013.6	2554.6
1973	45.7	297	1579	1058.6	2753.6
1974	59.5	324	1592.8	1081.6	2708.6
1975	82.5	328.8	1613.7	1077	2676.7
1976	98	331	1681.6	1138.1	2851.2
1977	121.2	338.7	1726.6	1171.7	2945.5
1978	140	365.2	1744.7	1215.6	3056.8
1979	144.7	383.3	1834.8	1293.5	3104.7
1980	161.1	390	1814.9	1307.1	2975.2
1981	189.4	397.7	1826.2	1322.4	2870.7
1982	207.4	414.2	1863.8	1327	2778.3
1983	233.2	433.6	1917	1340.4	2763
1984	281.7	447.4	2010	1458.6	2814.7
1985	335.4	456.9	2105.7	1584.2	2802.6
1986	361.2	462.6	2136	1522.6	2892
1987	393	470.5	2202.8	1599	2949
1988	428.4	482.1	2252.2	1672.7	3039.1
1989	440.5	482.5	2287.4	1753.9	3087.8
1990	453	498.2	2264.4	1794.5	3140.2
1991	474.9	510.7	2217.7	1829.2	3137.6
1992	478.5	509.2	2202.8	1836.2	3170.4
1993	495.3	537.3	2201.4	1867.8	3140.1
1994	504	541.8	2218.8	1875.6	3199.5
1995	526	570.3	2290.3	1936.6	3245
1996	544.9	580	2340.7	2030.1	3321.3
1997	541.2	588.6	2350.6	2023.4	3395.2
1998	550.5	596.1	2268.8	2058.1	3411.7
1999	571.3	601	2136.7	2107.2	3480.4
2000	584.7	616.3	2174.3	2198.7	3517.5
2001	601	584.7	2243.1	2219.5	3571.1
2002	610.6	592.1	2397.9	2282	3522.5

Fossil fuels are extremely versatile energy sources. They serve a wide diversity of purposes, including transportation, heating, electricity, industrial applications and feedstock for the chemical industry. Crude oil in particular can be separated and processed into a wide range of useful products. Its lightest fractions – consisting mainly of propane and butane – are in the gaseous state at ambient pressure and temperature, but may be liquefied at relatively low pressures. This liquefied petroleum gas (LPG) is suitable for a variety of applications: as feedstock for the chemical industry, as a convenient portable fuel for heating and cooking purposes and, increasingly, as automotive fuel. The gasoline fraction, comprising volatile hydrocarbons of low molecular weight which are naturally resistant to spontaneous ignition, turned out to be ideal for knock-free operation in spark-ignition engines. Kerosene, the next heavier fraction, consists of relatively involatile components comprising nine to fourteen carbon atoms. Formerly used as "lamp kerosene" for illumination, it has nowadays become the most important aviation fuel.

Gas oils are more complex molecules of 10 to 22 carbon atoms. Their relative ease of spontaneous ignition renders them suitable for operation in high-speed compression-ignition engines. The most important use for gas oils, however, is domestic and industrial heating. The residual components which cannot undergo distillation without thermal damage to the molecules are referred to as fuel oils. Fuel oils are used in industry and marine transportation. Due to their high viscosity, they require warming prior to transportation and combustion. Any heavy inorganic contaminants present in the parent crude are generally accumulated in this residual fraction. Furthermore, crude oil delivers bitumen for road construction and a large number of products serving as important feedstock for the petrochemical industry.

Natural gas is a more uniform product than crude oil. It is the least carbonaceous of all fossil fuels, an ideal transportation fuel and very clean in combustion. The main disadvantages of gaseous fuels are the difficulties associated with storage and transport. Natural gas can be shipped and marketed as compressed natural gas (CNG) or as a cryogenic liquid (liquid natural gas, LNG). Liquefaction is achieved by cooling the gas to produce a low-viscosity liquid with a boiling point of about $-160\,°C$ at atmospheric pressure. Expensive infrastructure requirements also represent the main drawbacks for the widespread use of natural gas as an automotive fuel.

The term "coal" is used for a very broad and divergent group of solid fossil fuels. The most important market for coal is power generation. More than 60 % of coal production is used for power generation worldwide. With a share of 38 %, coal is the most important energy source in power generation by far [6]. Coals have the highest carbon content of all fossil fuels. Consequently, they produce more CO_2 upon combustion and have a lower energy content. The solid nature of coal makes it less convenient for transportation and storage. On the other hand, the abundant presence of easily accessible coal reserves in many parts of the world guarantees that coal will remain one of the most important energy f sources worldwide.

3.1.3
Formation and Composition of Fossil Fuels

Fossil fuels are complex mixtures of hydrocarbons formed in a process lasting millions of years by decomposition of biological matter. Carbon and hydrogen are the main components, but other elements such as sulfur, nitrogen, oxygen and metals are also present in small amounts. At normal temperatures and pressures, these compounds may be gaseous, liquid or solid, depending on the complexity of their molecules.

All fossil fuels are derived from biological materials, either marine plankton or terrestrial vegetable matter. The formation of oil and natural gas was initiated in shallow ocean basins, under conditions that favored the growth of plankton and algae. When large numbers of dead marine organisms sank to the ocean floor and were subsequently covered by sediments, they were slowly transformed by processes that were partly due to the biological activity of anaerobic bacteria on the seafloor and partly chemical processes taking place at elevated temperatures and pressures. The sedimentary layer in which this transformation took place is generally referred to as source rock. Crude oil can only be formed within a precisely defined temperature and pressure window. If temperatures exceed a critical limit, the liquid hydrocarbon molecules break down to form natural gas. But also at low temperatures of up to about 80 °C, crude oil is often biologically degraded, over geological timescales, by micro-organisms that destroy hydrocarbons and other components to produce altered, denser heavy oils [7].

Due to their low density, freshly formed oil and natural gas deposits migrate upwards through porous strata. Eventually, a large proportion of the hydrocarbons that were formed in the geological past reaches the surface, is lost to the atmosphere and degraded in an aerobic environment. Only those fractions that were trapped beneath an impervious layer of sedimentary rock remain to form oil and gas reserves, captured within the microscopic pore spaces of so-called reservoir rocks.

Crude oils are liquid compounds composed mainly of saturated hydrocarbons. Sulfur is present in almost every crude oil at concentrations varying between less than 1 % and over 4 %. Further components include organic and inorganic nitrogen and oxygen species (both usually less than 0.5 %) and traces of sodium, potassium, vanadium and nickel [8]. The physical properties of crude oils differ widely. These variations are mainly due to the different proportions of various hydrocarbon compounds. Depending on their origin, oils may range from yellowish, mobile liquids to black, viscous semi-solids. Densities range from about 0.7 to 1.00 kg/l. Estimates suggest that about 9000 different crude oils can be found worldwide, of which only about 200 are traded.

Natural gas consists mainly of methane, but also contains higher hydrocarbons together with traces of nitrogen, carbon dioxide and hydrogen sulfide. The precise composition of natural gas varies greatly depending on the geological location and the history of the gas source. Natural gas may be found associated with crude oil as a gas-cap above the oil or on its own, unassociated with oil reserves. It can be derived directly from vegetable material, from the breakdown of crude oil at high temperatures, or from the decomposition of coal. This latter process results in a gas very rich in methane, containing only small quantities of other hydrocarbons.

Where terrestrial vegetable matter accumulates, peat is formed which, after burial, is transformed in various steps into lignite, bituminous coal and, eventually, anthracite. The coalification process is initiated by anaerobic bacteria and continues under the action of temperature and pressure over a period of millions of years. Upon maturation, the volatile gaseous and liquid hydrocarbons are progressively lost, leaving only solid carbonaceous compounds. Consequently, coals are solid compounds with a relatively high carbon and low hydrogen content.

3.1.4
Global Reserves and Production

Although reserves are finite, a shortage of fossil fuels is not to be expected in the foreseeable future. Accumulations of hydrocarbons occur in almost every part of the world, from conveniently located, easily accessible sites, to very hostile and remote environments such as the Polar Regions or offshore locations in deep water. Fossil fuel resources are usually subdivided into several categories. The well-explored shares of total resources that can be extracted with available technology at acceptable costs are referred to as reserves. This category may be divided into "proven reserves," and "probable reserves," based on exploration results and the degree of confidence in those results. The term "proven reserves" is thus used for the amount of oil, gas or coal that can be recovered economically under current market conditions. Advances in exploration and extraction techniques constantly transfer fossil fuels from the resource category to the proven reserves pool. Innovations such as horizontal drilling extended the extractable share, as they allowed production from many previously inaccessible reservoirs. The distribution of global proven oil, gas and coal reserves in 2002 is depicted in Figure 3.2 and Table 3.2. An additional distinction is made between conventional and nonconventional resources. The term "nonconventional" is used for reservoirs that cannot be produced at economic flow rates or that do not produce economic volumes of oil and gas without assistance from massive stimulation treatments or special recovery processes and technologies [9].

At the end of 2002, global proven oil reserves were estimated to be about 1.43×10^{11} tonnes. Two thirds of these reserves are located in the Middle East. Saudi-Arabia holds 25 % of the global proven reserves, followed by Iraq with 10.7 %, Kuwait and the United Arab Emirates with 9.3 and 9.2 %, respectively and Iran with 8.6 % [3]. As oil production in the North Sea and the USA shows signs of declining, the dependence of the industrialized world on the immense oil reserves in the Middle East and other politically sensitive regions will increase over time [5]. At present, however, the influence of the Middle Eastern countries and other OPEC members is not as strong as one might assume. In 2002, annual global oil production reached 3.56×10^9 tonnes. Of this amount, only 38.4 % was produced in OPEC-countries. Major non-OPEC oil producing countries include the Russian Federation, the United States, Mexico, China, Norway, the United Kingdom and Canada. In total, oil is produced in approximately 90 countries worldwide.

How long are the reserves going to last? This crucial question is as old as the oil industry itself. Warnings of an imminent physical shortage of global oil resources were issued as early as in the 1920s, when the US Geological Survey warned that the nation was running out of crude oil [2]. The subject of oil depletion has always been discussed very controversially [4, 10–12]. In the 1960s, oil geologist Marion King Hubbert proposed the Hubbert curve, which became the best known model of oil production. It implied that the discovery and production of oil over time would follow a single-peaked, symmetric bell-shaped curve with a peak in production when 50 % of the total oil in place had been extracted. Based on this

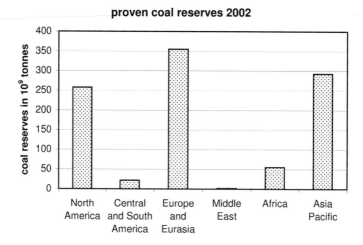
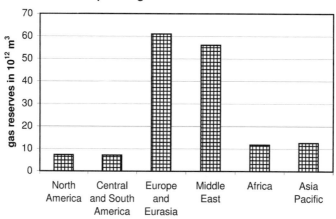
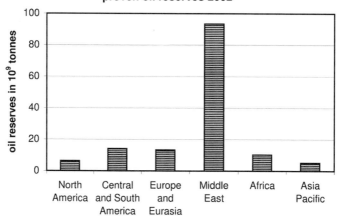

Fig. 3.2 Distribution of global proven oil, gas and coal reserves at the end of 2002. Data from [3].

Table 3.2 Oil, gas and coal reserves 2002.

	Coal in 10^9 tonnes	Gas in 10^{12} m^3	Oil in 10^9 tonnes
North America	257.8	7.2	6.4
Central and South America	21.8	7.1	14.1
Europe and Eurasia	355.4	61	13.3
Middle East	1.7	56.1	93.4
Africa	55.3	11.8	10.3
Asia Pacific	292.5	12.6	5.2

model, Hubbert predicted that the US oil production would peak some time in the late 1960s. His forecast was quite accurate, as the actual peak occurred in 1970.

The Hubbert curve was used by other authors to extrapolate future oil production on a global scale [4, 10, 12]. Most recently, Colin J. Campbell forecast the peak in conventional oil production to occur around 2010 [10]. The difficulty of these projections lies in the fact that, in reality, production curves are not likely to be symmetrical. The shape of the trailing leg of the Hubbert curve is a function of many variables including supply, price, access to environmentally sensitive areas and technology. Many curve-fitting models only poorly reflect the improvements in oil recovering technology and the – hardly predictable – influence of fluctuating oil prices. Interestingly, most recent results of curve-fitting methods show a consistent tendency to predict a peak within a few years, followed by a decline, no matter when the predictions were made [4]. This finding indicates the great amount of uncertainty that is associated with all estimates of oil reserves and future production.

The official statistics generally do not rely on curve-fitting methods. Instead, they use a very straightforward term, the proven reserves to production (R/P) ratio. The R/P ratio is based on present conditions and does not account for future changes in exploration and production technology, price and demand. In 2003, an R/P ratio of 40.6 years was reported for conventional oil reserves [3]. Interestingly, this ratio has altered very little during the last forty years. In fact, the R/P ratios of all fossil fuels were substantially higher in 2003 than a century earlier.

As global consumption of fossil fuels is expected to rise further in the coming years, nonconventional oil resources become increasingly more important. These include tar sands, shale oil, heavy oil and deep-water oil, which are more expensive and difficult to extract and process than conventional liquid crude oils. The world's nonconventional oil resources far exceed the proven conventional reserves. The largest accumulations of nonconventional oil are not found in the Middle East, but on the flanks of foreland basins on the American continent. Large oil shale deposits occur in the Western Unites States and Australia. Major tar sand reserves are found in Canada, in Venezuela and to a lesser extent in Madagascar. The Canadian tar sands contain more oil than the entire oil fields of Saudi Arabia. Production of nonconventional oil reserves is associated with considerable costs, technical difficulties and often a high environmental impact. However, the divide between conventional and nonconventional oil reserves is steadily shifting due to technical innovation. For example, oil is nowadays produced from the Athabascan tar sands in Northern Canada on an economical scale. In some statistics, the Athabascan tar sands have therefore been shifted to the proven reserves category, which led to a significant increase in total global oil reserves [13]. With further substantial improvements in technology and increasing oil prices, nonconventional oil reserves are likely to play an increasingly important role in the future.

World reserves of natural gas were about 1.56×10^{14} cubic meters at the end of 2002. 30.5 % of all reserves are located in the Russian Federation, 14.8 % in Iran, 9.2 % in Qatar, 4.1 % in Saudi Arabia and 3.9 % in the United Arab Emirates. Global annual production reached 2.53×10^{12} cubic meters in 2002, corresponding to 2.27×10^9 tonnes of oil equivalent. Based on these figures, the reserves to production ratio of the world's conventional gas reserves is estimated to lie around 61 years [3].

In recent years, the gas industry has developed an increasing interest in nonconventional gas reserves, such as coalbed methane, tight-gas sands and methane hydrates. Tight-gas sands and coalbed methane are already economically produced at certain locations, while energy recovery from methane hydrate is still far from being a commercial application. Coalbed methane is formed during the process of coalification and stored in the micropores of solid coal. It can be desorbed from the coal by lowering the pressure. However, only a minority of coalfields are suitable for commercial coalbed methane recovery, because economic production is only possible from coal beds with exceptional permeability [14].

The most extensive nonconventional natural gas reserves occur in the form of solid compounds of gas and water commonly called gas hydrates. These crystalline structures typically form when small "guest" molecules such as methane or carbon dioxide contact water at low temperatures and moderate to high pressures. The gas concentration in a hydrate is comparable to that of a highly compressed gas. Large methane hydrate resources exist in arctic permafrost undergrounds and in the deep ocean bottom. Estimates of the extent of these deposits vary widely. Some authors suggest that the amount of energy in hydrates is equivalent to twice that of all other fossil fuels combined [15]. However, the potential of these ice-like crystal structures for energy recovery is controversial, as the extraction of methane from the hydrate crystals is expensive and technically challenging. Nevertheless, several countries, including Japan, India and the United States, have set up research programs to examine the possibilities of methane hydrate production. Pilot drilling, characterization and production testing of hydrates have begun in arctic permafrost regions and in ocean drilling programs. Japanese and American programs predict that stand-alone hydrated energy recovery will begin by 2015 [15].

As far as coal is concerned, resource depletion is not an issue. The world's coal reserves will last for generations. At the end of 2002, BP reported global coal reserves of 9.84×10^{11} tonnes and an R/P ratio of more than 200 years. Almost half the world's reserves are located in OECD countries. The largest accumulations of coal are found in the United States (25 % of global proven reserves), the Russian Federation (15.9 %), China (11.6 %), India (8.6 %) and Australia (8.3 %). The main producers in 2002 were China (29.5 % of global production) and the United States (24 %) [3].

In the end, it becomes clear that resource exhaustion of fossil fuels will not be a matter of actual physical depletion. Or, as the former Saudi Arabian oil minister Sheik Yamani expressed it:

> The Oil Age will not end because the world runs out of oil, just as the Stone Age did not end for a lack of stones.

3.1.5
Environmental Impact

Exploration, drilling and extraction activities are known as the "upstream phase" of the fossil fuels industry. Depending on the nature and location of the extracted fuel, these processes may affect the local environment in various ways. Coal extraction, particularly from surface mines, disrupts the natural landscape and produces large quantities of waste. The same holds true for surface mining of nonconventional oil resources such as tar sands. Production of conventional oil and gas reserves usually has less visible impact on the natural landscape. Deforestation and erosion, disturbance of ecosystems and animal populations, local pollution because of accidental spills and human health and safety risks for neighboring communities are possible environmental consequences. The most important environmental issues during oil and gas production arise from storage, handling and discharge of processed water, drilling muds and chemicals. Modern oil production requires a large number of chemicals, including detergents used for enhanced oil recovery, hydraulic fluids, corrosion inhibitors, hydrate inhibitors and so forth. A major problem for any oil- or gas-producing company is the fact that even if the direct environmental impact of an operation is properly managed, there may be impact from other activities that follow oil and gas development and that are outside the control of the company. These problems may include land clearing, human colonization of pristine areas and introduction of invasive species.

Accidental oil spills during production, transportation and storage can have serious environmental consequences. The scale of these is influenced by a large number of factors, including the size of the spill, the weathering and buoyancy characteristics of the oil, the season of the spill and the vulnerability of local ecosystems. Some ecosystems, such as mangroves, salt marshes and coral reefs are particularly sensitive to oil spills and may take years to recover. Tight safety systems are an absolute necessity wherever large amounts of potentially hazardous goods are stored and transported. Throughout the world, oil spill preparedness and response systems are part of the regulatory framework of most countries. The International Tanker Owner Pollution Federation's worldwide oil spill statistics show good progress over the past two decades, but the prevention of spills is still a challenge for the industry [16]. Groundwater pollution from underground storage tanks has largely been overcome in Europe since the vast majority of underground storage tanks are double hulled, but it remains an issue in other parts of the world.

Fossil fuel production, processing and transportation may have considerable environmental impact at local and regional levels. On the whole, however, the most significant impact is caused by fossil fuel combustion. During this process, five primary air pollutants are formed: sulfur oxides (SO_x), nitrogen oxides (NO_x), volatile organic carbons (VOC), carbon monoxide (CO) and particulate matter. Combustion of fossil fuels is the largest source of anthropogenic SO_x and NO_x emissions. The eventual oxidation of these compounds in the atmosphere produces sulfates and nitrates, which are responsible for regional acid deposition. Atmospheric reactions of NO_x, CO and volatile organic compounds lead to the formation of ozone, a highly reactive oxygen species causing respiratory diseases, reduced crop yields and damaged forests. Particulate matter emissions have emerged as an issue of major concern. They appear to be responsible for various respiratory and cardiovascular diseases and may contain potentially carcinogenic compounds. A recent study suggests that black carbon particles may even enhance global warming [17].

Modern engine and burner technologies, improved fuel handling and storage facilities as well as continuous optimization of fuel quality have led to a dramatic reduction in air pollution in the industrialized world, though local air pollution remains a major problem in large cities in developing countries. The automotive industry has contributed to this improvement by developing optimized vehicle and engine technologies, encompassing control of fuel distribution and air–fuel ratio, combustion-chamber design incorporating stratified charge, temperature reduction by exhaust-gas recirculation and various exhaust catalyst devices. The mineral oil industry has made a strong effort to improve fuel quality. Stringent emission limits and product specifications limiting, for example, the amount of sulfur and aromatics, have led to significantly cleaner fuels.

Levels of sulfur dioxide emissions have dropped dramatically due to the rising levels of desulfurization of transport and heating fuels at the refineries and flue gas desulfurization at coal- and oil-fired power plants. Desulfurization to below 10ppm is now economically feasible and many countries are introducing ultra low sulfur fuels on a national scale. Limits on engine exhaust and evaporative emissions are becoming more and more stringent. Since the early 1970s automotive NO_x, CO and VOC emissions have been greatly reduced by installation of exhaust catalysts, though more vehicles, longer travel distances and more congested traffic have to some extent counteracted the effect of these controls. A new generation of exhaust treatment technologies, including particulate traps and de-NO_x-catalysts, promises significant further reductions of automotive emissions.

A dramatic reduction of air pollutant emissions can thus be achieved by optimizing both fuels and engines. A decrease in the levels of virtually all regulated emissions can now be observed in the industrialized world. At the same time, CO_2 emissions are rapidly increasing throughout the world. It is now widely believed that human activities, primarily the burning of fossil fuels, are modifying natural atmospheric processes and contributing to global warming. The impact of anthropogenic CO_2 emissions on global warming has emerged as the foremost global environmental concern arising from the combustion of fossil fuels. The first systematic measurements of rising background CO_2 concentrations began in 1958 at American observatories in Mauna Loa, Hawaii and the South Pole. In 2001, the Intergovernmental Panel on Climate Change published its third assessment report, stating that "the balance of evidence suggests a discernible human influence on global climate" [18]. Even if proof of the anthropogenic greenhouse effect cannot currently be established beyond any doubt, there is broad international consensus that reduction of greenhouse gas emissions will be the major environmental issue to tackle for coming generations.

3.1.6
Future Trends

The International Energy Agency expects world energy demand to rise by more than 50% until 2030 [5]. A substantial decoupling of energy and economic production has been observed in several studies, but energy and economic activity are still strongly connected in most industrialized and developing countries [4]. As economic development and population growth continue to push energy demand upward and concerns over the acceleration of global warming by combustion of fossil fuel are increasing, drastic changes in energy supply must occur someday, possibly in the second half of the century. In the nearer future though, fossil fuels will remain our most important energy sources. While alternatives are feasible for heating purposes, transportation is still very much dependent on fossil fuels and is likely to remain so in the next couple of decades. Consequently, the International Energy Agency predicts a further increase in global consumption of all three fossil fuels over the next 30 years [5].

Nonconventional fossil fuel resources such as tar sands, oil shale or methane hydrates will undoubtedly gain importance as conventional oil and gas reserves decrease. As production and processing of these resources is expensive and energy intensive, it will only become economic if oil and gas prices increase. The high CO_2 emissions arising from the production and consumption of nonconventional fossil fuels will call for effective ways of capturing CO_2. One possible way of reducing CO_2 emissions to the atmosphere is to sequester the emitted CO_2 and store it, for example, in underground caverns, depleted oil and gas fields and porous and permeable reservoir rocks or in the deep ocean. Carbon sequestration is currently a field of intense research. The necessary technology for underground storage of carbon dioxide exists and is commercially proven. The first large-scale commercial underground CO_2 sequestration project began at the Sleipner West gas field in the North Sea in 1996. On a smaller scale, CO_2 is re-injected into producing oil fields to enhance oil recovery. However, there is still great uncertainty regarding the potential of this method for atmospheric CO_2 reduction. Attempts to estimate global or regional underground CO_2 storage capacity have produced a wide range of figures. Furthermore, sequestration and safe storage of CO_2 are expensive, both in terms of money and energy. In the end, public perception of CO_2 storage will also play an important role. As yet, little data has been published on this question. As an end-of-pipe solution, CO_2 storage is unlikely to enjoy enthusiastic public support [14, 19].

Theoretically, sequestered CO_2 could also be disposed of in the depth of the oceans. However, concerns have been raised concerning the impact of increased oceanic CO_2 concentrations on marine life [20]. Absorption of carbon dioxide in the oceans may be associated with adverse effects on coral reefs and other organisms with a calcium carbonate skeleton shell. High CO_2 concentrations lead to lower pH values, which hinder the formation of carbonate structures.

It is obvious that CO_2 sequestration alone will not be sufficient to stabilize anthropogenic greenhouse gas emissions. Future energy supply will need to be diversified and provided by the wider use of renewable energies and hydrogen technologies, as well as highly efficient clean technologies for the use of fossil fuels. The overall conversion efficiency of advanced power production GTL systems is expected to increase as new turbine materials are developed and the CO_2 produced per megawatt of electricity generated declines [21].

Especially the conversion of coal, the most CO_2-intensive of all fossil fuels, has a huge potential for improvement. Clean coal technology (CCT) projects supported by the World Bank, the Global Environmental Facility, the Prototype Carbon Funds, the European Investment Bank and the European Union are carried out in several countries [22, 23]. Modern technologies offering improved conversion efficiency will enable coal fired power plants to provide at least the same amount of electricity while emitting less CO_2. This will lead not only to the reduction of significant amounts of CO_2, but also reductions in tar, phenol, benzene, particulate and other local emissions. The further development of CCT will be strongly dependent on the uncertain impact of environmental policies on coal demand. Concerns about future environmental regulations, including carbon-emission constraints, could deter investment in new coal mining projects and coal conversion technologies. In any case, the rate at which clean coal technologies are adopted and the scope at which they are put into place will be crucial for future coal use [18].

A shift to fossil fuels with lower carbon content, such as natural gas and liquefied petroleum gas (LPG) will occur in many places. Global demand for natural gas is expected to rise more strongly than for any other fossil fuel. The International Energy Agency forecasts a doubling in primary gas consumption until 2030 [5]. New power stations using combined cycle gas turbine technology will account for the largest part of this increase. Over the last two decades, natural gas and LPG have also emerged as alternative transport fuels. These fuels have a number of benefits over petroleum-derived fuels, but they also have a number of drawbacks limiting their market share. Main disadvantages are the requirement for costly vehicle modifications and the development of separate fuel distribution and vehicle fuelling infrastructure.

Gas to liquid (GTL) technology offers a long-term possibility of combining the advantages of natural gas and liquid fuels. Natural gas can be converted into syn-gas containing mainly CO and H_2, which in turn can be converted to methanol, ammonia or liquid hydrocarbon fuels. Methanol can be used as fuel for fuel cells or turned into gasoline for automobiles. Liquid hydrocarbon fuels are synthesized using the Fischer–Tropsch process. These sulfur- and aromatics-free "synfuels" can be burned in current diesel engines and lead to a substantial decrease in pollutant emissions. GTL technology is expensive, but currently achievable and marginally economic. Several major oil and gas companies have erected GTL plants or announced new projects at various locations in Africa, the Middle East and Southeast Asia. At present, GTL technology is no more than a niche market, but it has the potential to grow strongly in the next decades. Further technical development will be necessary to make the process cost-effective. In further development of GTL technology, the Fischer–Tropsch process could also be employed to produce liquid fuels from biomass ("sunfuels").

Liquid and gaseous fuels can also be made from coal. In fact, this is what the process developed around 100 years ago by Franz Fischer and Hans Tropsch was originally intended for. Several coal conversion technologies are being tested and applied today. Coal hydrogasification yields substitute natural gas (SNG) and high value chemicals. Coal liquefaction yields a synthetic crude oil, which can be used for the production of liquid fuels. However, it is highly questionable whether future environmental policies will encourage the use of coal for the production of transportation fuels.

In the short to medium term, a relatively low-cost and low-risk option for enhancing the diversity of fuel supply and reducing CO_2 emissions from fossil fuels is blending of conventional fuels with biomass-derived liquids. Agricultural surplus, energy plants, lignocellulose, sewage and domestic waste may be used to produce biofuels such as bioethanol, biodiesel and biogas. Use of pure biofuels requires engine modifications, but current vehicles without further modifications can safely be fuelled with conventional gasoline or diesel containing a low percentage of biofuels. Biofuels for transport are promoted and subsidized in many parts of the world. Pure and blended biofuels are currently available in a whole range of countries including Brazil, the USA and countries of the European Union. Co-firing of biomass in power plants is another relatively straightforward way of adding biomass capacity to the energy supply system [21].

In the more distant future, hydrogen and fuel cell technologies are expected to contribute to energy supply security and environmental protection. Most hydrogen is currently produced by steam reforming of hydrocarbons. In the long term, it is clear that only hydrogen produced from renewable energy sources can be truly sustainable. In the short and medium terms, however, fossil fuels are likely to continue to play a dominant role.

3.2
The Carbon Cycle and Biomass Energy
Samuel Stucki

3.2.1
The Carbon Cycle

Living systems are characterized by a continuous exchange of material and energy with their environment. The chemical energy responsible for maintaining life on the planet is produced by complex photochemical reactions involving the photochemical reduction of CO_2 with water to organic forms of carbon and molecular oxygen. The process is referred to as photosynthesis. The photosynthetic conversion of CO_2 and water to glucose and oxygen absorbs an energy equivalent of 2800 kJ mol^{-1}:

$$6H_2O + 6CO_2 \rightarrow C_6H_{12}O_6 + 6O_2$$

The reverse reaction, respiration, oxidizes the organic carbon with atmospheric oxygen. It releases energy as heat and creates a flow of carbon dioxide back to the atmosphere. On a global scale the material turnover of this reaction is huge: 120 Gton of carbon is converted annually by the photosynthetic reaction of plants on land alone (IPPC Report). This quantity is generally referred to as the gross primary production (GPP). GPP corresponds to a conversion of 4.7×10^{21} J of energy per year, which exceeds the annual consumption of fossil fuels by a factor of circa 10. The total marine GPP is of the same order of magnitude and is estimated to amount to 102 Gton per year, exchanging CO_2 between organisms and the water (dissolved inorganic carbon). The carbon exchange between the atmosphere and the hydrosphere (oceans) amounts to 90 Gton per year and is driven by photosynthesis and by transport processes between surface water and deep water. In a steady state all the fluxes from and to the atmosphere are well balanced, that is the photosynthetic production is balanced by an equal flux of carbon resulting from respiration processes. One differentiates between autotrophic respiration (i.e. the roughly 50 % of the GPP carbon flux which is released due to respiration processes of the photosynthetic organisms themselves) and heterotrophic respiration (respiration of organisms feeding on photosynthetic plants) and combustion ("natural" and anthropogenic combustion of biomass). The difference between GPP and autotrophic respiration is the so-called net photosynthetic production (NPP). Figure 3.3 summarizes the flows and reservoirs relevant to the global carbon cycle.

The carbon content of the atmosphere is small compared to the carbon content of the other reservoirs taking part in the exchange of carbon. Any deviation from a well balanced regime of carbon flows from and to the atmosphere will therefore lead to an accumulation or depletion of this reservoir within a short period of time. Although the flow of carbon to the atmosphere due to fossil fuel burning is less than 5 % of the terrestrial GPP, this flow seriously disturbs the balance of flows and has led to the well-documented increase in the CO_2 concentration in the atmosphere. The famous record of the CO_2 concentration over the past nearly 50 years at the site of Mauna Loa in Hawaii reflects the sensitivity of the atmosphere to imbalances in carbon fluxes. The steady increase in the curve is due to the net excess of CO_2 resulting from fossil fuels and deforestation, while the annual oscillations of the curve reflect a seasonal imbalance of the fluxes, due to the vegetation cycle and the fact that the land mass is not evenly distributed between the northern and the southern hemispheres. The observed fluctuations in the net increase in the CO_2 content of the atmosphere (e.g. by 1.9 Gton in 1992 and 6.0 Gton in 1998) have been ascribed to variations in land and ocean uptake [27].

A very small part of the photosynthetic net production is buried in sediments, isolated from the oxygen of the atmosphere and thus sequestered from the rapid annual carbon turnover between biosphere and atmosphere. Under favorable

Main components of the natural carbon cycle

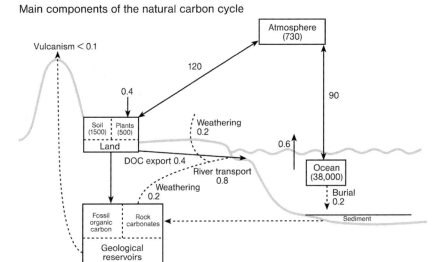

Fig. 3.3 The carbon flows between the important reservoirs in the geochemical cycle. From (Intergovernmental Panel on Climate Change (IPCC) (2001), Ref. [4]).

geological conditions, the accumulation and chemical transformation of part of this sequestered organic carbon in sediments has led to the formation of exploitable fossil fuel deposits over time. The rate of organic carbon sequestration in sediments is more or less balanced by oxidative weathering of old organic carbon deposits in sediment rocks. Estimates derived from isotopic analysis of carbon in sediment rocks reveal that the rate of deposition has varied over geological times. It ranges between 2 and 6×10^{18} mol Myr^{-1} [24], or, on average, about 5×10^{13} gC yr^{-1}, which corresponds to 0.1 % of the present annual net photosynthetic production (NPP). The rate of release of organic carbon from sediments by burning of fossil fuels is higher than the natural rate of weathering by a factor of 100. The "sudden" (on geological time scales) oxidation of sequestered organic carbon in the form of fossil fuels has led to the unprecedented surge in carbon dioxide concentration in the atmosphere, with its effect on global climate.

3.2.2
Biomass for Energy

Biomass is the oldest form of fuel utilized by humans. The discovery of fire is at the very beginning of civilizations and still today about 10 % of the primary energy needs globally are covered by so-called "traditional biomass," that is burning of fuel wood, dung and other agricultural residues, mainly for cooking and heating. Beyond the traditional use of biomass as a fuel, there has been increasing interest in using biomass as a resource in a sustainable energy system. Biomass has the advantage of being a storable form of solar energy and can easily be converted to heat and power, or to clean and commercially established secondary fuels, such as methane or liquid hydrocarbons, using more or less established processes derived from the utilization of fossil fuels (refineries, coal technology). Furthermore, biomass is widely available in a broad range of climates, and therefore contributes to secure local provision of energy. Its local character has further implications on the social dimension of sustainable development, as it secures the development of rural communities by providing local jobs and income.

Biomass is usually considered a carbon neutral form of energy, that is its photosynthetic production absorbs the amount of carbon released during its use as a fuel. This is, strictly speaking, only true as long as biomass energy systems which are in a steady state or at equilibrium are considered, that is if there are no basic changes in land use and/or harvesting practices involved and if annual production and use are balanced. There may, however, be important deviations from carbon-neutrality, if biomass utilization is to be increased significantly. Increased use of crop residues, for example in forestry, can temporarily lead to increased CO_2 emissions, as the use of these residues releases carbon faster than the natural decomposition of residues left on the forest floor, but the carbon stock will reach a new equilibrium after a relatively short period [25]. The establishment of biomass plantations and the consequent change in land use can lead to more dramatic changes in the carbon stock of the land which is being converted. These changes in the carbon budget of land can have a positive or a negative sign, depending on the type of plantation considered and the land use prior to conversion to a biomass plantation. Influencing carbon stocks on land by targeted plantations has become a hot topic discussions as to what extent tree plantations can be credited as carbon sinks in carbon emission trading or the clean development mechanism (CDM) of the Kyoto Protocol.

The potential of biomass residues and wastes to produce energy depends on the amount of biomass grown for food and fibers. It is a limited resource, which will not be able to cover the energy needs. The theoretical potential of residues and wastes has been estimated to be about 77 EJ, or about one third of today's global energy consumption. The potential estimates for biomass from dedicated energy plantations depend very much on to what extent sustainability criteria are considered. Sustainability requires that economic feasibility, social and environmental compatibility are considered. Many biomass energy potential studies have been published and, in a recent review of these studies [26], it was stated that the range of potentials for additional biomass energy use that have been estimated by different authors is huge. Estimates range from very modest potentials of 33 EJ to 1135 EJ, a multiple of today's world primary energy consumption. It is no wonder that there is such a great scatter of potential projections as they include difficult assumptions regarding sustainability constraints (availability of land for energy crops, competition with the production of food and materials, achievable carbon emission savings, etc.) (Table 3.3).

3.2.3
Biomass and Hydrogen

The chemical composition of biomass with respect to carbon, hydrogen and oxygen can be approximated by the formula $C_6H_9O_4$, that is biomass contains a little less

Table 3.3

	Pg C yr^{-1}	Energy conversion J yr^{-1}	
Terrestrial GPP	120	$4.67 \cdot 10^{21}$	based on glucose as primary product of photosynthesis
Terrestrial NPP	60	$2.33 \cdot 10^{21}$	Ref. [27]
Fossil fuel consumption	5.3	$3.44 \cdot 10^{20}$	Ref. [28]
Solar radiation hitting earth surface		$2.7 \cdot 10^{24}$	

oxygen than the primary product of photosynthesis, glucose($C_6H_{12}O_6$), but much more than fossil fuels (CH_2 to CH_4). The oxidation state of carbon in biomass is close to zero (while for natural gas it is $-IV$), which means that biomass can either be oxidized to $+IV$ (CO_2), or reduced to $-IV$ (methane). This fact offers a variety of processes for conversion of biomass, which are interesting in the context of hydrogen as an energy vector as there are reactions producing hydrogen, as well as reactions requiring hydrogen as reactant.

The formal overall chemical reactions involved in biomass conversion are listed in Table 3.4, together with the corresponding reaction enthalpies at standard conditions ($\Delta H°_{298}$).

Biomass *combustion* releases the stored energy in biomass as heat (heat of combustion) and the complete organic carbon as CO_2. The enthalpy change of 2668 kJ mol^{-1} is the maximum energy that can be recovered as heat.

Hydrogen production via *auto-thermal gasification and shift reaction*. Biomass is reacted with water and oxygen in such a ratio that over all no reaction enthalpy is released. In order to compensate for heat losses, practical systems usually run with higher amounts of oxygen and negative reaction enthalpies. This reaction is the base for thermo-chemical hydrogen production from biomass. Technically the reaction is carried out in two steps: gasification at high temperature (to avoid the

Table 3.4 Options for thermal conversion of biomass.

	$\Delta H°_{298}$ (kJ mol^{-1})
Combustion $C_6H_9O_4 + 6.25O_2 = 6CO_2 + 4.5H_2O_g$	-2668
Autothermal reforming (gasification)/shift reaction $C_6H_9O_4 + 5.5H_2O_{fl} + 1.25O_2 = 10H_2 + 6CO_2$	0
Methanation $C_6H_9O_4 + 1.75H_2O_{fl} = 3.125CH_4 + 2.875CO_2$	-83
Hydrogasification $C_6H_9O_4 + 11.5H_2 = 6CH_4 + 4H_2O_g$	-636

formation of methane), followed by conversion of the CO to H_2 and CO_2 using steam (water gas shift reaction). Essentially the same process is currently applied on a large scale to produce hydrogen from natural gas or from naphtha.

Methanation. The formation of methane and CO_2 from biomass is a classical example of chemical disproportionation of the zero-valence carbon in biomass. The reaction is mildly exothermic and is the natural decomposition reaction of wet biomass in the absence of oxygen (anaerobic digestion of biomass). The reaction also proceeds at elevated temperatures (up to 400 °C) in supercritical water as a reaction medium [29]. Alternatively the reaction can be carried out in a two-stage process of gasification of biomass to synthesis gas, followed by catalytic methanation at $T < 400$ °C (Biollaz).

Hydrogasification. Biomass can be reduced to methane using hydrogen as a reducing agent. The reducing reaction is referred to as hydrogasification [30], is exothermic and proceeds at temperatures of 800–1000°C. Hydrogasification converts all biomass carbon to methane fuel and emits no carbon in the form of CO_2. Due to the hydrogen consumed by the reaction, the energy content of the product gas is higher than the heating value of the biomass.

3.2.4
Use of Biomass for Hydrogen Production or as a Renewable Carbon Resource

As biomass is per se a form of chemically stored renewable energy, there is no need to convert it to hydrogen for storage purposes, as is the case for solar energy, wind- or hydro-power, where hydrogen is expected to play an important role as a storage medium. Hydrogen can be produced from biomass in very much the same way as it is currently produced from fossil fuels, that is by steam reforming of the carbon compounds constituting the biomass material. As with fossil fuels, it only makes sense to produce hydrogen fuel from biomass if the overall conversion efficiency to electricity, in for example a fuel cell or a combustion engine, can be made better and/or less polluting than the direct conversion. A strong alternative to hydrogen from biomass is the production of conventional fuels (synthetic gasoline/diesel fuel or synthetic natural gas). With the broad introduction of fuel cell cars, it is likely that a market for renewable hydrogen derived from biomass will develop as its production is expected to be considerably less costly than the production of hydrogen from renewable electric power via electrolysis, the only viable alternative for production of hydrogen from solar, wind or hydro-power today.

In the long term, biomass is predestined to perform the essential functions that petroleum-based products perform today. Among these are the use of petrochemicals as the basis for chemical industry and transportation fuels. The production of hydrocarbons from biomass via thermochemical gasification to synthesis gas and subsequent synthesis of hydrocarbons in the Fischer–Tropsch process is state of the art. In the very long term, when fossil fuels will be exhausted, or their use will be restricted for environmental reasons, biomass will play an important role as a source of organic carbon. Hydrogen derived from water will be essential for the chemical reduction of biomass to liquid or gaseous hydrocarbons with high heating value. Conversion reactions of the type of the hydration of biomass (e.g. the hydro-gasification reaction given by way of example in Table 3.2) will become important as a means of converting biomass to chemicals without wasting carbon to the atmosphere as CO_2. Renewable hydrogen could boost the potential of biomass to produce the hydrocarbons, which are hard to substitute by a factor of 2 to 3. This "marriage" of renewable hydrogen with biomass has also been discussed as a liquid medium for seasonal storage of hydrogen in the form of methanol, an energy carrier that can easily be converted back to CO_2 and hydrogen [31].

3.3
The Hydrogen Cycle
John D. Speight

3.3.1
Introduction

Global energy demand is forecast to double by 2050. By persisting in our use of current energy technologies most of this demand will be met by the increased use of our rapidly dwindling supplies of petrochemicals. In Europe alone, the oil import dependence is set to grow from 50 % currently, to 70 % or more, by 2025. Even more serious, the European Community is already dependent on oil for 90 % of its transport needs.

A central theme of this volume is that the Western economies must take the lead in moving away from fossil fuels towards the use of hydrogen to supply major parts of their demand for power, heat and transportation. Only by undertaking this transition in the usage of our energy sources will we avoid the threats of resource depletion and man-made(anthropogenic)carbon emission-induced climate change. In this chapter, the cyclic nature of existing fossil fuel usage will be compared to that offered by moving towards a hydrogen-based economy. Particular attention will be paid to highlighting how the transition to clean, sustainable energy supplies could stabilize climate change.

The role of cyclic phenomena in the Earth's biosphere began with the identification of the carbon cycle in 1783 by Senebier. Over the following century, the cycles associated with nitrogen (1886) and the water cycle (late nineteenth century) were recognized. The integration of these cycles and other environmental phenomena into one model, the Gaia hypothesis, was finally achieved by Lovelock [32] in 1979.

3.3.2
The Hydrogen Cycle

3.3.2.1 Replacing the Fossil Fuel/Carbon Cycles with Hydrogen
Senebier's carbon cycle in its original and simplest form, described the absorption of atmospheric carbon dioxide by plants, which are subsequently eaten by animals, who in turn return the carbon dioxide to the atmosphere by respiration and

Table 3.5 Global carbon budget.

	Gt C/yr
Sources	
Fossil fuel combustion	5.5
Tropical deforestation	1.6
	7.1
Sinks	
Retained in atmosphere	3.3
Ocean uptake	2.0
Uptake by Northern Hemisphere reforestation	0.5
	5.8
Net Imbalance	1.3

excrement. This concept and a more extended form of the carbon cycle, including fossil fuel usage, is shown in Fig. 3.4. Carbon-rich fossil fuels, oil, coal and natural gas, are extracted from the near surface of the earth and burned as energy sources. The combustion of fossil fuels produces large quantities of carbon dioxide, oxides of nitrogen and sulfur together with soot. Since the Industrial Revolution the proportion of these gases in the atmosphere has increased, accompanied by a corresponding rise in the global average near surface temperature; the greenhouse effect.[1] Typical carbon fluxes associated with the major components of the cycle are shown in Table 3.5. The only significant natural absorption mechanism for excess carbon dioxide is via vegetation and this is not able to keep pace with the rate of fossil fuel-generated emissions. Equally serious are the geological timescales needed for the plant life cycle to replace the original coal or oil resources. Thus, whilst similar in mechanisms, the traditional carbon cycle and the fossil fuel cycle, shown on opposite sides of Fig. 3.4, differ in cycle times by tens of thousands of years. This fundamental imbalance has a number of major consequences, the most important being:

1. Use of fossil fuels, particular petrol chemicals, is rapidly depleting a finite and nonrenewable resource.
2. Carbon dioxide emissions from fossil fuel combustion will accumulate and accelerate the greenhouse warming mechanisms underlying global warming.

To escape from these impending crises, a fuel system capable of providing ongoing energy sources in a nonpolluting manner is needed. Fortunately, the hydrogen cycle, which has a similar timescale to that of the carbon bio-system, can provide an answer.

---

[1] Solar heating of the Earth occurs via shortwave radiation, which is absorbed by the Earth's surface and then reflected back as longer wavelength radiation. A fraction of the long wavelength radiation can be trapped by "greenhouse gases" in the upper atmosphere, thus creating a blanket effect and raising global temperatures [33].

3.3 The Hydrogen Cycle

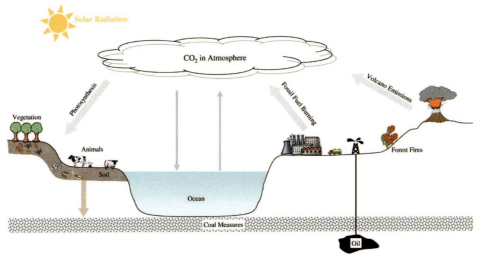

Fig. 3.4 The global carbon/fossil fuel cycle.

3.3.2.2 Key Elements of the Hydrogen Cycle

A common feature of the hydrogen and carbon cycles is that both are inaccurate descriptions of the phenomena they claim to depict. In neither case does elemental carbon or hydrogen play a major role in the underlying mechanisms of the cycle. In the case of the carbon cycle the predominant forms of carbon are carbon dioxide and complex petrochemicals. For the hydrogen cycle, the major engine of the cycle activity is provided by water rather than elemental hydrogen. A full understanding of the phenomena involved begins, therefore, with the hydrological cycle.

The hydrological cycle, as the name implies, describes the circulation of water between terrestrial and atmospheric systems. The key elements are shown in Fig. 3.5. Water enters the atmosphere by evaporation, predominantly from the ocean surfaces. Smaller degrees of evaporation occur via landlocked bodies of water and evapotranspiration of plant life across the continental land masses. The return of water from the atmosphere occurs in various forms of precipitation. Typical fluxes participating in the global water budget are also shown in Fig. 3.5.

The use of the hydrogen cycle as a practical source of energy is shown schematically in Fig. 3.6. Production of hydrogen by the electrolysis of water using renewable energy sources, such as solar, wind, biomass or hydro-electricity, allows the derived hydrogen to be stored, distributed and converted into useful energy. Hydrogen may be used directly in combustion processes in a similar manner to petrol or natural gas. However, as will be shown later, hydrogen fuel cells represent a carbon-free method of heat and electricity generation compatible with many domestic and industrial applications. During all these end of cycle conversion processes, hydrogen reacts with oxygen leading to water or steam, which is returned to the atmosphere or hydrological system. At the front end of the cycle, using renewable energy sources for hydrogen production would eliminate the carbon dioxide emissions associated

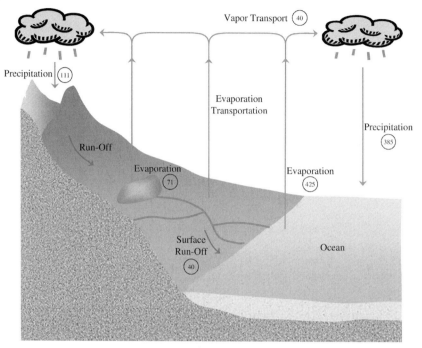

Fig. 3.5 The global water cycle [Fluxes in 10^3 cu·km/gr.].

with the currently used reformation technologies for hydrogen manufacture, thus reducing global warming. A further benefit of the hydrogen energy cycle is that, in contrast with electricity, it allows the possibility of large-scale storage. This facet of the cycle is of particular importance where hydrogen is derived from intermittent energy sources, such as solar or wind.

Thus, the hydrogen cycle may be regarded as a perturbation or subset of the larger scale hydrological cycle, in that hydrogen may be extracted from water by a variety of processes (these are described later in this chapter) and burned as a fuel from which the combustion product is water.

3.3.2.3 The Centrality of Water

The interlinking of the hydrogen and hydrological cycles implies that availability of water, for either electrolysis or other means of splitting it into its constituent elements, is a crucial factor in achieving sustainable, renewable and "clean" production routes for gaseous hydrogen.

The Earth is 70 % covered by water, but 97 % of this is saline, leaving around 2.5 % as directly electrolysable freshwater. Of the freshwater budget, only 1 % is available for direct use, as the balance is frozen in the global ice caps. Current IPCC assessments [34] note that many areas of the world are likely to experience water stress. It is estimated that up to 2.3 Bn people worldwide live in areas where the annual availability of water is less than 1700 m^3/year. To compound the problem, 30 % of

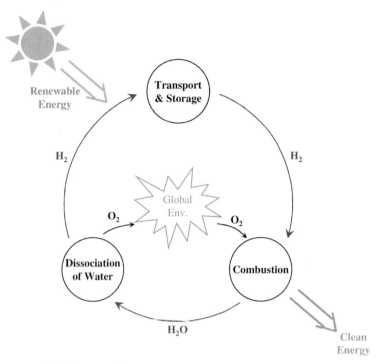

Fig. 3.6 The hydrogen cycle.

freshwater sources are aquifer-based. Such sources have long replenishment cycles and may not be renewable in practical timescales.

For the developed countries the emergence of extra water demand for hydrogen production is unlikely to be critical. The projected annual demand for hydrogen generated by migrating the US light transportation fleet to hydrogen fuel [35] is estimated as 150 Mtonnes/year, which is equivalent to 100 billion gallons of water/year. As domestic water usage in the US is around 4800 billion gallons/year and conventional power generation is 70 trillion, the amount needed for hydrogen generation would not be a significant perturbation. As noted above, for developing countries with existing water stress constraints, diversion of water to the production of hydrogen could be more problematic. The advent of new technologies may present a unique opportunity for third World countries to break the existing cycle of energy related indebtedness. Whilst poor in petrochemical resources many third World countries are rich in solar and other sustainable energy sources. Appropriate combinations of aid and technology support could thus alleviate poverty and energy dependence.

3.3.2.4 Other Cycles Active in the Global Environment

A central theme of the present volume is amelioration of global warming problems by the adoption of hydrogen as an alternative fuel. The dominant biosphere cycles underlying these problems are those associated with carbon and water. Other cycles operate in the biosphere but, whilst of importance, have a less marked influence on climate and the environment [37].

Sulfur Sulfur is an essential life element, being assimilated by living organisms and released as the end product of metabolism. Over the past two centuries, in a similar manner to carbon, the sulfur cycle has been perturbed by human activity. Currently anthropogenic emissions, mostly from transport and manufacturing, constitute 75 % of the global budget for sulfur. Removal of SO_2 from the atmosphere can occur in the form of acid rain and the consequential severe impacts on plant and animal populations. Vast quantities of sulfur dioxide are released into the upper atmosphere during volcanic eruptions (see Fig. 3.4). The gases are converted into sulfuric acid and aerosols. When large-scale eruptions occur the released sulfur compounds can lead to hemispherical global cooling lasting 2–3 years. Given that major eruptions occur about every decade means that volcanoes are a significant factor in climatic variability. However, sulfur injection into the atmosphere is relatively short-lived compared to carbon dioxide [37].

Halogens Chlorine, bromine, fluorine and iodine are all present in the atmosphere due to several natural and anthropogenic processes. The most important compounds are the chlorofluorocarbons (CFCs). Besides being greenhouse gases, CFCs were responsible also for the rapid depletion of the Antarctic ozone layer in the 1980s. As a result of international action, the concentration of atmospheric CFCs has declined, which should stabilize the ozone depletion.

Nitrogen Nitrogen is the most abundant (80 %) element in the terrestrial atmosphere. Molecular nitrogen is stable and has a minor role in the lower atmosphere. In contrast, minor constituents such as N_2O, NO, NO_2, nitric acid and ammonia are chemically active. All play important roles in environmental problems, such as acid rain, urban smogs, tropospheric aerosols and destruction of the ozone layer. Oxides of nitrogen (NO_x) generally, but most notably N_2O, are potent greenhouse gases. Large amounts are released in industrial combustion processes and exhaust emissions from petrochemical fueled transportation.

3.3.2.5 The Present Hydrogen Scenario

There is now almost unanimous international agreement on the need for urgent action to address the twin threats of petrochemical resource depletion and the onset of extreme climate change. With few exceptions, leading economies are evolving plans for the use of hydrogen and other petrochemical alternatives, particularly in the automotive sector.

Despite the clear advantages of basing our future energy needs on the hydrogen cycle, there are many technical, political and socio-economic barriers to be overcome before hydrogen becomes a viable option. The current status of hydrogen technology for the leading economies is similar. All have large hydrogen-oriented research and development programs, but actual usage of hydrogen as a fuel is limited to a small number of demonstrators. Implementation of the transition will require significant technical advances or breakthroughs in all aspects of the hydrogen cycle. The current status of three major aspects of the hydrogen cycle will be examined in the following sections:

- production of hydrogen
- distribution, transmission and storage of hydrogen
- converting hydrogen into useful power.

3.3.3
Exploiting the Hydrogen Cycle for Energy Production

3.3.3.1 Hydrogen Production
Non-Renewable Production Methods for Hydrogen

Natural Gas Reformation Large amounts of elemental hydrogen are currently produced by the petrochemical industry for use in the desulfurisation of diesel fuels, hydrogenation of edible oils and as a feedstock for many hi-tech manufacturing processes. The majority of today's hydrogen is made from natural gas reformation with steam, some oxygen and air. The equation for this process is:

$$CH_4 + 2\,H_2O = 4\,H_2 + CO_2 \tag{3.1}$$

Unfortunately, production of hydrogen from natural gas has the flaws of other fossil fuel sources:

1. Accelerating depletion of finite petrol chemical resources.
2. Carbon dioxide as a major by-product of the process.
3. Process heating by natural gas contributes further carbon dioxide.

Although simple, well-established, commercially viable and being capable of meeting the initial demand for hydrogen, production of hydrogen from natural gas cannot be regarded as other than an intermediate solution. If the major economies are to move towards large-scale use of hydrogen as a fuel, other low carbon production routes for hydrogen must be exploited.

Coal gasification has been explored as a possible mass production process for hydrogen, during previous petrochemical supply crises. The process has attractions in that it could exploit the large coal reserves known to exist in many parts of the world. Unfortunately, without sequestration of the resultant carbon dioxide outputs, the process would be environmentally unacceptable.

Nuclear power, especially in its advanced variants, has also been proposed as a means of providing high quality heat to the many possible production routes for hydrogen. In terms of ease of scaling-up, engineering efficiency and the benefits of 50 years of operating experience, persuasive arguments can be made in favor of this, versus the uncertainties of more novel approaches. Nuclear technology is, however, not strictly renewable and it brings the politically and economically unresolved problems of waste management.

Hence, other production routes for hydrogen must be exploited to avoid the current spiral of increasing pollution and fossil fuel dependence. Fortunately, there are numerous renewable production processes for hydrogen.

Renewable Processes for Hydrogen Production Supplies of hydrogen from nuclear and fossil fuels are commercially attractive in the short to medium term, but do not address the critical issues of geo-resource depletion and anthropogenic global warming. Ideally, sustainable processes for hydrogen production need to be established by the time petrochemical depletion begins to lead to unacceptably high fuel prices. Of the currently available methods, only water electrolysis seems capable of answering production needs. There are, however, emerging alternative and potentially sustainable processes, which will be considered in the following section.

Established Methods *Electrolysis.* Apart from the natural gas reforming processes, electrolysis of water is the most widely used method for hydrogen production. To electrolyse water, the minimum electrical energy required is determined by the free energy change of the underlying reaction:

$$H_2O = H_2 + 1/2 O_2$$

With a free energy change of 237 kJ mol^{-1}, a theoretical decomposition voltage of 1.23 V (25 °C and 1 bar pressure) is obtained. Under isothermal conditions, the cell voltage is 1.47 V and gives a process efficiency of 83 %. In practice, higher cell voltages (1.7 to 1.9 V) are used, giving lower energy utilisation and efficiencies (80 %). Practical efficiencies are in the range 70–75 %.

Volume production of hydrogen by electrolysis of water is achieved using one of the following process variants:

- *Water electrolysis* with aqueous alkaline electrolytes (30 % KOH) and asbestos electrodes is widely used in small- and medium-scale units (0.5–5.0 MW)
- *Solid polymer electrolyte water electrolysis* uses proton conducting ion exchange membranes as electrolyte and membrane. Commercial units with power ratings up to 100 Kw are available. Projected efficiencies are in the range 80–90 %.
- *High temperature steam electrolysis* exploits the marked decrease in operating cell voltage above 700 °C. Oxygen ion conducting membranes, operating at 700–1000 °C are used as the electrolyte. The water to be disassociated enters on the cathode side as steam, leading to a steam–hydrogen mixture. Production costs could be lower for this option, as process heat is cheaper than electrical power.

A significant consideration with all electrolytic options is the purity of the input water supply. Impure water can significantly reduce the lifetime of electrolytic cells and front–end purification plant will add to the final hydrogen production costs.

Table 3.6 Existing and potential sources of electrical power for electrolysis.

Technology	Power output	Input energy
Nuclear power	GW	Nuclear fuel
Hydropower	GW–kW	Water Heads
Solar power	MW	Solar radiation
Solar parabolic	GW–kW	Solar radiation
Wind turbine	GW–kW	On/off shore wind
Solar cell	GW–kW	Solar radiation

A wide variety of electrical energy sources are currently in use, or being developed, for electrolyzing water via one of the above options. These are summarized in Table 3.6. Of the options shown, only hydropower, wind turbine and solar cells may be regarded as mature technologies. Nuclear power offers an apparently attractive route to the enhanced efficiencies of high temperature electrolysis. In addition, the nuclear route could make electrical power for hydrogen production available at GW levels, independent of naturally varying phenomena such as wind and/or solar. A further advantage claimed is that nuclear plants could be sited near to points of use. Several design studies are being conducted in the US, Japan and the EU [38, 39]. Whatever the attractions, nuclear technology remains dependent on depletable resources and continues to be a contentious issue for public debate, especially regarding the safety and economic viability of long-term waste storage.

Emerging Processes Electrolysis is currently the sole source of sustainable and renewable hydrogen operating at commercial scales. Other renewable processes have, however, been known for many years. These processes have only recently been reconsidered for commercial production of hydrogen and are, in most cases, at early states of development.

(i) Biomass The use of biomass techniques to produce a variety of liquid and gaseous fuels which could ultimately be reformed into hydrogen is well established [40]. The main limitations to its exploitation are up-scaling of the feedstock supply to meet national and global demands. By definition, biomass production techniques are sustainable, but have the major limitations of not being able to supply sufficient hydrogen for the foreseen growth in demand in advanced economies. Biomass may, however, be relevant in specific geographic zones of western economies, where feedstock is readily available, for example Scandinavia and Canada. In the third World, the choice will be complex. Should the biomass crops be directed to alleviating hunger or fuel crises?

(ii) Thermochemical Water can be split by purely thermal energy at temperatures in excess of 2000 °C [41, 42]. Clearly, the engineering difficulties of designing a reactor with materials to withstand these temperatures over sustained periods are beyond our current capabilities. Even if such a reactor could be built, the

associated problems of thermal management and heat extraction would pose further insurmountable engineering problems. At present, the upper temperature limit for commercially viable processes is in the region of 100 to 1200 °C. Working knowledge of these conditions is available from the nuclear industry and solar furnace technologies. If an upper operating temperature of around 1000 °C is accepted, direct disassociation of water can only be achieved by multistep thermochemical processes.

The minimal process involves two steps:

$$H_2O + 2X = 2XH + \frac{1}{2}O_2$$

$$2XH = 2X + H_2$$

Of key importance is the nature and role of the X intermediate compounds. Many variants of possible reactions have been explored, involving from three up to twentyfive steps. Despite the intense activity in this field, none of the multistep processes has moved beyond laboratory scale demonstration. Of the possibilities, the iodine–sulfur reaction appears the most promising, but hydrogen sulfide, sulfuric acid–methanol and Br–Ca–Fe cycles are also being investigated [41, 42]. The past decade has seen a significant increase in research and development in this area. Typically, temperatures in the region of 880 °C are needed for efficient hydrogen production. Using nuclear plant as a source of heat, it is projected that production costs could be significantly lower than for room temperature electrolysis processes. However, a major barrier is, that even at the lower projected process temperatures, all the thermochemical production processes for hydrogen will require the development of improved high temperature materials for the reactor vessels and transfer functions.

(iii) Thermophysical   As noted above, the direct splitting of water into its constituent elements requires temperatures of 2000 °C or higher, which are impractical to produce or to contain under current commercial conditions.

Direct production of hydrogen at lower temperatures can, however, be achieved using technologies similar to that of a fuel cell (see later). Water is split into protons by a catalyst and these are transported by an ion conducting membrane. The key requirements are catalysts capable of achieving water splitting at around 1000 °C and a ceramic membrane capable of conducting both electrons and protons, which subsequently recombine into hydrogen on the other side of the membrane. The method is relatively simple, of high efficiency (around 50 %) and potentially cheap. The economics, as with fuel cells, are dependent on the availability of suitable membranes. The required temperatures are readily available via solar furnaces or high-temperature nuclear reactors. As with the other technologies discussed in this section, virtually no carbon dioxide emissions would be involved.

(iv) Photo-Electrochemical Methods   Other photo-induced water splitting processes are available besides the use of photo-electricity to manufacture hydrogen from the electrolysis of water. An intermediate absorption system is required in all these methods, as water molecules do not absorb radiation in the visible and near-UV ranges. The minimum energy required for the splitting of water molecules in the liquid phase is only 2.45 eV.

The basis of the processes is the absorption of incident solar photons leading to the generation of electron–hole pairs in a semiconductor in contact with an aqueous electrolyte. It is energetically favorable for the minority carriers (holes) to diffuse to the semiconductor/electrolyte interface, where recombination with electrons from the valence band of water can take place. The hydrogen ions migrate to the metal cathode and are reduced to hydrogen molecules.

The overall process may be shown as:

$$2H_2O + 4h\gamma = 2H_2 + O_2 + \Delta H$$

Theoretical efficiencies for the process are high ($\sim$25 %) but, in practice, an order of magnitude lower is observed. A major problem with the process is the evolution of oxygen at the semiconductor anode surface. To date, very few materials of the required band-gap (1.5–2 eV) have proved suitable [36, 43].

Overall, the method would seem to offer the potential for efficiencies in the range of 5–10 %. This range is close to that projected for practical solar cell/electrolysis of water systems. Given that these technologies are relatively well established, photo-electrochemical methods must aspire to significantly higher efficiencies or lower costs to be considered a viable option.

(v) Photo-Biological Processes Solar energy can be used to release hydrogen and oxygen from biological systems, such as green or purple algae. The mechanism in all these processes is enzymatic conversion of protons and electrons into molecular hydrogen.

Biophotolysis is essentially photosynthesis, that is splitting water into electrons, protons and oxygen. Green algae and cyano-bacteria may be used for the transfer of electrons to protons, to achieve reduction and thus form hydrogen. A notable advantage is that the starting material is only water.

Phototrophic bacteria (purple) are also able to reduce protons obtained from biomass and give off gaseous hydrogen. Production rates for hydrogen can be good via this approach (150 LH_2/h/kg biomass). Combined biophotolytic and phototrophic processes have also been explored.

Other biological methods such as *Photo and Dark Fermentation* can also produce hydrogen from organic substances with the addition of micro-organisms. A major advantage here is that the basic technologies of fermentation are well established and therefore amenable to scaling-up without further research and development. Against this, fermentation processes are relatively low yield and, in addition to hydrogen, produce carbon dioxide, methane and other troublesome organic by-products.

A recent study has pointed out the path forward for biological production of hydrogen is unclear [43]. In particular, much work remains to be done on current processes to achieve hydrogen production rates sufficient to power practical fuel cells, or to match the current electrolysis production rates for hydrogen ($\sim$1000 l/h). Despite these adverse features, the methods have the virtue of being simple to execute and are ecologically friendly.

The Path Forward The near-term outlook is good in that current hydrogen production methods should be adequate for the early stage demands of the hydrogen

Table 3.7 Hydrogen production methods: environmental, economic and scaling factors.

Method	Projected electricity price US$/GJ[a]	Renewable	Effect on hydrogen/ water cycle	Effect on other cycles	Effect on carbon cycle	Scaling potential[b]
Natural gas reformation	7–11	×	Low	High	High[c]	Good
Coal gasification	8–11	×	Low	High	High[c]	Good
Biomass gasification	10–18	✓	High		High[c]	Moderate
On-shore wind	17–23	✓	Low	—	Low	Moderate
Off-shore wind	22–30	✓	Low	—	Low	Moderate
Thermal solar/electrolysis	27–35	✓	Moderate	Low	Low	Good
Solar PV/electrolysis	47–75	✓	Moderate	Low	Low	Moderate
Nuclear electrolysis	15–20	×	Moderate	Low	Low	Good

[a] IEA 2003.
[b] Ability to meet projected rate of growth in global energy demand.
[c] Low if carbon sequestration included.

economy. Supply would, however, be largely via the reformation of natural gas with the associated carbon dioxide emission problems. For the medium and longer term, continued adoption of present production processes would be unacceptable; the associated carbon dioxide burden and escalation of fossil fuel prices brought on by resource depletion would present insurmountable barriers.

Hydrogen production costs will be a key factor in determining the migration away from natural gas reformation as the accepted means of hydrogen volume production. In Western economies, the current costs for the common fuels are $6–8/GJ for gasoline and $3–5/GJ for natural gas. The IEA [44] has recently published projected (2020) comparative hydrogen cost figures for a range of production technologies (see Table 3.7). For many of the techniques, the projections are encouraging. With the possible exception of solar PV, where large reductions in the cost of plant need to occur, the clean forms of traditional technologies (coal and natural gas) and wind and solar thermal seem capable of offering hydrogen prices approaching substitution levels with traditional fuels. For these techniques, reductions in production costs may be anticipated as market demand grows. This trend will also be applicable to electrolysis, where there appears to be limited scope for cost reductions through improved efficiency beyond 70–75 %, but savings may be possible via lower plant costs and improved reliability.

Whilst in the initial stages of the hydrogen economy the cost of hydrogen production will be crucial, other factors, such as those listed in Table 3.7 must also be considered as the penetration of hydrogen gathers pace. Turner [35] has estimated that if the enormous potential of solar and wind-powered production can be harnessed, electrolysis could supply all foreseen hydrogen needs. As Table 3.7 indicates, such a combination of production technologies would be an ideal implementation of the hydrogen cycle and would have little impact on any of the other major global cycles affecting the biosphere.

An even more optimistic scenario for the emergence of the hydrogen economy could emerge should rapid research and development progress be made on the novel processes being explored for cleaner and high efficiency of hydrogen production. Of the emerging processes considered in the present section, thermophysical and other high temperature and high efficiency techniques appear to be the most promising. Production costs for hydrogen via these processes have been estimated to be $\sim\$8/GJ$ [44]. Proposals have been made in the European Community states (notably France) to exploit the higher temperatures offered by the next generation of nuclear reactors to access this route. No consensus has formed on the nuclear option for hydrogen generation because of political barriers and unresolved issues associated with waste management. Longer term, major benefits could be derived from exploiting the third World's huge, and as yet untapped, solar resources to produce hydrogen.

3.3.3.2 Transmission, Distribution and Storage of Hydrogen

Of the elements needed for successful exploitation of the hydrogen cycle, the downstream technologies needed to reach the end user, that is transmission, distribution and storage of hydrogen, are likely to be the least problematic. Western industrial economies have over 150 years of experience in the technology requirements for coal, or town gas, which was largely based on hydrogen. More recently, similar problems have been solved successfully for the exploitation and distribution of natural gas. Direct experience also exists for the transmission and storage of hydrogen in specific industrial usages and sites. Much of the existing expertise has, however, evolved in sectors likely to play minor roles in the hydrogen economy of the twenty first century. In transport particularly, there will be far reaching demands on infrastructure as the change is made from fossil fuels to hydrogen.

Transmission If a migration towards a hydrogen economy gathered momentum, it would be necessary to transmit hydrogen from the point of production to major end users and distribution hubs by pipelines similar to those currently used for natural gas. Extensive pipelines for industrial usage of hydrogen already exist in the UK (Teeside), the US (NASA) and Europe (Germany – 210km linking 18 industrial sites). Transition to the transport of hydrogen via long distance pipelines seems unlikely to present major problems. Several earlier studies [45] have shown that by using higher pressures for hydrogen than natural gas, to compensate for the lower density, hydrogen pipelines are economically viable. Two technical problems seem relevant.

1. Seals and materials for pipelines would need careful choice given the very high mobility of hydrogen and its ability to permeate many low density solids.
2. Current compressors are likely to be inadequate for handling high volume, high pressure hydrogen systems.

The EC, US and Japan have large nationally funded programs addressing the problems of transmission of hydrogen by pipelines. An EC program is exploring the possibility of joint natural gas–hydrogen transmission with separation into constituent gases at the point of use [46]. It should be noted, however, that hydrogen is odorless and burns with an invisible flame. In widespread use it would be necessary to add odorants and colorants to the distributed hydrogen for detection and general safety.

Distribution The distribution of hydrogen from regional production centers or pipeline hubs, is likely to be based on the rail, road or marine transport of liquefied hydrogen. All these modes of distribution have been used for liquefied gases in the leading economies for the last quarter of a century. Very extensive experience has been acquired in the US via the NASA programs dependent upon liquid hydrogen fueled rocket launchers. Some of this working knowledge would be of direct relevance in the early stages of a hydrogen economy, when fueling stations would be dependent upon hydrogen supply by road or rail.

For optimum exploitation of renewable global resources, the predominant future sources of hydrogen should, ideally, be sited in areas of high insolation or wind assets. Transportation of hydrogen between continents would therefore be needed on a similar scale to that currently encountered with crude oil tankering. Several major studies on the tanker-scale marine transport of hydrogen have been made [45]. Tanker-scale cryogenic containers can be fabricated suitable for existing ships' holds. The lower heating value of hydrogen compared to that of natural gas implies that to deliver the same energy content would require approximately 2.5 times larger cargo volumes. Most other features of the marine architecture of liquid hydrogen tankers would be closely similar to current practice for natural gas. Australia has plans to exploit the high insolation resources of its western deserts. Hydrogen would be converted into ammonia (17% hydrogen) for marine transport, thus avoiding the need for cryogenic transportation techniques.

Storage of Hydrogen In current industrial usage, hydrogen is stored as a compressed gas, as a cryogenic liquid or, in a small number of instances, solid-state compounds, such as metal hydrides. The choice of storage for specific applications will be determined by volume requirements and cost limitations. Some of the main application areas are discussed in this section.

Large-Scale Storage of Hydrogen Systems for large-scale storage of hydrogen are needed typically at the point of hydrogen production, at pipeline hubs or termini. In general, it can be assumed that the hydrogen to be stored will be compressed or cryogenically reduced as part of the overall storage process. In studying the economics of large-scale stores, the cost of both reservoir and compressor plant are equally important.

Underground storage of hydrogen will probably evolve along similar lines to that used for natural gas. Several cavern sites are already used for this purpose in West Germany and in the UK, notably in Teeside [45]. Many factors need to be considered in choosing suitable cavern sites. Even if the sites are geologically and technically feasible, there may be limitations in storage capacity.

Above ground, large-scale storage of hydrogen is typically one to two orders of magnitude more expensive than underground cavern systems. This cost differential means that surface pressure/cryogenic storage systems are largely restricted to applications demanding short storage cycles, that is NASA cryogenic stores for rocket launch or storage systems at points of production and distribution.

Medium-Scale Storage Within this volumetric storage range, a number of commercially important application areas are emerging. Most prominent in this sector, but unlikely to be deployed elsewhere on costs grounds, is the one million gallon liquid hydrogen storage tank at Cape Kennedy. At somewhat lower volumes, there is wide ranging engineering expertise on the storage of coal and natural gas for local and domestic usage. In this sector, two current applications are prominent.

(i) Intermittent Storage Many of the energy sources capable of producing renewable hydrogen are intermittent in nature, for example solar, wind and tidal sources (see Table 3.6). To overcome this limitation, hydrogen can be produced in low demand periods, stored and used at peak periods or for other local energy needs. Many field study and demonstration projects are in progress exploring hydrogen storage in association with renewable energy sources [47, 48]. The preferred storage technologies are compression, typically up to 350 bar, or absorption of hydrogen in a solid state medium, such as a metallic hydride (see Section 3.2.3) Typical storage volumes are around 10 000 litres.

(ii) Storage of Hydrogen at Fuel Cell Vehicle Refueling Stations In the major economies, hydrogen is expected to become the leading fuel for transportation within the next 20 to 30 years. Energy suppliers and vehicle companies are conducting field trials on refueling facilities needed for city buses, defined route vehicles and, to a lesser extent, for fuel cell-powered cars. At present, there are around 30 experimental hydrogen fuel stations worldwide, mostly in Germany, USA and Japan. A further 30 to 50 are projected to emerge in the 2004/2006 period. Currently the fuel storage capacity at an urban fueling station is designed to replenish around 500 family saloons. If this were to be replicated for hydrogen, it would translate into fuel station storage for 2500 kg of hydrogen. Current field trials are at a lower scale, typically storing 1000 kg of hydrogen, produced locally by electrolysis. Nearly 60 % of current stations use compression at 350 bar as the storage system, with the balance being liquefied hydrogen (21 %) and solid-state stores. The technology for compression and liquefaction of hydrogen on this scale are well established. For solid state fuel station stores to become technically viable, the prohibitive costs of the storage medium and the need to standardize both the on-board and off-board hydrogen storage solutions must be addressed [49].

Small-Scale Hydrogen Storage The major fossil fuel dependence in western economies is for transport. At present, the biggest barrier to the adoption of hydrogen as a fossil fuel replacement, especially in light vehicles and cars, is a viable on-vehicle storage system for hydrogen.

A family saloon powered by fuel cells would need around 5 kg of hydrogen to achieve a range of 500 km [50]. A practical store must be about the same volume and weight as a current gasoline tank and be able to be recharged in the same time and under the same safety conditions as for gasoline. Needless to say, it must also be low cost. Both compressed gas and the other, though less widely accepted, possibility of liquefied hydrogen, would require bulky storage vessels for hydrogen to provide the target range. Pilot fuel cell vehicles operated with compressed gas and cryogenic hydrogen have been, and are being, successfully operated in the US, Japan and Germany.

Whilst compressed hydrogen at pressures of 1000 bar or higher is currently seen as the preferred option for the market entry of fuel cell cars, solid-state storage of hydrogen offers the most promising long-term solution for cars and light vehicles. Only via this approach can the target parameters of weight, volume and the desired range be achieved. Several trial vehicles have been operated successfully with solid-state stores. All the main operating requirements, apart from range, can be met. Unfortunately, with the presently known hydrogen absorbing compounds, stores capable of achieving the target range would be excessively heavy. For example, using titanium–iron alloys the weight would be 400 kg. A wide variety of materials is currently under investigation; the status is summarized in Table 3.8. For the conversion of light vehicles to renewable fuels to occur, a commercially viable storage system for hydrogen would seem to be the single most challenging technological hurdle.

The Path Forward As a result of our long working knowledge on the use of piped gases as energy carriers, the transmission, distribution and storage of hydrogen present lesser technical challenges than other aspects of hydrogen technology. In addition, this sequence of processes is the most efficient in the hydrogen cycle; overall, losses in this part of the cycle are around 10%, due to a combination of leakage and boil-off. More serious are the current problems associated with specific aspects of the processes, for example:

- In the liquefaction of hydrogen, the energy expended in compression is ~30% of the potential energy yield of the stored hydrogen.
- To gain equivalent energy yield for pipelined hydrogen to that of natural gas requires compression to higher pressures.
- Current compressor technology is optimized for heavier and less mobile gases than hydrogen, hence losses on hydrogen compression due to leakage and lower efficiencies are high.

Whilst these issues are not technically limiting, the losses represent significant cost additions in the evolution of a hydrogen infrastructure.

Table 3.8 Hydrogen storage technologies.

Technology	H$_2$ storage density mass %	volume (kgH$_2$/m^3)	Operating temperature (C)	Pressure (bar)	Status/Issues
High pressure composite cylinders	13	33	25	100	Current choice for on-board storage in demo vehicles.
Liquid hydrogen	100	71	−252	1	Boil off and bulk still issues. In use for large fleet demos.
High surface area materials (physisorpion)	2–5	20	−196	15–50	Based on well established phenomena of gas/surface interaction. Volume densities low.
Metal hydrides – low temperature	2	150	25	2–9	Used with small scale PEM FC and ICE units. Limited range/excess weight.
Metal hydrides – high temperature	7	—	~300	2–10	Limited practical store experience. Thermal management problems.
Complex hydrides	18	150	>100	1	Nonregenerative on-board vehicle. Recycling waste products.

3.3.3.3 Conversion of Hydrogen to Energy

Hydrogen's primary use is likely to be as a fuel in the transport sector, as this is the area in Western economies with the highest fossil fuel dependence, but more generally, hydrogen may be used in three main modes to generate power.

1. By direct combustion with air in conventional engines, that is internal combustion, gas turbine or steam engines.
2. Reaction with pure oxygen in rocket type combustors.
3. Through electrochemical conversion of hydrogen and air mixtures to electrical energy in a fuel cell.

Direct combination of hydrogen with oxygen is likely to remain confined to relatively small areas of application, such as space exploration vehicle launchers. In the current chapter, our emphasis will, therefore, be on modes 1 and 3 above.

Conventional Engines Hydrocarbon-fed gas turbines are widely used as sources for electrical power generation in the civil, transport and military sectors. In all these applications, hydrogen could be used as a substitute fuel. Many studies and practical trials have been carried out on this subject [45, 51]. A number of distinct benefits arise from the use of hydrogen in gas turbines.

1. Lower maintenance and longer life as sediments and fuel-associated corrosion are lower.
2. Hydrogen is readily available in pressurized form.
3. Higher operating temperatures with hydrogen give enhanced efficiency (+10 %).

On the debit side, the increased operating temperatures could lead to higher nitrogen oxide emissions (see Section 3.2.2.4). Moreover, hydrogen does not lead to any reduction in noise, which is a major drawback to the use of turbines in populated areas. Hydrogen offers a further advantage over petrochemical fuels in that no sulfur pollutants result from its combustion.

Internal combustion engines are readily converted to hydrogen fuel. As noted above, the combustion of hydrogen gives no carbon dioxide, sulfur dioxide, hydrocarbons and related compounds in the exhaust emissions, only low levels of nitrogen oxides. A major development of hydrogen fueled internal combustion engines is currently being carried out by the EEC [51].

BMW have a demonstrator fleet of 30 saloon cars with 7L engines fueled from a cryogenic hydrogen store. All the benefits of hydrogen as a low pollution fuel are being demonstrated in these trials. Although fuel cells are seen as the power source of choice for the transport sector, BMW maintain that the use of hydrogen in internal combustion engines will act as an intermediate step until fuel cells become commercially viable and a hydrogen fueling infrastructure is established.

Fuel Cells The development of fuel cells as sources of electrical power may represent the biggest change in power source technology since the invention of the steam engine in the late eighteenth century. Despite the principles of fuel cells being demonstrated as early as 1839 by Grove, its development as a power source, for transport and small- to medium-scale electricity generation, occurred only in the last decade of the twentieth century.

The principle of the fuel cell is the inverse of electrolysis. Hydrogen and oxygen in the form of air are recombined electrochemically to deliver electrical energy. A schematic of a hydrogen oxygen fuel cell is shown in Fig. 3.7. At the anode, hydrogen is dissociated into protons and electrons. The electrolyte separating the electrodes is chosen to favor proton transport over electrons, which are used to perform useful work in the external circuit. Protons and electrons are discharged at the cathode, giving only water and waste heat as the overall emissions. The electrochemical reaction of fuels and oxidizers to generate electricity is possible for a large number of reactions. However, one of the most amenable, through availability of reactants and room temperature operation, is the hydrogen–air mixture. For this reason, the future large-scale uses of hydrogen and fuel cells are intrinsically linked.

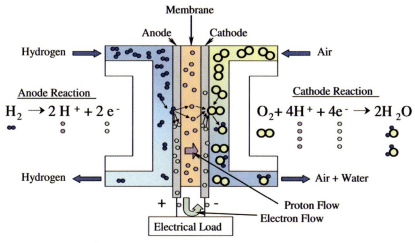

Fig. 3.7 PEM fuel cell.

Current development activity of fuel cells is centered around two main application areas.

1. Substitution of fuel cells for internal combustion engines for transportation, especially for cars and light vehicles.
2. Domestic and district generation of heat and electrical power.

In a fuel cell based on the reaction of hydrogen plus oxygen, about 80 % of the associated reaction enthalpy equal to 286 K J^{-1} mol^{-1} at 300 K is converted to electrical energy. In practice fuel cell efficiencies are nearer 60 %, the difference being due to waste heat.

Zero Emission Vehicles In the past decade, proton exchange membrane (PEM) fuel cells have emerged as the leading contender for this application. Many vehicle trials using PEM fuel cells, notably for urban buses, have been launched and successfully operated. A large-scale EC supported project, CUTE (Clean Urban Transport for Europe), involving the operation of fuel cell busses on commercial routes in 30 European cities is a notable example [52]. All the major automobile manufacturers in the US, Japan and Europe have prominent development programs and field trials aimed at the commercial introduction of fuel cell-powered light vehicles. For PEM fuel cells to achieve displacement of internal combustion engines in the light vehicle market, costs must be reduced from the current levels for fuel cells of around $1500 per kw to approach those acceptable for internal combustion engines around $100 per kw. A key area for cost reduction is the PEM membrane where minimization of expensive and relatively rare catalysts, notably platinum, must be achieved. Earlier in this chapter it was noted that hydrogen storage also presented a barrier to fuel cell adoption in vehicles. This is unlikely to be problematic for larger vehicles, such as busses, where there are few volume constraints on the storage of hydrogen and compressed hydrogen is the preferred option. Projections of the likely commercial entry date for fuel cell cars have become more pessimistic in the 2002–2007 period. In 2002, several automobile manufacturers were indicating timescales of 2003 to 2007 whereas more recent estimates are now indicating 2015 to 2020. Besides the above technical problems, other factors have contributed to the receding market horizons.

1. Lack of infrastructure for hydrogen refueling.
2. Auto companies see ICE-hybrid vehicles as the next and bridging step to fuel cell cars.
3. Many of the major automotive manufacturers are generating insufficient profit to fund a technology change of the order needed.
4. Diesel technology has emerged as a fuel-efficient option in light vehicles. Investment in diesel manufacturing plant will need to be recouped over the next decade.

Market entry of fuel cell vehicles may thus coincide with the emergence of high cost petrochemical fuels, that is around 2015. Unless the timescales are brought nearer, the light vehicle sector seems to likely to become progressively more exposed to higher priced and increasingly insecure fuel supplies. Fortunately, the scenario for other areas of fuel cell applications is more optimistic.

Power and Heating (CHP) Over the next two decades, fuel cells are expected to replace a large part of the current, oil and natural gas technologies used in power and heating. A more optimistic market penetration timescale than for automobiles is possible because the fuel cells in this sector are likely to be natural gas, rather than hydrogen fueled. It will therefore be possible to exploit existing fuel infrastructure.

At the intermediate level of electrical power generation ($\sim$100 MW), fuel cells are now regarded as the technology of choice. A major advantage of fuel cells over turbine and diesel engine-powered generators is the absence of noise and low carbon emissions. Operation in heavily populated neighborhoods is therefore possible. Some urgency has been added to these developments following the spectacular grid failures in the US and Europe during high demand periods in 2003. Natural gas fuel cells for domestic heat and power should make significant market penetration in the next decade. All the advanced economies have large projects studying the possible transition of today's energy supply towards a more decentralized and market oriented supply system. The ultimate scenario, by the twenty second century, would be locally distributed heat and power production, enhanced by grid power feeds from privately owned PEM fuel cell vehicles, with hydrogen as the common fuel [53]. In this scenario hydrogen fed fuel cells play a major role, whilst fuel cells powered by fossil derivatives such as natural gas or coal make dwindling and minor contributions. If the hydrogen feed stock for this scenario is produced by renewable techniques, a global transition towards a zero emission electricity system will be possible. Half of global electricity could be generated from mobile fuel cells [54]. At the close of the twenty first century hydrogen and electricity could become the main energy carriers giving cleaner, more flexible power generation and wider access.

The commercial and environmental opportunities associated with these changes are on a scale similar to those produced by the first industrial revolution or the rise of the information technology age. Many unforeseen synergies will no doubt emerge as the scenarios develop.

Path Forward All the applications touched on in this section will require considerable development work if hydrogen is to become accepted as the energy source of choice. For the well-established technologies of internal combustion engines and turbines the translation to hydrogen fueling will require developments in a number of critical areas:

ICE

- On-board vehicular storage for hydrogen must provide practical ranges at viable weight and cost.
- Injection systems will need optimizing for hydrogen
- Combustion zones and processes will need further development

Turbines

- Combustion zones will need modification and superior materials for hydrogen usage.
- Peripheral systems require further development.

Fuel cells will play a key role in introducing hydrogen fuel into both the automotive and stationary power sectors. Again there are specific areas needing development to achieve this:

Automotive Fuel Cells

- Stack operating temperature range needs to be extended both upwards and downwards.
- Membrane costs must be drastically reduced.
- Membrane operating reliability has to be established, together with the underlying physical degradation mechanisms.
- Current automotive design is ICE centered. Design for FC automobiles should fully exploit the flexibility and distributed control offered by an electrical power source.

3.3.4
Implementation of the Hydrogen Cycle: Technical Issues

3.3.4.1 Possible Scenarios
In Section 3.3.1, the key elements of the hydrogen cycle were identified as Production, Distribution/Storage and Power Conversion. Overall, the chain involves the use of renewable electricity to electrolyze water for hydrogen, which is used as an energy carrier or energy vector to various conversion devices, where it is transformed back into useful electricity for the end-user applications. This sequence is shown schematically in Fig. 3.8.

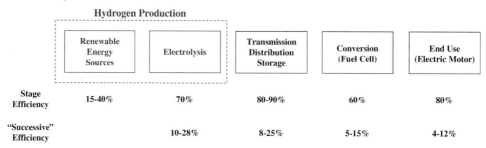

Fig. 3.8 The hydrogen cycle in practice.

Critics of the hydrogen economy have pointed out [55] that the sequence in Fig. 3.8 has many disadvantages; the key one being that it involves generating electricity twice. At each stage in the process losses occur, as shown, giving somewhat low overall efficiencies. Seemingly persuasive arguments can be made in favor of substituting nuclear generation, natural gas or coal-based generation for the "Production" front-end of Fig. 3.8. Though superficially attractive, and leaving aside the issue of the carbon emissions of the coal and natural gas generation processes, none of these options are viable long term because of their dependence on depletable, finite fuel resources.

Automobiles generate by far the largest demand for fossil fuels worldwide and give rise to the most troublesome emissions. Hence, the transport sector is likely to be decisive in deciding these arguments, and leading to a hydrogen cycle-based infrastructure as shown in Fig. 3.6. Hydrogen and fuel cells are the only realistic options capable of avoiding dire economic and climatic consequences. Whilst the efficiencies of the overall cycle appear to be low (4–12%), it should be recalled that the well to wheel efficiency for a gasoline car is only around 15%. Given that the raw materials needed to feed the hydrogen cycle are nondepletable, potentially secure and lead to more flexible transportation and power supply systems, a small percentage loss of efficiency would seem a small price to pay. Increased efficiencies for all parts of the hydrogen cycle are the key drivers for the large research and development Hydrogen Roadmap programs in the US, Japan and the EEC.

The most likely scenario for volume penetration of hydrogen fuel cells is, thus, in the transport sector, eventually leading to the wider adoption of hydrogen. Nearer term, nonhydrogen fuel cells will continue to grow steadily in the medium-scale standby-power and domestic heating markets. These early developments will provide invaluable operating experience of practical fuel cell systems.

As noted earlier, fuel cells will, also, be critical in the emergence of decentralized power generation systems. Wider availability of fuel cells as part of the personal transport or the domestic heat and power units will make this transition possible. To achieve this break from centralized, grid-based power supply, there are significant technical, but more importantly, economic and political, hurdles to be cleared. The emergence of distributed power generation will present major challenges for the power utility corporations in all leading economies. More than likely, the transition to decentralized power will follow a similar pattern to that seen in the telecommunications industry, where mobile telephones and computing technology were introduced as de facto products and, despite initial resistance, had to be absorbed by the telecommunication utilities.

3.3.4.2 Potential Hazards of the Hydrogen Economy

Effects on the Environment Until recently, it was assumed that migration to a hydrogen fuel economy would have no negative aspects for the environment. However, in mid-2002, it was proposed that the widespread use of hydrogen could lead to hitherto unknown environmental impacts due to hydrogen emissions [56], the central claim being there will be substantial leakage of hydrogen associated with the production, transportation and storage of hydrogen. It is estimated that the total anthropogenic emissions of hydrogen from these sources could be of the order of 60 to 120 million tonnes per year. Assuming the leaked hydrogen accumulates in the stratosphere, it could react with hydroxyl radicals to form water vapor. Hence, increasing leakages of molecular hydrogen could lead to increased levels of water vapor in the stratosphere. As a consequence, the lower stratosphere could cool and perturb the ozone chemistry of the polar regions, which depend upon reactions involving hydrochloric acid and chlorine nitrates on H_2O ices.

Other studies have noticed that enhanced leakage rates of molecular hydrogen could lead to an increase in the concentration of greenhouse gases [57]. In the troposphere hydroxyl radicals are effective getters of many pollutant species, including methane, a particularly effective greenhouse gas. Enhanced removal of hydroxyl radicals by hydrogen could thus be responsible for enhanced methane levels and the associated warming effects.

There is little uncertainty that the reaction of hydroxyl radicals and molecular hydrogen could occur in the troposphere and stratosphere. What is far less certain is the scale of these reactions, which in turn is dependent on the scale of man-made leakage. Debate continues about the proposed levels of leakage likely to occur as the use of hydrogen expands. Diminishing use of fossil fuels could lead, in itself, to reductions in hydrogen emissions, by eliminating the hydrogen produced from incomplete combustion. Leakage rates are also very dependent upon the nature of hydrogen storage technologies. Greater rates of leakage would probably apply more to liquid hydrogen storage than compressed gas systems and both would be inferior to solid-state stores from which leakage is negligibly small.

Further debate is currently centered on the absorption of molecular hydrogen by soils [58]. Hydrogen is a microbiological nutrient, hence one would expect increased partial pressures of hydrogen over soils to lead to increased uptake, but this possibility has been questioned by some workers.

Safety Issues in the Use of Hydrogen A convenient summary of the current status of the hydrogen economy is that the three biggest barriers to progress are the three "Ss": Storage, Supply and Safety. The majority of this chapter has been devoted to the first two of these issues, but safety and the adverse public perception of hydrogen as a fuel remain significant barriers. Despite the widespread use of coal gas (60% hydrogen content) for over 150 years in Europe and the USA, a negative public perception of hydrogen persists. Most common amongst the concerns are recollections of the Hindenburg airship disaster and the hydrogen bomb. Perceptions such as these are deep seated despite studies, such as the 1997 report of hydrogen safety by the Ford Motor Company [59], which concluded that the safety of a hydrogen vehicle would potentially be better than that of a gasoline or propane vehicle with proper engineering.

Large public programs are needed to alter these adverse perceptions and reassure the public that the use of hydrogen poses no greater risks than gasoline. Though at an early stage, national and international operating and safety codes for the whole of the hydrogen supply chain are in the process of development in the US, Japan and EEC countries.

3.3.5
Hydrogen in the Twenty First Century

During the twenty first century, the world pattern of energy usage has to be transformed. To continue our ever increasing dependence on fossil fuels is futile and will accelerate the onset of further extreme climatic events. Hydrogen has the potential to be a key factor in this transformation. This chapter has outlined the hydrogen cycle-based principles, by which the change from fossil fuel dominance to a sustainable global energy system can be achieved.

The penetration and growth of hydrogen technologies in the twenty first century is difficult to predict with any degree of certainty. However, a number of general features seem likely to be dominant:

- Use of solid fuels and petrochemical products will decline rapidly in the latter half of the century.
- By the end of the century, fuel cells and hydrogen fuels will have transformed the transport market and the domestic heat and power sector.
- Production of hydrogen will be initially via steam reformation of natural gas, with renewable sources such as solar, wind and biomass becoming significant in the later decades.
- There will be no single pattern of shift towards hydrogen. Some developed and emerging economies will have bigger contributions from nuclear power and coal-based systems. Others will stay closer to truly renewable methods of power generation [60].
- In developing countries, renewable technology will evolve directly, as there are no limits posed by availability of existing infrastructure and so on. The emergence of cost effective solar power technologies could have huge impacts on the ability of third world countries to adapt rapidly to hydrogen.

Many technological, socio-economic and political barriers will influence the course of the transition from the fossil/carbon to the hydrogen cycle-based energy system [60, 61]. Different economies will adopt strategies reflecting their access to energy resources. Security of energy supply is a universal concern, but the scale and imminence of global warming-induced climate change remains contentious for some [61]. A point of common agreement is the need to move away from the use of depletable resources.

Across the OECD nations a mix of legislation to reduce emissions and fiscal measures to incentivise early adoption of cleaner fuel, including hydrogen, are being introduced. A minority, with the US being notable, are relying on a mixture of business as usual and enhanced programs of research and development aimed at earlier introduction of clean fuels and ultimately a hydrogen economy. Unfortunately, the US is responsible for around 25 % of the global output of carbon dioxide In the next decade China and India will become similar contributors to the global carbon dioxide output as their industrialization gathers pace. Inevitably, the delay in remedying carbon outputs now will only lead to more serious problems in the coming decades.

Despite the lack of agreement on how best to achieve a transition from fossil fuels, there is a growing acceptance that hydrogen cycle-based fuel systems will have a role in major economy energy systems by the end of the twenty first century. How quickly this new age will dawn is very dependant on the views and resources of individual nations. It is worth recalling that the oil crisis of 1973 accelerated the development of nuclear power in many advanced economies and, as a result, produced a blip in global carbon dioxide production. The moral would seem to be that economic catastrophes can galvanize global action. Could the recent sequence of excessively hot summers in Europe of 2003 have a similar effect? On the political front, all plans will be compromised should there be further disruption in the security of petrochemicals from major Middle Eastern suppliers. Faced with these scenarios governments could be forced to adopt accelerated measures to combat global warming.

It would be complacent to believe that the geo-politics of energy resources which, in the recent past, have led to economic emergencies, the rise of terrorism and, on occasion, war, will disappear with the promises of the hydrogen economy. At present, the rule of international law is flouted by the dominant economies adopting unilateral policies to secure their energy supplies. A renewed agenda for energy and world trade is urgently needed if we are to aspire to peaceful use of resources. Of equal importance is equal access to energy supplies for emerging nations, rather than preferential supply to the biggest users being enforced by arms.

Many southern hemisphere nations have huge untapped resources of solar power, potentially harnessable to the production of hydrogen and power. Very few, however, have the technical and financial resources to exploit these latent assets.

For the 2 billion people without electricity, the emergence of hydrogen infrastructures, combined with fuel cells for the provision of local power could offer a chance for them to leapfrog into the twenty first century.

References

1. Sampson, A. (1975) *The Seven Sisters – The Great Oil Companies and the World they Shaped*, The Viking Press, New York.
2. Smil, V. (2000) *Annu. Rev. Energy Environ.*, **25**, 21–51.
3. BP (2003) *BP Statistical Review of World Energy 2003*, British Petroleum, London.
4. Hall, C., Tharakan, P. et al. (2003) *Nature*, **426**, 318–22.
5. International Energy Agency (2002) *World Energy Outlook 2002*, OECD/IEA, Paris.
6. Dach, G., Wegmann, U. (2002) *Energy Environ.*, **13**, 579–89.
7. Head, I.M., Jones, D.M., Larter, S.R. (2003) *Nature*, **426**, 344–52.
8. Royal Dutch/Shell (1983) *The Petroleum Handbook*, 6th edn, Elsevier Science Publishers B.V., Amsterdam.
9. Holditch, S.A. (2003) *J. Petroleum Technol.*
10. Campbell, C.J. (1997) *The Coming Oil Crisis*, Multi-Science Publishing Co. Ltd., Brentwood.
11. Campbell, C.J. (2002) *The Essence of Oil and Gas Depletion*, Multi-Science Publishing Co. Ltd, Brentwood.
12. Ryan, J. (2003) *Hubbert's Peak Déjà vu All Over Again*, 2nd quarter, IAEE Newsletter, Cleveland, pp. 9–12.
13. ExxonMobil (2003) *Oeldorado 2003*, ExxonMobil Central Europe Holding GmbH, Hamburg.
14. Holloway, S. (2001) *Annu. Rev. Energy Environ.*, **26**, 145–66.
15. Sloan, E.D. (2003) *Nature*, **426**, 353–9.
16. International Tanker Owner Pollution Federation (2003) *Oil Tanker Spill Statistics 2003*, ITOPF, London.
17. Jacobsen, M. (2001) *Nature*, **409**, 695–7.
18. Intergovernmental Panel on Climate Change (2001) *Climate Change 2001: Synthesis Report*. IPCC, Geneva.
19. Dresselhaus, M.S., Thomas, I.L. (2001) *Nature*, **414**, 332–7.
20. Caldeira, K., Wickett, M.E. (2003) *Nature*, **425**, 365.
21. International Energy Agency (2003) *Integrating Energy and Environmental Goals – Investment Needs and Technology Options*, OECD/IEA, Paris.
22. Masaki, T. (2003) *Energy Environ.*, **14**, 51–7.
23. Cabal, V.L. (2003) *Energy Environ.*, **14**, 3–15.
24. Berner, R.A. (2003) *Nature*, **426**, 323–6.
25. Schlamadinger, B., Grubb, M., Azar, C., Bauen, A., Berndes, G. (2001) Carbon sinks and biomass energy production – a study of linkkages, options and implications. Climate Strategies Report.
26. Hoogwijk, M., Faaij, A., Broek, R.V.D., Berndes, G., Gielen, D., Turkenburg, W. (2003) *Biomass Bioenergy*, **25**, 119–33.
27. Intergovernmental Panel on Climate Change (IPCC) (2001) Climate Change 2001: The Scientific Basis; Chapter 3. *The Carbon Cycle and Atmospheric Carbon Dioxide*, Coordinating Lead Author: I.C. Prentice, Geneva.
28. BP statistical review of world energy (2003).
29. Vogel, F., Hildebrand, F. (2002) Catalytic hydrothermal gasification of woody biomass at high feed concentrations. Paper 123, 4th International Symposium On High Pressure Process Technology and Chemical Engineering, EFCE, Venice Italy.
30. Steinberg, M., Grohse, E.W., Tung, Y.A. (1991) Feasibility study for the co-processing of fossil fuels with biomass by the hydrocarb process. Brookhaven National Laboratory Report BNL-46058, DE 91011971.
31. Stucki, S., Schucan, T.H. (1994) Speicherung und Transport von Wasserstoff in der Form organischer Verbindungen. Tagung Global Link des VDI, (VDI Berichte 1129), Okt. 94.
32. Lovelock, J.E. (1972) Gaia as seen through the atmosphere. *Atmospheric Environment*, vol. **6**, Elsevier, p. 579.
33. Bolin, B. et al. (1986) *The Greenhouse Effect, Climate Change and Ecosystems (SCOPE29)*, John Wiley & Sons, Chichester, UK.
34. Report of the International Scientific Steering Committee (2005) *International Symposium on the Stabilization of*

Greenhouse Gas Concentrations Hadley Centre, Met Office Exeter UK, May 2005.
35 Turner, J.A. (2004) *Science*, **305**, P972.
36 Nitsch, J., Voigt, C. (1988) in *Hydrogen as an Energy Carrier*, Springer Verlag, p. 314.
37 Brasseur, G.P. et al. (1999) *Atmospheric Change and Global Change*, Oxford University Press, p. 173.
38 Nuttall, W.J. (2005) *Nuclear Renaissance: Technologies and Policies for the Future of Nuclear Power*, IOP Publishing, Bristol, UK.
39 Forsberg, C. (2003) *Int. J. Hydrogen Energy*, **28**, 1073.
40 Iwasaki, W. (2003) *Int J. Hydrogen Energy*, **28**, 939.
41 Baykarra, Z. (2004) *Int. J. Hydrogen Energy*, **29**, 1451.
42 Lucchese, P. (2004) *Hydrogen Production 1st General Assembly Meeting EEC Jan 2004*.
43 Levin, D.B., Pitt, L., Love, M. (2004) *Int. J. Hydrogen Energy*, **29**, 173.
44 Winter, C.J. (2004) *Int. J. Hydrogen Energy*, **29**, 1095.
45 Carpetis, C. (1988) in *Hydrogen as an Energy Carrier*, Springer Verlag, p. 265.
46 EEC Sixth Framework Programme "NATURHALY": Project No SE6/CT/2004/502661 www.naturhaly.net
47 Haris, I.R., Book, D., Anderson, P., Edwards, P. (2004) *Fuel Cell Rev.*, 17.
48 Zuttel, A. (2003) *Mater. Today*, 24.
49 Neelis, M.L. (2004) *Int. J. Hydrogen Energy*, **29**, 537.
50 Ogden, J.M. (2002) *Phys. Today*, 69.
51 *HyICE EEC Sixth Framework Programme.*
52 *CUTE (Clean Urban Transport for Europe) EEC Sixth Framework Programme.*
53 Rifkind, J. (2003) *The Dawn of the Hydrogen Age*, European Union Hydrogen Conference Brussels.
54 Elam, C.C. et al. (2003) *Int. J. Hydrogen Energy*, **28**, 601.
55 Bossel, U. (2004) *Renewable Energy World*, 155.
56 Tromp, T.K. et al. (2003) *Science*, **300**, 1740.
57 Horowitz, L. (2003) National Oceanic and Atmospheric Administration Report.
58 Rahn, T., Kitchen, N., Eiler, J.M. (2002b) *Geophys Rev. Lett.*, **29**(18), 1888.
59 Ford Motor Co. (1997) Direct Hydrogen-Fuelled PEM Fuel Cell System for Transportation Applications: H2 Vehicle Safety Report DOE/CE/50389-502 Directed Technologies Inc., Arlington, Va., May 1997.
60 Pacala, S., Socolow, R. (2004) *Science*, **305**, 968.
61 Thorpe, A. (2007) *New Scientist*, 24, 17th March.

4
Properties of Hydrogen
Andreas Züttel, Andreas Borgschulte, Louis Schlapbach, Ib Chorkendorff, and Seijirau Suda

4.1
Hydrogen Gas
Andreas Züttel

4.1.1
Hydrogen Isotopes

Hydrogen [1] (hydrogenium, $υδωρ$ = the water, $γεννειν$ = to give birth) is the first element in the periodic table of the elements having the atomic number 1 and the electron configuration $1s^1$. Hydrogen was prepared many years before it was recognized as a distinct substance by Cavendish in 1766 and it was named by Lavoisier. Hydrogen is the most abundant of all elements in the universe and it is thought that the heavier elements were, and still are, built from hydrogen and helium. It has been estimated that hydrogen makes up more than 90 % of all the atoms or 75 % of the mass of the universe. It is found in the sun and most stars, and plays an important part in the proton–proton reaction and calin_rbon–nitrogen cycle, which accounts for the energy of the sun and stars. It is thought that hydrogen is a major component of the planet Jupiter and that at some depth in the planet's interior the pressure is so great that solid molecular hydrogen is converted into solid metallic hydrogen.

The ordinary isotope of hydrogen, H is known as *Protium* and has an atomic weight of 1.0078 (1 proton and 1 electron). In 1932, Urey [2] announced the preparation of a stable isotope, *Deuterium* (D) with an atomic weight of 2.0140 (1 proton, 1 neutron and 1 electron). Two years later an unstable isotope, *Tritium* (T), with an atomic weight of 3.0161 (1 proton, 2 neutrons and 1 electron) was discovered [3]. Tritium has a half-life of 12.5 years [4]. Deuterium is found in 0.017 % of all hydrogen isotopes [5]. Tritium atoms are also present in $\approx 10^{-18}$ % of all hydrogen isotopes as a result of natural processes in the atmosphere, as well as from fallout from past atmospheric nuclear weapons tests and the operation of nuclear reactors and fuel reprocessing plants (Table 4.1 and Figure 4.1).

Hydrogen as a Future Energy Carrier. Edited by A. Züttel, A. Borgschulte, and L. Schlapbach
Copyright © 2008 WILEY-VCH Verlag GmbH & Co. KGaA, Weinheim
ISBN: 978-3-527-30817-0

Table 4.1 Atomic data of hydrogen isotopes, from (Ref. [1])

CAS No.	^{1}H [12385-13-6]	^{2}D [16873-17-9]	^{3}T [15086-10-9]
Atomic mass [u]	1.007825	2.0140	3.01605
Natural abundance [%]	99.985	0.015	$\approx 10^{-18}$
Half life time [yr]			12.26
Ionisation energy [eV]	13.5989	13.6025	13.6038
Thermal neutron capture cross section [10^{-24} cm^2]	0.322	0.51×10^{-3}	$< 6 \times 10^{-6}$
Nuclear spin [$h/2\pi$]	$+1/2$	$+1$	$+1/2$
Nuclear magnetic moment, nuclear magnetons [μN]	$+2.79285$	$+0.85744$	2.97896

The atomic radius of the free hydrogen atom in the ground state, the Bohr radius is $a_0 = h^2/(\pi^2 \cdot e^2 \cdot m) = 0.529$ Å and the covalent atomic radius of hydrogen in crystal structures is between 30 and 35 pm.

4.1.2
Hydrogen Molecule

The isotopes H, D and T form diatomic molecules. The interaction potential energy of two hydrogen atoms goes through a minimum at a certain interatomic distance when two electrons form a singlet state $^1\Sigma_g^+$, namely, the state with a total electron spin equal to zero (the combination of two electrons with opposite spin). The energy of the triplet state $^3\Sigma_u^+$, having a total electron spin of unity, increases when two hydrogen atoms with parallel spin approach each other. Hydrogen atoms of opposite spin exist in equal numbers, and readily combine in pairs to form singlet state molecules. The formation of molecules can be suppressed by applying a strong magnetic field to an atomic hydrogen gas [6] (Figure 4.2).

Quantum mechanics requires that the wavefunction of a molecule must be antisymmetric with respect to the interchange of the space coordinates of two fermions (spin = $\frac{1}{2}$), and symmetric for the interchange of bosons (spin = 1). Thus the wavefunction of a H$_2$ (T$_2$) molecule should be antisymmetric on interchange of two protons (tritons), and that of D$_2$ symmetric on interchange of two deuterons. There are a total of $(2I + 1)2$ combinations of nuclear spin states for two identical

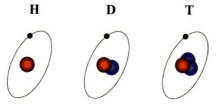

Fig. 4.1 Schematic representation of hydrogen isotopes: protium (H), deuterium (D) and tritium (T).

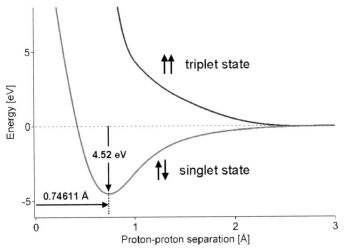

Fig. 4.2 The singlet state exhibits a minimum in the interaction potential energy for a proton–proton separation of 0.74611 Å. The binding energy is 4.52 eV, that is 2.26 eV per H atom (218.1 kJ mol^{-1} H) to separate the H—H bond. From (Häussinger, Lohmüller, and Watsin (1990), Ref. [1]).

nuclei of spin I, of which $(2I+1)(I+1)$ states are symmetric and $(2I+1)I$ are antisymmetric with respect to the interchange of the nuclei. Consequently there are two kinds of hydrogen molecules in the singlet state, namely, ortho-hydrogen (symmetric, with parallel nuclear spins ↑↑) and para-hydrogen (antisymmetric, with antiparallel nuclear spins ↑↓).

When the nuclei are fermions, as in the case of protons (tritons), the symmetric nuclear spin states must be coupled with the antisymmetric rotational states (odd J) and the antisymmetric nuclear spin states must be coupled with the symmetric rotational states (even J) (Figure 4.3).

The population ratio of ortho-hydrogen to para-hydrogen in thermal equilibrium is given by

$$\frac{N_o}{N_p} = \frac{(I+1) \cdot \sum_{J=\text{odd}} q_J^r}{I \cdot \sum_{J=\text{even}} q_J^r} \quad \text{with} \quad q_J^r = (2J+1) \cdot \exp\left\{-\frac{2 \cdot h^2}{8 \cdot \pi^2 M \cdot r_0^2} \frac{J(J+1)}{k \cdot T}\right\}$$
(4.1)

The following forms of hydrogen are distinguished: o-H$_2$ ortho-hydrogen, p-H$_2$ para-hydrogen, e-H$_2$ equilibrium hydrogen and n-H$_2$ normal hydrogen. e-H$_2$, e-D$_2$ and e-T$_2$ all correspond to the equilibrium concentration at definite temperatures. n-H$_2$ is a mixture of the equilibrium concentration of ortho- and para-hydrogen at a temperature of 293.15 K (20 °C), that is 25 mol% p-H$_2$ and 75 mol% o-H$_2$. At low temperature p-H$_2$, o-D$_2$ and p-T$_2$ can be present in a virtually pure state. The conversion [8] of o-H$_2$ to p-H$_2$ is exothermic and a function of

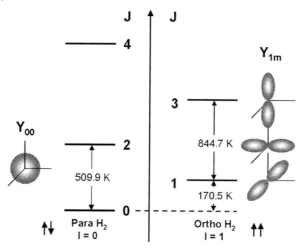

Fig. 4.3 Energies and wavefunctions of rotational states (J) of a para-hydrogen and an ortho-hydrogen molecule. The total nuclear spin I is 0 or 1, respectively and the rotational quantum number is 0, 1, 2... From (Silvera (1980), Ref. [7]).

temperature. When hydrogen is cooled from room temperature (RT) to the normal boiling point (nbp = 21.2 K) the ortho-hydrogen converts from an equilibrium concentration of 75 % at RT to 50 % at 77 K and 0.2 % at nbp (Figure 4.4).

The self conversion rate is an activated process and very slow, the half-life of the conversion is greater than one year at 77 K. The conversion reaction from ortho- to para-hydrogen is exothermic and the heat of conversion is also temperature dependent [9]. At 300 K the heat of conversion is 270 kJ kg^{-1} and increases as the temperature decreases, until it reaches 519 kJ kg^{-1} at 77 K. At temperatures lower than 77 K the enthalpy of conversion is 523 kJ kg^{-1} and almost constant. The enthalpy

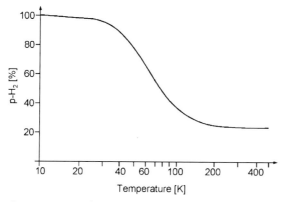

Fig. 4.4 Amount of para-hydrogen (p-H$_2$) in the equilibrium composition e-H$_2$ as a function of temperature.

Table 4.2 Physical properties of molecules of the hydrogen isotopes. From Ref. [1]

	n-H$_2$	n-D$_2$	n-T$_2$	HD	HT	DT
CAS No.	1333-74-0	7782-39-0	10028-17-8	13983-20-5	14885-60-0	14885-61-1
Molecular mass [u]	2.016	4.029	6.034	3.022	4.025	5.022
Bond length [pm]	74.6	74.2		74.1		
Dissociation energy [eV]	4.473	4.552		4.510		
Triple point						
Temperature [K]	13.96	18.73	20.62	16.60	17.63	19.71
Pressure [kPa]	7.3	17.1	21.6	12.8	17.7	19.4
Critical point						
Temperature [K]	32.98	38.35	40.44	35.91	37.13	39.42
Pressure [kPa]	1.31	1.67	1.85	1.48	1.57	1.77
Normal boiling point [K]	20.39	23.67	25.04	22.13	22.92	24.38
Density at n-b.p.						
ρ_L [kg m^{-3}]	70.811	162.50	260.17	114.80	158.62	211.54
ρ_V [kg m^{-3}]	1.316	2.230	3.136	1.802	2.310	2.694
Heat of vaporization [J mol^{-1}] at 25 K	825	1175	1400	1000	1100	1290

of conversion is greater than the latent heat of vaporisation ($\Delta H_V = 451.9$ kJ kg^{-1}) of normal and para-hydrogen at the nbp. If the unconverted normal-hydrogen is placed in a storage vessel, the enthalpy of conversion will be released in the vessel, leading to the evaporation of the liquid hydrogen. The transformation from ortho- to para-hydrogen can be catalyzed by a number of surface active and paramagnetic species, for example n-H$_2$ can be adsorbed on charcoal cooled with liquid hydrogen and desorbed in the equilibrium mixture (e-H$_2$). The conversion may take only a few minutes if a highly active form of charcoal is used. Other suitable ortho–para catalysts are metals such as tungsten, nickel, or any paramagnetic oxides [10] like chromium or gadolinium oxides. The nuclear spin is reversed without breaking the H—H bond (Table 4.2).

4.1.3
Physical Properties

In hydrogen, the interaction between molecules is weak as compared to other gases, therefore the critical temperature is low ($T_c = 33.0$ K). The melting curve, the solid–liquid boundary in a p–T diagram, has been determined by several groups [11] for p-H$_2$ and n-D$_2$. The following functions were determined by least-squares

Table 4.3 Vapor pressure and density of p-hydrogen at low temperatures

Temperature [K]	Vapor pressure [kPa]	Density [kg m^{-3}]		
		ρ_S	ρ_L	ρ_G
1	11 × 10^{-37}	89.024		
5	4.76 × 10^{-3}	88.965		
10	255.6	88.136		0.006
12	1837	87.532		0.037
13.803[a]	7.0	86.503	77.019	0.126
20	93.5		71.086	1.247
20.268[b]	101.3		70.779	1.338
30	822.5		53.930	10.887
32.976[c]	1293			31.43

[a] Triple point.
[b] 101.3 kPa.
[c] Critical point.

fitting:

$$p_m = -51.49 + 0.1702 \cdot (T_m + 9.689)^{1.8077} \quad \text{for} \quad H_2 \quad (4.2)$$

$$p_m = -51.87 + 0.3436 \cdot (T_m)^{1.691} \quad \text{for} \quad D_2 \quad (4.3)$$

where p_m is in MPa and T_m in K. The melting pressure for D_2 is about 4 % lower than for H_2 at a given temperature (Table 4.3 and Figure 4.5).

At zero pressure, hydrogen (H_2, D_2) solidifies in the hexagonal close-packed (hcp) structure. Data for p-H_2 are [13]: $a = 375$ pm, $c/a = 1.633$, molar volume

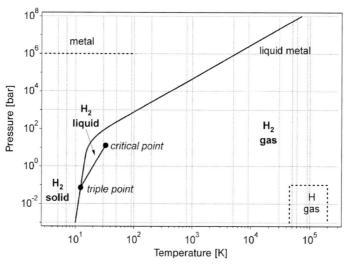

Fig. 4.5 Primitive phase diagram for hydrogen. (Leung, March, and Motz (1976), Ref. [12]).

Table 4.4 Physical properties of para-hydrogen (p-H$_2$) and normal-hydrogen (n-H$_2$) at the triple- and normal boiling point

	p-H$_2$	n-H$_2$
Triple point ($T = 13.803$ K, $p = 7.04$ kPa)		
Temperature [K]	13.803	13.957
Pressure [kPa]	7.04	7.2
Density (solid) [kg m^{-3}]	86.48	86.71
Density (liquid) [kg m^{-3}]	77.03	77.21
Density (vapor) [kg m^{-3}]	0.126	0.130
Heat of melting ΔH_m [J mol^{-1}]	117.5	
Heat of sublimation ΔH_V [J mol^{-1}]	1022.9	
Enthalpy $\Delta H°$ [J mol^{-1}]	−740.2	
Entropy $\Delta S°$ [J mol^{-1} K^{-1}]	1.49	
Thermal conductivity [W m^{-1} K^{-1}]	0.9	
Dielectric constant	1.286	
Boiling point at $p = 101.3$ kPa		
Temperature T_b [K]	20.268	20.39
Heat of vaporization ΔH_V [J mol^{-1}]	898.30	899.1

22.56 cm^3 mol^{-1}. The spherical $J = 0$ species (p-H$_2$, o-D$_2$) undergo a structural transition below 4 K where the rotational motion of the molecules is quenched, and the molecules are located on the fcc lattice (space group $Pa3$) with their axes oriented along the body diagonals [14].

At very high pressures (>2 × 10^{11} Pa), solid hydrogen is expected [15] to transform from a diatomic molecular phase to a monatomic metallic phase with a density >1000 kg m^{-3}. This phase may become a high-temperature superconductor [16] (Tables 4.4 and 4.5).

4.1.4
Equation of State

Among the most fundamental knowledge about the thermodynamic properties of hydrogen is the equation of state (EoS), namely, the volume as a function of pressure and temperature $V(p,T)$. Once the EoS is known, all the thermodynamic quantities can be calculated. The Gibbs energy $G(p,T)$ and entropy $S(p,T)$ can be obtained by integration

$$G(p, T) = G(p_0, T) + \int_{p_0}^{p} V \, dp \tag{4.4}$$

$$S(p, T) = S(p_0, T) - \int_{p_0}^{p} \left(\frac{\partial V}{\partial T}\right)_p dp \tag{4.5}$$

Table 4.5 Physical properties of liquid and gaseous para-hydrogen (p-H$_2$) and normal-hydrogen (n-H$_2$).

	p-H$_2$	n-H$_2$
Liquid phase at boiling point		
Density [kg m^{-3}]	70.78	70.96
Specific heat capacity c_p [J mol^{-1} K^{-1}]	19.70	19.7
Specific heat capacity c_V [J mol^{-1} K^{-1}]	11.60	11.6
Enthalpy $\Delta H°$ [J mol^{-1}]	−516.6	−548.3
Entropy $\Delta S°$ [J mol^{-1} K^{-1}]	16.08	34.92
Viscosity [mPa s]	13.2 × 10^{-3}	13.3 × 10^{-3}
Velocity of sound [m s^{-1}]	1089	1101
Thermal conductivity [W m^{-1} K^{-1}]	98.92 × 10^{-3}	100 × 10^{-3}
Compressibility factor	0.01712	0.01698
Vapor phase at boiling point		
Density [kg m^{-3}]	1.338	1.331
Specific heat capacity c_p [J mol^{-1} K^{-1}]	24.49	24.60
Specific heat capacity c_V [J mol^{-1} K^{-1}]	13.10	13.2
Enthalpy $\Delta H°$ [J mol^{-1}]	381.61	1447.4
Entropy $\Delta S°$ [J mol^{-1} K^{-1}]	60.41	78.94
Viscosity [mPa s]	1.13 × 10^{-3}	1.11 × 10^{-3}
Velocity of sound [m s^{-1}]	355	357
Thermal conductivity [W m^{-1} K^{-1}]	16.94 × 10^{-3}	16.5 × 10^{-3}
Compressibility factor	0.906	0.906
Critical point		
Temperature [K]	32.976	33.19
Pressure [MPa]	1.29	1.325
Density [kg m^{-3}]	31.43	30.12
Properties at STP ($T = 273.15$ K, $p = 101.3$ kPa)		
Density [kg m^{-3}]	0.0899	0.0899
Specific heat capacity c_p [J mol^{-1} K^{-1}]	30.35	28.59
Specific heat capacity c_V [J mol^{-1} K^{-1}]	21.87	20.3
Viscosity [mPa s]	8.34 × 10^{-3}	8.34 × 10^{-3}
Velocity of sound [m s^{-1}]	1246	1246
Thermal conductivity [W m^{-1} K^{-1}]	182.6 × 10^{-3}	173.9 × 10^{-3}
Dielectric constant	1.00027	1.000271
Compressibility factor	1.0005	1.00042
Prandtl number	0.6873	0.680

The enthalpy is then given by

$$H(p, T) = G(p, T) + T S(p, T) \tag{4.6}$$

At low gas density, all real gases tend to obey the ideal gas law

$$p \cdot V = n \cdot R \cdot T \tag{4.7}$$

where p is the absolute pressure, V the volume, n the number of gas molecules and R the gas constant (8.314 J mol^{-1} K^{-1}). In a real gas, however, the gas molecules occupy a certain volume – the "excluded volume" b – and the molecules interact by means of the Van der Waals force (dipole interaction), which is expressed with the "interaction strength" a. The Van der Waals equation for real gases is therefore

$$\left(p + \frac{a \cdot n^2}{V^2}\right)(V - b \cdot n) = n \cdot R \cdot T \tag{4.8}$$

with the appropriate parameters for hydrogen: $a(H_2) = 2.476 \times 10^{-2}$ m^6 Pa mol^{-2} and $b(H_2) = 2.661 \times 10^{-5}$ m^3 mol^{-1}. The pressure as a function of the volume of a real gas is given by

$$p(V) = \frac{n \cdot R \cdot T}{V - n \cdot b} - a \cdot \frac{n^2}{V^2} \tag{4.9}$$

The $p(V)$-isotherm of a Van der Waals gas exhibits, at the critical temperature T_K, a saddle point (p_K, V_K) when $dp/dV = 0$ and $d^2p/dV^2 = 0$.

$$T_K = \frac{8a}{27 \cdot R \cdot b}, \quad p_K = \frac{a}{27 \cdot b^2}, \quad V_K = 3 \cdot b \tag{4.10}$$

The critical parameters of a real gas are used to define the reduced variables T_r, p_r and V_r

$$T_r = \frac{T}{T_K}, \quad p_r = \frac{p}{p_K}, \quad V_r = \frac{V}{V_K} \tag{4.11}$$

which lead to the general Van der Waals equation independent of the type of gas

$$\left(p_r + \frac{3}{V_r^2}\right) \cdot (3 \cdot V_r - 1) = 8 \cdot T_r \tag{4.12}$$

A detailed analysis of the experimental thermodynamic data for hydrogen by Hemmes et al. [17] has led to the proposal of an equation of state based on the modified Van der Waals equation [18] for 1 mol of hydrogen:

$$\left(p + \frac{a(p)}{V_m^\alpha}\right)(V_m - b(p)) = R \cdot T \tag{4.13}$$

where $b(p)$ is a pressure dependent excluded volume, $a(p)$ a pressure dependent interaction term, α is approximately constant but not equal to two and V_m is the molar volume. The pressure dependences are described by

$$a(p) = \exp\left[a_1 + a_2 \cdot \ln(p) - \exp(a_3 + a_4 \cdot \ln(p))\right] \quad (p > 1 \text{ bar}) \tag{4.14}$$

$$b(p) = \begin{cases} \sum_{i=0}^{8} b_i \cdot \ln(p)^i & (p \geq 100 \text{ bar}) \\ b(100) & (p < 100 \text{ bar}) \end{cases} \tag{4.15}$$

Table 4.6 Coefficients for the calculation of $a(p)$, $b(p)$ and α in Eq. (4.13). From [17].

α_0	2.9315	b_0	20.285
α_1	-1.531×10^{-3}	b_1	-7.44171
α_2	4.154×10^{-6}	b_2	7.318565
		b_3	-3.463717
		b_4	0.87372903
a_1	19.599	b_5	-0.12385414
a_2	-0.8946	b_6	9.8570583×10^{-3}
a_3	-18.608	b_7	$-4.1153723 \times 10^{-4}$
a_4	2.6013	b_8	7.02499×10^{-6}

where p is measured in bar. For the weak temperature dependence of the power α a simple quadratic form was assumed.

$$\alpha(T) = \alpha_0 + \alpha_1 T + \alpha_2 T^2 \quad \text{for} \quad T \leq 300\,\text{K} \tag{4.16}$$

Above 300 K $\alpha(T) = \alpha(300)$ (Tables 4.6 and 4.7).

From the EoS the changes in the thermodynamic quantities can be determined. In order to calculate absolute values we have to know the properties of hydrogen for some reference state, for example the 1 bar reference isobar.

A large part of the experimental thermodynamic data on hydrogen is reproduced by the modified Van der Waals equation within 0.1 % and practically all data within 0.5 %.

The compressibility Z is often applied to correct for the nonideal behavior of real gases and is defined by

$$Z = -\frac{1}{V}\frac{\partial V}{\partial p} = \frac{p \cdot V_m}{R \cdot T} \tag{4.17}$$

$$p = \frac{R \cdot T}{V_m - b(p)} - \frac{a(p)}{V_m^\alpha} \tag{4.18}$$

With $p(V_m)$ and V_m given by the EoS, the compressibility can be calculated.

Table 4.7 Reference isobar ($p = 1$ bar) for the temperature range 100 to 1000 K

T [K]	V_m [cm^3 mol^{-1}]	H [J mol^{-1}]	G [J mol^{-1}]	S [J mol^{-1} K^{-1}]
100	8314.34	2999	-7072	100.71
200	16628.68	5687	-18184	119.36
300	24943.02	8506	-30724	130.77
400	33257.36	11402	-44237	139.10
500	41571.70	14311	-58474	145.57
600	49886.04	17221	-73305	150.88
700	58200.38	20131	-88622	155.36
800	66514.72	23039	-104357	159.25
900	74829.05	25947	-120456	162.67
1000	83143.39	28852	-136879	165.73

The Gibbs energy $G(p,T)$ is according to Equation (4.6) for an ideal gas

$$G(p, T) = G(p_0, T) + \int_{p_0}^{p} V\,dp = G(p_0, T) + n \cdot R \cdot T \cdot \ln\left(\frac{p}{p_0}\right) \quad (4.19)$$

In order to correct for the nonideal behavior of the real gas the pressure is replaced by the fugacity f. The Gibbs energy of a real gas is

$$G(p, T) = G(p_0, T) + \int_{p_0}^{p} V\,dp = G(p_0, T) + n \cdot R \cdot T \cdot \ln\left(\frac{f}{p_0}\right) \quad (4.20)$$

The fugacity coefficient ϕ is defined as $\phi = f/p$ and is independent of the standard pressure p_0 ($p_0 = 1.013 \times 10^5$ Pa). The logarithm of the fugacity coefficient can be calculated [19] by means of the virial coefficients (Tables 4.8 and 4.9).

$$\ln\left(\frac{f}{p}\right) = \ln(\phi) = \frac{1}{R \cdot T} \int_{0}^{p} \left(V_m - \frac{R \cdot T}{p}\right) dp = \sum_{i=1}^{} C_i \cdot \frac{p^i}{i} \quad (4.21)$$

The relation between the compressibility and the fugacity is therefore

$$\ln(\phi) = \int_{0}^{p} \frac{(Z-1)}{p}\,dp \quad (4.22)$$

4.1.5
Joule–Thomson Effect, Inversion Curve

The differential coefficient μ was first investigated by James Joule and William Thomson in the 1850s [23], before Thomson was elevated to the peerage, to become the first Lord Kelvin. So it is also referred to as the Joule–Kelvin coefficient.

$$\mu = \left(\frac{\partial T}{\partial p}\right)_H \quad (4.23)$$

It is a measure of the effect of the throttling process on a gas, when it is forced through a porous plug, or a small aperture or nozzle. The drop in pressure, at constant enthalpy H, has an effect on temperature. The enthalpy of hydrogen as a Van der Waals gas is given by the specific values (per mole) of the extensive variables H and V, written h and v.

$$h = u + pv = \frac{f}{2}RT - \frac{a}{v} + v\left(\frac{RT}{v-b} - \frac{a}{v^2}\right) = RT\left(\frac{f}{2} + \frac{v}{v-b}\right) - \frac{2a}{v} \quad (4.24)$$

where u is the internal energy, f is the degree of freedom (5 for H_2) and v is the molar volume of the gas. The change in enthalpy is given by the differential and is

Table 4.8 Coefficients C_i as a function of temperature

T [K]	C_1	C_2	C_3	C_4	C_5
60	-3.54561×10^{-4}	1.66337×10^{-7}	-2.99498×10^{-11}	2.42574×10^{-15}	
77	-1.38130×10^{-4}	4.67096×10^{-8}	5.93690×10^{-12}	-3.24527×10^{-15}	3.54211×10^{-19}
93.15	-3.86094×10^{-5}	1.23153×10^{-8}	9.00347×10^{-12}	-2.63262×10^{-15}	2.40671×10^{-19}
113.15	1.32755×10^{-5}	1.01021×10^{-8}	4.43987×10^{-13}		
133.15	3.59307×10^{-5}	5.40741×10^{-9}	4.34407×10^{-13}		
153.15	4.24489×10^{-5}	5.03665×10^{-9}	8.93238×10^{-14}		
173.15	4.29174×10^{-5}	5.56911×10^{-9}	-2.11366×10^{-13}		
193.15	4.47329×10^{-5}	3.91672×10^{-9}	-4.92797×10^{-14}		
213.15	4.34505×10^{-5}	3.91417×10^{-9}	-1.50817×10^{-13}		
233.15	4.45773×10^{-5}	2.18237×10^{-9}	5.85180×10^{-14}		
253.15	4.48069×10^{-5}	8.98684×10^{-10}	2.03650×10^{-13}		
273.15	4.25722×10^{-5}	9.50702×10^{-10}	1.44169×10^{-13}		
293.15	3.69294×10^{-5}	2.83279×10^{-9}	-1.93482×10^{-13}		
298.15	3.49641×10^{-5}	3.60045×10^{-9}	-3.22724×10^{-13}		
313.15	4.16186×10^{-5}	-5.28484×10^{-10}	2.73571×10^{-13}		
333.15	4.05294×10^{-5}	-7.21562×10^{-10}	2.52962×10^{-13}		

Table 4.9 The volume V, enthalpy H, Gibbs free energy G and entropy S of n-H_2 at a temperature ($T = 300$ K) and pressures ($p = 0.1$ MPa-100 GPa, i.e. 1×10^6 bar)[a]

p [bar]	V_m [cm^3 mol^{-1}]	H [J mol^{-1}]	G [J mol^{-1}]	S [J mol^{-1} K^{-1}]
1	24943.02	8506	−30724	130.77
2	12485.87	8507	−28994	125
5	5003.08	8510	−26704	117.38
10	2508.87	8515	−24968	111.61
20	1261.83	8526	−23224	105.84
50	513.73	8560	−20895	98.18
100	264.51	8620	−19091	92.37
200	140.09	8747	−17210	86.52
500	65.78	9176	−14454	78.77
1000	40.98	9954	−11924	72.93
2000	27.96	11529	−8615	67.15
5000	18.75	15896	−1962	59.53
10000	14.58	22319	6189	53.77
20000	11.56	33414	19013	48
50000	8.64	60517	48402	40.39
100000	6.84	94585	86094	28.31
200000	5.52	154070	146981	23.63
500000	4.1	293183	287378	19.35
1000000	3.23	472676	467424	17.51

[a] The values are taken from Hemme et al. [20]. The complete set of data for temperatures ($T = 100$–2000 K) can be found in Fukai [21] and Sugimoto [22].

zero for an isoenthalpic expansion

$$dh = \frac{\partial h}{\partial v} dv + \frac{\partial h}{\partial T} dT = 0 \tag{4.25}$$

$$dT = -dv \frac{\frac{\partial h}{\partial v}}{\frac{\partial h}{\partial T}} = \frac{\frac{T \cdot b}{(v-b)^2} - \frac{2a}{Rv^2}}{\frac{5}{2} + \frac{v}{v-b}} dv \tag{4.26}$$

In a pressure–temperature plot, for any gas, the locus of points at which the drop in pressure, at constant enthalpy H, has no effect on the temperature is called the inversion curve for that gas. So the inversion curve has the simple form $\mu = 0$. These curves are plotted by first finding a family of isoenthalpies (H = const.) on the T–p plane, then connecting their stationary points. Except along the inversion curve, throttling either heats the gas ($\mu < 0$), or cools it ($\mu > 0$). Cooling of gases, in order to subsequently liquefy them, is usually accomplished by throttling in a region where it causes cooling (Figure 4.6).

The specific values (per mole) of the extensive variables H, S and V, are written h, s and v. Then the specific enthalpy is [24]

$$dh = T \cdot ds + v \cdot dp \tag{4.27}$$

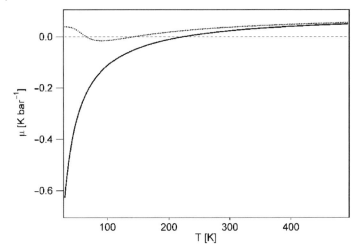

Fig. 4.6 The Joule–Thomson coefficient for H_2 in the Van der Waals approximation (Equation (4.26)) at a pressure of $p = 0.1$ MPa (solid line) and $p = 10$ MPa (dashed line).

and

$$T \cdot ds = c_p \cdot dT - T \cdot \left(\frac{\partial v}{\partial T}\right)_p \cdot dp \tag{4.28}$$

Thus it follows that the Joule–Thomson coefficient is

$$\mu = \frac{1}{c_p}\left[T\left(\frac{\partial v}{\partial T}\right)_p - v\right] = \frac{T^2}{c_p}\frac{\partial}{\partial T}\left(\frac{v}{T}\right)_p \tag{4.29}$$

The first thing to notice about the Joule–Thomson coefficient is that it vanishes for an ideal gas. The inversion "curve" is everywhere for an ideal gas! For real gases, the isoenthalpies have to be determined experimentally and the line connecting the stationary points is the inversion curve.

The simplest equation of state for a gas which predicts an inversion curve is the Van der Waals equation. It yields an inversion curve, that is fairly close, but not exactly correct. The Van der Waals equation in reduced variables (Equation (4.14)) used in the equation for the Joule–Thomson coefficient (Equation (4.29)) leads for $\mu = 0$ and substituting index r by i to

$$T_i = \frac{3 V_i (V_i - 1)^2}{4 V_i^3} \tag{4.30}$$

Substituting V_i with p_i one finds (Figure 4.7)

$$T_i = \frac{1}{12}\left(45 - p_i \pm 12\sqrt{9 - p_i}\right) \tag{4.31}$$

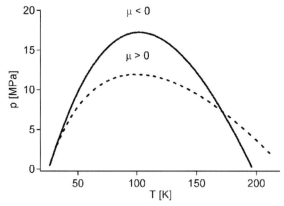

Fig. 4.7 Inversion curve of n-hydrogen and calculated inversion curve (Equation (4.31)) using the critical pressure $p_k = 1.325$ MPa and the critical temperature $T_k = 33.19$ K for hydrogen.

The two branches join at $p_i = 9$, where $T_i = 3$ and $V_i = 1$. They connect this point with the T_i axis, where p_i vanishes, at the two points $T_i = \frac{3}{4}$ and $T_i = 27/4$. And of course, inside this curve is the region in which the Van der Waals equation predicts that throttling any gas will produce cooling.

4.1.6
Chemical Properties and Diffusion

Hydrogen is the element with the highest diffusion capacity because of its small size and small mass. Some diffusion coefficients [25] of hydrogen in gases and liquids are listed in (Table 4.10). In metals with a high hydrogen solubility a high hydrogen diffusion rate is also usually found. The diffusion coefficients are in the same range as for hydrogen ions in water (10^{-4} cm^2 s^{-1} at 25 °C).

Table 4.10 Diffusion coefficients of hydrogen in gases ($p = 101.3$ kPa) and liquids

Diffusion in	D [cm^2 s^{-1}]	T [°C]
N_2	0.674	0
O_2	0.701	0
H_2 (selfdiffusion)	1.285	0
H_2O, vapor	0.759	0
H_2O, liquid	4.8×10^{-5}	25
Iron, smelting	5.64×10^{-3}	1600
Al, smelting	1.28×10^{-5}	960
Palladium	5.0×10^{-7}	25
Vanadium	5.0×10^{-5}	25

Hydrogen has the electron configuration 1s^1. The first electron shell can be filled with a maximum of two electrons. Therefore, the chemistry of hydrogen depends mainly on four processes (Figure 4.8):

1. donate the valency electron to form the hydrogen ion, H$^+$
2. accept an electron to form the hydride ion H$^-$
3. share the electron with a partner atom to form a pair bond (covalent bond) H—H
4. share the electron with an ensemble of atoms to form a metallic bond H^0.

The degree of covalent to ionic character of the covalent bond depends on the electronegativity of the element to which hydrogen is attached. The electronegativity of hydrogen according to Pauli is 2.1. Except in the hydrogen molecule itself, where the bond is homopolar all H—X bonds will possess some polar character. Hydrogen is not exceptionally reactive, although hydrogen atoms react with one another and with all other elements with the exception of the noble gases (with helium only short-lived unstable forms such as HeH$^+$ and HeH$_2^{2+}$ are known). Hydrogen oxidizes less electronegative elements (e.g. alkali and alkaline earth metals), and reduces more electronegative ones (e.g. halogens, oxygen, nitrogen, carbon). The strength of the H—X bond in covalent hydrides depends on the electronegativity and size of the element X. The bond strength decreases in a group with increasing atomic number and generally increases across any period. The most stable covalent bonds are those formed between two hydrogen atoms (or isotopes), or with halogens, oxygen, carbon and nitrogen (Table 4.11).

Halogens react with hydrogen to yield hydrogen halides, with increasing reactivity in the sequence iodine, bromine, chlorine and fluorine. At $-210\,°$C liquid fluorine ignites immediately in hydrogen; solid fluorine explodes violently in the presence of liquid hydrogen; similarly, combined hydrogen reacts more or less violently with fluorine. Oxygen reacts with hydrogen according to

$$H_2 + {}^1\!/_2\, O_2 \rightarrow H_2O;\ \Delta H^\circ_{298} = -285\,\text{kJ} \cdot \text{mol}^{-1},\ S^\circ = 70\,\text{J}\,\text{K}^{-1} \cdot \text{mol}^{-1}$$

Above 550 °C the reaction occurs with flame propagation, explosion or detonation (oxyhydrogen reaction). The flame temperature is limited by the thermal dissociation of water vapor and attains a maximum of 2700 °C.

At elevated temperatures in the presence of suitable catalysts, hydrogen will react with nitrogen to form ammonia. Industrially this is one of the most important reactions.

$$3\,H_2 + N_2 \rightarrow 2\,NH_3;\quad \Delta H^\circ_{298} = -46\,\text{kJ} \cdot \text{mol}^{-1},\quad S^\circ = 192\,\text{J} \cdot \text{K}^{-1} \cdot \text{mol}^{-1}$$

Hydrogen reacts with carbon at high temperatures to yield methane. The equilibrium of the reaction is on the methane side at low temperatures, and on the hydrogen and carbon side at high temperatures.

$$2\,H_2 + C \rightarrow 2\,CH_4;\quad \Delta H^\circ_{298} = -75\,\text{kJ} \cdot \text{mol}^{-1},\quad S^\circ = 186\,\text{J} \cdot \text{K}^{-1} \cdot \text{mol}^{-1}$$

1	2	3	4	5	6	7	8	9	10	11	12	13	14	15	16	17	18
H 2.20																	He
LiH 0.97	BeH_2 1.47											BH_3 2.01	CH_4 2.50	NH_3 3.07	H_20 3.50	HF 4.10	Ne
NaH 1.01	MgH_2 1.23											AlH_3 1.47	SiH_4 1.74	PH_3 2.06	H_2S 2.44	HCl 2.83	Ar
KH 0.91	CaH_2 1.04	ScH_2 1.20	TiH_2 1.32	VH VH_2 1.45	CrH (CrH_2) 1.56	Mn 1.60	Fe 1.64	Co 1.70	$NiH_{<1}$ 1.75	CuH 1.75	ZnH_2 1.66	(GaH_3) 1.82	GeH_4 2.02	AsH_3 2.20	H_2Se 2.48	HBr 2.74	Kr
RbH 0.89	SrH_2 0.99	YH_2 YH_3 1.11	ZrH_2 1.22	(NbH_2) 1.23	Mo 1.30	Tc 1.36	Ru 1.42	Rh 1.45	$PdH_{<1}$ 1.35	Ag 1.42	(CdH_2) 1.46	(InH_3) 1.49	SnH_4 1.72	SbH_3 1.82	H_2Tc 2.01	HI 2.21	Xe
CsH 0.86	BaH_2 0.97	LaH_2 LaH_3 1.08	HfH_2 1.23	TaH 1.33	W 1.40	Re 1.46	Os 1.52	Ir 1.55	Pt 1.44	(AuH_3) 1.42	(HgH_2) 1.44	(TlH_3) 1.44	PbH_4 1.55	BiH_3 1.67	H_2Po 1.76	HAt 1.90	Rn
Fr	Ra	AcH_2 1.00															

CeH_3 1.06	PrH_2 PrH_3 1.07	NdH_2 NdH_3 1.07	Pm	SmH_2 SmH_3 1.07	EuH_2 1.01	GdH_2 GdH_3 1.11	TbH_2 TbH_3 1.10	DyH_2 DyH_3 1.10	HoH_2 HoH_3 1.10	ErH_2 ErH_3 1.11	TmH_2 TmH_3 1.11	(YbH_2) YbH_3 1.06	LuH_2 LuH_3 1.14
ThH_2 1.11	PaH_2 1.14	UH_3 1.22	NpH_2 NpH_3 1.22	PuH_2 PuH_3 1.22	AmH_2 AmH_3 1.2	Cm	Bk	Cf	Es	Fm	Md	No	Lr

Fig. 4.8 The different hydrides formed with the elements and their electronegativity.

Table 4.11 Average bond dissociation energies of representative covalent hydrogen bonds at 298 K. From [1].

Bond	$\Delta H°_{298}$ [kJ mol^{-1}]	Bond	$\Delta H°_{298}$ [kJ mol^{-1}]
H—H	428.2	H—N	314
H—D	440	H—N=	435.4
D—D	443.4	H—O	428.2
T—T	446	H—OH	498.1
H—CH$_3$	435.3	H—S	344.5
H—C=	452	H—Cl	431.6
H—C≡	536	H—F	568.9

The equilibrium constant, using graphite as a basis, is 2×10^5 bar at 150 °C and 9.4×10^{-3} bar at 1000 °C. Carbon monoxide and carbon dioxide can react with hydrogen in the presence of a catalyst. Depending on the reaction conditions, catalyst and CO : H$_2$ ratio a variety of products can be formed. Unsaturated hydrocarbons are converted to saturated or partially saturated hydrocarbons by hydrogenation. In the petroleum industry hydrogenation and hydrocracking reactions are of importance as well as desulfurization and selective hydrogenation (dienes, monoolefins).

Many metals react with hydrogen to form hydrides. Depending on the nature of the hydrogen bond, hydrides are classified into four principal categories: covalent, ionic, complex and metallic.

The saline hydrides or ionic hydrides are formed when the strongly electropositive alkali metals and the alkaline earth metals (calcium, strontium, barium), all usually in the elemental form, react with hydrogen at high temperatures. Salt-like hydrides contain the hydride ion, H$^-$; they are crystalline, have high heats of formation and high melting points. They are reactive, powerful reducing agents, all of them reacting with water to liberate hydrogen.

Covalent hydrides may be either solid, liquid or gaseous; typical are the hydrides of boron, aluminum, silicon. germanium and tin. Beryllium and magnesium hydride seem to represent a transition between ionic and covalent hydrides. The bond between hydrogen and the atom is of the nonpolar electron-sharing type. Molecular stability is often provided by means of three-centered two-electron bridge bonds (e.g. in diborane, B$_2$H$_6$, an electron-deficient type molecule), by hydride anion formation (BH$_4^-$, AlH$_4^-$), or formation of polymeric structures [polymeric (AlH$_3$)$_x$].

Metallic hydrides are formed by the transition metals. Transformation of a pure metal into a hydride occurs through continuous solution of hydrogen in the metal with subsequent abrupt phase transition at defined stoichiometric hydride phases (ZrH$_2$, PdH. VH, VH$_2$).

Substances containing hydrogen coupled to the most electronegative elements exhibit properties that are best explained by assuming that the hydrogen atom of an H—X bond still has a small but significant affinity for other electronegative atoms although it remains strongly bonded to the original atom. This relatively weak secondary bond is called a hydrogen bond, normally depicted by the notation

X—H⋯Y. Hydrogen bonds result in molecular association. NH_3, H_2O and HF have much higher boiling points and heats of vaporization than the corresponding homologues PH_3, H_2S and HCl. Carboxylic acids such as acetic or benzoic acids are present as dimers with the configuration

$$-C\underset{O-H\cdots O}{\overset{O\cdots H-O}{\diagup\diagdown}}C-$$

The presence of hydrogen bonds also causes formation of crystalline hydrates such as $NH_3\ H_2O$, $SO_2\ H_2O$ hydrocarbon hydrates (e.g. with natural gas components) as well as the zig-zag structure of the $(HF)_n$ polymer.

The existence of the hydrogen bond can be shown by infrared spectroscopy. When the hydrogen of the X—H group forms a hydrogen bond with the atom Y, the X—H stretching frequency is lowered and the absorption band characteristically broadened. In NMR spectroscopy a proton which participates in a hydrogen bond, produces a definite paramagnetic shift. The energies of hydrogen bonds are in the range 4–40 kJ mol^{-1}, which is small compared to 200–400 kJ mol^{-1} for ordinary bonds. The average bond length is 200–300 pm (O—H⋯O in ice 276 pm).

Atomic hydrogen is much more reactive than the molecular form. The dissociation reaction

$$H_2 \rightarrow 2\,H; \quad \Delta H^\circ_{298} = 432.2\,\text{kJ mol}^{-1}$$

is highly endothermic. Atomic hydrogen can be produced thermally by supplying sufficient energy, for example in electric arcs with a high current density, electrical discharges at low pressures, irradiation with UV light as well as electron bombardment (10–20 eV). The hydrogen atom has a short half-life (circa 0.3 s) and reacts even at room temperature with nonmetallic elements such as halogens, oxygen, sulfur and phosphorus to form the corresponding hydrogen compounds. At room temperature it reduces many oxides to the elements. The heat of recombination of hydrogen molecules from hydrogen atoms leads to very high temperatures (Langmuir flare, 4000 K). Atomic hydrogen can be absorbed in the metallic structure of some elements of Group 8–10 (Fe, Co, Ni) of the periodic system. The suitability of many catalysts for reactions involving hydrogen is based on the dissociation and solubility of hydrogen in the atomic form. Hydrogen formed upon dissolution of metals in acids is very reactive at the moment of formation (nascent hydrogen), much more so than normal hydrogen. This strong reduction capability is often explained by a short term formation of atomic hydrogen. Another explanation is the intermediate formation of a hydride-type compound which is very similar to Grignard compounds.

The ionization potential of the reaction

$$H_{(g)} \rightarrow H^+_{(g)} + e^-; \quad \Delta H^\circ_{298} = 1310\,\text{kJ mol}^{-1}$$

which is even higher than the first ionization potential of the noble gas xenon. The hydrogen ion can be formed only in a medium which solvates the protons. The solvation process provides the energy required for bond rupture. The order of magnitude can be seen in the solvation reaction with water:

$$H^+_{(g)} + x\,H_2O \rightarrow H^+_{(aq)}; \quad \Delta H^\circ_{298} = 1070\,\text{kJ}\,\text{mol}^{-1}$$

In aqueous systems, H_3O^+ is the form implied by the customary designation hydrogen ion. The structure of the H_3O^+ ion is that of a rather flat triangular pyramid with a H—O—H bond angle of circa 110°, O—H bond distances of 102 pm and a H—H distance of 172 pm. Normally, it is coordinated with four water molecules not all of which are equivalent, so that the hydrated hydrogen ion is best represented as $H_9O_4\cdot H_2O$. The lifetime of an individual H_3O^+ ion is exceedingly short, 10^{-13} s, because the protons are rapidly exchanged among the water molecules.

In aqueous solutions the hydrogen ion activity is given in terms of the pH value, defined as $-\log(a(H^+))$, where $a(H^+)$ is the hydrogen ion activity. By definition, the potential of the redox system $\tfrac{1}{2}\,H_2(g) \rightarrow H^+(aq) + e^-$ is zero ($E = 0.000$ V) at all temperatures, when an inert metallic electrode is immersed in a solution of unit activity (i.e. pH = 0) in equilibrium with hydrogen gas at 100 kPa pressure. The electrode material is a small platinum foil or a wire, coated with finely dispersed platinum black and is placed partly in the solution and partly in the hydrogen atmosphere. This standard hydrogen electrode is the reference electrode for all other oxidation–reduction systems. The relation between the electrochemical potential E and the hydrogen partial pressure p_{H_2} is

$$\Delta G = -n \cdot F(E - E^0) = R \cdot T \cdot \ln\left(\frac{p_{H_2}}{p_0}\right) \tag{4.32}$$

4.1.7
Ignition and Detonation Performance

Hydrogen reacts, when the ignition energy (thermal activation energy) of ≈ 0.02 mJ is provided, violently with oxidizing agents such as oxygen (air), fluorine or chlorine and N_2O. Combustion, deflagration or detonation may occur, depending on the conditions. The ignition and detonation properties of hydrogen–air mixtures are particularly important from the safety aspect. The flammability limits (i.e. the minimum and the maximum concentration of hydrogen in air) are exceptionally wide for hydrogen.

Table 4.12 shows a comparison of safety-relevant thermo-physical and combustion properties of hydrogen with those of methane, propane and gasoline [26]. The flammability limits are affected by temperature, as shown in Figures 4.9 and 4.10, so that a preheated mixture has considerably wider limits for coherent flames [27]. An increase in pressures up to 10 kPa has only a small effect. Water vapor has a strongly inhibiting influence on the oxyhydrogen reaction.

The flammability limits of hydrogen in pure oxygen at room temperature are 4.65–93.9 vol% hydrogen, those of deuterium in pure oxygen 5.0–95 vol%. In general, the chemical properties of H, D and T are essentially identical. Small

Table 4.12 Combustion and explosion properties of hydrogen, methane, propane and gasoline

	Hydrogen	Methane	Propane	Gasoline
Density of gas at standard conditions [kg m^{-3}(STP)]	0.084	0.65	2.42	4.4^a
Heat of vaporisation [kJ kg^{-1}]	445.6	509.9		250–400
Lower heating value [kJ kg^{-1}]	119.93×10^3	50.02×10^3	46.35×10^3	44.5×10^3
Higher heating value [kJ kg^{-1}]	141.8×10^3	55.3×10^3	50.41×10^3	48×10^3
Thermal conductivity of gas at standard conditions [mW cm^{-1} K^{-1}]	1.897	0.33	0.18	0.112
Diffusion coefficient in air at standard conditions [cm^2 s^{-1}]	0.61	0.16	0.12	0.05
Flammability limits in air [vol%]	4.0–75	5.3–15	2.1–9.5	1–7.6
Detonability limits in air [vol%]	18.3–59	6.3–13.5		1.1–3.3
Limiting oxygen index [vol%]	5	12.1		11.6^b
Stoichiometric composition in air [vol%]	29.53	9.48	4.03	1.76
Minimum energy for ignition in air [mJ]	0.02	0.29	0.26	0.24
Autoignition temperature [K]	858	813	760	500–744
Flame temperature in air [K]	2318	2148	2385	2470
Maximum burning velocity in air at standard conditions [m s^{-1}]	3.46	0.45	0.47	1.76
Detonation velocity in air at standard conditions [km s^{-1}]	1.48–2.15	1.4–1.64	1.85	1.4–1.7^c
Energy^d of explosion, mass-related [gTNT g^{-1}]	24	11	10	10
Energy^d of explosion, volume-related [gTNT m^3(STP)]	2.02	7.03	20.5	44.2

^{a}100 kPa and 15.5 °C.
^{b}Average value for a mixture of C_1—C_4 and higher hydrocarbons including benzene.
^{c}Based on the properties of n-pentane and benzene.
^{d}Theoretical explosive yields.

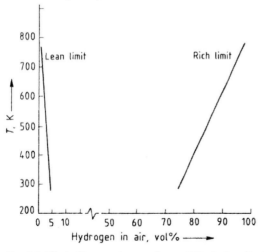

Fig. 4.9 Effect of temperature on flammability limits of hydrogen in air (pressure 100 kPa).

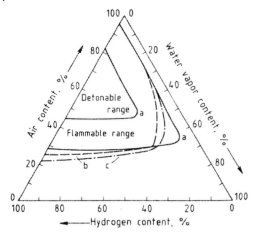

Fig. 4.10 Flammability and detonability limits of the three-component system hydrogen–air–water vapor [28] at (a) 42 °C, 100 kPa; (b) 167 °C, 100 kPa, (c) 167 °C, 800 kPa.

differences can be found in processes where the atomic mass has a significant influence, for example the thermal dissociation of a D—X bond requires a greater energy than that of the comparable H—X bond [29].

For comprehensive compilations of safety characteristics and hazards on handling hydrogen see Ref. [30]. The envisaged large-scale introduction of hydrogen as an energy carrier requires a comparative evaluation of safety aspects with regard to fossil fuels. Volumetric leakage's of hydrogen gas will be 1.3–2.8 times as large as gaseous methane leakage and approximately four times that of air under the same conditions (rule: "airproof is not hydrogen-proof"). Released hydrogen disperses rapidly by turbulent convection, drift and buoyancy, thus shortening the hazard duration. The prompt dispersion, however, favors the formation of gas mixtures within the wide flammability and detonability limits; the lower limit is the vital one in most applications and is comparable to that of other fuels. The minimum energy for ignition of hydrogen–air mixtures is extremely low. However, the ignition energy inherent in virtually every source is also more than sufficient for ignition of any other fuel–air mixture. Ignition by catalytic action is also possible.

4.1.7.1 Fire Hazards

Hydrogen flames are nearly invisible in daylight. Hydrogen fires last only one fifth to one tenth of the time of hydrocarbon fires and the fire damage is less severe because of several characteristics: (i) high burning rate resulting from rapid mixing and high propagation velocity, (ii) high buoyant velocity, (iii) high rate of vapor generation of liquid hydrogen. Although the maximum flame temperature is not much different from that of other fuels, the thermal energy radiated from the flame is only a part of that of a natural gas flame. Smoke inhalation, one of the major causes of injury and, therefore, a main parameter of fire damage, is considered less serious in the case of hydrogen because the sole combustion product is water vapor and some nitrogen oxides (depending on the temperature of the flame).

4.1.7.2 Explosive Hazards

High laminar burning velocity as well as the high laminar flame speed of hydrogen makes the transition to turbulent flame speeds exceeding $800\,\mathrm{m\,s^{-1}}$ up to several $\mathrm{km\,s^{-1}}$ easy. Hence, hydrogen is more sensitive to deflagration to detonation transition [31] (DDT) than hydrocarbons. Hydrogen has by far the widest limits of detonability; detonation of stoichiometric hydrogen–air mixtures in confined spaces will produce a static pressure rise of circa 15:1. If a DDT process proceeds, that is unburned hydrogen–air mixtures are precompressed, a pressure rise ratio of 120:1 could result.

Explosions are rated in terms of the amount of energy released, commonly expressed as an equivalent quantity of trinitrotoluene (TNT). The theoretical maximum values of explosive potentials for fuels (TNT equivalents) are recorded in Table 4.7. Experimental data indicate, that real yield factors of 10% are considered reasonable. Note that hydrogen is most potent on a mass basis and least potent on a volumetric basis. It also has the least theoretical explosive potential, when equivalent energy storage is taken into account.

4.1.7.3 Preventive Measures

In safety concepts, distinctions are made between primary, secondary and tertiary measures. Primary safety precautions aim at the exclusion of causative risks such as leakage, formation of explosive mixtures by proper conceptual design (inertization, open-air installation, flame arrestors, etc.). Secondary measures consist mainly in the avoidance of ignition sources of any kind (electrostatically or mechanically generated sparks). Tertiary measures should minimize dangerous results in case fire or explosion occurs. This is achieved by installation of explosion-proof or explosion relief systems, hydrogen process shut-down systems and suitable fire extinguishing systems. Safety Regulations (mandatory) and standards (mandatory or nonmandatory) apply for the safe production, storage and handling of hydrogen. They are mostly concerned with transportation; other operations are covered by more general regulations.

4.1.7.4 Future Outlook

Considering the possible use of hydrogen in a future hydrogen energy carrier, no major safety problems are anticipated. Hydrogen has been handled successfully for decades in the process industries. Handling as a liquid is also routine. In comparison with other inflammable gases or liquids, the safety risk of hydrogen as a fuel is not significantly different.

4.1.7.5 Toxicology

Hydrogen does not show any physiological effect. It is nonpoisonous. Inhalation of the gas leads to sleepiness and a high-pitched voice. A danger of asphyxiation exists if the oxygen content sinks below 18 vol% because of hydrogen accumulation in the air. Direct skin contact with cold gaseous or liquid hydrogen leads to numbness and a whitish coloring of the skin and to frostbite. The risk is higher than, for example with liquid nitrogen because of the greater temperature differences and the higher thermal conductivity of hydrogen.

4.2
Interaction of Hydrogen with Solid Surfaces*
Andreas Borgschulte and Louis Schlapbach

The first step in the formation of metal hydrides and solid solutions from molecular hydrogen occurs on the surface of the host metal. This chapter contains a description of the properties of the surface and of the interaction of that surface with hydrogen, that is physisorption, dissociation, chemisorption and solution on surface, near-surface and bulk sites. The interaction ranges from the sticking of hydrogen with a reaction probability of 0.99 on the Pd(111) surface, to the other extreme case, the complete absence of any reaction of H_2 with previously air exposed and thus oxide-covered Mg at room temperature. The chapter begins with a description of effects and phenomena occurring on clean surfaces (structural, thermodynamic and electronic properties) and on precovered surfaces (activation, poisoning and segregation). Then results on the hydrogen adsorption on clean surfaces of metals are summarized followed by a description of hydrogen effects on oxides. Subsequently, the impact of hydrogen interactions on thin films is highlighted by typical examples. The chapter ends with a discussion of the relevance of surface reactions for bulk hydrides.

4.2.1
Introduction

The particular importance of surface effects in hydrogen adsorption and absorption by metals, for getters, permanent magnets, in catalytic reactions, battery electrode reaction, H embrittlement and plasma-wall interaction in fusion stems from two facts: The first relates to the surface itself. The sharp discontinuity of matter with electric charges and potentials of electrons and atom cores at the surface together with the loss of periodicity in the direction orthogonal to the surface leads to

- structural properties
- electronic and magnetic properties
- a chemical composition of the uppermost layer and of near-surface layers and
- dynamical properties of surface and near-surface atoms.

which are quite different from those of the bulk. It belongs to the general goals of solid state and surface sciences to describe these phenomena, which, of

---

* Adapted from L. Schlapbach, in: L. Schlapbach (Ed.), Surface Properties and Activation, Hydrogen in Intermetallic Compounds, vol. II, Springer, Berlin, 1992.

course, determine the reactivity of the surface. The second important factor is the interaction of H with that surface. Gaseous hydrogen (mainly H_2, but also H) and its isotopes (D,T) and H^+ (protons) in an electrolyte interface with a surface in the process of adsorption and may or may not penetrate it. The most common reaction proceeds from the physisorption of gaseous H, via dissociative adsorption of H and subsurface H to H dissolved in the bulk. The surface is considered to comprise not only the top atomic layer of the substrate, but also the first few layers which differ from the bulk. In the case of a clean single crystalline substrate the top three or possibly four monolayers constitute the surface. On oxidized and contaminated substrates, for example multiphase alloys, however, the surface or surface layers may be as much as 10 nm thick. Usually these covered surfaces are not reactive towards hydrogen or much less so; they have to be activated. The activation process is of particular importance for H absorption by intermetallic compounds, in catalysis and for getters. A chapter on surface effects should, of course, describe the phenomena observable in a clean well-defined simple H-metal system as well as the activation of more complex H-metal systems, for example H, on an oxidized surface. General references to surface science are given in Ref. [32]. H adsorption on single crystalline substrates of elemental d-transition metals has been studied very extensively as a prototype reaction and because of the dominant role of H in heterogeneous catalysis [33–35].

An overview of these studies and the underlying physical models is given in Section 2.2. In view of the need for light-weight hydrides, Li, Na, Mg, Al, B and Ca are promising starting materials because of their abundance, low density, low price and high H content in their hydrides. Studies of surface effects have recently started on these elements and their compounds [36] (Mg [37, 38]; $NaAlH_4$ [39, 40] Al(Ti) [41]).

Studies of the reaction of H with surfaces of binary alloys began at the end of the 1970s with the growing interest in bimetallic catalysts and with the first successful explanation of the astonishingly high reactivity of some intermetallics towards H by a surface segregation model [42–44]. Surface segregation occurs in vacuum [45], but was also found to depend on the adsorbed gas species [46] and must, therefore, always be considered if the desired material has to function in a gaseous atmosphere, that is as a catalyst in heterogeneous chemical reactions or as a hydrogen storage material [47].

Hydrogen storage is restricted to pure surface effects in adsorption materials only, for example hydrogen storage by carbon nanotubes and metal organic frameworks. Hydrogen storage in metal hydrides is a bulk phenomenon, which is mediated by the surface of the metal (hydride). While the final states are determined by the thermodynamic properties of the bulk, the intermediate states are controlled by the bulk as well as the surface properties. The simplest approach neglects any interactions between surface and bulk and assumes one transition step to be rate limiting, for example diffusion or dissociation. This particular one is then treated independently of the other steps. Most recent results give hints that this treatment is successful in extreme cases only and from this model empirically derived parameters, for example the apparent activation energy, can lead to wrong interpretations [48]. A typical hint is the so-called Constable–Cremer relation, which is an indication of a competition between two processes, here most likely between surface and bulk processes [49].

4.2.2
A Survey of Effects, Phenomena and Models

The adsorption of H is conveniently described in terms of simplified one-dimensional potential energy curves for an H plus a molecule and for two H atoms on a clean metal surface (Figure 4.11). Far from the surface the two curves are separated by the heat of dissociation $E_D = 218$ kJ (mol H)$^{-1}$ (4.746eV [51]). The first attractive interaction of the hydrogen molecule approaching the metal surface is the Van der Waals force leading to the physisorbed state ($E_{Phys} \approx -5$ kJ (mol H)$^{-1}$) approximately one hydrogen molecule radius ($\approx$0.2 nm) from the metal surface [50]. Closer to the surface the hydrogen has to overcome an activation barrier for dissociation and formation of the hydrogen metal bond. The height of the activation barrier depends on the surface elements involved. The top of the barrier is sometimes called the transition state, given by the nearest distance possible to a molecule approaching the surface before dissociation. A survey of the involved electronic interactions and their calculation is given in Section 4.3.1. Hydrogen atoms sharing their electron with the metal atoms at the surface are called chemisorbed ($E_{Chem} \approx -50$ kJ (mol H)$^{-1}$) [34]. The chemisorbed single H atoms may have a high surface mobility. They interact with each other at sufficiently high coverage and form surface phases [34]. The sketched energy surface in Figure 4.11 can be extended to multi-dimensional energy surfaces derived from DFT-calculations, which enables model-independent description of the molecule/atom surface interactions,

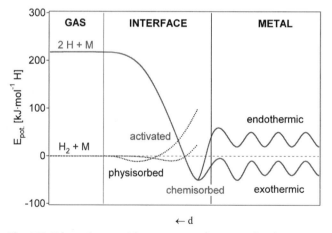

Fig. 4.11 Schematic potential energy curves for activated and non-activated chemisorption of hydrogen on a clean metal surface and exothermic or endothermic solution in the bulk. A more pronounced minimum just below the surface allows for subsurface hydrogen (one-dimensional Lennard-Jones potential, Somorjai (1987) Ref. [33]).

the electron distribution and dissociation probabilities [52]. For a survey, we refer to Section 4.3.1. Many surface properties such as heat of chemisorption and sticking probability become coverage dependent. In further steps the chemisorbed H atoms penetrate the surface and are dissolved exothermically or endothermically in the bulk, where a hydride phase may nucleate and grow. The H-metal bonding is electronic in nature both in the bulk and at the surface. An understanding of the surface and of the adsorption of H on that surface requires a description of the structural properties, thermodynamic properties (heat of adsorption, equilibrium composition, phases), electronic and magnetic properties (nature of the chemical bond) as well as dynamic and kinetic properties (sticking, vibration, rotation, diffusion), all of which are, of course, interrelated.

Experimental and theoretical studies of H adsorption are mostly performed on idealized surfaces, that is single crystals. Adsorption on real surfaces is a more complex phenomenon, mainly because of the physical and chemical nonuniformity of the surfaces. Surface defects (steps, kinks, grain boundaries) and the presence of impurity atoms strongly affect the adsorption [53]. Surfaces of alloys and intermetallic compounds show additional phenomena related to the fact that the chemical composition at the surface may differ from that in the bulk [47].

4.2.3
Surface Structure

In order to minimize the free energy of a crystal the equilibrium position of surface atoms can be different from that given by the lattice periodicity of the bulk. Very often the interatomic spacing between the top atomic layers differs from that deeper in the bulk, without noticeable change in the lateral symmetry. This effect is called surface relaxation. Often it amounts to a 5–10 % contraction between the first and second layer and a smaller, but measurable expansion between the second and third, and third and fourth atomic planes. In a few cases clean metal surfaces are reconstructed [54, 55] (e.g. the (100) surface of Ir, Pt, Au), that is the lateral symmetry of the surface differs from that of the bulk. Usually additional diffraction spots indicate a lowering of the symmetry. Generally the crystallographically open surfaces [fcc (110), bcc (111) and (211), hcp (1010)] are more susceptible to relaxation and reconstruction than the close-packed surfaces [fcc (111), bcc (110)] [54]. The notation used to describe surface reconstructions is elucidated, for example in Ref. [56]. Some illustrative examples are shown in Figure 4.12. The positions of adsorbed H atoms on the substrate surface are called atop (on top of a substrate atom), bridge or twofold (above the centre of two substrate atoms), threefold and fourfold (Figure 4.12(c)).

Upon increasing the coverage, adsorbed H atoms form disordered or, at lower temperatures, ordered surface phases. Adsorption itself can induce relaxation or reconstruction of the substrate surface or may even lift the relaxation or reconstruction of the clean substrate surface, for example Ref. [34]. The van der Waals type bonding of physisorbed H_2 however is too weak to cause noticeable displacement of substrate atoms. There is experimental and theoretical evidence [34, 36, 57–60]

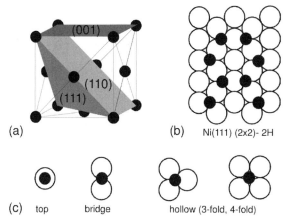

Fig. 4.12 (a) Definition of lattice planes in a fcc-lattice; (b) hydrogen adsorbed on Ni(111); (c) high symmetry sites of low index surfaces.

that chemisorbed H does not necessarily only occupy sites on top of the first metal atom layer, but also sites between and below top layer metal atoms. Subsurface H is generally accompanied by a strong surface reconstruction (surface hydride formation) and can be considered as an intermediate stage between adsorbed H and bulk hydride formation. The H–H (and D–D) equilibrium distances are comparable to those between bulk interstitial sites, that is 2.2 Å. The distance between planes of surface H and subsurface H (e.g. on Pd (111) [61]) can be shorter. However, it is unlikely that a surface H (or D) sits on top of a subsurface H (or D) in thermal equilibrium. On alloy substrates the reconstruction of the clean surface includes not only a geometrical rearrangement of the surface atoms, but also a redistribution of atoms of the alloy components. It results in an alloy composition at the surface different from that of the bulk and is known as surface segregation. The important experimental methods for surface structural analysis are not only the most widely used LEED and the very powerful He diffraction, but also STM, surface EXAFS, photoelectron spectroscopy and photoelectron diffraction, and neutron scattering. Results of the structure analysis of H layers on transition metals are summarized in Ref. 3 [Tables 5, 10–12].

4.2.4
Thermodynamic Properties, Surface Segregation

Although the H-metal bond is electronic in nature, a thermodynamic description of some adsorption properties is adequate. The relevant thermodynamics, with emphasis on surface problems, can be found, for example, in Ref. [62]. A weak van der Waals attraction is the origin of the physisorption of molecular H_2 [63, 64]. Accordingly, heats of physisorption are very small (> -5 kJ (mol H)$^{-1}$), and experimental physisorption studies have to be performed at low temperatures. The growing interest in physisorption has been stimulated by the successful use of

Table 4.13 Experimental ((111)-planes) and theoretical ((211) plane) heats of chemisorption compared to heats of solution in transition metals in kJ (mol H)$^{-1}$

	Cu	Ni	Ag	Pd
ΔH_{chem} (Exp)	−21 [77]	−52 [77]	<0 [77]	−50 [75]
ΔH_{chem} (Theo)	−27.9 [78]	−79.0 [78]	+51.0 [78]	−75.4 [78]
ΔH_{sol}	+42.4 [79]	+16.4 [79]	+68.5 [79]	−9.6 [79]

molecular beam scattering as a probe of surface structure. It is well established that stable physisorbed H_2 can exist on the noble metal surfaces Cu [65, 66], and Ag [67]. Low energy molecular beam studies show sharp diffraction patterns due to physisorbed H_2 and have allowed the determination of the interaction potential. The H_2–Cu physisorption well, for example, was determined to be 23 meV deep [68]. Studies of H_2 physisorption on simple metals were initiated both theoretically [64, 69] and experimentally [70]. Adsorption of molecular H, at temperatures well above 20 K may occur on special sites of lower coordination, for example on edge sites of stepped Ni (100) [71]. An important application of physisorption is hydrogen storage by adsorption on nanostructures, for example carbon nanotubes [72], zeolites [73] and metal organic frameworks [74]. Due to the weak interaction, a significant physisorption and therefore storage capacity is here also only observed at low temperatures (<273 K).

Chemisorption of atomic hydrogen occurs through electronic bonding states formed by H 1s and substrate metal states [75]. The heat of chemisorption is of the order of 20–60 kJ (mol H)$^{-1}$ and varies smoothly with the number of d-electrons within the 3d-transition metal series [75]. Values of the heats of chemisorption of H are compared to the heat of hydrogen solution in the bulk (Table 4.13). The trend of both enthalpies is similar due to a similar bonding between hydrogen and metal states. Moreover, the difference between ΔH_{sol} and ΔH_{chem} is constant for most metals. Accordingly the heat of solution can be used to estimate the heat of chemisorption [76].

The transition from physisorbed H_2 to chemisorbed 2H involves dissociation. Dissociative chemisorption turns out to be one of the most important surface-related steps in the formation of hydrides of intermetallic compounds [42]. Dissociation and chemisorption are called activated if the H_2 molecule approaching the surface has to overcome an activation barrier (Figure 4.11). Activation barriers of up to 1 eV were found on s-electron metals (Mg [37]); d-electrons tend to reduce the barrier [80].

On irregular surfaces with high defect concentrations (e.g. catalysts, clusters), H is preferentially adsorbed at sites of high coordination and near steps [34]. This leads to an adsorption energy which may be initially several kJ (mol H)$^{-1}$ higher and shows that H absorption can be used as a probe for surface defects.

If a new surface is created by the fracture of a metal (in absolute vacuum) the chemical potential of a surface atom μ^s is different from that of a bulk atom

μ^B. Thermodynamic equilibrium requires $\mu^s = \mu^B$ so that surface relaxation and reconstruction are induced. Alloy surfaces have an additional degree of freedom, the composition. Thermodynamic equilibrium requires that the chemical potential of the components A and B be uniform right up to the surface,

$$\mu_A^s = \mu_A^b, \quad \mu_B^s = \mu_B^b \tag{4.33}$$

and induces a surface composition which generally differs from that of the bulk, the phenomenon of surface segregation.

Surface segregation of binary alloys is often described [46] in the form

$$\frac{c_A^s}{c_B^s} = \frac{c_A^b}{c_B^b} \exp(\Delta H / kT) \tag{4.34}$$

where c stands for concentration. The enthalpy of segregation ΔH results from the differences in the chemical potentials

$$\Delta H = \left(\mu_A^s - \mu_A^b\right) \pm \left(\mu_B^s - \mu_B^b\right) \tag{4.35}$$

In the absence of chemisorption ΔH can be approximated [46] by the difference in surface enthalpy of the constituents A and B.

Adsorbates that interact strongly with the substrate (e.g. oxygen, sulfur) may induce an additional surface segregation which is generally much stronger than the segregation expected or observed in (absolute) vacuum. Different bond strengths between the adsorbate and the different components act as driving forces. Thus differences in the heats of adsorption of the adsorbate with the different components have to be added to the heat of segregation in (4.34) [46]. Surface segregation induced by the adsorption or absorption of H is much weaker and therefore only observed experimentally in a few cases studied so far (e.g. RhPt-clusters [81]).

However, oxygen-induced segregation effects are strong and are very important in the activation and poisoning of intermetallics for hydrogen absorption, in bimetallic catalysts, getters, and in surface magnetism [82–87]. Cu–Ni and Au–Ni are the best investigated solid solutions [87–90]. Strong Cu enrichment was observed on the topmost surface layer of Ni–Cu alloys containing up to 84 % Ni. On the Ni-rich side of the alloy surface segregation of Ni extending over the top surface layer and near surface layers was observed. H chemisorption was found, experimentally and theoretically [90], to reduce the Cu segregation. Alloy formation and subsequent oxygen-induced surface segregation were found to inhibit hydrogen uptake in Pd-capped yttrium layers [91].

Two special forms of surface segregation have been observed in binary and ternary hydrides and in H solid solution phases. First, the hydrogen-to-metal ratio in the surface or near-surface region differs from that in the bulk. The fact that the distance between the first and second metal atom planes can be adjusted more easily than between planes in the bulk, generally leads to enhanced hydrogen content in this subsurface region. Nb and Pd, for example, are known to exhibit strongly enhanced near-surface solubility for H [92–94]. In rare earth hydrides REH_{2+x} the opposite occurs: at low temperatures some surface hydrogen seems to diffuse into the bulk leading to a H-depleted REH_2 surface [95]. The depth concentration profile or near-surface H is strongly affected by the presence of impurities like oxygen [93]. Second, the presence of hydrogen in an intermetallic compound or alloy (hydride or solid solution) may induce a redistribution of the surface metal atoms and affect the surface composition. The effect is very weak or absent in all hydrides of intermetallics studied so far by surface-sensitive techniques, with the exception of a few Cu and Ca alloys. Strong modifications of the alloy composition upon hydrogen absorption were noticed on Pd–Cu [96] and recently on Ca–Pd [97], although one has to keep in mind that it is not always an easy task to distinguish clearly between artifacts induced by, for example oxygen and other impurities. Numerous theoretical studies have predicted the surface composition of ordered and disordered alloys from bulk thermodynamic properties and from microscopic electronic theories [88, 90, 98–100]. They describe quite successfully the surface concentration ratio as a function of the bulk concentration ratio at a fixed temperature, which is mostly room temperature or zero temperature. However, they do not describe the temperature dependence of segregation. Experimentally strong variation of the surface concentration ratio and even cross-overs were observed in chemisorption-induced segregation as a function of temperature.

Surface segregation is normally studied experimentally by the surface-sensitive techniques of photoelectron spectroscopy (ESCA, XPS), Auger electron spectroscopy and atom probe.

4.2.5
Chemisorption on Metal Surfaces

The electronic properties of elemental metals, ordered and disordered alloys and metal hydrides, and the experimental and theoretical methods to study them are described in Chapter 4.3 by I. Chorkendorf. Here we restrict ourselves to a short overview.

The isolated H_2 molecule possesses an occupied $1\sigma_g$, bonding level well below the bottom of most metal bands and a $1\sigma_u$, antibonding level above E_F. At large distances from the metal surface, the electronic structure of H_2 is little affected by the presence of the surface. The physisorption well is determined by the dynamic polarization properties of H_2 (van der Waals attraction) and the steep rise in energy due to Pauli repulsion as the separation is reduced. H_2 acts as a neutral but polarizable adsorbate. A physisorbed state of that nature is expected on all simple and noble metal surfaces. The corresponding potential energy curves were calculated for the simple metals Al, Mg, Li, Na and K and for the noble metals Cu, Ag and Au [101–106]. They compare well with the few available experimental results [107, 108].

The 4.7 eV binding energy of H_2 is comparable to the cohesive energy of metals. The bond length of 1.4 a.u. is much smaller than a typical metal lattice constant. Thus dissociative chemisorption requires hard work to increase the proton–proton (deuteron–deuteron) distance. A change in the orbital structure with sharing of electrons between H and the substrate and the formation of new electronic configurations are needed. H acts as a reactive adsorbate. For simple and noble metals the antibonding σ_u level of H_2 mixes with metal (M) levels of the same symmetry forming H–M bonding and antibonding levels.

The H–M bonding and H–M antibonding levels move downwards as H_2 approaches the metal surface. If the H–M bond wins the competition dissociative chemisorption occurs. On transition metal surfaces an additional effect comes into play as a result of the unfilled d-band. The d-orbitals are quite strongly localized and thus overlap little with the H_2 orbitals. However, when H_2 orbitals approach the surface, substrate s-electrons can be transferred to the d-band, avoiding the penetration of the $H_2\sigma_u$, orbital and thus weakening the Pauli repulsion. The functioning can be depicted discussing the model system H_2 approaching Cu and Ni [109, 110]. Cu and Ni have a high and low dissociation barrier, respectively. The Cu–H_2 potential energy curves display strong Pauli repulsion and the potential energy rises steadily to a maximum value = dissociation barrier of around 1 eV. In Ni–H_2, however, the gradual filling of d-holes lowers the barrier to around 0.1 eV. In the chemisorption region strong M–H bonds are formed without a complete break of the H–H and M–M bonds. The total energy falls −1.25 eV relative to that of isolated Ni and H_2. Due to the weakening of the Pauli repulsion the activation barrier may be reduced practically to zero and the physisorbed state may cease to exist, allowing direct entry of H_2 into the chemisorption region.

This quasiclassical model summarized correctly describes the observed fact that most transition metals adsorb H_2 dissociatively whereas most simple and noble metals do not. It does not, of course, account for detailed variations of the H adsorption properties, for example within transition metals series, but has eliminated earlier controversy about the role of s- and d-electrons in the H–M bonding.

For simple systems, the potential energy curves, activation energy and chemisorption energy can be calculated quite accurately, and satisfactory approximations are available for clean surfaces of simple and noble metals and for clean transition metal surfaces [37, 52, 111, 112].

4.2.6
Chemisorption on Metal Oxides

The surfaces of metal oxides and their H_2 chemisorption characteristics have been far less studied than the surfaces of elemental metals and semiconductors [113, 133]. Cation surface states are formed on ideal oxide surfaces at about 2 eV below the bottom of the conduction band. The charge of the surface ions is found to be reduced compared with that of the bulk ions and this leads to an enhanced covalency at the surface. The reduction amounts to less than 10 % for oxides of simple metals such as MgO and to 20–30 % for transition metal oxides. Cluster and slab calculations reveal that special surface state bands with metallic character can be formed on polar surfaces by charge compensation effects. To what extent the metallic band accounts for special catalytic activity is not yet known [114].

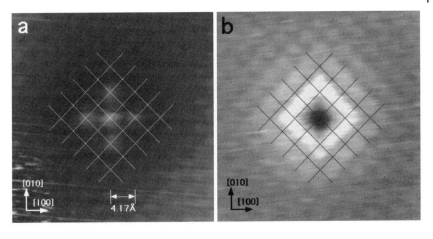

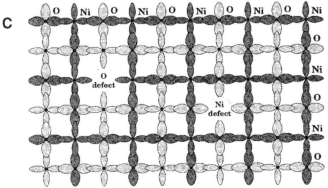

Fig. 4.13 (a) Empty-states (+1.12 V sample bias, 1.0 nA tunnelling current); (b) filled-states (−1.3 V sample bias, 1.0 nA tunnelling current) STM images of a single point defect. In (a) the defect appears bright and is surrounded by four bright SNN Ni sites that are in registry with the Ni (1 × 1) sublattice (indicated by the overlaid mesh). In (b) the defect center again lies on the overlaid mesh, which is identical to that shown in (a). The bright spots seen away from the defect are therefore on the oxygen sublattice. For illustration see (c). From (Castell, Wincott, Condon, Muggelberg, Thornton, Dudarev, Sutton and Briggs (1997), Ref. [115]).

A lattice defect on the oxide surface usually introduces deep localized states around it. Figure 4.13 illustrates the electronic structure changes of a NiO (100) surface as measured by STM [115]. Such defects play a crucial role as active centers for surface reactions of otherwise inert surfaces (NiO: steam reformation [116], TiO_2: photocatalysis [117], etc.). Chemisorption on defect sites is generally much stronger than on normal sites. The photocatalytic activity of the rutile surface is greatly enhanced by the introduction of oxygen vacancies [118]. Studies of the adsorption of H_2 and H on single crystal oxides of TiO_2 [116, 119], $SrTiO_3$, WO_3, ZnO and NiO paint a consistent picture of the behavior of oxide surfaces towards hydrogen: most perfect or nearly perfect oxide surfaces are essentially inert to H_2.

However, atomic H has been shown to adsorb on stoichiometric oxides, for example on $WO_3(001)$ and ZnO [120]. NiO(100) surfaces cleaned by ion bombardment and annealing – and thus probably slightly reduced [121] – become further reduced when exposed to H_2 at 400–600 K.

The defect-free surface of $TiO_2(110)$ is inert. Below 370 K, adsorption of H was observed on $TiO_2(110)$ surfaces having O-vacancy point defects. H_2 was postulated to dissociate at the defect site and bond to the adjacent, partially coordinated, cations resulting in Ti^{4+}–H-hydride bonds [119]. Results of earlier studies [122] of H_2 adsorption on Ar^+-bombarded surfaces were interpreted in terms of surface OH formed by dissociated H, with H bonding to surface oxygen ions. Nonetheless, recent publications report on hydrogen adsorbed on *neutral* surface oxygen sites on perfect TiO_2 [123]. Though H_2 does not dissociate on perfect TiO_2, the diffusion of atomic H across single crystalline TiO_2 is fast: the crystal structure of rutile TiO_2 is characterized by large open channels parallel to the *c*-axis which allow rapid diffusion of light interstitials [124]. For H the diffusion is described by $D = 1.8 \times 10^{-3} \exp(-0.59 \,\mathrm{eV}/kT)$ cm^2 s^{-1} and amounts to about 2×10^{-13} cm^2 s^{-1} at 300 K [125]. Lattice defects sharply inhibit H diffusion. Accordingly TiO_2-coating effectively prevents H-embrittlement [126].

H_2 and H adsorption on simple metal oxides has been studied even less than that on transition metal oxides. H_2 adsorption onto various defects on the MgO(100) surface has been treated theoretically using defect lattice techniques, including the relaxation of the lattice around the defects. Surface defects (F- and V-centers and self-trapped holes) were all found to activate dissociative chemisorption, resulting in the formation of OH radicals [127, 128]. This is in general agreement with the observed catalytic activity of MgO after the creation of F-centers by X-rays [129].

Recently, the debate on the dissociation properties of oxides has received considerable attention due to the fact that transition metal oxides enhance hydrogen sorption of MgH_2 [130]. The exact phase of the additive, its role, and the corresponding mechanisms have remained unclear, sparking controversy about the origin of the effect. By an analysis of the desorption kinetics Barkhordarian *et al.* tried to evaluate the critical kinetic step of the reactions at different temperatures and catalyst contents. The authors concluded that oxide catalysts lower the surface barrier during desorption [131]. It could be shown that the best additive, Nb_2O_5, has considerable dissociation properties [134]. However, Hanada *et al.* found pure 3d-elements to be good additives [132]. In this study, nanocrystalline Ni was found to have the highest catalytic activity. Clean surfaces of 3d-elements, in particular Ni, are well known for their catalytic activity for hydrogen dissociation (Ref. [52], see Chapter 4.3). Their dissociation activity is orders of magnitude higher than that of the transition metal oxides used by Barkhordarian *et al.*, including Nb_2O_5, while these oxides are the better additives for hydrogen uptake [133, 134]. Thus, additional models of the origin of oxide-catalyzed kinetics of MgH_2 are discussed, such as: transition metal oxides may enhance diffusion properties of MgH_2 [135], oxide interfaces may kinetically destabilize the hydride [136] and oxides may support the milling process and stabilize the grain/particle size [137].

4.2.7
Effect of Thin Film Coating and Precoverage on H_2 Dissociation and H Adsorption and Absorption

The ability to deposit monolayers of metal atoms on single crystal metallic substrates and the formation of multilayer structures initiated a surge of activity in surface and interface science, which has also had an impact on H sorption and permeation studies. The modification of structure, electronic structure and magnetism at the interface and in the overlayer are widely studied properties [138–142]. Strongly enhanced magnetic moments were found for some transition metal overlayers, as were induced magnetic moments at the interface of otherwise nonmagnetic elements. Epitaxial constraints limit lattice expansion due to H absorption in Nb–Ta superlattices and depress superconductivity [143]. A survey of magnetic effects in hydrogenated superlattices is given in Chapter 7.1.

It was noticed [144] that the deposition of more than two monolayers of Pd or Pt dramatically enhances the H uptake rate of Nb or Ta substrates. It is the variation in the surface atomic and electronic structure and the weakening of subsurface bonding, not just protection from surface oxidation, that are at the root of the enhanced uptake [145]. The use of "catalytic overlayers" was also successfully implemented in switchable mirrors by depositing a thin Pd capping layer on top of the optically active metal hydride. The switching kinetics of such a switchable mirror – and consequently its hydrogen uptake rate – is mainly determined by the thickness of the catalytic Pd cap layer [146, 147]. In particular, it was observed that a minimum Pd thickness is required for a sufficient hydrogen uptake. The effect was explained by an encapsulation of the clusters by a reduced yttrium oxide layer after exposure to hydrogen [148], the so-called Strong Metal Support Interaction (SMSI) [149]. There is evidence that this explanation holds also for similar switchable mirror systems [150], and its impact (slowing down of the kinetics) might be reduced by blocking the interdiffusion of the layers [151, 152].

It is well known from catalysis that electropositive (e.g. Na, Cs, K) and electronegative (e.g. S, O, C, Cl) adatoms decrease or increase the reaction rate and thus poison or promote the reaction, respectively [153–155]. Alkali-metal influenced adsorption on transition metals was reviewed by Bonzel [154]. Coadsorption of alkali metals and H, or D, on Al(100) revealed that the sticking coefficient and dissociation rate are extremely weak ($\sim 10^{-4}$ at all alkali coverages [156]). Upon exposing alkali-covered metal substrates to a beam of atomic H or D, alkali hydride formation was observed.

The effect of oxygen precoverage of transition metals on the H_2 chemisorption increases with the degree of oxide formation and was investigated in relation to the inhibition of H sorption and embrittlement and SMSI in catalysis. Monolayer amounts of oxygen reduce the H_2 adsorption and desorption on single and polycrystalline transition metals by orders of magnitude [42, 157, 158]. On polycrystalline metals oxygen precoverage seems to deteriorate somewhat less than on monocrystals: oxidation of polycrystalline Ti caused a decreased D_2 absorption rate; complete inhibition of the reaction occurred at oxide thicknesses exceeding 2 nm

[159]. Ko and Gorte observed a complete suppression of H_2 adsorption even for TiO [160]. At elevated temperatures some substrate metals dissolve surface oxygen, a mechanism which is particularly important in the activation of getter alloys [85, 161]. From detailed volumetric studies Fromm and coworkers [162] conclude that thin suboxide layers, formed during the initial stages of oxidation or by partial oxide reduction in H_2, do not impede the H_2 reaction drastically. This was found in particular for Ni overlayers, where a strong inhibition of hydrogen absorption occurs only for fully oxidized, that is NiO layers [163]. A decrease in the H_2 reaction probability of typically one order of magnitude was observed for a Ti substrate film covered with 3 nm thick metal overlayers after precoverage of the overlayers with 10 monolayers of oxygen. The results indicate that H_2 dissociation is strongly impeded if the oxygen precoverage exceeds a critical value which is given by the maximum amount of oxygen that can be adsorbed with sticking probability of one in the initial stage of oxidation. H_2 on TiO_2 dissociates in neither the rutile nor anatase structure modifications [119]. TiO_2 is used, as mentioned earlier, as an effective coating to prevent embrittlement [126]. Various authors have reported that transition metal overlayers on the oxide layer restore the original uptake rate of the substrate [159, 162, 164]. This clearly indicates that oxidation reduces the dissociation rate and is in agreement with the observed rather fast diffusion of atomic H across TiO channels [124, 125] mentioned earlier.

The adsorption of sulfur on clean transition metal surfaces completely inhibits H_2 dissociation and H adsorption (from gas phase and cathodic in aqueous acid medium) [165–168]. A major effect seems to be the blocking of H adsorption sites, possibly even the blocking of several H sites by one S atom. Marcus and Protopopoff [167] conclude from S and H adsorption experiments on Pt(110) and Pt(111) that one S atom blocks approximately 8 and 12 H adsorption sites, respectively. Self-consistent linearized APW calculations of the electronic structure perturbations induced by S on the Rh(001) surface reveal a substantial reduction in the density of states at the Fermi level, even at nonadjacent sites [169]. Thus blocking is explained by local electronic structure arguments. Using density functional theory calculations of sulfur-poisoned Pd surfaces, this argument could be made precise. Wilke and Scheffler showed that adsorbed sulfur builds up energy barriers which hinder the dissociation and that, because these energy barriers are located in the entrance channel of the dissociation-reaction pathway of hydrogen, this hindrance is particularly effective [170]. Surface poisoning by adsorption of sulfur species is known to inhibit H desorption from Pd hydride [171].

In an effective medium approach to poisoning and promotion by Norskov et al. [172] evidence is given for the importance of empty (antibonding) states near the Fermi level. Competition between the closed-shell kinetic energy repulsion and the attraction due to the gradual filling of the antibonding level determines the height of the molecular adsorption barrier, the depth of the molecular wall and the height of the dissociation barrier in the reaction path. Possible ways of altering the balance between the two competing terms are the addition of a partially empty d-band around the Fermi level (transition metal atoms) or changes in the work function of the surface.

The close relation between inhibitors of H embrittlement and H_2 dissociation poisons was noticed early by Berkowitz et al. [173]. The source of hydrogen determines whether, for example surface S, which slows down the $H_2 \Leftrightarrow 2H$-reaction in both directions, inhibits or promotes embrittlement. For dissolved H in the host metal, for example during electrochemical treatment, S prevents recombinative desorption and thus promotes embrittlement. In contrast, S prevents dissociation and solution of gaseous H_2, for example in a H_2-pipeline tube, and thus inhibits H embrittlement under these circumstances.

4.2.8
Relevance of Surface Reactions for Bulk Absorption

Hydrogen absorption in metal hydrides involves two main steps: dissociation of the hydrogen molecule and transport of the chemisorbed hydrogen toward the subsurface and adjacent diffusion in the bulk (see Figure 4.11). According to diffusion data, hydrogen transport inside metals or metal hydrides (with the important exception of MgH_2 [174]) is fast enough to provide high hydrogen absorption and desorption rates at room temperature. However, in most systems only slow rates are observed, which is one clue that the surface properties of the metal grains determine their H sorption kinetics, that is, the dissociation of the hydrogen molecule as discussed in the previous sections. This is, in particular, true for so-called inactivated metal hydrides. Here, the grains are covered with a dissociatively inactive oxide skin, which blocks hydrogen sorption [175]. The passivation is highly desirable in some ways, for example to avoid corrosion of stainless steel or H embrittlement, but it is very annoying in all cases where a reactive surface is needed: that is for catalysts, getters and hydride-forming elemental metals and alloys. An activation process is needed to make the surface reactive again.

Nonetheless, surface and bulk processes are not completely decoupled, because the coverage of the surface, and therefore the number of free active sites for dissociation, depends also on the concentration of hydrogen in the bulk (see previous section) [92, 176]. During hydrogen sorption, the concentration in the bulk, as well as at the surface, changes until it reaches its equilibrium [176]. The experimental observation of such changes is extremely difficult because most surface science techniques to measure the surface concentration need high vacuum, while absorption in metal hydrides takes place at high pressures (>1 bar). Therefore, most publications model the sorption kinetics to identify the rate-limiting processes involved, that is, to distinguish between surface and bulk processes [49, 131, 176–178]. Interestingly, obtained kinetic data are scattered extensively, even for similar systems. For example from empirical data calculated apparent activation energies are often scattered [48]. Despite this scattering, a clear correlation was found between the model parameters apparent activation energy and its connected prefactor, which suggested the existence of a universal physical effect. However, the origin of this so-called compensation effect is the simultaneous determination of the apparent prefactor and apparent activation energy from an Arrhenius analysis. Thus, the observed compensation effect is probably an artifact of the data analysis rather than a physical phenomenon [48]. The Arrhenius analysis will only deliver reliable results, if the kinetic constraint is based on only one condition-independent barrier [179]. Conversely, the existence of a Constable effect hints towards a competition between two processes, in this case most likely between dissociation and diffusion [49].

The relevance of surface processes to the sorption properties of complex hydrides is, however, under discussion. Within this material class, doped sodium alanate [180] is the archetypical hydrogen storage material for mobile applications. It decomposes in two steps, giving a total theoretical capacity of 5.6 mass% hydrogen [181]:

$$NaAlH_4 \Leftrightarrow \tfrac{1}{3}Na_3AlH_6 + \tfrac{2}{3}Al + H_2 \Leftrightarrow NaH + Al + \tfrac{3}{2}H_2$$

Generally, for hydrogenation, the hydrogen has to be (i) transported to the surface of the material, (ii) be adsorbed, (iii) dissociated and (iv) transported in the bulk of the material. At this stage, the desired hydride nucleates (v) and grows (vi). A peculiarity of nearly all complex hydride systems, including $NaAlH_4$, is that despite the transport of hydrogen substantial mass transfer of metal atoms also has to occur (vii), since the reactants (here Na_3AlH_6, NaH and Al) are spatially separated over several hundred atom layers. Considering the number of coexisting elementary steps (i)–(vii) involved in the total reaction, it is not surprising that a final answer of the question on the reaction mechanisms has not yet been found. Interestingly, the reaction is only reversible under technically applicable conditions, if it is doped with transition metal compounds, most efficiently with titanium compounds [180, 182]. Here, surface reactions might be reinforced [40, 41]. A more detailed discussion is given in chapter 6.6.

4.3
Catalysis of Hydrogen Dissociation and Recombination
Ib Chorkendorff

The hydrogen molecule is basically the simplest system one can consider to investigate the interaction between a solid surface and molecules. Dissociation/adsorption of hydrogen has been investigated widely as it is an essential step in many large-scale catalytic processes such as the synthesis of ammonia, the steam reforming process, the methanol synthesis reaction and so on; see for instance Ref. [183] and references therein. Dissociation of hydrogen also plays a crucial role in both low temperature fuel cells and the formation of metal hydrides. In the following we shall give a brief overview of the interaction of hydrogen with metals, describing some of the essential trends of reactivity for hydrogen and explaining them in terms of a relatively simple model formulated by Nørskov and coworkers [184]. After this theoretical overview we will give some examples of hydrogen adsorption/adsorption and of how essential parameters like sticking coefficients and activation energies/adsorption energies can be measured experimentally using copper as an example. Finally, we will present some results on hydrogen interaction with the so-called free electron metal, which constitutes the most interesting metal for hydrogen storage.

4.3.1
The Theoretical Approach

When a H_2 molecule approaches a metal surface, as indicated in Figure 4.14 it will first feel the van der Waals interaction. This is a weak interaction due to polarization effects, which typically amounts to only $10\,kJ\,mol^{-1}$ for H_2 and thus only leads to adsorption at very low temperatures or high pressures.

When approaching even closer there may be an exchange between the hydrogen electrons and the electrons in the metal leading to associative chemisorption, that is the H_2 molecule remians intact. The typical situation for hydrogen is, however, that this associative chemisorption is also very weak and the hydrogen only interacts strongly with the metal if it is allowed to dissociate. Whether this happens is naturally dependent on the height of the barrier E_a indicated in Figure 4.14. If no barrier is found the molecule will dissociate straightforwardly, as is the case for most transition metals where the probability for dissociative sticking of hydrogen is always close to one. If the barrier is large, as indicated by the full line there will be an activation energy for dissociation and the sticking coefficient will drop accordingly. This barrier is typical for the late transition metals and especially for the noble metals and the so-called free electron metals where there is no d-electron to facilitate the reaction. If the H_2 molecule dissociates, the proton may, dependent on the type of metal and its potential energy diagram, enter into the metal, where

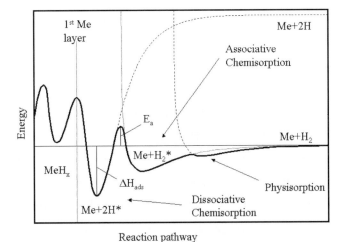

Fig. 4.14 Schematic potential energy diagram for a hydrogen molecule approaching a metal surface. In this case the metal hydride formation is endothermic. See the text for details.

it can either occupy interstitial sites or form a regular metal hydride, but that is beyond the scope of this chapter and will be discussed in chapter 6. Thus the potential energy diagram shown in Figure 4.14 basically sets the scene for the hydrogen dissociation and recombination since that is given by the height of the activation energy barrier E_a and the depth of the chemisorption well of the adsorbed hydrogen atoms ΔH_{ads}. The adsorption energy will then be $E_{ads} = \Delta H_{ads} + E_a$.

The individual elements of this potential energy diagram are readily explained: When an atom or molecule approaches a metal surface with free electrons, image charges will be introduced through polarization. This leads to an attractive interaction, the so-called Van der Waals interaction, which is proportional to d^{-3}, where d is the distance between the atom/molecule and the surface. However, if the electrons in the surface and the molecule do not adjust there will be a strong repulsion proportional to e^{-d} since this is the manner in which the electronic wavefunctions decay far from the surface and the atom/molecule. The dotted line going very steeply upwards indicates this repulsive behavior.

The chemisorption is due to an interaction between the states of the H_2 molecule and the electrons in the metal. The simple picture is given by the Newns–Anderson model where an adsorbate state is lowered and broadened when interacting with a sea of valence electrons in a metal, as indicated in Figure 4.15. Basically, all metals have a broad band of sp-electrons that are populated up to the Fermi level. Since these metals can be considered more or less as free electrons their density of states (DOS) is usually assumed to be proportional to $\sqrt{E}$, as indicated in Figure 4.15. In the case of associative H_2 chemisorption the interaction is rather weak

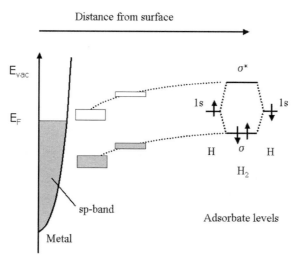

Fig. 4.15 The hydrogen molecule far away from a metal surface is formed by overlap between the two 1s electrons of the individual atoms forming a bonding and antibonding orbital. When the molecule approaches a surface these orbitals become broadened and lowered due to the interaction with the metal sp-band.

and further approach to the surface results again in strong Pauli repulsion, as indicated by the dotted line, unless dissociation takes place. If we took a hydrogen atom and approached it to the surface its energy would be much higher since it requires 4.55 eV to dissociate a H_2 molecule. This is indicated by the slashed curve. The atom also feels a van der Waals attraction but that is negligible on this scale where the hydrogen atoms are simply falling into a chemisorbed state. Where the crossover occurs determines the size of the barrier for dissociation. Since all metals basically have this feature how do the differences arise? The d-electrons have not yet been considered and they are the key to the difference. The d-electrons form a considerably narrower d-band since these orbitals are more localized than the sp-orbitals. Remember the width of the band is proportional to the overlap of the orbitals. The effect of the d-band is illustrated schematically in Figure 4.16 where we now let our hydrogen molecule approach a transition metal with a partially filled d-band, which naturally also has a broad sp-band.

Just as in Figure 4.15 the interaction with the sp-band will lead to a lowering and broadening of the molecular levels, that is the molecular chemisorption. But in addition to this interaction there will also be a coupling to the narrow d-band. Just as we considered a simple interaction between the two 1s orbitals in the atomic hydrogen leading to the bonding and antibonding orbitals in the hydrogen molecule in Figure 4.15 we may now consider the interaction between each of the lowered (and broadened) σ orbitals of the hydrogen molecule interacting with the narrow d-band in a similar manner. This will lead to a splitting of each of the σ orbitals into a bonding and antibonding orbital, as indicated in Figure 4.16. The effect of this interaction is basically the holy grail of catalysis. Notice how the antibonding orbital of molecular hydrogen σ^* is being pulled below the Fermi level and thus

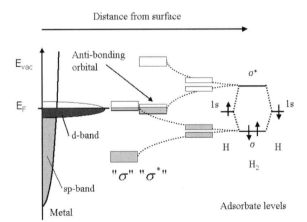

Fig. 4.16 The hydrogen molecule is as in Figure 4.15 interacting with the metal sp-band becoming lowered and broadened but now the molecular levels also interacts with a narrow d-band, resulting in a splitting of the two molecular orbitals into bonding and antibonding orbitals. Notice how the internal bonding of the hydrogen molecule becomes weakened due to the filling of the otherwise empty σ^* state.

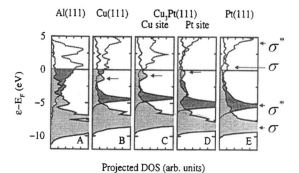

Fig. 4.17 Projection of the H$_2$ states onto the metals states. The dark shaded states are the projection of the down-split σ^* while the light shaded is the down-split σ state. See text for details. Adapted from (Hammer and Nørskov (1995), Ref. [185]).

filled with electrons. This increases the bonding to the surface by weakening the internal bonding of the molecule since it is an antibonding orbital that is being filled. Thus the internal bonding is weakened and hence its dissociation becomes much easier – that is the activation energy for dissociation E_a shown in Figure 4.14 is reduced and dissociation of the molecule becomes much easier, forming adsorbed atomic hydrogen on the surface. This hydrogen may now diffuse into the metal forming interstitials or metal hydrides, dependent on the energy potential, as indicated to the far left in Figure 4.14, or it may interact with other surface intermediates forming useful compounds, as is the case in catalysis.

The above strongly idealized picture captures the main effects of molecules approaching a surface and is also applicable to other molecules like CO, N$_2$ and O$_2$. Nevertheless, when turning to the real density functional theory (DFT) calculations [185], which describe the approach much more accurately, the picture becomes considerably more complex, as is seen from Figure 4.17. In the left panel are the bonding σ and the antibonding σ^* orbitals of the Al(111) surface. Notice that the levels are broadened. In the next panel is shown the result of the interaction with a d-band metal Cu(111). We still have the two bands, but it is now clear that an additional splitting due to the interaction with the d-band is occurring. The last panel shows the interaction with pure Pt(111) and here the additional splitting due to the d-band becomes even clearer. The downshifted antibonding orbital is shifted way below the Fermi level and is being filled, weakening the internal bond of the hydrogen molecule. This picture corresponds well with the experimental observation that the free electron metals are not good at dissociating hydrogen. When entering the transition metal series dissociation happens readily while it becomes difficult again when going to the far right, that is the noble metals, due to the filling and lowering of the d-band.

To understand the trends in reactivity we must consider the reaction pathway for dissociating hydrogen and, in particular, the detail of the transition state where the

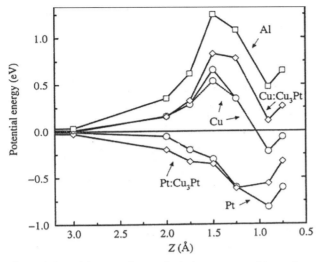

Fig. 4.18 Potential energy diagram for hydrogen approaching various metal surfaces as indicated. For further details please consult the text and the literature. Adapted from (Hammer and Nørskov (1995), Ref. [185]).

hydrogen is dissociating. The DOS shown in Figure 4.17 is for the situation where the hydrogen molecule is in the transition state and the potential energy diagram is shown in Figure 4.18. Notice the very high barrier for dissociating hydrogen on Al(111) where there is no interaction with a d-band. The barrier simply disappears over the d-band metal, but start to reappears when going to the noble metals such as Cu(111) where the d-band is filled and basically only results in a repulsive interaction with the molecular orbitals.

This effect is even more pronounced if going to the even more noble metals like Ag and Au and has been treated in great detail by Nørskov et al. [186].

4.3.2
The Experimental Approach

Huge amounts of literature dealing with the experimental investigation of hydrogen interaction with metals exist (see, for example Christmann [187]) both in terms of thermodynamics, kinetics, and dynamics and it is beyond the scope of this chapter to deal with all of this in detail. Instead, we will concentrate on a few examples illustrating how information on the thermal stability and kinetics can be obtained and also correlated with the much more detailed dynamics.

Hydrogen interaction with copper is an interesting border case since it processes a substantial barrier for the adsorption process and still is an interesting metal from a catalytic point of view. The catalyst used in the low temperature water gas shift reaction is based on metallic copper on a Zn/Al_2O_3 support whereby CO and water are converted into hydrogen and CO_2 [188]. The same type of catalyst is also used for the methanol synthesis where CO_2 and hydrogen are converted into methanol [189].

$$CO_2 + 3H_2 \xrightleftharpoons{Cu} CH_3OH + H_2O \qquad (4.36)$$

Therefore it is desirable to have detailed information of the hydrogen behavior under reaction conditions so it is possible to model the overall reaction by, for example, constructing microkinetic models, which has been done in two cases [188, 189].

If hydrogen is the only gas over a surface it is straightforward to set up the differential equation determining the surface coverage

$$\frac{d\theta_H}{dt} = k_H^+ P_{H_2} \theta_*^2 - k_H^- \theta_H^2 = k_H^+ P_{H_2}(1-\theta_H)^2 - \frac{k_H^+}{K'_{H_2}} \theta_H^2, \qquad (4.37)$$

where we have introduced the forward rate k_H^+ and the backward rate k_H^- and the equilibrium constant for the adsorption process is $K_{H_2} = K'_{H_2} P_0$ where $P_0 = 1$ bar is the standard pressure. θ_H and θ^* describe the occupied and unoccupied sites respectively and P_{H_2} the pressure of H_2 over the crystal.

If we want to determine the rate constants we need to design our experiments so that the adsorption and desorption can be measured independently, that is we must dose hydrogen at such low temperatures that the desorption rate can be neglected or get hydrogen on the surface at low temperature and then remove the hydrogen gas so that further adsorption can be neglected during the following heating and desorption process. Let us look into the latter case first.

Measuring atomic hydrogen on surfaces is not trivial since the standard methods in surface science like Auger electron spectroscopy (AES) and X-ray electron spectroscopy (XPS) are blind to hydrogen. Vibrational methods like high resolution energy loss spectroscopy (HREELS) can be used, but the best method is to use temperature programmed desorption (TPD). Here any hydrogen that is adsorbed will desorb when the crystal is ramped up in temperature with a heating rate β and can easily be detected by a mass spectrometer with high sensitivity – less than 1 % of a monolayer – as shown in Figure 4.19. Here a Cu(100) surface was exposed to increasing dosages of predissociated hydrogen. As we saw in Figure 4.18 the barrier for hydrogen dissociation on Cu is substantial and, therefore, it is not possible to get any hydrogen to dissociate under UHV conditions where the hydrogen pressure cannot be taken beyond 1×10^{-5} mbar. Because of that the predissociation is carried out by having a hot tungsten filament in front of the crystal so that equilibrium between atomic and molecular hydrogen is established [190].

The feature observed from 250 to 330 K corresponds to 0.5 ML and can be filled, as we shall see, by dissociating molecular hydrogen. The sharp feature growing in at 235 K is only observed when dosing atomic hydrogen and carries the typical fingerprint of a surface reconstruction. It is also observed to give rise to a $c(2 \times 2)$

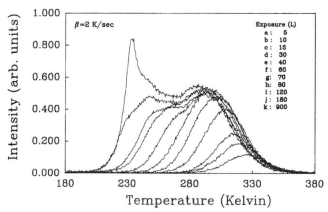

Fig. 4.19 TPD spectra of hydrogen desorbing from a Cu(100) surface for increasing exposures of predissociated H$_2$. The symmetric feature appearing for a dose of 60 l (curve f) corresponds roughly to 0.5 ML, which is the maximal amount of hydrogen observed if the hydrogen is not predissociated. Predissociation leads to adsorption of yet another 0.5 ML, which is followed by a characteristic reconstruction of the Cu surface atoms. Adapted from (Chorkendorff and Rasmussen (1991), Ref. [191]).

clock reconstruction of the surface, which can be investigated by low energy electron diffraction (LEED) [191]. It is, in this context, only of academic interest as the Cu surface seldom will be exposed to atomic hydrogen. The advantage of these TPD curves is that it is possible to evaluate the energy required to desorb hydrogen from this surface. The desorption rate is, in its simplest form, given by the following expression:

$$\frac{d\theta_H}{dt} = -k_H^- \theta_H^2 = -2\nu e^{\frac{-E_{des}}{RT}} \theta_H^2 \quad \text{and} \quad T = \alpha + \beta t \tag{4.38}$$

Both the prefactor ν and the desorption energy E_{des} may depend on the coverage. The parameters can be extracted by a number of more or less reliable methods [183]; here we shall restrict ourselves to the simplest possible. By reading the temperature at which the adsorption rate is a maximum $T_m = 290$ K for the first 0.5 ML and by assuming that the prefactor $\nu = 1 \times 10^{13}$ s^{-1} we find by analyzing Equation (4.38) that the following relation must be fulfilled:

$$E_{des} = RT_m \ln\left(\frac{RT_m^2 \nu}{E_{des}\beta 2\theta}\right) \tag{4.39}$$

This transient equation can easily be solved and we find in this simple picture that the adsorption energy is $E_{des} = 76$ kJ mol^{-1}.

If we want to estimate how much hydrogen there is on the surface under realistic conditions we must also find out how fast the hydrogen is adsorbed. In this case we must perform our experiments under such a condition that the adsorption rate can be neglected, that is below the adsorption temperature. The uptake rate of hydrogen under such conditions is given by:

$$\frac{d\theta}{dt} = k_H^+ P_{H_2} \theta_*^2 = \frac{2 S(T) P_{H_2}}{N_0 \sqrt{2\pi m_{H_2} k_B T}} \theta_*^2 \quad \text{where} \quad S(T) = S_0 e^{\frac{-E_a}{RT}} \quad (4.40)$$

where $S(T)$ is the sticking probability for dissociating hydrogen, m_{H_2} is the mass of a hydrogen molecule, k_B is Boltzman's constant, θ_* is the number of free sites where hydrogen can dissociate and is, in this case where only hydrogen is adsorbed, related to θ_H as $\theta_* = 1 - \theta_H$.

This means it is quite easy to solve this differential equation resulting in

$$\theta_H(t) = \frac{2 S(T) F t / N_0}{1 + 2 S(T) F t / N_0} \quad F = \frac{P_{H_2}}{\sqrt{2\pi m_{H_2} k_B T}} \quad (4.41)$$

where N_0 is the number of free sites per area and F is the incoming hydrogen flux. Thus by dosing hydrogen for a certain time t the dosage can be found. The amount that has been adsorbed is then determined by a subsequent TPD experiment and the result, as a function of dosage, is shown in Figure 4.20 for different surface temperatures. Notice that the sticking coefficient is very low so rather high pressures of hydrogen are used (1.8 bar). Here it is very important to perform the experiment in such a manner that the crystal and the gas are in thermal equilibrium. For most gases this is not a problem and will happen already around 1–10 mbar, but hydrogen is in this respect special due to the large quantities of energy that are

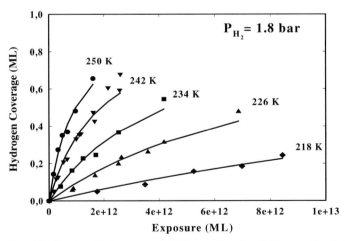

Fig. 4.20 Uptake of hydrogen on Cu(100) as a function of dosage (time × pressure) at a pressure of 1.8 bar of H_2 at a different temperature below the desorption temperature. The data points are fitted with Equation (4.5) where the sticking coefficient $S(T)$ is the only fitting parameter. The fit is shown as full drawn lines. Adapted from (Rasmussen, Holmblad, Christoffersen, Taylor and Chorkendorff (1993), Ref. [192]).

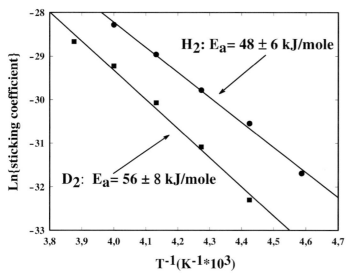

Fig. 4.21 The $S(T)$ obtained from Figure 4.20 is plotted in an Arrhenius plot and the activation energies for the adsorption process E_a can be extracted, as indicated in the Figure. Notice that the same experiment was also performed for D_2. Adapted from (Rasmussen, Holmblad, Christoffersen, Taylor and Chorkendorff (1993), Ref. [192]).

stored in the internal coordinates, especially in vibrations, and may not equilibrate well before 100–500 mbar [189]. Having the uptake curves and Equation (4.6) it is straightforward to extract the sticking coefficient $S(T)$ by fitting the curves using $S(T)$ as the fitting parameter. The results are shown as the full drawn lines in Figure 4.20.

Since the $S(T)$ can be obtained for a number of different temperatures it is possible to extract the activation energy for the adsorption process by plotting the data in an Arrhenius plot, as shown in Figure 4.21 [192]. The sticking coefficients of D_2 are also shown in the same plot, showing a higher activation energy primarily due to the lower ground state energy of D_2.

Now having both the prefactors and the activation energies for both adsorption and adsorption is it possible to estimate the equilibrium constant for hydrogen adsorption. Under equilibrium conditions Equation (4.37) will be reduced to:

$$\frac{d\theta_H}{dt} = 0 \Rightarrow k_H^+ P_{H_2}(1-\theta_H)^2 = \frac{k_H^+}{K'_{H_2}}\theta_H^2 \Rightarrow \theta_H = \frac{\sqrt{K_{H_2}\left(\frac{P_{H_2}}{P_0}\right)}}{1+\sqrt{K_{H_2}\left(\frac{P_{H_2}}{P_0}\right)}} \quad (4.42)$$

allowing us to predict the surface coverage of hydrogen for different pressures and temperatures as long as our assumptions are fulfilled.

Table 4.14 Essential parameters for the calculation of the partition functions of H_2 and H^*

Species	Rotations/Vibrations (cm^{-1})	E_g (kJ mol^{-1})
H_2	$B = 60.9$, $\nu = 4405$	-35
H^*	$\nu = 1121$, $\nu_{||}(2) = 928$	-34

The method can, however, be refined since we have made some serious approximations concerning the prefactors for the adsorption process. This is also the method typically used for microkinetic modeling of catalytic reactions. Here the dissociation of hydrogen is typically in quasi-equilibrium and, therefore, we can neglect the transition state and only concentrate on the initial and final state of the adsorption process. If we have equilibrium the chemical potentials for molecular hydrogen will be equal to the chemical potential of the two resulting adsorbed hydrogen atoms

$$\mu_{H_2} = 2\mu_H \qquad (4.43)$$

The chemical potentials can easily be estimated from the respective partition functions for the hydrogen molecule and the adsorbed atomic hydrogen. The partition function for molecular hydrogen is easily found since all the energy states of the involved degrees of freedom are well-known and measured in great detail and can be found in Table 4.14, see also, for example Ref. [183, 193] for details on the respective partition functions. The chemical potential for N_g molecules of hydrogen is given by

$$\mu_{H_2} = -k_B T \frac{\partial \ln(Q_{H_2})}{\partial N_g} \quad \text{where} \quad Q_{H_2} = \frac{(q_{H_2})^{N_g}}{N_g!} \quad \text{and}$$
$$q_{H_2} = q_{\text{trans}} q_{\text{vib}} q_{\text{rot}} q_{\text{el}} q_{\text{nucl}} \qquad (4.44)$$

In a similar manner the partition function for adsorbed hydrogen atoms can be estimated.

$$\mu_H = -k_B T \frac{\partial \ln(Q_H)}{\partial N_a} \quad \text{where} \quad Q_H = \frac{N!}{(N-N_a)! N_a!} (q_H)^{N_a} \quad \text{and}$$
$$q_H = q_{\text{vib}}^H q_{\text{el}}^H q_{\text{nucl}}^H \qquad (4.45)$$

Here the adsorbed atomic hydrogen is assumed to be localized on a specific site leading to N_a out of N sites being occupied. In this case the three degrees of freedom of the adsorbed hydrogen atom are reduced to three frustrated vibrations which, in principle, can be measured by HREELS [191] or alternatively estimated by DFT. Anyway this is typically the case for the two dipole forbidden frustrated vibrations parallel to the surface since they are very weak in a HREELS experiment.

It is, nevertheless under certain conditions, possible to measure them, as was done for instance for H on tungsten [194]. The nuclear partition function is unity and the problem left is to estimate the electronic partition function of the adsorbed atom since that now includes in this scheme the adsorption energy of the atom. Inserting the expression for the partition functions from Equations (4.41) and (4.10) in Equation (4.43) we find straightforwardly, utilizing the ideal gas law and Stirlings approximation:

$$N_g = \frac{P_{H_2} V}{P_0} \ln(N!) \cong N\ln(N) - N; \theta_H = \frac{N_a}{N} \quad (4.46)$$

that the equilibrium constant can be expressed as:

$$K'_{H_2} = \frac{K_{H_2}}{P_0} = \frac{q_H^2}{\left(\frac{q_{H_2}}{V}\right) k_B T} = \frac{(q_{vib}^H q_{el}^H)}{\frac{k_B T q_{trans}}{V} q_{vib} q_{rot} q_{el} q_{nucl}} \quad (4.47)$$

where q_{el}^H is the only unknown since the binding energy to the surface is unknown.

This can now be found utilizing that the forward rate of the equilibrium constant has been determined from our uptake experiment. Thus by expressing the desorption rate in terms of the equilibrium constant:

$$k_H^- = \frac{k_H^+}{K_{H_2}} = \frac{2S(T)}{\sqrt{2\pi n_{H_2} k_B T} K_{H_2}}$$

we can fit the TPD spectra of Figure 4.19 and find, using the data in Table 4.14, that the electronic ground energy of adsorbed hydrogen is $E_g = -34 \text{ kJ mol}^{-1}$. Notice that the calculated adsorption energy estimated from Figure 4.18 by assuming a fixed prefactor compares reasonably well with that estimated from the more elaborate scheme $E_{des} = E_a - \Delta H_{ads} = E_a - (E_g^{H_2} - 2E_g^H) = 80.8 \text{ kJ mol}^{-1}$ (see Figure 4.14) where a more realistic and temperature dependent prefactor appears. It is now straightforward to estimate the hydrogen coverage under different pressures and temperatures, as indicated in Figure 4.22 where the hydrogen coverage on Cu(100) is estimated using the above procedure as a function of temperature for different pressures.

It is interesting that the sticking coefficient is so low despite the fact that it is not particularly activated. This is due to the fact that the prefactor is very low – only 5.9×10^{-3} – reflecting the loss of entropy the H_2 molecule encounters if it is going from the gas phase directly into a localized transition state on the surface. Theoretical expression for the sticking coefficient can be found using standard transition state theory combined with detailed information on the transition state from DFT calculations. Understanding the detailed dynamics of the hydrogen adsorption/adsorption has obtained substantial interest for the last decade and the hydrogen Copper systems has been one of the test systems. Today most features of the Cu(100) and Cu(111) system has been understood in great detail, see for instance Ref. [195]. The dynamics have been investigated in great detail for this system using

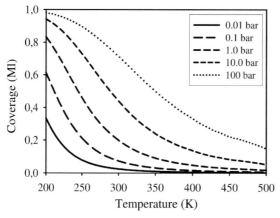

Fig. 4.22 Equilibrium coverage of atomic hydrogen on Cu(100) for a variety of pressures, calculated by the method described in the text.

super sonic molecular beams [196]. Also the refined details of enhanced sticking by population of the internal coordinates such as vibrations and rotations [197, 198] and references therein have been investigated in great detail.

One of the major results is shown in Figure 4.23, which is reproduced from the elegant work of Hayden et al. [197]. Here the sticking coefficient of pure molecular hydrogen is measured on Cu(110) as a function of translational energy using a

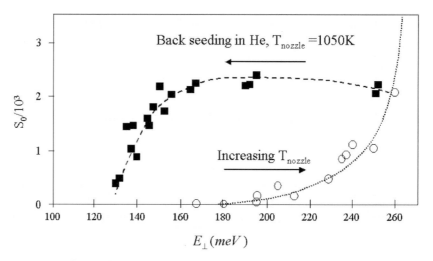

Fig. 4.23 Sticking coefficient of molecular hydrogen on Cu(110) as a function of translational energy (open symbols). By measuring the sticking coefficient for a fixed nozzle temperature at 1100 K and reducing the translational energy by back-seeding, it is demonstrated that the vibrational energy contributes substantially to the sticking as the sticking is constant over a broad range (filled symbols). Adapted from (Hayden and Lamont (1989), Ref. [197]).

supersonic molecular beam (open symbols). Increasing the temperature of the nozzle from which the beam is expanded controls the translational energy.

It is seen how the sticking takes off at relative low energies (substantially below the 48 kJ mol^{-1} or 0.5 eV mentioned above). This is because this sort of translational energy requires high temperatures where also the first vibrational state of the molecules will become populated. This is demonstrated by the back-seeded-beam where the translational energy is lowered for a fixed nozzle temperature. For example this is done by adding a heavier inert component into the expanding hydrogen gas – in this case He was used. Hereby the translational energy is lowered, but the population of the vibrational excited molecules can be kept constant, demonstrating that vibrational energy helps to overcome the barrier for dissociation.

The dynamics of the sticking and the gas–surface dynamics continue to be intriguing areas of research, but it is beyond the scope of this chapter to pursue this any further.

4.3.3
Relation to Metal Hydride Formation

The above example of hydrogen interacting with Cu(100) demonstrates the main features of metal–hydrogen adsorption/desorption and the various related phenomena. However, if we go beyond catalysis we may be also interested in forming metal hydrides for, for example forming media for hydrogen storage. Again we can use the Cu(100) system as an illustration of a system that will definitely not work as a storage medium since copper does not form stable metal hydrides, as indicated in Figure 4.14. Nevertheless, it will still be able to take up some hydrogen, as shown in Figure 4.24, for sheer entropy reasons.

Here the same crystal was exposed to high pressures of hydrogen at elevated temperatures and then cooled down in the gas. After removing the hydrogen gas in the high-pressure cell and taking the crystal back in the vacuum chamber TPD could be performed, as shown in Figure 4.24. All the features below or around 300 K are related to the adsorbed atomic hydrogen, discussed above, which is a necessity for forming any hydrides. The broad features at higher temperature show that while exposed to hydrogen at 2.1 bar at high temperature atomic hydrogen diffuses into the bulk of the crystal, equilibrating with the surface hydrogen. When cooling the crystal fast enough we are able to catch some of the hydrogen in the crystal, which is revealed by the TPD experiment.

The metals that are capable of forming metal hydrides with sufficiently high hydrogen contents (6.5 mass% according to Ref. [199]) are the free electron metals like Mg and Al and/or combinations of these with some of the other light elements. But as we have discussed above hydrogen dissociation is not particular easy on these metals since there are no d-bands available. Unfortunately, there have not been very many studies investigating the dissociation of hydrogen on such surfaces in great detail although this might be a serious kinetic limitation for utilizing such metals for hydrogen storage. In the following we shall demonstrate a few examples of the

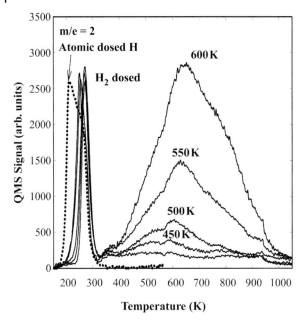

Fig. 4.24 Hydrogen released in a TPD experiment from a Cu(100) crystal that has been exposed to high pressures of hydrogen at elevated temperature (indicated above each curve) and then cooled in the gas. The TPD feature of a crystal dosed with atomic hydrogen is also indicated as a dotted curve. Notice that the adsorption feature below 300 K is basically constant while the amount of hydrogen dissolved in the crystal increases with temperature. Adapted from (Rasmussen, Holmblad, Christoffersen, Taylor and Chorkendorff (1993), Ref. [192]).

problems encountered in investigating the surface kinetics of such surfaces, but we will not consider the details of the bulk formation, since metal hydride formation will be treated in detail elsewhere.

The major problems are that the dissociative sticking is very low and that the atomic hydrogen form very stable or even the metal hydrides are volatile which complicates the TPD method substantially. This is, for example the case for Al, which has been investigated both using supersonic molecular beams [200] and atomic hydrogen [201, 202]. The sticking of molecular hydrogen on Al has a very high activation energy and is only observed for nozzle temperatures above 2000 K where atomic hydrogen may also be formed [200]. When using atomic hydrogen, substantial coverage may be built up and volatile aluminum hydrides (Al_2H_6 or AlH_3) are formed. These could be observed to desorb already at 340 K as a zero-order desorption feature, see Figure 4.25 [201]. The rapid low-temperature kinetics and high-energy density make AlH_3 an unusual and promising hydrogen storage medium for a number of applications [203], however the low stability is as yet an unsolved problem. The alanates $XAlH_4$, where X is an alkali metal, are more stable and are interesting for hydrogen storage (see chapter 6.6).

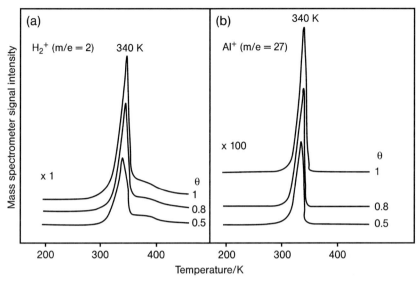

Fig. 4.25 TPD spectra following mass 2 (H_2) and 27 (Al) when dosing Al(111) with atomic hydrogen, indicating that volatile alane has been formed. Adapted from (Hara, Domen, Onishi, Nozoye, Nishihara, Kaise, and Shindo (1991), [201]).

Turning to a more stable metal hydride like Mg similar problems are encountered. Plummer et al. have investigated this system by dosing atomic hydrogen and measuring the HREELS spectrum of the adsorbed atoms [204]. However, this does not give any insight into the dynamics and kinetics of the surface. The activation energy for adsorption was estimated by DFT calculations to be rather high 0.37 eV [205] and the metal hydride formation energy is so high that the hydrogen cannot be released from the crystal in a TPD experiment [200] since it will vanish into the bulk and is first released at a temperature where the Mg also sublimes, making this approach impossible. Although, magnesium forms too stable metal hydrides, which first gives a reasonable hydrogen pressure (1–5 bars) at temperatures of 550–600 K it is still interesting since one may find routes to lower their stability by, for instance, making alloys with other metals. It has been shown that alloying with Ni can lower the temperature, see, for example, Ref. [205] but not sufficiently, moreover there is not much scope for adding heavier elements since then the hydrogen content by weight will be too low. The kinetics of the surface reaction can also be improved by adding more catalytically active metals to the magnesium. This approach has, for instance, been applied by several authors, adding palladium and studying the overall and bulk kinetics of the metal hydride formation [206, 207].

Understanding the key parameters of the surface requires the possibility of performing detailed measurements of the adsorption and desorption process, as was done for copper, obtaining some value for the activation and desorption energies involved. As mentioned above, in the magnesium case it is not possible to work with a single crystal, which is usually the preferred model system in catalysis.

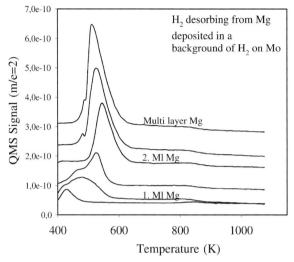

Fig. 4.26 Hydrogen as a function of the amount of Mg deposited in a background of H_2 of 1×10^{-6} mbar. The little sharp feature growing in at 500 K is an artifact due to a multiplayer of magnesium desorbing into the mass spectrometer. Adapted from (Ostenfeld, Davies, Vegge and Chorkendorff (private communication), Ref. [208]).

Instead we have to use thin layers of magnesium deposited on stable metals, like for instance molybdenum, as also was used in the study of aluminum hydrides [201]. Very thin layers will naturally be influenced by the substrate, as shown in Figure 4.26, but for thicker layers this influence can be neglected and it is then possible to perform experiments where the surface kinetics can be studied in greater detail, both with respect to alloys destabilizing the magnesium hydride and with surface additive in very small amounts that may lower the high activation energy barriers for adsorption/desorption [208]. This may be a route for improving and getting input for the current efforts to find new and better metal hydrides in place of the trial and error approach which, to a large extent, has been prevalent over the years, just as it has in the field of catalysis.

4.3.4
Summary

A qualitative account of the reactivity of metals towards molecular hydrogen has been given. It is seen that the transition metals are good catalysts for the dissociation of hydrogen and a method has been given for experimentally investigating the potential energy diagram, extracting the important parameters such as activation energies for adsorption and desorption. Detailed knowledge of these parameters opens the possibility for predicting the hydrogen coverage under realistic conditions. This quantity has proven very useful in the field of catalysis for identification of the slowest steps in a series of surface reactions so that it is possible to identify which parameters should be modified in order to improve the overall process. The same parameters are crucial for understanding in detail the processes involved in the low-temperature fuel cell where there are strong demands for new electrode catalysts to replace the precious metal Pt, which incidentally is very susceptible to CO impurities in the gas, and for development of new metal hydrides that can solve the essential hydrogen storage problem for mobile units.

4.4
The Four States of Hydrogen and Their Characteristics and Properties
Seijirau Suda

4.4.1
Introduction

There are four states of hydrogen that have been applied practically in various devices and processes. The aim of this section is to introduce these states from technological and engineering viewpoints:

- *Molecular hydrogen* in the gaseous state has been used widely as a chemical resource and is now becoming a significantly important energy source, especially as the fuel source for the *Proton Exchange Membrane Fuel Cell* (PEMFC).
- *The Proton* or hydrogen ion is encounteed in various electrochemical processes such as electrolysis and PEMFC. It exists in the form of H^+ in acid solutions and proton exchange membranes such as Nafion are well known.
- *Protium* is a state of hydrogen, which can exist independently in the interstitial site of a crystalline lattice in metal hydrides. Metal hydrides have been developed for more than three decades as one of the major hydrogen storage materials. However, the importance of metal hydrides varies from their function as hydrogen storage materials to their chemical and electrochemical functions as catalysts, electrode and reducing agents. Protium in metal hydrides is introduced briefly with regard to the nickel–hydrogen rechargeable battery.
- *Protide*, also called "hydride ion", is one of the states of hydrogen that can not exist independently but forms binary saline hydrides and hydrogen–metal complex ions in combination with alkaline and alkaline earth metals, mostly in aluminum or boron complex compounds.

In the past few years, $NaBH_4$ has been investigated extensively as a hydrogen storage material as it generates hydrogen easily and by catalytic hydrolysis. It dissociates to BH_4^- (borohydride complex ion) in aqueous alkaline solution. The BH_4^- ion is applied as a source of gaseous hydrogen and also as a source of a liquid type of fuel supplied directly to a new class of fuel cell, the *Direct Borohydride Fuel Cell* (DBFC).

In an advanced process for producing $NaBH_4$ that applies $NaB_4O_7 \cdot 10H_2O$ (borax as the natural resource) and $NaBO_2 \cdot 4H_2O$ (sodium borate as the "spent fuel"), Mg

Table 4.15 The four states of hydrogen

H_2: molecular hydrogen (dihydrogen) ($r^a = 0.3$ Å)	$[H°:H°]$
H^+: proton ($r = 1.5 \times 10^{-5}$ Å)	$[H^+ = H° - e^-]$
$H°$: protium (monatomic hydrogen) ($r = 1.0$ Å)	$[H° = H^- - e^-]$
H^-: protide (hydride ion) ($r = 2.08$ Å)	$[H^- = H^+ + 2e^-]^b$

$^a r$ denotes the radius of hydrogen as classified.
$^b H^-$ in MgH_2: 1.3 Å and H^- in $NaBH_4$: 2.0 Å. Values are taken from Cotton and Wilkinson [209]).

is used as H^--donor and O-acceptor. Protide in Mg is used as an active reducing agent to produce $NaBH_4$ from $NaBO_2$.

4.4.2
Classification of the Four States of Hydrogen

The state of hydrogen is classified into four states as shown in Table 4.15.

In Figure 4.27, the four states of hydrogen are illustrated as a triangular diagram in which the three states of proton, protium and protide are placed at the corners and molecular hydrogen is placed at the center. In the figure, individual relations between the states of hydrogen are illustrated by typical chemical and

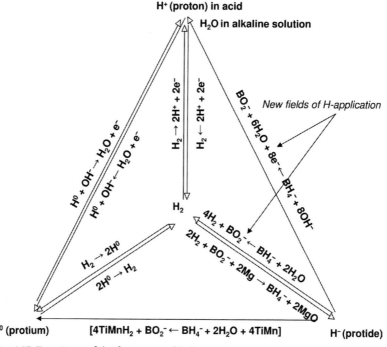

Fig. 4.27 Transitions of the four states of hydrogen.

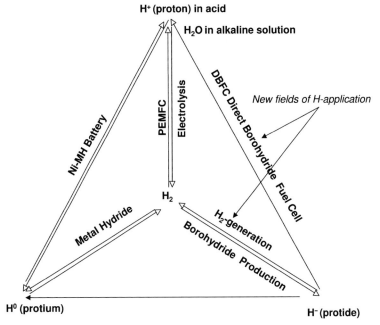

Fig. 4.28 Applications of the four states of hydrogen.

electrochemical rations. The principles of several applications currently available or under development are shown in Figure 4.28.

4.4.3
Molecular Hydrogen (H_2: Diatomic Hydrogen)

Hereinafter, we simply call "molecular hydrogen" "hydrogen" for the sake of simplicity and to distinguish clearly from the three other states of hydrogen.

Hydrogen in the gaseous state has long been investigated as a clean energy source for internal combustion engines and PEMFC. [Note PEMFC is sometimes simply called *Polymer Electrolyte Fuel Cell* (PEFC)].

Hydrogen is used as a fuel of high energy density when it is combusted properly with air. Nowadays, hydrogen is employed as the fuel source of PEMFC and it represents a typical example of hydrogen application. In PEMFC, hydrogen is supplied to the anode (negative electrode) as the fuel source and air is supplied to the cathode (positive electrode) to transfer electrons to generate electric power. At the anode side, hydrogen is converted to protium in contact with the surface of anode which then splits into two protons (H^+) and two electrons (e^-), see Chapter 5.2.

The anode plays a catalytic role first converting hydrogen to protium and then splitting the protium into a proton and an electron. Hydrogen does not convert directly to proton (H^+) by transferring two electrons to the cathode (see Equation

(4.48)). As the anode is composed of a Pt-based catalyst, it acts first at the surface as a hydrogen-absorbing material to convert hydrogen to protium. In many textbooks and references this initial stage from hydrogen to protium is abbreviated or sometimes overlooked.

$$H_2 \rightarrow 2H° \rightarrow 2H^+ + 2e^- \tag{4.48}$$

4.4.4
Proton (H^+: Hydrogen Ion)

This state of hydrogen is frequently encountered in chemical and electrochemical reactions. One typical example among many proton-based applications is PEMFC in which protium splits into a proton and electron, as described in Section 4.4.2.

Another application is electrolysis in which hydrogen is generated by water splitting (Equation (4.46)). Proton usually exists in acid solution in equilibrium with H_3O^+, the hydroxonium ion (Equation (4.47)).

$$2H^+ + 2e^- \rightarrow H_2 \tag{4.49}$$
$$H_3O^+ \leftrightarrow H^+ + H_2O \tag{4.50}$$

4.4.5
Protium ($H°$: Monatomic Hydrogen)

Protium is a state of monatomic hydrogen that exists independently in arcs at high current density, in discharge tubes at low pressure, and by ultraviolet irradiation of hydrogen at high temperature. It is chemically reactive, reducing various oxides either to lower oxides or to the pure metals and alloys. It has an extremely short half-life of 0.3 (s) [209].

It should be noted that protium is very reactive when it leaves the outer surface of metal hydride particles. However, protium can be stored internally in a reversible manner under stable conditions in hydrogen storage materials to form metal hydrides and the protium is released as hydrogen at the surface of the metal hydrides.

Internal reversibility of $H°$ in metal hydrides should be strictly distinguished from "external reversibility" because the internal reversibility means that it accompanies the endothermic and exothermic thermal exchanges between the hydrogen–metal hydride system and the surroundings in hydrogenation and dehydrogenation reactions. The heat release (output) and supply (input) are required according to the enthalpy changes (ΔH) during hydrogenation and dehydrogenation. These reactions are "internally reversible".

It does not form a bond such as $M-H°$ with metals (M: transition metals such as Ti and Zr, and lanthanides such as La and Sm) but it exists as protide in the interstitial site of metallic elements and mostly in the crystalline lattice of hydride-forming alloys [210].

Typical examples are hydride-forming alloys such as FeTi and LaNi$_5$:

In lanthanum nickel alloy: LaNi$_5$ + 2H$_2$ ↔ LaNi$_5$H$_4$ ↔ $LaNi_5 \cdot 4(H°)$
In iron titanium alloy: FeTi + H$_2$ ↔ FeTiH$_2$ ↔ FeTi · 2(H°)

In a general form;

$$2M + xH_2 \leftrightarrow 2MH_x \leftrightarrow 2M \cdot x(H°) \tag{4.51}$$

($-\Delta H$ in hydrogenation and $+\Delta H$ in dehydrogenation reactions)

The maximum numbers of protide sited in the lattice may vary freely between 0 and x depending on the pressure and temperature. The number given in Eq. (4.51) is $x = 4$ for LaNi$_5$H$_x$ and $x = 2$ for FeTiH$_x$, respectively. This illustrates the number of protium per metal atom as indicated as H/M (H: number of protium and M: number of metal atoms) that is simply representing the maximum value when fully hydrided. It should be noted that hydrogen is in the state of protium in metal hydrides except for metal–hydrogen complex compounds, as introduced later in this section.

Nickel–metal hydride (Ni–MH) secondary batteries are a typical example of H° applications and have been widely used as rechargeable batteries for CD cameras, cordless tools, and hybrid cars. The Ni–MH battery is based on the internal reversibility between the electrochemical charge and discharge of protium in aqueous KOH solutions;

$$H° + OH^- \leftrightarrow H_2O + e^- \ (\rightarrow \text{for discharging and} \leftarrow \text{for charging}) \tag{4.52}$$

Such an expression as H° ↔ H$^+$ + e$^-$ is misleading with regard to the electrochemical principles of the Ni–MH battery. The reversible transformation of protium should be expressed as shown in Equation (4.52).

4.4.6
Protide (H$^-$: Hydride Ion)

Protide is anionic hydrogen and is also called a hydride ion. It is one of the states of hydrogen that cannot exist independently but exists in binary saline hydrides of alkaline and alkaline earth metals such as NaH, CaH$_2$, MgH$_2$, Mg(AlH$_4$)$_2$, and ternary hydrogen–metal complex compounds LiAlH$_4$ and NaAlH$_4$ with Al, and also with boron that are classified as a family of semiconductor or half-metals to form LiBH$_4$, KBH$_4$, NaBH$_4$ and Mg(BH$_4$)$_2$ [see Chapters 6.4 and 6.5].

Most of these hydrides are reactive and generate hydrogen violently when exposed to moist air and water. However, in recent years, they have been attracting strong interest as hydrogen storage materials because of their high hydrogen contents [211]. Among such complex compounds, NaBH$_4$ is the only material that generates hydrogen by catalytic hydrolysis in a controllable manner and it has been investigated for the past few years as a new class of hydrogen storage material. NaBH$_4$ can be stabilized in aqueous NaOH solution in which it exists as borohydride complex ions (BH_4^-).

4.4.7
New Fields of Protide Applications

4.4.7.1 Hydrogen Generation

The protide transition to hydrogen has been successfully developed as a hydrogen storage device for PEMFC (see Chapter 6.1).

The BH_4^- complex hydride ion releases hydrogen catalytically at ambient pressure and temperature conditions as shown in Equations (4.50) and (4.54). The B–H bonding in BH_4^- is easily broken with the aid of catalysts such as Ru, Raney-Co and Raney-Ni, and the fluorinated Mg_2NiH_4 (see Chapter 6.8).

The generation of hydrogen is considered to proceed as shown below;

$$NaBH_4 \rightarrow Na^+ + BH^{4-} \rightarrow Na^+ + B^{3+} \cdot 4(H^-) \text{ (in alkaline solution)} \quad (4.53)$$

Followed by

$$B^{3+} \cdot 4(H^-) + 2H_2O \rightarrow 4H_2 + BO_2^- \text{ (in alkaline solution)} \quad (4.54)$$

4.4.7.2 Borohydride Fuel Cell

The borohydride complex ion, BH_4^-, is used not only to generate hydrogen but also for a new class of fuel cell by applying it directly as the liquid source of protide. Protide in BH_4^- can release two electrons per protide individually and a total of eight electrons when it is transformed to proton:

$$H^- \rightarrow H^+ + 2e^- \quad (4.55)$$

In contrast, hydrogen transfers two electrons per mole through $1/2 H_2 \rightarrow H° \rightarrow H^+ + e^-$ in the PEMFC.

The protide transition to proton has also been successfully applied to the development of a new type of fuel cell, the *Direct Borohydride Fuel Cell* (DBFC). The principle of the DBFC is briefly summarized by the following equation:

$$BH_4^- + 8OH^- \rightarrow BO_2^- + 6H_2O + 8e^- \quad (4.56)$$

This reaction is based on the electrochemical oxidation of BH_4^- to BO_2^- via protide. (Details of the DBFC will be introduced in Chapter 5.1).

4.4.7.3 Behavior of Protide on the Outer Surface of Mg

As described in Chapter 6.7, $NaBH_4$ can be synthesized by providing a process that is simply expressed by $NaBO_2 + 2H_2 + 2Mg \rightarrow NaBH_4 + MgO$. It is well known that Mg forms a binary hydride MgH_2 under high pressure and temperature conditions.

However, MgH_2 is not formed when Mg is mixed with $NaBO_2$ under hydrogen atmospheres if the temperature is raised rapidly, the hydrogen is transformed directly to protide as the transitional form of $Mg \cdot 2(H^-)$ (Equation (4.57)).

$$Mg + H_2 \leftrightarrow Mg \cdot 2(H^-) \tag{4.57}$$

Recently, H^- transformed from hydrogen at the outer surface of Mg was found to be quite useful in producing $NaBH_4$ from $NaBO_2$ where protide is exchanged with O^{2-} in $NaBO_2$ as shown in Equation (4.58) [212].

$$2Mg + 2H_2 + NaBO_2 \rightarrow 2Mg \cdot 2(H^-) + NaBO_2 \rightarrow 2MgO + NaBH_4 \tag{4.58}$$

Protide has an apparent distance in Mg–H bonding of 1.3–1.5Å and the effective radius of the free ion is 2.08 Å [209]. This is surprisingly larger than the proton radius of 0.4 Å and the 1.0 Å bond length of the H–H bond of hydrogen.

4.4.8
Summary

In this chapter, the four states of hydrogen are described briefly from practical viewpoints and the transition of states is summarized as follows:

(a) *Molecular hydrogen in PEMFC*

$$H_2 \rightarrow 2H° \rightarrow 2H^+ + 2e^- \tag{4.59}$$

(b) *Proton in electrolysis*

$$2H^+ + 2e^- \rightarrow H_2 \tag{4.60}$$

(c) *Protium*

   (c-1) Protium in metal hydrides

$$2M + xH_2 \leftrightarrow 2MH_x \leftrightarrow 2M \cdot x(H°) \tag{4.61}$$

   (c-2) Protium in Ni–MH rechargeable battery

$$H° + OH^- \leftrightarrow H_2O + e^- \quad (\rightarrow \text{for discharge and for charge}) \tag{4.62}$$

(d) *Protide in metal–hydrogen complexes*

   (d-1) Protide in $NaBH_4$ as a hydrogen storage material

$$NaBH_4 \rightarrow Na^+ + BH_4^- \rightarrow Na^+ + B^{3+} \cdot 4(H^-) \tag{4.63}$$

$$B^{3+} \cdot 4(H^-) + 2H_2O \rightarrow 4H_2 + BO_2^- \tag{4.64}$$

(d-2) Protide in DBFC

$$H^- \rightarrow H^+ + 2e^- \tag{4.65}$$

$$BH_4^- + 8OH^- \rightarrow BO_2^- + 6H_2O + 8e^- \tag{4.66}$$

(d-3) Protium in the reprocessing of $NaBH_4$ from $NaBO_2$

$$Mg + H_2 \leftrightarrow Mg \cdot 2(H^-) \tag{4.67}$$

$$2Mg + 2H_2 + NaBO_2 \rightarrow 2Mgo2(H^-) + NaBO_2 \rightarrow 2MgO + NaBH_4 \tag{4.68}$$

4.5
Surface Engineering of Hydrides
Seijirau Suda

4.5.1
Introduction

The surface of hydrogen-absorbing materials provides the contact point for interactions between gas–solid and liquid–solid phases. A typical example of the role and function of the surface is the transition of gaseous hydrogen at the metal hydride surface where it dissociates to monatomic hydrogen which is then located in the interstitial sites of the crystalline lattice (see Chapter 4.4).

Engineering applications such as hydrogen storage in metal hydrides, the nickel–metal hydride rechargeable battery (Ni–MH), and the proton exchange membrane fuel cell (PEMFC) are basically dependent on the surface properties and characteristics.

In this chapter, the transition of H_2 (*gaseous hydrogen*) to H^+ (*proton*), H^+ to $H°$ (*protium*), and H^- (*protide*) to H^+ is described from the surface interaction viewpoints in engineering applications. These four states of hydrogen are described in Chapter 4.4.

4.5.2
Various Types of Surface Interactions

Transition of the four different states of hydrogen, as described in Chapter 4.4, occurs at the gas–solid and liquid–solid interface as illustrated below:

At gas–solid interface:

(a) H_2 (hydrogen*)–$H°$ (protium):
  (a-1) hydrogen storage in metal hydrides,
  (1-2) $NaBH_4$ synthesis via Mg-particle surface
(b) H_2 (hydrogen)–$H°$ (protium)–H^+ (proton): anode (negative electrode) in PEMFC

At liquid–solid interface:

(c) H^+ (proton)–$H°$ (protium): Ni–MH rechargeable battery
(d) H^- (protide)–H_2 (hydrogen): hydrogen storage in sodium borohydride solution ($NaBH_4$)
(e) H^- (protide)–H^+ (proton): direct borohydride fuel cell (DBFC)

(Note: hereinafter, H_2 (gaseous hydrogen) is denoted simply hydrogen.)

4.5.2.1 Hydrogen to Protium

Hydrogen Storage in Metal Hydrides Interfacial reactions of the hydrogen–protium transition at the surface of a metal hydride are explained by the successive transformation from hydrogen, that is, dissociation to monatomic hydrogen, diffusion into the crystalline lattice, and formation of protide by hydrogenation [213]. Hydrogen dissolves first at the metal hydride surface, and then, decomposes to two protiums [$H_2 \leftrightarrow 2H°$]. Finally, protium diffuses through the crystalline lattice to form a metal hydride by a hydrogenation reaction [see Chapter 4.4].

In metal hydrides, hydrogen can be hydrogenated and dehydrogenated in an internally reversible manner but not in an externally reversible manner so long as the surface maintains cleanliness and durability against impurities.

It is well known that hydrogen uptake capability is significantly reduced when the surface is exposed to the air or impurity gases such as CO, H_2S, N_2 and other contaminants. The surface forms oxides and hydroxides easily when in air and loses hydrogen uptake capability [214, 215]. The surface poisoning of the Pt-based electrode in the PEMFC is known to be caused during long term use by a trace of CO contained in the hydrogen.

However, the surface durability against impurities can be improved by modifying the surface configuration such as composition and structure. Chemical plating, alkaline treatment, and fluorination have been attempted as means to improve the hydrogen uptake capability, the electrochemical properties, and the durability. Among these treatments, a fluorination technique will be described in the latter part of this chapter in relation to the role and function of a Ni-enriched surface and a fluoride layer.

(Note: The term "*reversible*" is misleading about the nature of "back and forth" reactions between hydrogen and metal hydrides. The hydrogenation and dehydrogenation are *not externally reversible* because they are accompanied by *exothermic* and *endothermic* heat transfer between the system and the surroundings [216]. The hydrogen–protium transition is a *non-equilibrium phenomenon* but it is a *transitional and dynamic phenomena* that accompanies the heat transmission.)

$NaBH_4$ Synthesis via a Mg-Particle Surface Protide obtained from hydrogen at the surface of Mg particles was found to be useful as the protide donor and the oxygen ion (O^{2-}) acceptor in the production of $NaBH_4$ from $NaBO_2$ (anhydrous sodium metaborate) through the reaction

$$NaBO_2 + 2Mg + 2H_2 \rightarrow NaBH_4 + 2MgO$$

Hydrogen at the surface of Mg particles is considered to exist instantaneously in the form Mg·2(H$^-$) when the reaction proceeds under transitional conditions where the temperature and pressure conditions are transferred from hydrogenation to dehydrogenation regions by a rapid increase in temperature at a constant rate.

The surface plays a significantly important role in the transitional reaction to generate protide as the extremely reactive state of hydrogen at the Mg-particle surface (see Chapter 6.8). Protide is delivered directly at the contact surface between particles of Mg and NaBO$_2$ to form NaBH$_4$ in exchange with O^{2-}, where O^{2-} in NaBO$_2$ moves from the surfaces of NaBO$_2$ to Mg to form MgO. The reaction must be clearly distinguished from the reaction; $2Mg + O_2 \rightarrow 2MgO$ [217].

4.5.2.2 Hydrogen to Proton via Protium

In the PEMFC, also known as polymer electrolyte fuel cell (PEFC), hydrogen dissolves first at the anode surface of the Pt-based catalyst where it dissociates to protium, and then to proton to generate one electron per proton ($H_2 \rightarrow 2H° \rightarrow 2H^+ + 2e^-$). The magnitude of the specific surface area that provides the sites for the transition of hydrogen to protium is the key factor for the anode surface in a PEMFC.

4.5.2.3 Proton to Protium

In the case of electric charge and discharge in Ni–MH secondary batteries, the following reactions proceed at the interface between the liquid (electrolyte such as aqueous KOH solution) and the solid (metal hydride anode):

On charging:

In electrolyte, $H_2O \rightarrow H^+(+OH^-)$ and then in anode, $H^+ + e^- \rightarrow H°$

On discharging:

In anode, $H° - e^- \rightarrow H^+$ and then in electrolyte, $H^+(+OH^-)] \rightarrow H_2O$

Proton in H$_2$O in the electrolyte is transformed to protium in contact with the anode surface by receiving one electron during charging [$H^+ + e^- \rightarrow H°$], and protium returns to proton by releasing one electron and then forming H$_2$O in contact with OH$^-$ during discharging [$H° - e^- \rightarrow H^+$ (discharge)]. These electrochemical reactions occur at the surface of the anode material, metal hydride.

Protium is stored in the interstitial sites in the crystalline lattice of the metal hydride and the amount of protium varies in accordance with the levels of charge and discharge. In these electrochemical reactions, the hydrogen states between proton and protium change at the surface of the metal hydride. This is the typical interfacial transition of hydrogen between liquid and solid phases.

4.5.2.4 Protide to Hydrogen

Protide (H$^-$) is the most reactive state amongst the four states of hydrogen that possesses one excess electron over protium (H°) and two excess electrons over proton (H$^+$). It forms metal–hydrogen complex ions such as AlH$_4^-$ in NaAlH$_4$ and

BH_4^- in $LiBH_4$ and $NaBH_4$. Hydrogen is liberated from these materials by thermal decomposition or hydrolysis.

Although protide does not exist independently but only in binary saline hydrides or Al- or B-based complex hydrides of alkaline and alkaline earth metals, its strong reactivity has led to its application as a reducing reagent for many years. Recently, protide has been considered as the state of hydrogen that is suitable for storing hydrogen in the form of metal–hydrogen complex ions because of their high hydrogen capacities compared with many other materials [218, 219].

Hydrogen is generated by thermal decomposition or hydrolysis. For thermal decomposition some doped-catalytic reagents are applied to reduce the decomposition temperature to practical levels. Hydrolysis uses catalysts to generate hydrogen under ambient conditions. [See Chapter 6.5 for thermal decomposition and Chapter 6.8 for hydrolysis]. In these reactions, the aid of surface reactants is needed to liberate hydrogen at the gas–solid or liquid–solid interfaces.

On the other hand, protide in BH_4^- was found quite recently to be a convenient protide source in a new class of fuel cell, the DBFC. In the DBFC, H^- in BH_4^- that exists in an aqueous solution of NaOH delivers effectively two electrons per H^- (eight electrons per BH_4^-) through electrochemical reactions, as shown in Equations (4.69) to (4.70).

Anode side (negative electrode):

$$BH_4^- + 8OH^- \rightarrow B(OH)_4^- + 2H_2O + 8e^- \quad (E^\circ = -1.24 \text{ V}) \quad (4.69)$$

Cathode side (positive electrode):

$$2O_2 + 2H_2O + 8e^- \rightarrow 8OH^- \quad (E^\circ = 0.40 \text{ V}) \quad (4.70)$$

Overall reaction:

$$BH_4^- + 2O_2 \rightarrow B(OH)_4^- \quad (E^\circ = 1.64 \text{ V}) \quad (4.71)$$

For simplicity, Equation (4.69) can be expressed as Equation (4.69) with respect to the anode surface;

$$4H^- \text{ in } BH_4^- \rightarrow 4H^+ \text{ in } 8OH^- \text{ (where 8 electrons are released)} \quad (4.72)$$

Protide in BH_4^- to proton in $8OH^-$ occurs at the anode surface that is composed of a fluorinated intermetallic compound (see Section 4.4.4). Details of the DBFC are given in Chapter 6.1.

4.5.3
Significance of Surface Properties and Characteristics

Much effort has been made in the past to improve the properties and characteristics of the surface of intermetallic compounds, particularly, of hydrogen-absorbing materials in order to:

1. Create a hydrogen interactive surface
2. Increase the specific surface area
3. Increase the electrically conductive surface
4. Increase the number of catalytic sites
5. Increase the reaction kinetics at the surface
6. Create a fresh surface
7. Protect the growth of the oxide layer
8. Provide an impurity-resistant surface.

The surface reactivity must be maintained in any applications of hydrogen–metal systems with regard to the four different states of hydrogen, that is, hydrogen (H_2), proton (H^+), protium ($H°$), and protide (H^-). The surface sensitivity and durability are reduced considerably when exposed to air, forming an inactive surface composed of oxides and hydroxides. Inactive surfaces are also formed from contact with impurities such as CO, CO_2, H_2S, NO_2 N_2 and others [214].

With regard to reactivity and sensitivity, the chemically and electrochemically interactive sites on the surface need to be increased but, at the same time, the surface must be durable against air and water, impurity gases, and alkaline solutions.

An interactive surface is provided by (i) increasing the specific surface area, (ii) creating nanosized structures, (iii) chemical treatment. We discuss these specific topics briefly in the following sections.

4.5.4
Improvement of Surface Properties and Characteristics

Hydrogen-absorbing materials easily lose their hydrogen-uptake capability mostly by: (i) degradation of crystalline structures to amorphous states, which causes a decrease in the number of interstitial sites in the crystalline lattice available for protium occupation; (ii) disproportionation of crystalline compositions, which results in a decrease in hydrogen uptake capability; (iii) surface contamination, which causes the formation and growth of inactive surface layers [214].

4.5.4.1 Size Reduction of Particles
Excessive size reduction to micrometer (μm) or nanometer (nm) ranges by using a high-density ball milling technique is unfavorable due to the degradation of crystalline structure caused by the strong mechanical forces.

The size reduction of particles in order to improve the capacity of hydrogen-absorbing materials has not been successful in the past, any attempts have resulted in poor hydrogen capacity.

4.5.4.2 Creating Nanosized Surface Structures
The specific surface area can be increased by modification of the surface structure. Chemically treated surfaces should provide a reactive surface that is insensitive

to impurities [220–222]. The fluorination technique is the only technique that has been successful in solving the problems 1–8 listed in Section 4.5.4. A number of surface-treatment techniques for the preparation of an active surface for hydrogen uptake have been reported but these have all led to the sacrifice of other important properties and characteristics.

4.5.4.3 Fluorinated Surfaces

The fluorination technique was found to be effective by a series of long-term studies to improve the surface properties and characteristics of hydrogen absorbing materials as listed in Section 4.5.4.

Typical examples of materials with a nanostructured surface are shown as scanning electron microscopy (SEM) and electron microprobe analysis (EPMA) micrographs in Figure 4.29 for fluorinated Mg_2Ni (a), (c) and (e) and $LaNi_{4.7}Al_{0.3}$ (b) and (d).

Fluorination is effective not only for protecting surfaces from impurities but also for giving catalytic function and improving both the chemical and electrochemical characteristics of metal hydrides. For example, the surface of Mg_2Ni created by a fluorination treatment was found to be effective as a catalyst for the catalytic generation of hydrogen from an aqueous alkaline solution of $NaBH_4$ by hydrolysis (see Chapter 6.8).

It forms a Ni-enriched top-surface above the sub-surface layer of MgF_2 that is followed by the ordinal phase of Mg_2NiH_4 by a fluorination treatment and the MgF_2 surface works effectively in hydrolysis to break the B—H bond to generate hydrogen ($BH_4^- + 2H_2O \rightarrow 4H_2 + BO_2^-$).

A series of mischmetal nickel alloys are known as typical hydrogen-absorbing materials (see Chapter 6.3) and are used as the negative electrode in Ni–MH rechargeable batteries (see Chapter 8.4). By the fluorination treatment of mischmetal nickel alloys the electric conductivity is significantly improved by the replacement of La-oxide and La-hydroxide layers on the Ni-enriched LaF_2 surface. The layer was also found to protect the surface from KOH alkaline solution as electrolyte and to improve the long-term durability.

These fluorination techniques, which have been developed since 1991 to now are given in the Appendix as a literature list for the readers' convenience.

4.5.5
Concluding Remarks

In this chapter, the role of the surface is described from engineering viewpoints as summarized below:

1. The transitions between the four states of hydrogen at the surface are shown to be the basis of hydrogen devices and processes.

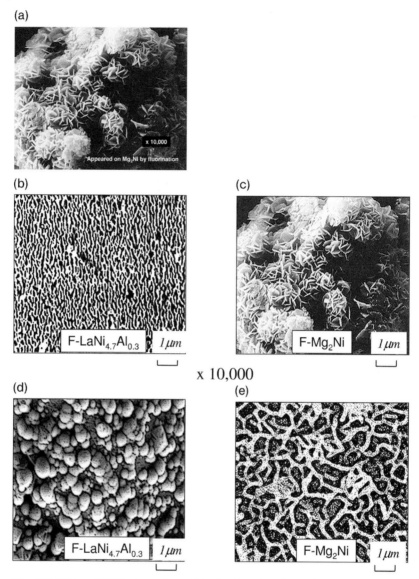

Fig. 4.29 SEM and EPMA observations of the fluorinated surfaces of (a), (c) and (e) Mg_2Ni and (b) and (d) $LaNi_{4.7}Al_{0.3}$.

2. The surface is the interface between gas–solid and liquid–solid phase reactions for both chemical and electrochemical applications.
3. The nanostructured surface created by the fluorination technique is briefly reviewed with respect to the improvement in hydrogen uptake capability and durability of hydrogen storage materials for both chemical and electrochemical applications.

References

1. Häussinger, P., Lohmüller, R., Watsin, A.M. (XXXX) Hydrogen. *Ullmann's Encyclopedia of Industrial Chemistry*, Fifth, Completely Revised Edition, volume A13: High-Performance Fibers to Imidazole and Derivatives, Wiley-VCH Verlag GmbH, pp. 297–442.
2. Urey, H.C., Brickwedde, F.G., Murphy, G.M. (1932) *Phys. Rev.*, **40** (1), 1–15.
3. Oliphant, M.L.E., Harteck, P., Rutherford, L. (1934) *Proc. R. Soc. London, Ser. A*, **144**, 692.
4. Novick (1947) *Phys. Rev.*, **72**, Letters to the Editor, 972.
5. Weast, R.C. (ed.) (1976) *Handbook of Chemistry and Physics*, 57th edn, CRC Press, Cleveland, Ohio.
6. Silvera, F., Walraven, J.T.M. (1980) *Phys. Rev. Lett.*, **44**, 164.
7. Silvera, F. (1980) *Rev. Mod. Phys.*, **52**, 393.
8. Massie, S.T., Hunten, D.M. (1982) *Icarus*, **49**, 213–26.
9. Sullivan, N.S., Zhou, D., Edwards, C.M. (1990) *Cryogenics*, **30**, 734.
10. Yucel, S. (1989) *Phys. Rev. B*, **39**, 3104.
11. Diatschenko, V., Chu, C.W., Liebenberg, D.H., Young, D.A., Ross, M., Mills, R.L. (1985) *Phys.Rev. B*, **32** (1), 381–9.
12. Leung, W.B., March, N.H., Motz, H. (1976) *Phys. Lett.*, **56A** (6), 425–6.
13. Mucker, K.F., White, D. et al. (1965) *Phys. Rev. Lett.*, **15**, 586.
14. Silvera, F. (1980) *Rev. Mod. Phys.*, **52**, 393.
15. Wigner, E., Huntington, H.B. (1935) *J. Chem. Phys.*, **3**, 764.
16. Asheroft, N.W. (1968) *Phys. Rev. Lett.*, **21**, 1748.
17. Hemme, H., Driessen, A., Griessen, R. (1986) *J. Phys. C*, **19**, 3571–85.
18. Hemme, H., Driessen, A., Griessen, R. (1986) *Physica*, **139/140B**, 116–18.
19. Zhou, L., Zhou, Y. (2001) *Int. J. Hydrogen Energy*, **26**, 597–601.
20. Hemme, H., Driessen, A., Griessen, R. (1986) *J. Phys. C*, **19**, 3571–85.
21. Fukai, Y. (1993) *The Metal-Hydrogen System Basic Bulk Properties*, Springer Verlag, Heidelberg, ISBN 3-540-55637-0.
22. Sugimoto, H., Fukai, Y. (1992) *Acta Metall.*, **40**, 2327.
23. Joule, J.P., William, T. XV (1854) *Philos. Trans.*, **114**, 321–64; Joule, J.P., William, T. XI (1856) *Proc. R. Soc.*, **7**, 127–30; Joule, J.P., William, T. XXV (1862) *Philos. Trans.*, **152**, 579–89.
24. Zemansky, M.W., Dittman, R.H. (1997) *Heat and Thermodynamics*, 7th edn, McGraw-Hill.
25. Holleck, G. (1970) *J. Phys. Chem.*, **74**, 503.
26. Hord, J. (1978) *Int. J. Hydrogen Energy*, **3**, 157.
27. Drell, L., Belles, F.E. (1952) *Survey of Hydrogen Combustion Properties*, NASA Report 1383, Washington D.C.
28. Fischer, M., Eichert, H. (1981) *DFVLR-Nachr.*, **34**, 33.
29. Wiberg, K.B. (1955) *Chem. Rev.*, **55**, 713.
30. NASA Lewis Research Center (1968) Hydrogen Safety Manual, NASA Tech. Memo TM-X 52454; Kalyanam, K.M., Hay, D.R. (1987) *Safety Guide for Hydrogen*, National Research Council Hydrogen Safety Committee, Ottawa; Gregory, F.D. (1997) Safety standard for hydrogen and hydrogen systems. Guidelines for Hydrogen System Design, Materials Selection, Operations, Storage, and Transportation, Office of Safety and Mission Assurance, Washington, DC 20546, National Aeronautics and Space Administration NASA.
31. Eichert, H., Fischer, M. (1986) *Int. J. Hydrogen Energy*, **11**, 117.
32. Somorjai, G.A. (1981) *Chemistry in Two Dimensions, Surfaces*, Cornell University Press, Ithaca; Bare, S.R., Somorjai, G.A. (1987) "Surface Chemistry," in *Encyclopedia of Physical Science and Technology*, vol. **13**, Academic Press, New York, p. 526.
33. Paal, Z., Menon, P.G. (1988) *Hydrogen Effects in Catalysis*, M. Dekker, New York.

34 Christmann, K.R. (1988) *Surf. Sci. Rep.*, **9**, 1, and references therein.
35 Burch, R. (1979) "The Adsorption and Absorption of Hydrogen by Metals," in *Chemical Physics of Solids and Their Surfaces*, vol. **9**, Chemical Society, London, p. 1.
36 Jacobsen, K.W., Nørskov, J.K. (1987) *Phys. Rev. Lett.*, **59**, 2764; Jacobsen, K.W. (1988) *Comments Condens. Matter Phys.*, **14**, 129.
37 Vegge, T. (2004) *Phys. Rev. B*, **70**, 035412.
38 Ostenfeld, C.W., Chorkendorff, I. (2006) *Surf. Sci.*, **600**, 1363–8.
39 Frankcombe, T., Lovvik, O.M. (2006) *J. Phys. Chem. B*, **110** (1).
40 Fu, Q.J., Ramirez-Cuesta, A.J., Tsang, S.C. (2006) *J. Phys. Chem. B*, **110**, 711–5.
41 Chaudhuri, S., Muckerman, J.T. (2005) *J. Phys. Chem. B*, **109** (15), 6952.
42 Schlapbach, L., Seiler, A., Stucki, F., Siegmann, H.C. (1980) *J. Less-Common Met.*, **73**, 145.
43 Siegmann, H.C., Schlapbach, L., Brundle, C.R. (1978) *Phys. Rev. Lett.*, **40**, 972.
44 Waldkirch, T.V., Zurcher, P. (1978) *Auvl. Phys. Lett.*, **33**, 689; Zhu, L., Wang, R., Kingy, T.S., DePristo, A.E. (1997) *J. Catal.*, **167**, 408–11.
45 Ruban, A.V., Skriver, H.L., Nørskov, J.K. (1999) *Phys. Rev. B*, **59**, 15990–16000.
46 Tománek, D., Mukherjee, S., Kumar, V., Bennemann, K.H. (1982) *Surf. Science*, **114**, 1.
47 Greeley, J., Mavrikakis, M. (2004) *Nature Mater.*, **3**, 810.
48 Andreasen, A., Vegge, T., Pedersen, A.S. (2005) *J. Phys. Chem. B*, **109**, 3340.
49 Borgschulte, A., Westerwaal, R.J., Rector, J.H., Schreuders, H., Dam, B., Griessen, R. (2006) *J. Catal.*, **239**, 263–71.
50 Lennard-Jones, J.E. (1932) *Trans. Faraday Soc.*, **28**, 333.
51 Muscat, J.P., Newns, D.M. (1978) *Progr. Surf. Science*, **9**, 1.
52 Hammer, B., Scheffler, M., Jacobsen, K.W., Nørskov, J.K. (1994) *Phys. Rev. Lett.*, **73**, 1400–3; and Kroes, G.J., Gross, A., Baerends, E.J., Scheffler, M., McCormack, D.A. (2002) *Acc. Chem. Res.*, **35**, 193–200.
53 Rendulic, K.D. (1988) *Appl. Phys. A*, **47**, 55.
54 Christmann, K. (1987) *Z. Phys. Chem. NF*, **154**, 145.
55 Müller, K. (1986) *Ber. Bunsenges. Phys. Chem.*, **90**, 184.
56 Kittel, C. (1986) *Introduction to Solid State Physics*, John Wiley & Sons, Inc., New York.
57 Li, Y., Erskine, J.L., Diebold, A.C. (1984) *Phys. Rev. B*, **34**, 5951.
58 Nicol, J.M., Rush, J.J., Kelley, R.D. (1987) *Phys. Rev. B*, **36**, 9315.
59 Chou, M.Y., Chelikowsky, J.R. (1987) *Phys. Rev. Lett.*, **59**. 1737.
60 Greeley, J., Mavrikakis, M. (2005) *J. Phys. Chem. B*, **109**, 3460.
61 Felter, T.E., Foiles, S.M., Daw, M.S., Stulen, R.H. (1986) *Surf. Sci.*, **171**, L379.
62 Zangwill, A. (1988) *Physics at Surfaces*, Cambridge University Press, Cambridge.
63 Nordlander, P., Holmberg, C., Harris, J., (1985) *Surf. Sci.*, **152/153**, 702.
64 Johansson, P.K. (1981) *Surf. Sci.*, **104**, 510.
65 Andersson, S., Wilzen, L., Persson, M. (1988) *Phys. Rev. B*, **38**, 2967; Andersson, S., Wilzen, L., Persson, M., Harris, J. (1989) *Phys. Rev. B*, **40**, 8146.
66 Lapujoulade, J., Perreau, J. (1983) *Phys. Scr. T*, **4**, 138.
67 Avouris, P., Schmeisser, D., Demuth, J.E. (1982) *Phys. Rev. Lett.*, **48**, 199; Llisca, E. (1991) *Phys. Rev. Lett.*, **66**, 667.
68 Harris, J., Leibsch, A. (1983) *Phys. Scr. T*, **4**, 14.
69 Nordlander, P., Holmberg, C., Harris, J. (1986) *Surf. Sci.*, **175**, L753.
70 Murphy, R.B., Mundnar, J.M., Tsuei, K.D., Plummer, E.W. (1988) *Bull. Am. Phys. Soc.*, **33**, 655.
71 Martensson, A.S., Nyberg, C., Andersson, S. (1986) *Phys. Rev. Lett.*, **57**, 2045.
72 Züttel, A., Sudan, P., Mauron, P., Kyiobaiashi, T., Emmenegger, C., Schlapbach, L. (2002) *Int. J. Hydrogen*

Energy, **27**, 203; Nijkamp, M.G., Raaymakers, J.E.M.J., van Dillen, A.J., de Jong, K.P. (2001) *Appl. Phys. A*, **72**, 619.

73 Langmi, H.W., Walton, A., Al-Mamouri, M.M., Johnson, S.R., Book, D., Speight, J.D., Edwards, P.P., Gameson, I., Anderson, P.A., Harris, I.R. (2003) *J. Alloys Compd.*, **356–357**, 710–5.

74 Rosi, N.L., Eckert, J., Eddaoudi, M., Vodak, D.T., Kim, J., O'Keeffe, M., Yaghi, O.M. (2003) *Science*, **300**, 1127.

75 Nørskov, J.K. (1984) *Physica*, **B127**, 193; *Phys. Rev. B*, **26**, 2875 (1982); Nørskov, J.K., Besenbacher, F. (1987) *J. Less-Common Met.*, **130**, 475.

76 Tanaka, K.-I., Tamaru, K. (1963) *J. Catal.*, **2**, 366–70.

77 Lee, G., Plummer, E.W. (1995) *Phys. Rev. B*, **51**, 7250.

78 Bligaard, T., Nørskov, J.K., Dahl, S., Matthiesen, J., Christensen, C.H., Sehested, J. (2004) *J. Catal.*, **224**, 206–17.

79 Griessen, R., Riesterer, T. (1988) "Heat of Formation Models," in *Topics in Applied Physics*, vol. **63**, Springer-Verlag, Berlin, Heidelberg, p. 219.

80 Harris, J. (1988) *Appl. Phys. A*, **47**, 63.

81 Zhu, L., Wang, R., King, T.S., DePristo, A.E. (1997) *J. Catal.*, **167**, 408–11.

82 Sinfelt, J.H. (1986) *J. Phys. Chem.*, **90**, 4711; Sinfelt, J.H. in *Bimetallic Catalysts*, John Wiley & Sons, Inc., New York (1983).

83 Schlapbach, L. (1986) in *Hydrogen in Disordered and Amorphous Solids* (eds G. Bambakidis, R.C. Bowman), NATO AS1 B 136, Plenum, New York, p. 397.

84 Ichimura, K., Matsuvama, M., Watanabe, K. (1987) *J. Vac. Sci. Technol. AS*, **220**; Ichimura, K., Ashida, K., Watanabe, K. (1985) *Vac. Sci. Technol. A*, **3**, 346.

85 Vedel, I., Schlapbach, L. (1989) *Structure and Reactivity of Surfaces* (eds C. Morterra, A. Zecchina, G. Costa), Elsevier, Amsterdam, p. 903.

86 Dowben, P.A., Miller, A. (eds) (1990) *Surface Segregation Phenomena*, CRC Press, Boca Raton, Florida.

87 Sakurai, T., Hashizume, T., Jimbo, A., Sakai, A., Hyodo, S. (1985) *Phys. Rev. Lett.*, **55**, 514; Sakurai, T., Hashizume, T., Kobayashi, A., Sakai, A., Hyodo, S., Kuk, Y., Pickering, H.W. (1986) *Phys. Rev. B*, **34**, 8379.

88 Foiles, S.M. (1985) *Phys. Rev. B*, **32**, 7685.

89 Durham, P.J., Jordan, R.G., Sohal, G.S., Wille, L.T. (1984) *Phys. Rev. Lett.*, **53**, 2038.

90 Modak, S., Khanra, B.C. (1986) *Phys. Rev. B*, **34**, 5909.

91 Borgschulte, A., Rode, M., Jacob, A., Schoenes, J. (2001) *J. Appl. Phys.*, **90**, 1147.

92 Strongin, M., Colbert, J., Dienes, G.J., Welch, D.O. (1982) *Phys. Rev. B*, **26**, 2715.

93 Pick, M.A., Hanson, A., Jones, K.W., Goland, A.N. (1982) *Phys. Rev. B*, **26**, 2900.

94 Wicke, E. (1985) *Z. Phys. Chem. NF*, **143**, 1.

95 Schlapbach, L., Burger, J.P., Thiry, P., Bonnet, J., Petroff, Y. (1986) *Phys. Rev. Lett.*, **57**, 2219; *Surf. Sci.*, **189/190**, 747 (1987).

96 Oelhafen, P., Lapka, R., Gubler, U., Krieg, J., DasGupta, A., Guntherodt, H.J., Mizoguchi, T., Hague, C., Kubler, J., Nagel, S.R. (1982) in *Rapidly Quenched Metals IV* (eds T. Masumot, K. Suzuki), Japan Institute of Metals, Sendai, p. 1259.

97 Burger, J.P., Schlapbach, L., Vedel, I., Maier, U. (1989) *Z. Phys. Chem. NF*, **163**, 569.

98 Muscat, J.P. (1985) *Progr. Surf. Sci.*, **18**, 59; *Surf. Sci.*, **152/153**, 684 (1985).

99 Sparnaay, M.J. (1984) *Surf. Sci. Rep.*, **4**, 101.

100 Miedema, A.R. (1978) *Z. Metallkunde*, **69**, 455.

101 Nordlander, P., Holmberg, C., Harris, J. (1985) *Surf. Sci.*, **152/153**, 702.

102 Johansson, P.K. (1981) *Surf. Sci.*, **104**, 510.

103 Andersson, S., Wilzen, L., Persson, M. (1988) *Phys. Rev. B*, **38**, 2967; Andersson, S., Wilzen, L., Persson, M., Harris, J. (1989) *Phys. Rev. B*, **40**, 8146.

104 Nordlander, P., Holmberg, C., Harris, J. (1986) *Surf. Sci.*, **175**, L753.

105 Harris, J., Andersson, S. (1985) *Phys. Rev. Lett.*, **55**, 1583.

106 Harris, J., Andersson, S., Holmberg, C., Nordlander, P. (1986) Phys. Scr. T, 13, 155.
107 Andersson, S., Wilzen, L., Persson, M. (1988) Phys. Rev. B, 38, 2967; Andersson, S., Wilzen, L., Persson, M., Harris, J. (1989) Phys. Rev. B, 40, 8146.
108 Eberhardt, W., Cantor, R., Greuter, F., Plummer, E.W. (1982) Solid State Commun., 42, 799.
109 Harris, J., Andersson, S. (1985) Phys. Rev. Lett., 55, 1583.
110 Castro, G.R., Drakova, D., Grillo, M.E., Doyen, G. (1996) J. Chem. Phys., 105, 9640.
111 Kroes, G.J., Gross, A., Baerends, E.J., Scheffler, M., McCormack, D.A. (2002) Acc. Chem. Res., 35, 193–200.
112 Gross, A., Scheffler, M. (1998) Phys. Rev. B, 57, 2493–506.
113 Henrich, V.E. (1985) Rep. Prog. Phys., 48, 1481; Prog. Surf. Sci., 9, 143 (1979).
114 Tsukada, M., Shima, N. (1987) Phys. Chem. Miner., 15, 35.
115 Castell, M.R., Wincott, P.L., Condon, N.G., Muggelberg, C., Thornton, G., Dudarev, S.L., Sutton, A.P., Briggs, G.A.D. (1997) Phys. Rev. B, 55, 7859.
116 Oefner, H., Zaera, F. (1997) J. Phys. Chem. B, 101, 9069.
117 Diebold, U. (2003) Surf. Sci. Rep., 48, 53–229.
118 Linsebigler, A.L., Lu, G., Yates J.T. Jr. (1995) Chem. Rev., 95, 69.
119 Göpel, W., Rocker, G., Feierabend, R. (1983) Phys. Rev. B, 28, 3427.
120 Dulub, O., Diebold, U., Kresse, G. (2003) Phys. Rev. Lett., 90, 016101.
121 Malherbe, J.B., Hofmann, S., Sanz, J.M. (1986) Appl. Surf. Sci., 27, 355.
122 Lo, W.J., Chung, Y.W., Somorjai, G.A. (1978) Surf. Sci., 71, 199.
123 Fujino, T., Katayama, M., Inudzuka, K., Okuno, T., Oura, K., Hirao, T. (2001) Appl. Phys. Lett., 79, 2716.
124 Bates, J.B., Wang, J.C., Perkins, R.A. Phys. Rev. B, 19, 4130 (1979).
125 Johnson, O.W., Paek, S.H., Deford, J.W. (1975) J. Appl. Phys., 46, 1026.
126 Nelson, J.G., Murray, G.T. (1984) Met. Trans. A, 15, 597.
127 Colburn, E.A., Mackrodt, W.C. (1982) Surf. Sci., 117, 571.
128 Pope, S.A., Guest, M.F., Hillier, I.H., Colburn, E.A., Mackrodt, W.C., Kendrick, J. (1983) Phys. Rev. B, 28, 2191.
129 Ibach, H., Lehwald, S., Voigtlinder, B. (1987) J. Electron Spectrosc. Relat. Phenom., 44, 263.
130 Barkhordarian, G., Klassen, T., Bormann, R. (2004) J. Alloys Compd., 364, 242.
131 Barkhordarian, G., Klassen, T., Bormann, R. (2006) J. Alloys Compd., 407, 249–255.
132 Hanada, N., Ichikawa, T., Fujii, H. (2005) J. Phys. Chem. B, 109, 7188.
133 Henrich, V.E., Cox, P.A. (1994) The Surface Science of Metal Oxides, Cambridge University Press, New York.
134 Borgschulte, A., Rector, J.H., Dam, B., Griessen, R., Züttel, A. (2005) J. Catal., 235, 353–8.
135 Friedrichs, O., Sanchez-Lopez, J.C., Lopez-Cartes, C., Klassen, T., Bormann, R., Fernandez, A. (2006) J. Phys. Chem. B, 110, 7845–50.
136 Borgschulte, A., Bösenberg, U., Barkhordarian, G., Dornheim, M., Bormann, R. (2007) Catal. Today, 120, 262.
137 Huhn, P.-A., Dornheim, M., Klassen, T., Bormann, R. (2005) J. Alloys Compd.
138 Durr, W., Taborelli, M., Paul, O., Germar, R., Gudat, W., Pescia, D., Landolt, M. (1989) Phys. Rev. Lett., 62, 206.
139 Fu, C.L., Freeman, A.J., Oguchi, T. (1985) Phys. Rev. Lett., 54, 2700; Brodsky, M.B. (1984) J. Phys., 45, C5–349; J. Magn. Magn. Mater., 35, 99 (1983).
140 Paul, O., Taborelli, M., Landolt, M. (1989) Surf. Sci., 211/212, 724.
141 Amiri-Hezaveh, A., Jennings, G., Joyner, D.J., Willis, R.F. (1984) J. Phys., 45, C5–371; Shinjo, T., Hosoito, N., Kawaguchi, K., Takada, T., Endoh, Y. (1984) J. Phys., 45, C5–361.
142 Martensson, N., Stenborg, A., Bjorneholm, O., Nilsson, A., Andersson, J.N. (1988) Phys. Rev. Lett., 60. 1731.
143 Uher, C., Cohn, J.L., Miceli, P.F., Zabel, H. (1987) Phys. Rev. B, 36, 815.

212 Suda, S., Iwase, Y., Morigasaki, N., Li, Z.-P. (2004) *Advanced Materials for Energy Storage II* (eds D. Chandra, R.G. Bautista, L. Schlapbach), TMS, pp. 123–133.
213 Sandrock, G., Suda, S., Schlapbach, L. (1992) *Hydrogen in Intermetallic Compounds II, Topics in Applied Physics*, vol. **67**, Ch. 5. Springer Verlag, Berlin, p. 197.
214 Suda, G. Sandrock (1994) *Z. Phys. Chem.* **183**, 149–56.
215 Suda, S. (2002) Hydrogen–Metal Systems: Technological and Engineering Aspects, in *Encyclopedia of Materials –Science and Technology*. Elsevier Science Ltd., Amsterdam, pp. 3970–6.
216 Cengel, Y.A., Boles, M.A. (2003) *Thermodynamics – An Engineering Approach*, 4th edn, McGraw Hill, New York, p. 844.
217 Suda, S., Iwase, Y., Morigasaki, N., Li, Z.-P. (2004) *Advanced Materials for Energy Storage II* (eds D. Chandra, R. G. Bautista, L. Schlapbach), TMS, USA, pp. 123–33.
218 Schlapbach, L., Züttel, A. (2001) *Nature*, **414**, 353–8.
219 Kung, H. (2003) Report on Hydrogen Storage Panel Findings in DOE-BES Sponsored Workshop on Basic Research for Hydrogen Production, Storage and Use, May 13–15, DOE, WA, USA.
220 Liu, F.-J., Suda, S. (1995) *J. Alloys Compd.* **231**, 742–50.
221 Liu, F.-J., Suda, S. (1996) *J. Alloys Compd.* **232**, 204–11.
222 Liu, F.-J., Sandrock, G., Suda, S. (1992) *J. Alloys Compd.*, **190** (1), 57–60.

List of literature related to the fluorination of intermetallic hydrogen storage materials

Fluorination Techniques

1 Liu, F.-J., Suda, S. (1995) *J. Alloys Compd.*, **231**, 742–50.
2 Liu, F.-J., Suda, S. (1996) *J. Alloys Compd.*, **232**, 204–11.
3 Liu, F.-J., Sandrock, G., Suda, S. (1992) *J. Alloys Compd.*, **190** (1), 57–60.
4 Li, Z.-P., Sun, Y.-M., Liu, B.-H., Gao, X.-P., Suda, S. (1998) Proceedings of the MRS Spring Meeting, San Francisco, CA, USA, vol. 513, pp. 25–36.
5 Sun, Y.-M., Gao, X.-P., Araya, N., Higuchi, E., Suda, S. (1999) *J. Alloys Compd.*, **293–295**, 364–8.
6 Sun, Y.-M., Suda, S. (2002) *J. Alloys Compd.*, **330–332**, 627–631.

Fluorination Effects

7 Wang, X.-L., Suda, S. (1993) *J. Alloys Compd.*, **191** (1), 5–7.
8 Wang, X.-L., Suda, S. (1993) *J. Alloys Compd.*, **194**, 73–6.
9 Wang, X.-L., Suda, S., Wakao, S. (1994) *Z. Phys. Chem.*, **183**, 297–302.
10 Wang, X.-L., Suda, S. (1994) *Z. Phys. Chem.*, **183**, 385–90.
11 Liu, F.-J., Sandrock, G., Suda, S. (1994) *Trans. Mater. Res. Soc. Jpn.*, **18** (B), 1237–40.
12 Wang, X.-L., Iwata, K., Suda, S. (1994) *Trans. Mater. Res. Soc. Jpn.*, **18** (B), 1245–8.
13 Wang, X.-L., Suda, S. (1994) *Trans. Mater. Res. Soc. Jpn.*, **18** (B), 1241–4.
14 Sun, Y.-M., Suda, S. (1995) *J. Alloys Compd.* **231**, 417–21.
15 Liu, F.-J., Kitayama, K., Suda, S. (1996) *Vacuum*, **47** (6–8), 903–6.
16 Liu, F.-J., Suda, S. (1996) *J. Alloys Compd.*, **232**, 232–7.
17 Wang, X.-L., Hagiwara, H., Suda, S. (1996) *Vacuum*, **47** (6–8), 899–902.

18 Gao, X.-P., Liu, B.-H., Imai, M., Ohta, H., Suda, S. (1998) *Hydrogen Energy Prog.*, **I**, 1075–84.

19 Suda, S., Imai, M., Uchida, M., Komazaki, Y., Higuchi, E. (1999) *J. Alloys Compd.*, **293–295**, 391–5.

Applications of Fluorination Techniques

20 Wang, X.-L., Suda, S. (1994) *Trans. Mater. Res. Soc. Jpn.*, **18** (B), 1249–52.

21 Liu, F.-J., Suda, S. (1994) *Trans. Mater. Res. Soc. Jpn.*, **18** (B), 1233–6.

22 Li, Z.-P., Yan, D.-Y., Suda, S. (1994) *Trans. Mater. Res. Soc. Jpn.*, **18** (B), 1209–1212.

23 Wang, X.-L., Haraikawa, N., Suda, S. (1994) *Trans. Mater. Res. Soc. Jpn.*, **18** (B), 1253–6.

24 Liu, F.-J., Suda, S. (1995) *J. Alloys Compd.*, **231**, 411–6.

25 Wang, X.-L., Suda, S. (1995) *J. Alloys Compd.*, **227**, 58–62.

26 Wang, X.-L., Hagiwara, H., Suda, S. (1995) *J. Alloys Compd.*, **231**, 376–9.

27 Wang, X.-L., Suda, S. (1995) *J. Alloys Compd.*, **231**, 380–386.

28 Yan, D.-Y., Sun, Y.-M., Suda, S. (1995) *J. Alloys Compd.*, **231**, 387–391.

29 Li, Z.-P., Suda, S. (1995) Proceedings of the Symposium on Hydrogen & Metal Hydride Batteries, Battery Division Proc, vol. 94-27, pp. 78–84.

30 Yan, D.-Y., Suda, S. (1995) *J. Alloys Compd.*, **231**, 565–72.

31 Liu, F.-J., Suda, S. (1995) *J. Alloys Compd.*, **231**, 666–9.

32 Wang, X.-L., Iwata, K., Suda, S. (1995) *J. Alloys Compd.*, **231**, 829–34.

33 Wang, X.-L., Iwata, K., Suda, S. (1995) *J. Alloys Compd.*, **231**, 860–864.

34 Suda, S. (1996) *Proceedings of the International AB-Sorption Heat Pump Conference, Montreal, Quebec, Canada*, pp. 107–126.

35 Suda, S., Iwata, K., Sun, Y.-M., Komazaki, Y., Liu, F.-J. (1997) *J. Alloys Compd.*, **253–254**, 668–72.

36 Liu, F.-J., Ota, H., Okamoto, S., Suda, S. (1997) *J. Alloys Compd.*, **253–254**, 452–8.

37 Sakashita, M., Li, Z.-P., Suda, S. (1997) *J. Alloys Compd.*, **253–254**, 500–5.

38 Sun, Y.-M., Iwata, K., Chiba, S., Matsuyama, Y., Suda, S. (1997) *J. Alloys Compd.*, **253–254**, 520–4.

39 Iwata, K., Komazaki, Y., Uchida, M., Suda, S. (1998) *Hydrogen Energy Prog.*, **I**, 647–60.

40 Li, Z.-P., Sun, Y.-M., Liu, B.-H., Gao, X.-P., Suda, S. (1998) *Mater. Res. Soc. Symp. Proc.*, **513**, 25–36.

41 Iwata, K., Sun, Y.-M., Suda, S. (1999) *Int. J. Hydrogen Energy*, **24**, 251–6.

42 Li, Z.-P., Higuchi, E., Liu, B.-H., Suda, S. (1999) *J. Alloys Compd.*, **293–295**, 593–600.

43 Gao, X.-P., Sun, Y.-M., Higuchi, E., Suda, S. (1999) *J. Alloys Compd.*, **293–295**, 707–11.

44 Liu, B.-H., Li, Z.-P., Higuchi, E., Suda, S. (1999) *J. Alloys Compd.*, **293–295**, 702–6.

45 Gao, X.-P., Sun, Y.-M., Higuchi, E., Toyoda, E., Suda, S. (1999) *J. Alloys Compd.*, **293–295**, 707–11.

46 Li, Z.-P., Higuchi, E., Liu, B.-H., Suda, S. (2000) *Electrochim. Acta*, **45**, 1773–9.

47 Gao, X.-P., Sun, Y.-M., Toyoda, E., Higuchi, E., Nakagima, T., Suda, S. (2000) *Electrochim. Acta*, **45**, 3099–3104.

48 Liu, B.-H., Li, Z.-P., Okutsu, A., Suda, S. (2000) *J. Alloys Compd.*, **296**, 148–51.

49 Suda, S., Sun, Y.-M., Liu, B.-H., Zhou, Y., Morimitsu, S., Arai, K., Tsukamoto, N., Uchida, M., Candra, Y., Li, Z.-P. (2001) *J. Appl. Phys. A*, **72**, 209–12.

50 Suda, S., Sun, Y.-M., Uchida, M., Liu, B.-H., Morimitsu, S., Arai, K., Zhou, Y., Tsukamoto, N., Candra, Y., Li, Z.-P. (2001) *Met. Mater. Int.*, **7** (1), 73–5.

51 Higuchi, E., Toyoda, E., Li, Z.-P., Suda, S., Inoue, H., Nohara, S., Iwakura, C. (2001) *Electrochim. Acta*, **46**, 1191–4.

52 Higuchi, E., Miyoshi, H., Li, Z.-P., Suda, S., Inoue, H., Nohara, S., Iwakura, C. (2001) *ITE Lett. Batteries, New Technol. Med.*, **2** (1), 69–74.

53 Suda, S., Sun, Y.-M., Liu, B.-H., Zhou, Y., Morimitu, S., Arai, K., Uchida, M., Li, Z.-P. (2001) *J. Appl. Phys.*, **A 72**, 209–12.

54 Liu, B.-H., Li, Z.-P., Kitani, R., Suda, S. (2002) *J. Alloys Compd.* **330-2**, 825–30.
55 Higuchi, E., Hidaka, K., Li, Z.-P., Suda, S., Nohara, S., Inoue, H., Iwakura, C. (2002) *J. Alloys Compd.*, **335**, 277–80.
56 Higuchi, E., Li, Z.-P., Suda, S., Nohara, S., Inoue, H., Iwakura, C. (2002) *J. Alloys Compd.*, **335**, 241–5.
57 Li, Z.-P., Liu, B.-H., Hitaka, K., Suda, S. (2002) *J. Alloys Compd.*, **330-2**, 777–81.

Mg-Treatments

58 Liu, F.-J., Sandrock, G., Suda, S. (1994) *Z. Phys. Chem.*, **183**, 163–7.
59 Liu, F.-J., Sandrock, G., Suda, S. (1994) *Trans. Mater. Res. Soc. Jpn.*, **18** (B), 1229–32.
60 Liu, F.-J., Suda, S. (1995) *J. Alloys Compd.*, **230**, 58–62.
61 Wang, X.-L., Haraikawa, N., Suda, S. (1995) *J. Alloys Compd.*, **231**, 397–402.
62 Liu, F.-J., Suda, S. (1995) *J. Alloys Compd.*, **231**, 696–701.
63 Liu, F.-J., Suda, S. (1996) *J. Alloys Compd.*, **232**, 212–7.
64 Suda, S. (2002) *Task 12 in IEA Agreement on the Production and Utilization of Hydrogen*, pp. 65–71.

5
Hydrogen Production
Andreas Borgschulte, Andreas Züttel, and Ursula Wittstadt

5.1
Hydrogen Production from Coal and Hydrocarbons
Andreas Borgschulte and Andreas Züttel

Hydrocarbons are the main source for the production of hydrogen on an industrial scale. The many production routes for hydrogen, however, are by no means of equal economic importance. Most of the hydrogen for industrial use is produced from natural gas and oil, either as a main product or as a byproduct from a process involving a chemical conversion. In many cases, the chemical processes releasing hydrogen are intermediate steps in a large process chain. Some of the main reactions were invented a hundred years ago and can be summarized by the general term "gasification." Principally, hydrogen can be extracted from any hydrocarbon. Accordingly, the number of possible raw materials is large:

- Coal
- Heavy oil
- Light oil
- Methane
- Biomass

Due to the various raw materials, the specific processes differ, in particular the technical realizations. We will restrict ourselves to a short review of the physico-chemical fundamentals starting with the underlying reaction equations.

5.1.1
Physico-Chemical Fundamentals of Hydrogen Production from Fossil Fuels*

The reaction characteristics of gasification can be subdivided into four groups (ΔH refers to 0°C and 0.1 MPa as the reference state):

---

* This chapter was adapted from Ref. [4].

Hydrogen as a Future Energy Carrier. Edited by A. Züttel, A. Borgschulte, and L. Schlapbach
Copyright © 2008 WILEY-VCH Verlag GmbH & Co. KGaA, Weinheim
ISBN: 978-3-527-30817-0

1. Reactions with molecular oxygen (combustion) [4]

$$C + 1/2 O_2 \Leftrightarrow CO \quad \Delta H = -111 \text{ kJ mol}^{-1} \quad (5.1)$$

$$CO + 1/2 O_2 \Leftrightarrow CO_2 \quad \Delta H = -283 \text{ kJ mol}^{-1} \quad (5.2)$$

$$H_2 + 1/2 O_2 \Leftrightarrow H_2O \quad \Delta H = -242 \text{ kJ mol}^{-1} \quad (5.3)$$

$$C_n H_m + (n+m)/4 O_2 \Leftrightarrow n CO_2 + m/2 H_2O \quad (5.4)$$

For example the combustion of methane gives:

$$CH_4 + 2 O_2 \Leftrightarrow CO_2 + 2 H_2O \quad \Delta H = -802 \text{ kJ mol}^{-1} \quad (5.4a)$$

For partial combustion of hydrocarbons to carbon monoxide and steam,

$$C_n H_m + (n/2 + m/4) O_2 \Leftrightarrow n CO + m/2 H_2O \quad (5.5)$$

The reaction enthalpy can be calculated from the enthalpies of Eqs. 5.4 and 5.2 as follows:

$$\Delta H(\text{Eq. (5.5)}) = \Delta H(\text{Eq. (5.4)}) - n\, \Delta H(\text{Eq. (5.2)})$$

2. Reactions with steam

$$C + H_2O \Leftrightarrow CO + H_2 \quad \Delta H = +131 \text{ kJ mol}^{-1} \quad (5.6)$$

$$CO + H_2O \Leftrightarrow CO_2 + H_2 \quad \Delta H = -41 \text{ kJ mol}^{-1} \quad (5.7)$$

also called homogenous water gas reaction and shift conversion, or, generally:

$$C_n H_m + n H_2O \Leftrightarrow n CO + (m/2 + n) H_2 \quad (5.8)$$

Again, for methane:

$$CH_4 + H_2O \Leftrightarrow CO + 3 H_2 \quad \Delta H = +206 \text{ kJ mol}^{-1} \quad (5.8a)$$

The enthalpy of the reaction of fuels with steam to produce carbon dioxide and hydrogen

$$C_n H_m + 2n H_2O \Leftrightarrow n CO_2 + (m/2 + 2n) H_2 \quad (5.9)$$

can be calculated by

$$\Delta H(\text{Eq. 5.9}) = \Delta H(\text{Eq. 5.8}) + n\, \Delta H(\text{Eq. 5.7})$$

3. Reactions with carbon dioxide

$$C + CO_2 \Leftrightarrow 2 CO \quad \Delta H = +173 \text{ kJ mol}^{-1} \quad (5.10)$$

(Boudouard reaction)

$$C_nH_m + n\,CO_2 \Leftrightarrow 2n\,CO + m/2\,H_2 \tag{5.11}$$

For our example:

$$CH_4 + CO_2 \Leftrightarrow 2\,CO + 2H_2 \quad \Delta H = 247\,\text{kJ mol}^{-1} \tag{5.11a}$$

Generally, the reaction enthalpies for hydrocarbon gasification with carbon dioxide are obtained from the enthalpies of Eqs. (5.8) and (5.7) as follows:

$$\Delta H(\text{Eq. 5.11}) = \Delta H(\text{Eq. 5.8}) - n\,\Delta H(\text{Eq. 5.7})$$

4. Decomposition of hydrocarbons (soot reactions)
   Hydrocarbon decomposition is described by

$$C_nH_m \Leftrightarrow n\,C + m/2\,H_2 \tag{5.12}$$

For example, the reactions

$$CH_4 \Leftrightarrow C + 2H_2 \quad \Delta H = +75\,\text{kJ mol}^{-1} \tag{5.12a}$$

$$C_2H_6 \Leftrightarrow 2\,C + 3H_2 \quad \Delta H = +85\,\text{kJ mol}^{-1} \tag{5.12b}$$

may be considered the reverse of hydrocarbon formation from its elements and are essential for the gasification of hydrocarbons. The reaction enthalpy can be determined from the combustion enthalpies of the reactants:

$$\Delta H(\text{Eq. 5.12}) = \Delta H(\text{Eq. 5.4}) - n\,\Delta H(\text{Eq. 5.1}) - n\,\Delta H(\text{Eq. 5.2})$$
$$- m/2\,\Delta H(\text{Eq. 5.3})$$

The gasification reactions (5.1)–(5.12) are never complete in either of the two possible directions but tend to reach an equilibrium expressed by the equilibrium constant K_p. For Eq. (5.8a), for example, this constant is given by

$$K_p = \frac{p_{CO} \cdot p_{H_2}^3}{p_{CH_4} \cdot p_{H_2O}} = \frac{x_{CO} \cdot x_{H_2}^3}{x_{CH_4} \cdot x_{H_2O}} P^2 = f(T)$$

Where x_i represents the mole fraction and p_i the partial pressure of each of the four components, P is the total pressure and T the temperature.

For Eq. (5.7), the carbon monoxide shift conversion (which in nearly every case has to be considered), the equilibrium composition of the shift conversion is not influenced by the total pressure, because the number of molecules (reactants) on the left- and right-hand sides is identical. However, the equilibrium constant depends on temperature. According to the laws of thermodynamics, the equilibrium constant can be calculated by means of the Gibbs energy of the formation energies of the reactants, which depend on temperature. Strictly speaking, the equilibrium constant K_p in the above form applies only to ideal gases. At increased pressure (process conditions) and near the boiling points of the liquids, the appropriate partial fugacities should be used instead of partial pressures. However, in practice, partial pressures can be used with sufficient accuracy in most cases. In heterogeneous reactions, partial pressures of the solids are not taken into account because they are regarded as a function of temperature alone, which is included in the equilibrium constant K_p.

The three reactions described by Eqs. (5.7), (5.8a) and (5.10) are almost always sufficient to calculate the theoretical product gas composition of various gasification processes. As can be seen from Figure 5.1, the equilibrium constant for endothermic reactions increases with temperature and that for exothermic reactions decreases. A general assessment of tendencies is facilitated by the Le Chatelier principle that states that a system in equilibrium tries to evade a change forced upon it [1]. For example, an increase in total pressure results in increased methane formation because the number of molecules on the right-hand side of Eq. (5.8a) is higher than that on the left. On the other hand, a pressure increase in the reaction proceeding without a change in the number of molecules, such as the homogenous water gas reaction Eq. (5.7), has almost no effect.

However, no statements can made on the basis of thermodynamics regarding the time dependence of the reaction and, therefore, the estimate of the production yield of the real reforming process. For this, calculations must include reaction

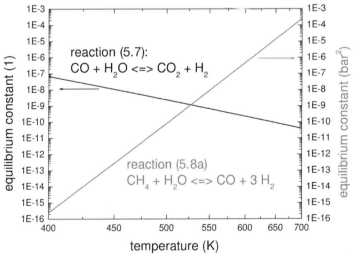

Fig. 5.1 Equilibrium constants for an exothermic reaction (5.7) and an endothermic reaction (5.8a). (Data from Häussinger, Lohmüller and Watsin (1995), Ref. [5]).

kinetics. Some processes must even employ catalysts to accelerate the reaction and enhance the reaction kinetics in order to approach equilibrium. The reasons are as follows: in general, the reaction kinetics is sufficiently high at high temperatures, which implies the use of these high temperatures. However, the temperature also fixes the equilibrium constant. For example, at high temperatures, the equilibrium concentration of an exothermic, reversible reaction is low. Important examples are methane production (the reversed reaction (5.8a)) and the technically more important synthesis of methanol from carbon dioxide and hydrogen [2]. Thus, for a high formation rate a low process temperature has to be used considering only thermodynamics, while the kinetic considerations require high temperatures. Figure 5.2 sketches the two contrary relations. For maximum yield, an optimum temperature can be found. The higher the reaction kinetics at low temperatures, the higher is the effective yield, that is the production rate of the sought products. These considerations are the basis of the engineering and design of chemical reactors. An acceleration of the reaction at low temperatures can be achieved by catalysts enhancing both the effective yield and reducing thermal losses and is therefore one of the hottest subjects in chemical engineering. In particular, catalysts for methanol formation from carbon dioxide and hydrogen are the prototypes for catalysts in chemical engineering (see also Chapter 4.3).

A second point is that selective catalysts can promote desirable reactions at the expense of undesirable ones (e.g. carbon formation according to Boudouard). Finally, while reaction rates increase with increasing temperature, in many cases the overall thermodynamic efficiency decreases with temperature (due to entropy production). An example worthy of mention is the methanol reformer for mobile applications,

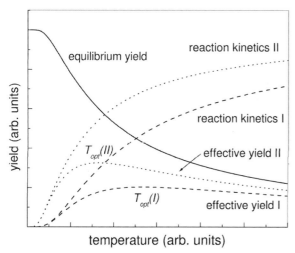

Fig. 5.2 Temperature dependence of the equilibrium yield and reaction kinetics of an exothermic, reversible reaction (e.g. methanol synthesis). The effective yield corresponding to the real production of the sought products is a compromise of both relations, giving an optimum temperature T_{opt}.

which could reach efficiencies near one if room temperature conditions could be used. However, a working catalyst has not yet been found [3].

In what follows is a description of economically important processes to produce hydrogen from hydrocarbons.

5.1.2
Hydrogen from Carbon and Water

Coal can be converted at temperatures of 800–1100 °C and pressures of 1–40 bar into syngas (CO + H_2) and methane (CH_4), also known as town gas, via coal gasification according to reactions (5.6) and (5.7) [4]. The heart of gasification-based systems is the gasifier. A gasifier converts carbon feedstock into gaseous components by applying heat under pressure in the presence of steam. A gasifier differs from a combustor in that the amount of air or oxygen available inside the gasifier is carefully controlled so that only a relatively small portion of the fuel burns completely. This "partial oxidation" process (Eqs. (5.1)–(5.4)) provides the heat. Rather than burning, most of the carbon-containing feedstock is chemically broken apart by the gasifier's heat and pressure, setting into motion the chemical reactions that produce syngas.

5.1.3
Hydrogen from Fossil Hydrocarbons and vice versa

Commercial bulk hydrogen is usually produced by the steam reforming of natural gas. At high temperatures (700–1100 °C), steam (H_2O) reacts with methane (CH_4) to yield syngas (reaction (5.8a)). The heat required to drive the process is generally supplied by burning some portion of the natural gas (reaction (5.4a)). Technical realizations have to consider also the side reactions, for example the Boudouard reaction (5.10). A typical example of such a gasifier is the Tubular Steam Reformer [5]. As suggested by reaction (5.8), in principle any hydrocarbon can be steam reformed into hydrogen and carbon monoxide. Because there are only small differences in the cost of hydrocarbons on the basis of heating values, the investment costs for the production of hydrogen from light hydrocarbons are lower and yields from light hydrocarbons are higher, hydrogen is preferably produced from light hydrocarbons, if available [5]. However, the idea and principle of the gasification process or related principles can be applied to biomass and to the disposal/recycling of organic waste and, to some extent, to all carbon-containing waste matter. Tubular steam reformers reforming methane work on a production scale of more than $10^6 \, m^3 H_2 \, h^{-1}$ [6]. If the co-produced CO_2 is sequestered, this technology is currently the most efficient and green hydrogen production (see also Ref. 7) and might, therefore, be the intermediate step towards a completely sustainable energy economy based on reversible energy vectors.

In principle, all reactions can be read in either direction, that is one can also produce hydrocarbons from hydrogen and carbon (monoxide). Historically, the so-called Lurgi-process was of great importance [4]. Nazi-Germany was cut off from any oil resources, which stimulated the invention of a process to synthesize synthetic fuel (i.e. liquid hydrocarbons) from town gas generated by coal gasification.

This idea has now again received considerable attention. The most promising process is the Fischer–Tropsch process using special catalysts to reduce the temperature and enhance the reaction kinetics. Typical catalysts used are based on iron and cobalt. The principal purpose of this process is to produce a synthetic petroleum substitute for use as synthetic lubrication oil or as synthetic fuel. The original Fischer–Tropsch process is described by Eq. (5.9). The resulting hydrocarbon products are refined to produce the desired synthetic fuel. There are several companies developing the Fischer–Tropsch process to enable practical exploitation of so-called stranded gas reserves. It is expected by geologists that supplies of natural gas will peak 5–15 years after oil does. A combination of biomass gasification (BG) and Fischer–Tropsch (FT) synthesis is a very promising route to produce renewable or "green" transportation fuels.

5.2
Electrolysis: Hydrogen Production Using Electricity
Ursula Wittstadt

Water, a huge "resource" on earth, can be split into its constituents hydrogen and oxygen by means of electrical energy. This process is called water electrolysis. The general concept, together with the most important thermodynamic relations, will first be given. Then two techniques of electrolysis – alkaline and solid polymer electrolysis – will be discussed. Finally, the use of electrolysis in renewable energy systems will be considered.

5.2.1
Water Electrolysis – General Concept

In all different technologies for water electrolysis the underlying general process is the same: water is supplied to an electrochemical cell where hydrogen evolves at the cathode and oxygen at the anode when supplied with a sufficiently high voltage level – above the so called zero-current cell potential E_0. Ions are transported through an electrolyte and a diaphragm ensures the separation of the two evolving gases. The principle of an electrolysis cell is shown in Figure 5.3.

The minimum energy that is required to split water is given by the Gibbs energy ΔG_R of the reaction taking place (see Eq. (5.13)).

$$H_2O \rightarrow H_2 + 0.5\,O_2 \tag{5.13}$$

At standard conditions (298.15 K; 101.3 kPa) ΔG_R has a value of -237.19 kJ mol^{-1} [8]. The zero-current cell potential E_0 can be derived as follows (see Eq. 5.14):

$$E_0 = \frac{\Delta G_R}{nF} \tag{5.14}$$

with n being the number of electrons exchanged per mole of water split and F the Faraday constant (96 485 C mol^{-1}). Thus, the standard zero-current cell potential $E_{0,\text{stand}}$ for the evolution of hydrogen and oxygen from water at standard conditions is 1.23 V. As the Gibbs energy is a function of both temperature and pressure, the

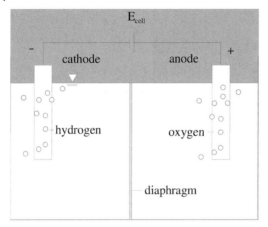

Fig. 5.3 Principle of an electrolysis cell.

zero-current cell potential is also dependent on these parameters. With increasing temperature E_0 decreases whereas a higher pressure causes an increase in cell potential.

The energy that is bound in one mole of water is given by its enthalpy of formation. It differs from the reaction Gibbs energy by the product of the thermodynamic temperature and the entropy of reaction. According to the second fundamental theorem of thermodynamics, a part of the enthalpy of reaction can be applied as thermal energy with a maximum of $\Delta Q_R = T \Delta S_R$ which is the amount of energy that corresponds to the entropy of reaction ΔS_R at the thermodynamic temperature T (see Eq. (5.15)).

$$\Delta H_R = \Delta G_R + T \Delta S_R \tag{5.15}$$

Thus, the total energy required for the reaction can be supplied by a combination of electricity and heat. The electrical power required can be reduced by operating at higher temperature. This is usually desirable, because heat is often available as a by-product and has a lower exergic value than electricity [9].

Up to now, reversible thermodynamics has been assumed. The electrical energy demand for water electrolysis under real conditions is significantly higher than the theoretical minimum energy derived above. The total voltage of an electrolysis cell under operation depends on the current in the cell, the voltage drop caused by Ohmic resistance and the anodic and cathodic overvoltages (see Eq. (5.16)).

$$E_{cell} = E_0 + iR + \left| E_{cath}^{ov} \right| + \left| E_{an}^{ov} \right| \tag{5.16}$$

E_{cath}^{ov} and E_{an}^{ov}, the anodic and cathodic overvoltages (also called oxygen and hydrogen overvoltages), represent the surplus of electrical energy that is necessary to "activate" the electrode reactions and to overcome the concentration gradients [10]. They are thus a measure of the reaction kinetics of each electrode.

iR, the ohmic voltage drop, is a function of the conductivity of the electrolyte and the electrodes, the distance between the electrodes, the conductivity of the diaphragm or the membrane and the contact resistances between the cell components.

The efficiency of an electrolysis process is often defined as the ratio of the energy content of the hydrogen produced per unit time and the electrical power needed for its production (see Eq. (5.17)). The lower heating value of hydrogen is most often used as a reference here. Commercial electrolysers reach efficiencies between 65 % and 75 % [11].

$$\varepsilon = \frac{\Delta H_R \cdot \dot{n}_{H_2}}{P_{el}} \tag{5.17}$$

Apart from this overall approach, the efficiency of electrolysis can be described in more detail. First there is the cell efficiency which compares the actual cell voltage to the minimum voltage possible (voltage) at operating conditions (see Eq. (5.18)).

$$\varepsilon_{cell} = \frac{E(T, p)}{E_0(T_0, p_0)} \tag{5.18}$$

The current efficiency or so called Faraday efficiency $\varepsilon_{Faraday}$ is the ratio of gas produced and the theoretical amount of hydrogen produced according to the current that flows through the cell. It takes into account parasitic currents in the cell and also includes the fact that the produced gases mix to a certain degree and therefore some hydrogen is lost as it diffuses from the cathode to the anode. Using the correlation between current and moles of gas produced, the Faraday efficiency for perfect gases can be written as

$$\varepsilon_{Faraday} = \frac{\dot{V}_{H_2, real}}{\dot{V}_{H_2, ideal}} \quad \text{with} \quad \dot{V}_{H_2, ideal} = \frac{I}{n \cdot F} \cdot v_{mol} \tag{5.19}$$

The Faraday efficiency typically reaches values of over 90 % [12]. To calculate the overall efficiency ε_{total} of an electrolyser, system losses due to the power supply of peripheral devices have to be included (see Eq. (5.18)).

$$\varepsilon_{total} = \varepsilon_{cell} \cdot \varepsilon_{Faraday} \cdot \varepsilon_{peripheral} \tag{5.20}$$

Having discussed the fundamentals of water electrolysis, the following sections will provide an overview of the technical realization of the electrolyser concept based on the electrolyte.

5.2.2
Alkaline Electrolysis

Alkaline electrolysis is a mature technique and is widely used. The electrolyte is an alkaline solution. Most electrolyser plants use a 20–40 % potassium hydroxide aqueous solution as an electrolyte. The gas sides are divided by a diaphragm which, until recently, used to be made from asbestos. Nowadays, materials such as polysulfone polymers or nickel oxide are used as substitutes for asbestos. Typical operating temperatures are between 80 and 100 °C. The electrode material is in most cases Raney-nickel [13].

Single cells are connected to form electrolyser units. For a simple installation unipolar concepts are used where cells are connected electrically in parallel and are open to the ambient. High currents have to be handled and operation at elevated pressure is not possible. Nowadays, only a few manufacturers offer this concept (see Table 5.1). Most industrial electrolysers have a bipolar connection of the cells where one electrode serves as anode and cathode at the same time. This arrangement of cells is called a stack, as single cells are stacked on each other. Thus, the plants can be built in a compact design. Currents are low but high voltages have to be handled. As the distance between electrodes governs the ohmic losses in an electrolyser stack, a so-called "zero gap" design has been developed where electrodes are placed directly onto the diaphragm [13].

Apart from the stack, an electrolyser plant consists of quite a number of peripheral units. The most important ones will be explained shortly. The quality of the feed water is guaranteed by deionization in order to prevent fouling in the system. It is necessary to maintain the concentration of the alkaline solution at a constant level through a process control system that adds as much water to the solution as is removed through water decomposition. In addition to this, the produced gases drain some electrolyte from the stack, a lye management unit controls the concentration of the alkaline electrolyte within the electrolyser system. As in most cases energy is available in the form of an AC power supply, a transformation station is necessary to supply direct current at the voltage level needed.

Table 5.1 Manufacturers of alkaline electrolyser systems [15]

Manufacturer	Hydrogen production capacity ($Nm^3\ h^{-1}$)	
GHW, Germany	12–500	
Hydrogen Systems, Belgium	1–100	
Idroenergy, Italy	0.4–64	
Norsk Hydro ASA, Norway	10–485	
Stuart Energy Systems, Canada	1–60	
The Electrolyser Cooperation, Canada	0.05–65.7	Unipolar cells
Wasserelektrolyse Hydrotechnik, Germany	1–250	

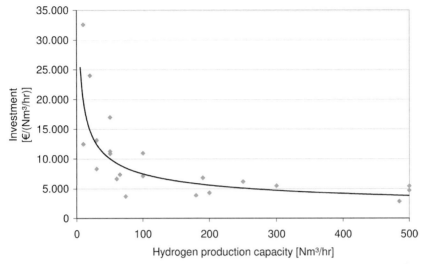

Fig. 5.4 Investment costs for alkaline electrolysers (calculated from requests for quotes from different suppliers in the year 2002).

Some of the manufacturers of alkaline electrolysers are listed in Table 5.1. Investment costs and efficiencies as a function of the hydrogen production capacity calculated from requests for quotes from various manufacturers are shown in Figures 5.4 and 5.5, respectively. Investment costs per production capacity are highly dependent on the size of the electrolyser unit, efficiencies are almost the same for all system sizes.

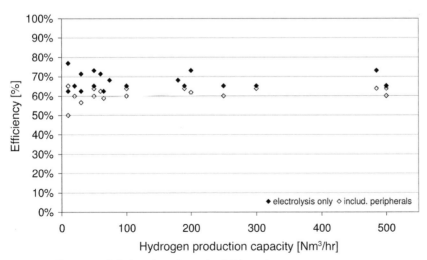

Fig. 5.5 Efficiencies of alkaline electrolysers for different plant sizes.

5.2.3
Solid-Polymer Electrolysis

Attempts have been made to develop solid electrolytes to replace liquid alkaline or acidic solutions. The advantages of the solid electrolytes are reduced corrosion, a constant electrolyte concentration (and therefore no need for concentration control) and the ability to use the electrolyte simultaneously as a diaphragm. Possible electrolyte materials for this are ion exchange membranes.

General Electric developed solid polymer electrolysis (SPE) in 1967. They used a special perfluorinated sulfuric acid polymer. The anode and cathode are applied directly on the membrane as a thin layer, and deionized water circulates the cell. Water decomposes at the anode into hydrogen ions (protons) and oxygen. The protons migrate as hydrated compounds through the membrane using fixed SO_3^--groups for transportation. Hydrogen is then formed at the cathode.

Other than for alkaline electrolysis, platinum (cathode) and iridium (anode) are used as catalysts. To reduce costs, carbon supported platinum is used to catalyze the hydrogen evolution reaction. Unfortunately, this is not possible on the anode side. The evolving oxygen at a potential of 1.7 V to 2 V would corrode the carbon material in a short time.

Compared to alkaline electrolysis, material costs are high for SPE electrolysers. The advantage of the SPE technology lies in a higher gas quality – especially in part load operation. Using a solid electrolyte also makes higher pressure possible as the absence of alkaline solution provides better sealing options. Some suppliers of SPE electrolysers are listed in Table 5.2.

Nowadays, SPE electrolyser systems cover the range of low production capacity. Fields of application are mainly on-site hydrogen supply and aerospace applications. The efficiency of commercially available SPE electrolysers is below that of the advanced alkaline technique. However, efficiencies of 85 to 93% have been reported and thus the potential for further improvement is expected to be high [14].

Table 5.2 Manufacturers of SPE electrolyser systems [15]

Manufacturer	Hydrogen production capacity ($Nm^3\ h^{-1}$)
Giner, Inc., USA	4–12.8
H2-interpower GmbH, Germany	0.02–0.04
Mitsubishi Corp., Japan	—
Proton Energy, USA	0.5–1
Shinko Pantec Kobe, Japan	0.5–2
Space Systems International, Inc., USA	50–100

5.2.4
Electrolyser Plants Using Renewable Energy – From Small to Large Scale

Only a small amount (3 % worldwide, [15]) of the world's hydrogen production is produced via electrolysis. As an energy source of high exergetic value (electricity) is necessary, electrolysis is used where electricity prices are low or where very pure gases are needed. Some hydrogen also comes from plants where it is produced as a by-product (e.g. in chloralkaline electrolysis). The produced hydrogen is mainly used as a chemical raw material and only very rarely as an energy carrier. The idea of an energy supply by means of renewable energy like wind and solar power offers a new field for hydrogen production from electrolysis. In that way, hydrogen can be used as energy storage where intermittent power supply from renewables has to be collected.

In the following sections two fields of application will be considered in more detail: small electrolyser systems for autonomous power supply and large units for large-scale renewable power plants in off-grid areas.

Autonomous power supply for off-grid applications can be, for example, for housing, huts, telecommunication stations or railway crossings. The power that has to be supplied is usually quite low (100 W to a few 10 kW [16]) but has to be available at any time. Using intermittent renewable energy sources means that it is necessary to install some means of storing energy. Batteries are an appropriate solution for short term storage (hours to some days), however, seasonal storage might be useful in order to avoid oversizing of the energy supply system. As an example, this is the case in areas where the solar radiation in summer is much higher than during winter.

The size of an electrolyser for such a system has to correspond to the size of the power supply in order to guarantee that all surplus energy can be stored. A major requirement is intermittent operation at changing loads. For small systems SPE electrolysers are favored. Compared with alkaline electrolysers they have no notable change in gas quality in intermittent operation and lye management is not necessary. Only deionized water has to be supplied, which can be easily done using ion exchange cartridges. Low maintenance is another important prerequisite as the systems are often not easy to access. Dynamic behavior has to be addressed thoroughly: most electrolysers work at a temperature between 333.15 and 353.15 K. The time of start-up has to be below the time constants of the energy supply. Direct coupling with a DC power supply, for example from solar panels, is possible, but leads to a highly coupled system behavior. A DC/DC converter provides the possibility to operate the electrolyser at an optimum voltage level and thus leads to a better overall system design. With wind power, an AC/DC converter is necessary, as operation with direct current is not possible and an additional component has to be provided anyway.

Apart from this niche market, hydrogen production could play an important role if the potential of renewable energy in regions far away from the user is exploited. It is important to mention, that the direct transport and use of electricity is, in many

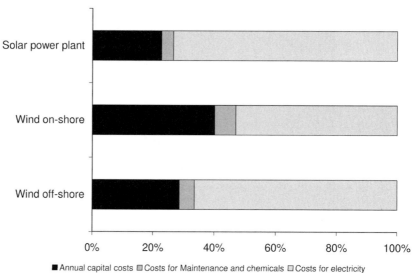

Fig. 5.6 Annual costs for hydrogen from renewable energies.

cases, more efficient than its conversion into hydrogen. Therefore, a thorough comparison of the two possibilities should be carried out. The power range of these applications is far above those for autonomous power supplies. A typical size for a solar power plant in regions with high solar radiation (e.g. Mexico) is between 50 and 100 MW [17]. Wind power plants range from 100 MW for onshore up to 250–500 MW for offshore applications [18, 19]. Up to now, electrolyser plants have not been built in this power range. A study for the design of an electolyser plant with a power of 300 MW recommends a plant with four electrolyser groups, each consisting of 24 single stacks [20].

As prices are of high importance in the energy market, it is interesting to consider what determines the cost for hydrogen produced from the power plants considered above. In general about a third of the annual costs are capital costs and costs for maintenance, two thirds are costs for electricity. Figure 5.6 shows the percentage for the three different power sources: solar energy, onshore and offshore wind energy. Prices for electricity that are used for the calculations are given in Table 5.3. Higher investment for an electrolyser plant that is installed offshore results in a higher

Table 5.3 Price for electricity from renewable energy

	Price (€ Ct/kWh)	Source
Wind power (onshore)	4.1	[19]
Wind power (offshore)	8.0	[18]
Solar power plant	12.5	[17]

percentage of capital costs (40%). The influence of the price of electricity can be clearly seen for hydrogen production from a solar power plant: The percentage of cost for electricity rises to nearly 75%.

5.2.5
Summary and Outlook

Hydrogen production by electrolysis is a mature technique which has been used for over a hundred years. Most commercial electrolysers are based on the principle of alkaline electrolysis, but solid polymers are also used as an electrolyte. Electrolyser efficiencies range between 65 and 75%. Units with a hydrogen production of liters to several 10.000 cubic meters per hour are available.

Only a very small percentage (3%) of the world's hydrogen is produced by electrolysis [15]. Even if hydrogen is used as a chemical raw material rather than as an energy carrier today, this might change in the future. If hydrogen is considered to play a major role among future energy carriers, electrolysis using energy from renewable sources might become a part of tomorrow's energy supply chain.

References

1 Atkins, P.W. (1994) *Physical Chemistry*, Oxford University Press, Oxford.
2 Fitzer, E., Fritz, W., Emig, G., Gerrenz, H. (1995) *Technische Chemie*, Springer Verlag, Berlin.
3 Steele, B.C.H., Heinzel, A. (2001) *Nature* **414**, 345–52.
4 Hiller, H., Reimert, R., "Gas Production", in *Ullmann's Encyclopedia of Industrial Chemistry*, 5th Completely Revised Edn., vol. **A12**, pp. 178–82.
5 Häussinger, P., Lohmüller, R., Watsin, A.M. (1995) "Hydrogen", in *Ullmann's Encyclopedia of Industrial Chemistry*, 5th Completely Revised Edn., Wiley VCH, vol. **A13**, pp. 333–40.
6 Rostrup Nielsen, J. (2002) *Encyclopedia of Catalysis*, Wiley and Sons, Chichester, New York.
7 Rostrup-Nielsen, J.R., Rostrup-Nielsen, T. (2002) *Large-scale Hydrogen Production*, Wiley and Sons, Chichester, New York.
8 Mortimer, C.E. (1986) *Chemistry*, 4th edn, Belmont.
9 Kroschwitz, J.I. (ex. ed.) and Howe-Grand, M. (ed.) (1995) *Kirk-Othmer: Encyclopedia of Chemical Technology*, vol. **13**, 4th edn, Wiley VCH, New York.
10 Wendt, H. (1990) *Electrochemical Hydrogen Technologies*, Elsevier, Amsterdam, pp. 1–14.
11 Wurster, R., Schindler, J. (2003) "Solar and Wind Energy Coupled with Electrolysis and Fuel Cells," in *Handbooks of Fuel Cells – Fundamentals, Technology and Applications* (eds W. Vielstich, H.A. Gasteiger, A. Lamm), John Wiley & Sons Ltd, Chichester, New York.
12 Hug, W., Divisek, J., Mergel, J., Seeger, W., Steeb, H. (1992) *Int. J. Hydrogen Energy*, **17** (9), 699–705.
13 Divisek. J. et al. (1985) *Neue Technologie zur Wasserstoffproduktion durch fortgeschrittene alkalische Wasserelektrolyse. DECHEMA Monographien Band 98 – Technische Elektrolysen*; Weinheim, Deerfeld Beach: Verlag Chemie 1985, p. 389
14 Crockett, R.G.M., Newborough, M., Highgate, D.J. (1997) *Solar Energy* **61** (5), 293–302.
15 Suresh, B. et al. (2001) "CEH Product Review: HYDROGEN," in

Chemical Economics Handbook, SRI Consulting.
16 Vakaraki, E. et al. (2003) *J. Power Sources*, **118**, 14–22.
17 Trieb, F. (ed.) (2002) *Concentrating Solar Power Now–Clean Energy for Sustainable Development*. Federal Ministry for the Environment, Nature Conservation and Nuclear Safety (BMU).
18 European Wind Energy Association/ Greenpeace (2001) Wind Force 12: A blueprint to achieve 12 % of the world's electricity from wind power by 2020.
19 Teske, S. (ed.) (2000) *North Sea Offshore Wind – A Powerhouse for Europe*, Study, Greenpeace e.V., Germany.
20 Clouman, A., d'Erasmo, P., Halvorsen, B.G., Stevens, P. (May, 1993) Analysis and Optimization of Equipment Cost to Minimize Operation and Investment for a 300 MW Electrolysis Plant.

6
Hydrogen Storage
Andreas Züttel, Michael Hirscher, Barbara Panella, Klaus Yvon, Shin-ichi Orimo, Borislav Bogdanović, Michael Felderhoff, Ferdi Schüth, Andreas Borgschulte, Sandra Goetze, Seijirau Suda, and Michael T. Kelly

6.1
Hydrogen Storage in Molecular Form
Andreas Züttel

Hydrogen can be transported in pipelines similarly to natural gas. There are networks for hydrogen already operating today, a 1500 km network in Europe and a 720 km network in the USA. The oldest hydrogen pipe network is in the Ruhr area in Germany and has operated for more than 50 years. The tubes, with a typical diameter of 25–30 cm, are built using conventional pipe steel and operate at a pressure of 10 to 20 bar. The volumetric energy density of hydrogen gas is 36% of the volumetric energy density of natural gas at the same pressure. In order to transport the same amount of energy the hydrogen flux has to be 2.8 times larger than the flux of natural gas. However, the viscosity of hydrogen (8.92×10^{-6} Pa s) is significantly smaller than that of natural gas (11.2×10^{-6} Pa s). The minimum power P required to pump a gas through a pipe is given by

$$P = 8 \cdot \pi \cdot l \cdot v^2 \cdot \eta \tag{6.1}$$

where l is the length of the pipe, v the velocity and η the dynamic viscosity of the gas. The transmission power per energy unit is therefore 2.2 times larger for hydrogen than for natural gas. The total energy loss during the transportation of hydrogen is about 4% of the energy content. Because of the great length, and therefore the great volume, of piping systems, a slight change in the operating pressure of a pipeline system results in a large change in the amount of hydrogen gas contained within the piping network. Therefore, the pipeline can be used to handle fluctuations in supply and demand, avoiding the cost of onsite storage.

Hydrogen as a Future Energy Carrier. Edited by A. Züttel, A. Borgschulte, and L. Schlapbach
Copyright © 2008 WILEY-VCH Verlag GmbH & Co. KGaA, Weinheim
ISBN: 978-3-527-30817-0

6.1.1
Hydrogen Storage

The ordinary isotope of hydrogen, H, is known as protium and has an atomic weight of 1 (1 proton and 1 electron). In 1932, the preparation of a stable isotope was announced, deuterium (D) with an atomic weight of 2 (1 proton and 1 neutron plus 1 electron). Two years later an unstable isotope, tritium (T), with an atomic weight of 3 (1 proton and 2 neutrons plus 1 electron) was discovered. Tritium has a half-life of about 12.5 years [1]. One atom of deuterium is found in about 6000 ordinary hydrogen atoms. Tritium atoms are also present but as a much smaller proportion. All the isotopes of hydrogen react together and form, due to the single electron in the atom, covalent molecules like H_2, D_2 and T_2, respectively. Hydrogen has a very ambivalent behavior towards other elements, it occurs as an anion (H^-) or cation (H^+) in ionic compounds, it participates with its electron in the formation of covalent bonds, for example with carbon and it can even behave like a metal and form alloys at ambient temperature.

The hydrogen molecule H_2 can be found in various forms, depending on the temperature and the pressure, which are shown in the phase diagram (Fig. 6.1). At low temperature hydrogen is a solid with a density of 70.6 kg m^{-3} at −262 °C and is a gas at higher temperatures with a density of 0.089886 kg m^{-3} at 0 °C and a pressure of 1 bar. A small zone starting at the triple point and ending at the critical point exhibits liquid hydrogen with a density of 70.8 kg m^{-3} at −253 °C. At

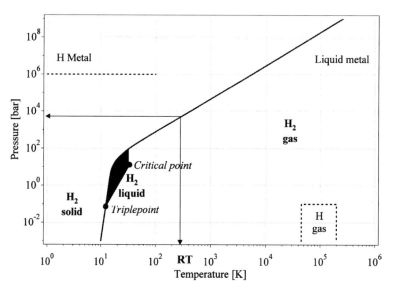

Fig. 6.1 Primitive phase diagram for hydrogen. Liquid hydrogen only exists between the solidus line and the line from the triple point at 21.2 K and the critical point at 32 K (Leung, March, and Motz (1976), Ref. [2]).

ambient temperature (298.15 K) hydrogen is a gas and can be described by the Van der Waals equation:

$$p(V) = \frac{n \cdot R \cdot T}{V - n \cdot b} - a \cdot \frac{n^2}{V^2} \qquad (6.2)$$

where p is the gas pressure, V the volume, T the absolute temperature, n the number of mols, R the gas constant ($R = 8.314$ J K^{-1} mol^{-1}), a is the dipole interaction or repulsion constant ($a = 2.476 \times 10^{-2}$ m^6 Pa mol^{-2}) and b is the volume occupied by the hydrogen molecules ($b = 2.661 \times 10^{-5}$ m^3 mol^{-1}) [1]. The strong repulsion interaction between hydrogen molecules is responsible for the low critical temperature ($T_c = 33$ K) of hydrogen gas.

Hydrogen storage basically implies the reduction of the enormous volume of the hydrogen gas. One kilogram of hydrogen at ambient temperature and atmospheric pressure takes a volume of 11 m^3. In order to increase the hydrogen density in a storage system work must either be applied to compress the hydrogen, or the temperature has to be decreased below the critical temperature or, finally, the repulsion has to be reduced by the interaction of hydrogen with another material. The second important criterion of a hydrogen storage system is the reversibility of the hydrogen uptake and release. This criterion excludes all covalent hydrogen carbon compounds as hydrogen storage materials because the hydrogen is only released from carbon hydrogen compounds if they are heated to temperatures above 800 °C or if the carbon is oxidized. There are basically six methods of reversibly storing hydrogen with a high volumetric and gravimetric density (Fig. 6.2). The following chapters focus on these methods and illustrate their advantages and disadvantages.

6.1.1.1 High Pressure Gas Cylinders

The most common storage systems are high pressure gas cylinders with a maximum pressure of 20 MPa. New lightweight composite cylinders have been developed which are able to withstand pressure up to 80 MPa and so the hydrogen can reach a volumetric density of 36 kg m^{-3}, approximately half that in its liquid form at the normal boiling point. The gravimetric hydrogen density decreases with increasing pressure due to the increasing thickness of the walls of the pressure cylinder. The wall thickness of a cylinder capped with two hemispheres is given by the following equation:

$$\frac{d_w}{d_o} = \frac{\Delta p}{2 \cdot \sigma_v + \Delta p} \qquad (6.3)$$

where d_w is the wall thickness, d_o the outer diameter of the cylinder, Δp the overpressure and σ_v the tensile strength of the material. The tensile strength of materials varies from 50 MPa for aluminum to more than 1100 MPa for high quality steel [3]. Future, developments of new composite materials have the potential to

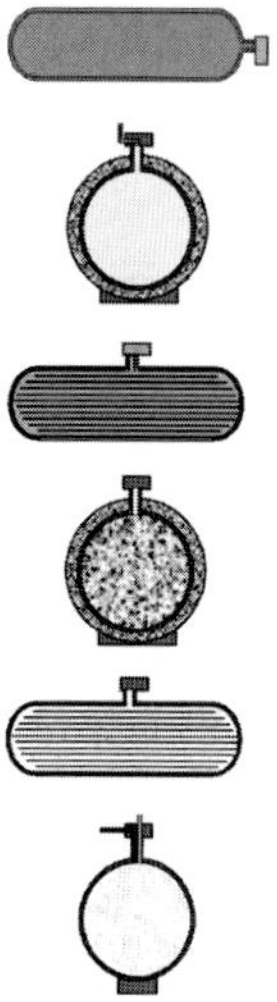

Storage Media	Volume	Mass	Pressure	Temperature	
	max. 33 kg $H_2 \cdot m^{-3}$	13 mass%	800 bar	298 K	Composite cylinder established
	71 kg $H_2 \cdot m^{-3}$	100 mass%	1 bar	21 K	Liquid hydrogen
	max. 150 kg $H_2 \cdot m^{-3}$	2 mass%	1 bar	298 K	Metal hydrides
	20 kg $H_2 \cdot m^{-3}$	4 mass%	70 bar	65 K	Physisorption
	150 kg $H_2 \cdot m^{-3}$	18 mass%	1 bar	298 K	Complex hydrides reversibility ?
	>100 kg $H_2 \cdot m^{-3}$	14 mass%	1 bar	298 K	Alkali + H_2O

Fig. 6.2 The six basic hydrogen storage methods and phenomena. The gravimetric density ρ_m, the volumetric density ρ_v, the working temperature T and pressure p are listed. RT stands for room temperature (25 °C). From top to bottom: compressed gas (molecular H_2) in a lightweight composite cylinder (tensile strength of the material is 2000 MPa); liquid hydrogen (molecular H_2), continuous loss of a few percent per day of hydrogen at RT; physisorption (molecular H_2) on materials, for example carbon with a very large specific surface area, fully reversible; hydrogen (atomic H) intercalation in host metals, metallic hydrides working at RT are fully reversible; complex compounds ($[AlH_4]^-$ or $[BH_4]^-$), desorption at elevated temperature, adsorption at high pressures; chemical oxidation of metals with water and liberation of hydrogen, not directly reversible.

increase the tensile strength above that of steel with a material density less than half that of steel [3].

Most pressure cylinders today use austenitic stainless steel (e.g. AISI 316 and 304 and AISI 316L and 304L above 300 °C to avoid carbon grain-boundary segregation), copper, or aluminum alloys which are largely immune to hydrogen effects at ambient temperatures. Many other materials are subject to embrittlement and should not be used, for example alloy or high strength steels (ferritic, martensitic and bainitic), titanium and its alloys and some nickel-based alloys [3].

Figure 6.3 shows the volumetric density of hydrogen inside the cylinder and the ratio of the wall thickness to the outer diameter of the pressure cylinder for stainless steel with a tensile strength of 460 MPa. The volumetric density of hydrogen increases with pressure and reaches a maximum above 1000 bar, depending on the tensile strength of the material. However, the gravimetric density decreases with increasing pressure and the maximum gravimetric density is found for zero overpressure! Therefore, the increase in volumetric storage density is sacrificed with the reduction of the gravimetric density in pressurized gas systems (see Fig. 6.4).

6.1 Hydrogen Storage in Molecular Form | 169

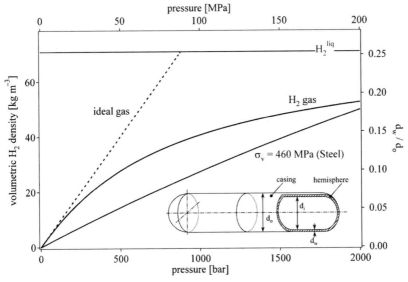

Fig. 6.3 Volumetric density of compressed hydrogen gas as a function of gas pressure including the ideal gas and liquid hydrogen. The ratio of the wall thickness to the outer diameter of the pressure cylinder is shown on the right-hand side for steel with a tensile strength of 460 MPa. A schematic drawing of the pressure cylinder is shown as an inset.

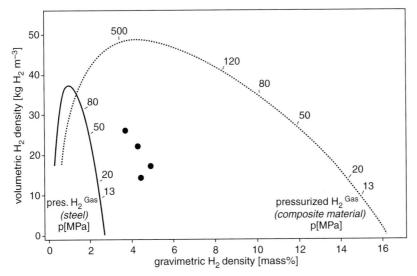

Fig. 6.4 Volumetric and gravimetric hydrogen storage density for pressurized gas. Steel (tensile strength $\sigma_V = 460$ MPa, density 6500 kg m^{-3}) and a hypothetical composite material ($\sigma_V = 1500$ MPa, density 3000 kg m^{-3}). Markers stand for pressure cylinders from Dynetek.

The safety of pressurized cylinders is an issue of concern, especially in highly populated regions. Future pressure vessels are envisaged to consist of three layers: an inner polymer liner, overwrapped with a carbon-fiber composite (which is the stress-bearing component) and an outer layer of an aramid-material capable of withstanding mechanical and corrosion damage. The target that the industry has set for itself is a 70 MPa cylinder with a mass of 110 kg resulting in a gravimetric storage density of 6 mass% and a volumetric storage density of 30 kg m^{-3}.

Hydrogen can be compressed using standard piston-type mechanical compressors. Slight modifications of the seals are sometimes necessary in order to compensate for the higher diffusivity of hydrogen. The theoretical work necessary for the isothermal compression of hydrogen is given by the following equation:

$$\Delta G = R \cdot T \cdot \ln\left(\frac{p}{p_0}\right) \tag{6.4}$$

where R stands for the gas constant ($R = 8.314$ J mol^{-1} K^{-1}), T for the absolute temperature and p and p_0 for the end pressure and the starting pressure, respectively. The error of the work calculated with Eq. (6.3) in the pressure range 0.1 to 100 MPa is less than 6%. The isothermal compression of hydrogen from 0.1 to 80 MPa therefore consumes 2.21 kWh kg^{-1}. In a real process the work consumption for compression is significantly higher because the compression is not isothermal. Metal hydrides can be used to compress hydrogen from a heat source only. Compression ratios of greater than 20:1 are possible with final pressures of more than 100 MPa [4].

The relatively low hydrogen density together with the very high gas pressures in the systems are important drawbacks of the technically simple and, on the laboratory scale, well established high pressure storage method.

6.1.1.2 Liquid Hydrogen

Liquid hydrogen is stored in cryogenic tanks at 21.2 K at ambient pressure. Due to the low critical temperature of hydrogen (33 K) liquid hydrogen can only be stored in open systems, because there is no liquid phase existent above the critical temperature. The pressure in a closed storage system at room temperature could increase to about 104 bar. The volumetric density of liquid hydrogen is 70.8 kg m^{-3}, slightly higher than that of solid hydrogen (70.6 kg m^{-3}). The challenges of liquid hydrogen storage are the energy efficient liquefaction process and the thermal insulation of the cryogenic storage vessel in order to reduce the boil-off of hydrogen. The hydrogen molecule is composed of two protons and two electrons. The combination of the two electron spins only leads to a binding state if the electron spins are antiparallel. The wavefunction of the molecule has to be antisymmetric in view of the exchange of the space coordinates of two fermions (spin = $^1/_2$). Therefore, two groups of hydrogen molecules exist according to the total nuclear spin ($I = 0$, antiparallel nuclear spin and $I = 1$, parallel nuclear spin). The first group with $I = 0$ is called para-hydrogen and the second group with $I = 1$ is called ortho-hydrogen. Normal hydrogen at room temperature contains 25 % of the para-form and 75 %

of the ortho-form. The ortho-form cannot be prepared in the pure state. Since the two forms differ in energy, the physical properties also differ. The melting and boiling points of para-hydrogen are about 0.1 K lower than those of normal hydrogen. At 0 K, all the molecules must be in a rotational ground state, that is in the para-form.

Liquefaction Process When hydrogen is cooled from room temperature (RT) to the normal boiling point (nbp = 21.2 K) the ortho-hydrogen converts from an equilibrium concentration of 75 % at RT to 50 % at 77 K and 0.2 % at nbp. The self-conversion rate is an activated process and very slow, the half-life time of the conversion is greater than one year at 77 K. The conversion reaction from ortho- to para-hydrogen is exothermic and the heat of conversion is also temperature dependent. At 300 K the heat of conversion is 270 kJ kg^{-1} and increases as the temperature decreases, where it reaches 519 kJ kg^{-1} at 77 K. At temperatures lower than 77 K the enthalpy of conversion is 523 kJ kg^{-1} and almost constant. The enthalpy of conversion is greater than the latent heat of vaporization ($H_V = 451.9$ kJ kg^{-1}) of normal and para-hydrogen at the nbp. If the unconverted normal hydrogen is placed in a storage vessel, the enthalpy of conversion will be released in the vessel, which leads to the evaporation of the liquid hydrogen. The transformation from ortho- to para-hydrogen can be catalyzed by a number of surface active and paramagnetic species, for example normal hydrogen can be adsorbed on charcoal, cooled with liquid hydrogen and desorbed in the equilibrium mixture. The conversion may take only a few minutes if a highly active form of charcoal is used. Other suitable ortho–para catalysts are metals such as tungsten, nickel, or any paramagnetic oxides such as chromium or gadolinium oxides. The nuclear spin is reversed without breaking the H—H bond (Fig. 6.5).

The simplest liquefaction cycle is the Joule–Thompson cycle (Linde cycle). The gas is first compressed, and then cooled in a heat exchanger, before it passes through a throttle valve where it undergoes an isenthalpic Joule–Thomson expansion, producing some liquid. The cooled gas is separated from the liquid and returned to the compressor via the heat exchanger [5]. The Joule–Thompson cycle works for gases, such as nitrogen, with an inversion temperature above room temperature. Hydrogen, however, warms upon expansion at room temperature. In order for hydrogen to cool upon expansion its temperature must be below its inversion temperature of 202 K. Therefore, hydrogen is usually precooled using liquid nitrogen (78 K) before the first expansion step occurs. The free enthalpy change between gaseous hydrogen at 300 K and liquid hydrogen at 20 K is 11 640 kJ kg^{-1}. The necessary theoretical energy (work) to liquefy hydrogen from RT is $W_{th} = 3.23$ kWh kg^{-1}, the technical work is about 15.2 kWh kg^{-1} almost 40 % of the higher heating value of the hydrogen combustion [6].

Storage Vessel The boil-off rate of hydrogen from a liquid hydrogen storage vessel due to heat leaks is a function of the size, shape and thermal insulation of the vessel. Theoretically, the best shape is a sphere, since it has the least surface to volume ratio and because stress and strain are distributed uniformly. However, large

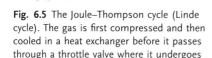

Fig. 6.5 The Joule–Thompson cycle (Linde cycle). The gas is first compressed and then cooled in a heat exchanger before it passes through a throttle valve where it undergoes an isenthalpic Joule–Thomson expansion, producing some liquid. The cooled gas is separated from the liquid and returned to the compressor via the heat exchanger.

size, spherical containers are expensive because of their manufacturing difficulty. Since boil-off losses due to heat leaks are proportional to the surface to volume ratio, the evaporation rate diminishes drastically as the storage tank size increases. For double-walled vacuum-insulated spherical dewars, boil-off losses are typically 0.4% per day for tanks which have a storage volume of $50 \, m^3$, 0.2% for $100 \, m^3$ tanks, and 0.06% for $20\,000 \, m^3$ tanks. Low temperature para-hydrogen requires the use of materials, which retain good ductility at low temperatures. Austenitic stainless steel (e.g. AISI 316L and 304L) or aluminum and aluminum alloys (Serie 5000) are recommended. Polytetrafluoroethylene (PTFE, Teflon) and 2-chloro-1,1,2-trifluoroethylene (Kel-F) can also be used.

6.1.1.3 Comparison of Pressure Storage and Liquid Storage

The gravimetric and volumetric hydrogen density depend strongly on the size of the storage vessel since the surface to volume ratio decreases with increasing size. Therefore, only the upper limit is defined. The large amount of energy necessary for the liquefaction, that is 40% of the upper heating value, makes liquid hydrogen not an energy efficient storage medium. Furthermore, the continuous boil-off of hydrogen limits the possible applications for liquid hydrogen storage systems to cases where the hydrogen is consumed in a rather short time, for example air and space applications Fig. 6.6.

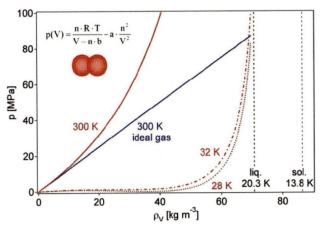

Fig. 6.6 Hydrogen density for compressed hydrogen, liquid and solid hydrogen.

6.2
Hydrogen Adsorption (Carbon, Zeolites, Nanocubes)
Michael Hirscher and Barbara Panella

6.2.1
Adsorption Phenomena

6.2.1.1 Physisorption

The physical adsorption of a gas on a surface is caused by weak Van der Waals forces between the adsorbate and the adsorbent. The consequent gas–solid interaction is composed of an attractive term which decreases with the −6 power of the distance between gas and substrate and a repulsive term, which decreases with the −12 power of the distance. The attractive interaction originates from long-range forces produced by fluctuations in the charge distribution of the gas molecules and of the atoms on the surface giving rise to dipole–image-dipole attraction. However at small distances the overlap between the electron cloud of the gas molecule and of the substrate is significant and the repulsion increases rapidly. The combination of these two terms gives rise to a shallow minimum in the potential energy curve at a distance of approximately one molecular radius of the gas molecule. This energy minimum corresponds typically to 1 to 10 kJ mol^{-1}. The small enthalpy change, which arises during the adsorption process, is not sufficient to cause bond breaking and the gas is adsorbed in its molecular form. As the forces involved in the interaction between adsorbate and adsorbent are very weak, physisorption usually takes place only at low temperature. Typically, there is no energy barrier to prevent the molecule approaching the surface from entering the physisorption well. Therefore the process is nonactivated and fast kinetics is characteristic for physical adsorption.

Because of the limited pore dimensions of microporous adsorbents the tendency of the gas molecules is to form a simple monolayer on the surface of the solid even

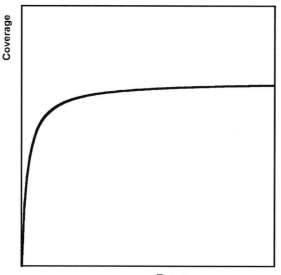

Fig. 6.7 Langmuir type isotherm.

at temperatures close to the liquefaction temperature of the gas. This behavior can be qualitatively described by a Langmuir isotherm, also called type I isotherm in the International Union of Applied Chemisty (IUPAC) classification (Fig. 6.7) and can be described by Eq. (6.5) [7]

$$\theta = \frac{KP}{1 + KP} \tag{6.5}$$

where K is the equilibrium constant, P the gas pressure and θ the coverage, expressed as the ratio of occupied adsorption sites to the total number of sites. For low pressure, the isotherm reduces to Henry's law, where the coverage is proportional to the pressure.

The saturation of all adsorption sites on the solid surface ($\theta = 1$) is characterized by a plateau in the isotherm. The Langmuir adsorption isotherm is based on the following assumptions: the surface is uniform and every adsorption site is equivalent to the others, the substrate surface is saturated when all adsorption sites are occupied and monolayer formation has occurred, there are no interactions between the adsorbed particles. In general, physisorption isotherms show various shapes. These are, according to the IUPAC classification, types II and III which describe adsorption on nonporous or macroporous adsorbent with strong and weak gas–solid interaction, respectively. Type IV and V are adsorption isotherms which show typically capillary condensation with hysteresis loops and type VI isotherm shows stepwise multilayer adsorption.

Hydrogen storage by physisorption is typically applied between 77K and room temperature. Since the temperature is considerably higher than the critical temperature of H_2 no multilayer adsorption takes place and type I or Henry-type isotherms are obtained for porous materials. The highest volumetric hydrogen density which can be physisorbed on a surface is limited by the density of liquid H_2. In fact, from the microscopic point of view, the minimum distance between hydrogen molecules in the adsorbed monolayer is determined by the intermolecular distance in the liquid phase. Liquids are nearly incompressible therefore higher densities imply chemical reaction with hydrogen and formation of new phases.

If the adsorption of hydrogen on a substrate has physical character, the interaction is nonspecific and the amount of hydrogen that can be stored is mainly dependent on the specific surface area of the adsorbent and on the operating conditions, like hydrogen pressure and temperature. Therefore materials with very high specific surface area seem to be very promising for hydrogen storage, at least in systems in which cooling is not a problem.

Through molecular simulations serious progress has been made in the last years in understanding adsorption phenomena on microporous surfaces.

The physical adsorption isotherms on carbon materials have been studied theoretically using Grand Canonical Monte Carlo simulations and an effective classical potential [8], or using Feynmann path formalism in conjunction with the Monte Carlo method to take into account the quantum effects [9]. To simulate hydrogen adsorption accurately at low temperature, these quantum effects have to be included. In this last case hydrogen is considered as a quantum fluid. The basic idea of Feynman path integral formalism is to look at the possible paths that a particle can take to move from one point to another.

In the case of Grand Canonical Monte Carlo simulation the adsorption isotherms are determined considering that the interaction between hydrogen molecules and carbon material is simply physical and can be described through a classical empirical potential. No chemical interaction is taken into account in these type of calculations. The pair wise interaction energy between two particles (hydrogen–hydrogen and hydrogen–carbon) is described through the Lennard–Jones potential (6.6)

$$U(s) = 4\varepsilon \left[\left(\frac{\sigma}{s}\right)^{12} - \left(\frac{\sigma}{s}\right)^{6} \right] \tag{6.6}$$

where s is the distance and σ and ϵ are the potential parameters, which are different for the hydrogen–hydrogen and the hydrogen–carbon interaction.

6.2.1.2 Chemisorption

In chemisorption the gas particles interact with the surface atoms of the adsorbent forming a chemical bond, typically of covalent character. The enthalpy involved in chemical adsorption is greater than for physisorption and of the order of magnitude of $100\,\text{kJ}\,\text{mol}^{-1}$. Sometimes the value of the energy of adsorption is used to discriminate between chemical and physical interaction with the adsorbate, but often this criterion is not absolute and most commonly spectroscopic techniques are necessary to identify the species adsorbed on the surface and the type of bonding which occurs. Without spectroscopic techniques it is very difficult to distinguish between strong physisorption and chemisorption.

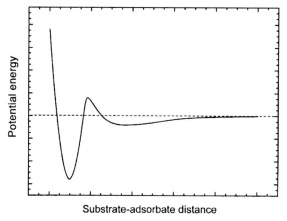

Fig. 6.8 Potential energy curve of the adsorbate as a function of the distance from the adsorbent.

The potential curve of an adsorbate approaching the solid surface typically shows two energy minima with different depths; one at greater distance from the adsorbent corresponds to physical adsorption, which can be the precursory state of chemisorption, the other occurs at smaller distances and has a deeper minimum, corresponding to the real chemical interaction (Fig. 6.8).

As the first step towards chemisorption can be physical adsorption, the amount of atoms chemisorbed can be determined by the amount of molecules physisorbed.

Chemisorption is typically an activated process, that means that an activation energy is required to break the chemical bonds in the gas molecules, when the adsorption is dissociative, or to rearrange the existing bonds when the gas particles are approaching at small distances to the surface. Therefore this type of adsorption can need higher temperatures and has a slower kinetics than physisorption. The energy can be supplied, for example by high temperature or through activation of the gas molecules.

Because of the existing dangling bonds on the surface of the solid, ideally every atom can act as an adsorption site for the gas particles and saturation is obtained with a monolayer formation when all accessible sites are occupied. This process is very well described through the Langmuir isotherm and, in the case of dissociative chemisorption, the equilibrium value of the surface coverage is proportional to the square root of pressure. Equation (6.5) is therefore modified to give Eq. (6.7)

$$\theta = \frac{(KP)^{1/2}}{1 + (KP)^{1/2}} \tag{6.7}$$

The release of chemically bonded hydrogen requires a high energy supply in order to break the covalent bonds between the atoms on the surface and the gas atoms. Even this process is activated, that means that fast desorption can occur only at high temperature.

Many theoretical calculations have been performed in order to explain and predict hydrogen adsorption on carbon nanomaterials, both considering chemisorption of hydrogen atoms on the substrate.

Monte Carlo simulations using classical potentials are not adequate to describe chemical processes like bond formation and bond breaking which occur in chemisorption.

In order to study the formation of chemical bonds between carbon and hydrogen atoms many authors have applied *ab initio* calculations [10].

6.2.2
Measuring Techniques

6.2.2.1 Volumetry or Sievert's Method

The volumetric technique is, because of the apparent simplicity of the set-up, one of the most used methods for measuring hydrogen storage capacity. In this case the pressure change due to adsorption or desorption from the sample is measured in a constant known volume. The experimental set-up consists of an adsorption cell containing the sample and a reservoir or expansion cell, both of calibrated volume. From the pressure difference, the volume of the system, the volume of the sample and the temperature, it is possible to calculate the quantity of hydrogen adsorbed by the sample, applying, at low pressures, the ideal gas equation and, at high pressures, the real gas equation. This technique seems to be very simple, in reality there are many difficulties in measuring the hydrogen uptake of small samples. One major problem is the evaluation of the volume of samples with very high specific surface area. Usually this volume is evaluated by expanding He gas in the sample holder containing the nanostructured material. Malbrunot *et al.* show that the helium density measurements can be in some cases erroneous due to a non-negligible effect of He adsorption [11].

Another problem arises from temperature changes due to gas expansion or due to external fluctuations. If only a small quantity of the sample is available, as usual for nanostructured materials, the error in the evaluation of the hydrogen storage capacity due to temperature changes can be of the same order of magnitude as the measured value. Temperature fluctuations are even more serious if the measurements are executed at 77 K.

Great errors, especially at high hydrogen pressures, can arise if leaks are present in the apparatus. To avoid this problem the system should be calibrated by measuring a well known metal hydride. Another way to detect leaks is to make sure that the hydrogen pressure is stable over a long time. Volumetric measurements are not a selective technique, that means that it is necessary to use high purity hydrogen gas to avoid the pressure drop being due to moisture condensation or adsorption of other gases on the surface of the substrate. Sievert's method involves multiple dosing, with an associated error which is summed at each pressure point; otherwise to evaluate the adsorption isotherm it is necessary to heat and evacuate the sample after each adsorption in order to start from the unloaded sample.

6.2.2.2 Thermal Desorption Spectroscopy (*TDS*)

The sample, is previously loaded with hydrogen at controlled pressure and temperature. Then it is heated in vacuum and the desorbed gas is selectively analyzed through a mass spectrometer. In order to convert the intensity of the peaks in hydrogen concentration it is necessary to make a calibration with a well known metal hydride, usually TiH_2. TDS is a very sensitive method, therefore it is possible to analyze small samples which is a great advantage considering the small quantity of nanostructured purified samples available.

6.2.2.3 Thermogravimetry

Hydrogen adsorption and desorption is measured with a microbalance by monitoring the mass change of the sample while it is subjected to a controlled temperature program under hydrogen pressure. The experimental set-up consists of a very sensitive microbalance in a vacuum-pressure vessel introduced into a furnace. Thermogravimetry is not a selective method because the mass change could also be induced by the adsorption of gases other than hydrogen. Therefore the apparatus has to be very clean and high purity hydrogen must be used. It is possible to evaluate a secondary effect, like thermomolecular flow induced by pressure gradients, by executing an experiment with an inert gas [12]. The thermogravimetric technique permits measurement of the storage capacity of samples with very low mass of about 10 mg in specially designed devices.

6.2.2.4 Electrochemical Method

The sample, e.g. carbon materials, can be loaded with hydrogen by electrochemical reactions. The carbon-containing electrode is prepared by mixing the adsorbing material with conductive powder (e.g. gold). The carbon based electrode and the counterelectrode are immersed in a KOH electrolyte solution. During charging the water is reduced on the negative electrode made from the analyzed material and part of the developed hydrogen is introduced into the sample. In the following discharge hydrogen and oxygen recombine to give water. Charge and discharge take place at constant current and the electrode potential at equilibrium conditions is determined. By determining the total electric charge in the galvanostatic set-up it is possible to determine the amount of desorbed hydrogen.

6.2.3
Carbon Materials

Carbon nanomaterials are very attractive candidates for hydrogen storage because of an ensemble of positive properties like high specific surface area, microporosity, low mass and good adsorption ability.

In these materials carbon is in the sp^2 hybridization state and has one nonhybridized free p-electron per atom, perpendicular to the sp^2-bonding. Because of their extended π-electron cloud these materials have very versatile properties and are called π-electron materials.

Activated carbon can be produced from a great number of different carbonaceous raw materials, like coconut shells, coal or lignin, which are processed in a carbonized form and subsequently activated through steam treatment and oxidation or through chemical activation. This processing is responsible for the formation of a very porous structure with a high specific surface area (from 1000 to 3000 $m^2\,g^{-1}$). Activated carbon is considered up to now to be the best carbonaceous adsorbent for hydrogen storage based on physisorption.

Since their discovery in 1991 by Iijima [13] carbon nanotubes have attracted much interest from the scientific society. Carbon nanotubes can be considered as rolled graphene sheets with an inner diameter of one to several diameters and length of 10 to 100 µm. Usually the nanotubes are closed at the two ends with caps, which have a fullerene-like structure. These tubules can be classified as single-walled carbon nanotubes (SWCNTs) and multi-walled carbon nanotubes (MWNTs). The MWNT consists of up to 50 graphitic layers with diameters typically ranging from 15 to 50 nm and interlayer distances close to that of graphite ($\approx$0.34 nm). SWCNTs consist of one layer and are therefore much thinner. Depending in which direction the graphite sheet is rolled with respect to the lattice vectors, three types of SWCNT can exist: zig-zag, armchair and chiral.

There are different potential sites where hydrogen adsorption can take place in these nanostructures.

SWCNTs exhibit a large free volume inside the tube and apart from this, the curvature of the graphene sheet and the channels between the tubes in a bundle can be sites for new interactions with hydrogen.

Large hydrogen storage in MWNTs between different concentric tubes seems to be impossible since then the strong carbon–carbon bond of the graphite sheets has to be stretched.

In recent years a new type of carbon nanostructure has been synthesized from catalytic decomposition of hydrocarbons. These fibrous materials are graphitic nanofibers (GNFs).

Graphitic nanofibers consist of graphite platelets stacked together in various orientations to the fiber axis with an interlayer distance similar to that of graphite. Depending on the orientation angle three distinct structures can exist: tubular, platelet and herringbone (Fig. 6.9).

6.2.3.1 Room Temperature

Many researchers have been spending time and effort to find new materials capable of storing reversibly large amounts of hydrogen at room temperature and moderate pressure. A storage medium with these characteristics would be ideal for mobile application, because low energy loss would be involved in adsorption and in desorption.

Dillon *et al.* [14] measured with TDS a reversible hydrogen uptake of 0.01 wt% for sonicated SWCNTs at room temperature. The sample contained only 0.1 wt% of nanotubes, therefore they estimated for pure nanotube samples a hydrogen storage capacity of approximately 5 to 10 wt%. A high-power ultrasonic treatment was applied to open the caps of the closed nanotubes, and cut the bundles in shorter

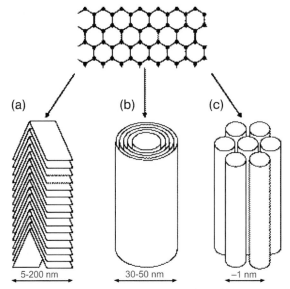

Fig. 6.9 Carbon nanostructures obtained from different assembly of graphene sheets: (a) graphitic nanofibers, (b) MWCNT and (c) a bundle of SWCNTs.

pieces in order to give easy access for hydrogen to the interior and interstitial sites of the bundles. It was later demonstrated that the high hydrogen adsorption could be attributed to titanium particles deriving from the tip used for the sonication treatment and not to the nanotubes [15]. Liu et al. measured the hydrogen storage capacity on SWCNTs volumetrically. They reported a storage capacity at room temperature of 4.2 wt% and at 10 MPa for SWCNTs with a large diameter of about 1.85 nm. The purity of the material was said to be 50 % and the authors claim that the storage capacity is repeatable [16]. However, it was never reproduced in any laboratory. Researchers from Japan have performed volumetric measurements of different carbon materials for hydrogen storage at room temperature. They obtained the highest uptake of 0.43 ± 0.03 wt% for purified HiPCO (High Pressure CO conversion) SWCNT [17].

Extremely high hydrogen uptake of herringbone GNFs, up to 67 wt%, was reported in 1998 by Rodriguez et al. They subjected the sample to a hydrogen pressure of 11.2 MPa in a constant volume and measured over a period of 24 h a great pressure drop [18]. However, in this case also, no laboratory has yet been able to reproduce this result. Using the same measuring technique, Ahn et al. found for GNFs with a significant fraction of herringbone structure a hydrogen storage capacity that does not exceed values of 0.2 wt% [19]. Recently, the group of Rodriguez and Baker measured a maximum adsorption of only 3.3 to 3.8 wt% for herringbone GNFs after hydrogen pretreatment at high temperature, 700 °C [20]. In 1998 a hydrogen uptake of 10–13 wt% for tubular formed GNFs was claimed by a group at the Chinese Academy of Science [21]. This result was obtained at a pressure of 11 MPa after boiling the nanofibers in hydrochloridric acid. However, in another publication [22], the same group reduced the hydrogen storage capacity of GNFs by a factor of 2. Up to now these results have still not been confirmed by any other group.

Very low hydrogen uptake values for carbon nanofibers are reported by both a Canadian group [23] and by Hirscher *et al* [24]. They obtained at around 11 MPa, respectively 0.7 and 0.1 wt%. Ritschel *et al.* [25] and Tibbetts *et al.* [26] confirm with their measurements insignificant or very low hydrogen storage capacity for GNFs.

Using thermogravimetric analysis, Ströbel *et al.* measured the hydrogen adsorption of carbon nanofibers and activated carbon at high pressure [27]. At a pressure of 12 MPa they observed a maximum weight increase corresponding to a hydrogen uptake of 1.6 wt% in activated carbon and 1.2 wt% in GNFs. Similar results are reported by Harutyunyan *et al.* for tubular and herringbone type carbon nanofibers. They obtained approximately 1 wt% at room temperature and a H_2 pressure of 12 MPa [28].

A Spanish group from the University of Alicante performed adsorption measurements on activated carbon materials at room temperature using a mixed gravimetric and volumetric method. The highest value of hydrogen adsorption they reported was close to 1 wt% at 10 MPa for an activated carbon obtained from anthracite and with a specific surface area of 1058 $m^2 g^{-1}$ [29].

The data concerning hydrogen adsorption in carbon materials at room temperature are scattered over a wide range. The reasons for these discrepancies can be attributed to the difficulty in measuring the hydrogen uptake and to the big differences in the sample quality. Unfortunately it seems obvious that all the reproducible results concern maximum storage capacities of approximately 1 wt% at 298 K far less than required for technical applications.

6.2.3.2 Cryogenic Temperature

The amount of hydrogen adsorbed on a solid surface depends on the chemical potential of the gas, which is correlated to temperature and pressure, and on the adsorption potential energy of H_2 on the solid, which is related to the pore structure, the specific surface area and the composition of the material. It is possible to vary the chemical potential of H_2 and enhance the storage capacity by lowering the temperature. Therefore many experiments on hydrogen adsorption have been performed around liquid nitrogen temperature (77 K).

Chahine and Bose have reported the hydrogen storage capacities of different types of activated carbon at cryogenic temperature [30]. The authors compare regular grade activated carbon with AX-21 activated carbon, which is produced by reaction with KOH and has cage-like porosity and very high surface area, of the order of 3000 $m^2 g^{-1}$. AX-21 has a hydrogen storage capacity of higher than 5 wt% at 77 K and 20 atm, twice as much as regular grade activated carbon with surface area between 700 and 1800 $m^2 g^{-1}$. These data are in very good agreement with the results obtained by volumetric measurements [31] on a similar type of activated carbon with a specific surface area of 2500 $m^2 g^{-1}$. The authors obtained, at 77 K and a pressure of 40 bar, a hydrogen uptake of approximately 4.5 wt%, and at RT and 65 bar one of 0.5 wt%.

Ye *et al.* carried out volumetric measurements on purified and DMF sonicated SWCNT at cryogenic temperatures. The authors obtained a high storage value of

8 wt% at 12 MPa at 80 K. Instead of reaching saturation, the hydrogen adsorption isotherms show a kink and a steep slope at high pressure. The authors suggest that the material undergoes a transition which involves attenuation of the van der Waals forces between tubes and decohesion due to the high chemical potential of hydrogen at high pressure [32].

Gravimetric measurements were executed by Pradhan *et al.* on SWNTs with different postsynthesis treatments. They reported hydrogen storage capacities at approximately 0.2 MPa ranging from approximately 1 to 6 wt%, depending on the processing of the material. The sample was first oxidized in dry air to remove the amorphous carbon, then refluxed in HCl or HNO_3 and finally annealed at different pressure.

The annealing treatment seems to be determinant for the hydrogen adsorption as the hydrogen storage capacity increases from 0.32 wt% for the sample annealed at 250 °C to 6.4% for the sample subjected to the same chemical treatment in HNO_3 but annealed at 1000 °C. The authors attribute the high storage capacities to local roughening of the surface and support this hypothesis through molecular dynamics studies on defective nanotubes [33].

It has been demonstrated experimentally by thermal desorption mass spectrometry that the adsorption mechanism of hydrogen on purified SWCNTs at low temperature is due to physical adsorption.

The SWNTs produced by the HiPCO method were loaded at 77 K with a equimolar mixture of H_2 and D_2. After evacuation the sample was degassed by increasing the temperature. Only H_2 and D_2 TDS signals at 100 K could be observed during desorption and the HD signal was comparable to the background. This means that hydrogen and deuterium are adsorbed on the SWCNTs in the molecular form, no dissociation and recombination of the molecules occurs on the surface of the sample. Purified SWCNT samples store therefore hydrogen isotopes via physisorpion at low temperature [34].

This hypothesis is also supported by inelastic neutron scattering spectra on hydrogen adsorbed on SWCNTs. The low intensity of the elastic peak (zero energy transfer) compared to the inelastic peak due to rotational transition of the adsorbed H_2 molecule, indicates that there is no significant amount of atomic hydrogen on the surface of the sample and that hydrogen maintains its molecular character when adsorbed on the sample [35].

Harutyunyan *et al.* reported for tubular and herringbone type carbon nanofibers 1.8 wt% at 77 K and a H_2 pressure of 12 MPa [36]. Very low storage capacity, less than 1 wt% at 0.45 MPa, is obtained through desorption measurements by Ahn *et al.* on similar samples.

It seems obvious that the phenomenon involved in the reversible adsorption of hydrogen at low temperature is physisorption.

It is possible to estimate from theoretical approximation the amount of hydrogen physically adsorbed in a monolayer per specific surface area on carbon substrates using Eq. (6.8) [37].

$$m_{ads}/S_{spec} = 2.27 \times 10^{-3} \text{ mass\% m}^{-2} \text{ g}^{-1} \tag{6.8}$$

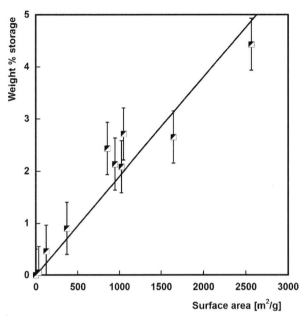

Fig. 6.10 Correlation between saturation hydrogen storage capacity and specific surface area of the adsorbent at 77 K [31].

Different carbon nanomaterials have been investigated with respect to their storage capacity at low temperature and, independently of their structure, the hydrogen uptake seems to be exclusively correlated to the specific surface area or pore volume (Fig. 6.10) [31]. The experimental relation obtained is $m_{ads}/S_{Spec} = 1.9 \times 10^{-3}$ mass% m^{-2} g^{-1} at 77 K and approximately 40 bar. Similar correlation between surface area and hydrogen uptake is also obtained by M.G. Nijkamp *et al.* at 1 bar [38]. This feature is very important because, depending only of the surface area, it is possible to choose as adsorbent material the one which is economically more advantageous to produce.

6.2.3.3 High Temperature Adsorption

Some investigations have been made on hydrogen adsorption in carbonaceous materials at high temperature or under energetic conditions. Typically, at high temperature the activation barrier for hydrogen bond breaking is overwhelmed, H_2 molecules dissociate on the surface of the carbon material and a covalent C—H bond is formed. Evard *et al.* have investigated the sorption capacity of nanoporous carbon exposed to hydrogen gas at temperature ranging from 520 to 970 K and at variable pressure. The highest uptake obtained is 1.3 wt% at 9300 kPa. The release of hydrogen takes place between 370 and 1470 K with a maximum desorption rate at approximately 800 K [39].

A ball milling process can introduce high local temperature in the sample. Hydrogen uptake during high energy ball milling of nanostructured graphite has been studied by Orimo et al. They showed that the hydrogen concentration increases with increasing milling time and after 80 h and under a H_2 pressure of 1 MPa they obtain a final storage capacity of 7.4 wt%. From neutron diffraction measurements the authors suggest that half of the hydrogen atoms form covalent bonds, which are favored because of the formation of dangling bonds during ball milling [40]. Two desorption peaks are obtained from the deuterided samples at 750 and 1000 K, indicating the possible existence of two carbon–deuterium coordinations [41]. In similar experimental conditions (graphite ball milled for 48 h under a hydrogen pressure of 0.8 MPa) Hirscher et al. obtain a hydrogen uptake greater than 5 wt%. Hydrogen desorption starts at 600 K and has a maximum at around 863 K accompanied by methane desorption. At 1066 K a second intense hydrogen desorption peak is obtained. The authors show that after desorption reloading the graphitic sample in hydrogen atmosphere is impossible [42].

Graphite and carbon materials are thought to be attractive candidates for plasma facing components in fusion reactors, that is as first wall materials. Atsumi analyzed the chemisorption properties of these materials to estimate the hydrogen recycling and tritium inventory. For this scope both untreated and highly defective materials, which were irradiated with ions or neutrons, have been investigated. The samples were exposed at 1273 K under a hydrogen pressure of 0.02 to 40 kPa. From his studies the author supposes the existence of two different trapping sites; both are carbon dangling bonds but with different energy. The rate determining step in hydrogen adsorption is suggested to be, in this case, molecular diffusion to the trapping sites and, in desorption, the rate is controlled by a sequence of detrapping and retrapping processes which occur during diffusion to the surface. The highest desorption rate is observed at 1220 K [43]. The samples which were irradiated with neutrons showed, because of the high defect density, higher hydrogen uptake than the unirradiated samples, but still less than 0.3 wt%. The authors suggest that the two desorption peaks observed by Orimo are ascribed to hydrogen detrapping at high temperature and recombination at lower temperature.

Hydrogen chemisorption on fullerenes has been investigated by several authors. Brosha et al. report an irreversible storage capacity of 2.6 wt%, which corresponds to a stoichiometric formula of $C_{60}H_{18.7}$, at 673 K and 103 bar H_2. While the samples can be directly hydrogenated through exposure to the gas phase, desorption is irreversible and accompanied by collapse of the fullerene structure to graphite-like species [44].

The high temperature necessary for hydrogen release and the high degree of irreversibility involved in chemical hydrogenation of carbon materials are very severe drawbacks from the viewpoint of technical storage application.

6.2.4
Zeolites

Zeolites are three-dimensional silicate structures, with isomorphous replacement of Si^{4+} ions with Al^{3+} and with SiO_4 tetrahedrons sharing all four corners. To maintain electroneutrality for every Si^{4+} substituted with an aluminum ion there is an additional extraframework metal ion adsorbed in the structure. The additional

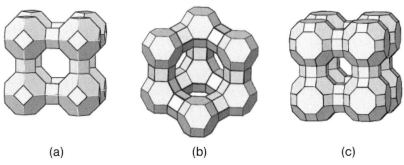

Fig. 6.11 Skeleton block of Zeolite A (a), Zeolite X and Zeolite Y (b) and of Zeolite RHO. The corners represent Al or Si atoms and at the middle of every line is one oxygen atom bridging two neighboring Si or Al atoms [47].

cations are usually larger metal ions such as K^+, Na^+, Ca^{2+}, or Ba^{2+}. The smaller ions do not occur in zeolites because the cavities in the lattice are too large. The ion exchange capacity of these materials depends on the extent of isomorphous substitution in the tetrahedral framework. Replacement of one quarter or one half of the Si atoms is quite common, giving structures $M^I[AlSi_3O_8]$ and $M^{II}[Al_2Si_2O_8]$.

Zeolites have a very open microporous structure with different framework types depending on the assembly of the building units (Fig. 6.11); the anion skeleton is penetrated by channels giving a honeycomb-like structure with high specific surface area. These channels are large enough to allow them to exchange certain ions or adsorb water and small molecules without the structure breaking down.

Because of these attractive properties, zeolites have been intensively investigated for hydrogen adsorption capacities [45]. The presence of strong electrostatic forces inside the channels and pores should enable hydrogen retention in the free volumes of the zeolite. The electric field is produced by the additional metal ions and it increases with increasing charge and decreasing size.

In spite of these exceptional characteristics, like high specific surface area, high porosity and the possibility to modulate the electric field inside the channels through ion exchange, very low hydrogen storage capacity has been reported for different types of zeolites.

Bose and Chaine compare hydrogen adsorption in activated carbon, in zeolites Linde 5A and Linde 13X at 77 K. In the three-dimensional alluminosilicates the storage is at 20 atm around 1 wt%, 5 times less than in activated carbon AX-21 [46].

For zeolite type ZSM5 Nijkamp et al. obtained the highest storage capacity, around 0.7 wt%, at 77 K and 1 bar [38].

Several types of zeolites with different cations have been analyzed for their hydrogen uptake by Langmi et al. [47] They report the highest H_2 uptake (1.8 wt%) for NaY at $-196\,°C$ and 15 atm; at room temperature the measured storage capacities are less than 0.3 %. The authors suggest that the adsorption is due to physisorption, because no hysteresis was measured between adsorption and desorption isotherms and the curve they obtained is a type I isotherm. Otero Areán et al. were able to measure the stretching mode of the H_2 molecule in the cages of Li-ZSM-5 zeolite with IR spectroscopy [48]. The electric field produced by the Li^+ ions induces a perturbation of the hydrogen molecule rendering this stretching mode IR active and shifting it to lower energies compared to the gas phase (where the vibrational mode is IR inactive but Raman active). Though the H—H bond is perturbed when hydrogen is adsorbed in zeolites, the existence of the stretching mode indicates that hydrogen is still adsorbed in the molecular form. Kazansky *et al.* found a relation between the amount of adsorbed hydrogen and the number of sodium ions in the zeolite structure, from which they assume that sodium ions are the adsorption sites. They think that the interaction is also influenced by the basicity of the zeolite [49]. The greatest storage capacity at 77 K and pressure <1 bar is obtained in this case for X zeolite with silicon to aluminum ratio of 1.05 and is approximately 1.3 wt%. All the results reported on zeolites are very congruent with each other and can be summarised by saying that the storage capacity for this inorganic material is less than 2 wt% at liquid nitrogen temperature and less than 1 wt% at room temperature.

Although zeolites have very high specific surface area and microporosity, these materials did not show the expected high adsorption capacities for hydrogen.

6.2.5
Metal–Organic Frameworks (MOFs)

MOFs are a new wide class of extremely porous polymeric structures which consist of metal ions linked together through organic ligands. Of particular interests is the crystalline material designed and synthesized by Yaghi and his colleagues, also called nanocubes because of their cubic crystalline structure [50].

They consist of $[OZn_4]^{6+}$ building blocks assembled in a uniform cubic lattice by organic connectors to give a three-dimensional structure with very high specific surface area. The $[OZn_4]^{6+}$ tetrahedrons are held together by carboxylate groups of the linkers resulting in a supertetrahedron cluster. Each supertetrahedron is bridged to the six neighboring ones through the organic chains in three dimensions. The authors report pore densities which can be even four times greater than for typical zeolites [51]. James explains in a recent paper the important aspects which influence the structure of MOFs during their synthesis. Among them the choice of rigid ligands is essential to reduce their degree of freedom and therefore the number of possible framework geometries which can be formed. Moreover the bridging anion is very important because of its coordinating abilities and the capacity to accommodate in the crystal structure [52].

Because of their apparently extraordinary properties, these porous materials have been studied with regard to their hydrogen storage capacity [53]. The sorption measurements were performed on MOF-5, IRMOF-6 and IRMOF-8 (Fig. 6.12) which have as organic linkers, respectively, benzenedicarboxylate, cyclobutylbenzenedicarboxylate and naphthalenedicarboxylate. These three structures essentially differ in the length of the linkers and consequently in pore size. Hydrogen adsorption isotherms were reported for MOF-5 at both room temperature and 78 K. At RT the authors found a linear relationship between hydrogen storage and pressure with 1 wt% storage at 20 bar, without reaching any saturation pressure in the range from 5 to 20 bars. At 78 K almost at very low pressure saturation occurs with a maximum

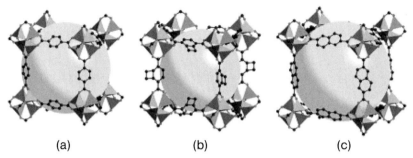

Fig. 6.12 MOF-5 (a), isoreticular MOF-6 (b) and IRMOF-8 (c) [53].

uptake of 4.5 wt% at 0.8 bar. The adsorption curve at cryogenic temperature is not a type I isotherm (saturation curve) as the authors claim but rather a step-like function.

For the other two materials, IRMOF-6 and IRMOF-8, the authors claim even doubled and quadrupled, respectively, specific H_2 uptake relative to MOF-5. Through inelastic neutron scattering mainly two types of different hydrogen adsorption sites in MOF-5 are identified. One site is associated to Zn and the other, which splits into four slightly different sites, is associated with the BDC linkers. From this information the authors suggest that further increase in H_2 sorption capacity can be achieved by employing larger linkers.

Férey et al. measured hydrogen adsorption in nanoporous metal-benzenedicarboxylates, where the metal is trivalent chromium or aluminum. Also in this case the material has a framework structure with high specific surface area ($1100 \, m^2 \, g^{-1}$). The authors report for these samples type I adsorption isotherms with hysteresis. The maximum storage capacity obtained for the chromium compound is 3.1 and 3.8 wt% for the aluminum compound at 1.6 MPa and 77 K [54].

Changing the organic linkers and the metal gives the possibility to investigate an infinite number of new structures with tuneable pore volume and surface area. Pan et al. developed a microporous metal–organic material with pore dimension comparable to the length scale of the molecular diameter of hydrogen. In this case two copper atoms share four carboxylate groups of the ligand, which is 4,4'-(hexafluoroisopropylidene) bis (benzoic acid). Each of these building units is connected to four others to give a 2D network. Two layers of two-dimensional structure are bound to give a three-dimensional network with open channels. At room temperature the reported storage capacity is 1 wt% at 48 atm [55].

6.2.6
Conclusion

Hydrogen storage through physical adsorption has many interesting aspects of great advantage for mobile application. The reversibility of the hydrogen uptake process is essential for a successful storage technology as all the fuel introduced in the system can be made easily available and with low energy lost. Moreover physisorption is a very fast process that means that refueling time is short compared to the case of the metal hydride storage medium. This is an essential factor for widespread use of hydrogen as an alternative fuel. Adsorption does not require very high hydrogen pressure, which is a positive factor from the point of view of safe storage.

Materials with high surface area and high potential adsorption capacity include a great variety of structures which can be modified and optimized for technical applications. Chemical engineering is very important for the breakthrough of these materials in hydrogen storage. Control of nanosize dimension gives the opportunity to develop new structures with higher porosity and surface area than the existing structures. These factors give the opportunity to use physical adsorption as a reversible process in hydrogen storage. The low temperatures needed limit at present the application of physisorption in nanostructured materials to systems with low cost cooling like satellites.

6.3
Metal Hydrides
Andreas Züttel

6.3.1
Binary and Intermetallic Hydrides

Metals, intermetallic compounds and alloys generally react with hydrogen and form mainly solid metal–hydrogen compounds. Hydrides exist as ionic, polymeric covalent, volatile covalent and metallic hydrides. The demarcation between the various types of hydrides is not sharp, they merge into each other according to the electronegativities of the elements concerned. This chapter focuses on metallic hydrides, that is metals and intermetallic compounds which with hydrogen form metallic hydrides. Hydrogen reacts at elevated temperature with many transition metals and their alloys to form hydrides. The electropositive elements are the most reactive, that is scandium, yttrium, the lanthanides, the actinides and the members of the titanium and vanadium groups (Fig. 6.13).

The binary hydrides of the transition metals are predominantly metallic in character and are usually referred to as metallic hydrides. They are good conductors of electricity and possess a metallic or graphite-like appearance [56]. Many of these compounds (MH_n) show large deviations from ideal stoichiometry ($n = 1, 2, 3$) and can exist as multi-phase systems. The lattice structure is that of a typical metal with atoms of hydrogen on the interstitial sites; for this reason they are also called interstitial hydrides. This type of structure has the limiting compositions MH, MH_2 and MH_3; the hydrogen atoms fit into octahedral or tetrahedral holes in the metal lattice, or into a combination of the two types. The hydrogen carries a partial negative charge, depending on the metal, an exception is, for example, $PdH_{0.7}$ [57]. Only a small number of the transition metals are without known stable hydrides. A considerable "hydride gap" exists in the periodic table, beginning at Group 6 (Cr) up to Group 11 (Cu), in which the only hydrides are palladium hydride ($PdH_{0.7}$), the very unstable nickel hydride ($NiH_{<1}$) and the poorly defined hydrides of chromium

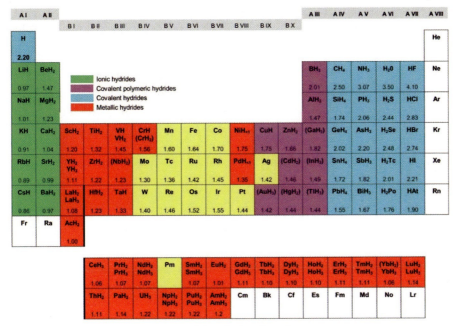

Fig. 6.13 Table of the binary hydrides and the Allred-Rochow electronegativity (Taken from: Huheey (1983), Ref. [58] and Allred and Rochow, Ref. [59]).

(CrH, CrH_2) and copper (CuH). In palladium hydride, the hydrogen has high mobility and probably a very low charge density.

In the finely divided state, platinum and ruthenium are able to adsorb considerable quantities of hydrogen, which thereby becomes activated. These two elements, together with palladium and nickel, are extremely good hydrogenation catalysts, although they do not form hydrides [60]. Especially interesting are the metallic hydrides of intermetallic compounds, in the simplest case the ternary system AB_xH_n, because the variation of the elements allows one to tailor the properties of the hydrides (Table 6.1).

The A element is usually a transition metal located left of the hydride gap in the periodic table or a rare earth metal and tends to form a stable hydride. The B element is often a transition metal from the right side of the hydride gap and forms only unstable hydrides. Some well defined ratios of B to A in the intermetallic compound $x = 0.5, 1, 2, 5$ have been found to form hydrides with a hydrogen to metal ratio of up to two.

6.3.2
Hydrogen Absorption Process

The reaction of hydrogen gas with a metal is called the absorption process and can be described in terms of a simplified one-dimensional potential energy curve (one-dimensional Lennard-Jones potential, Fig. 6.14) [61].

Table 6.1 The most important families of hydride-forming intermetallic compounds including the prototype and the structure

Intermetallic compound[a]	Prototype	Hydrides	Structure
AB_5	$LaNi_5$	$LaNiH_6$	Haucke phases, hexagonal
AB_2	ZrV_2, $ZrMn_2$, $TiMn_2$	$ZrV_2H_{5.5}$	Laves phase, hexagonal or cubic
AB_3	$CeNi_3$, YFe_3	$CeNi_3H_4$	Hexagonal, $PuNi_3$-type
A_2B_7	Y_2Ni_7, Th_2Fe_7	$Y_2Ni_7H_3$	Hexagonal, Ce_2Ni_7-type
A_6B_{23}	Y_6Fe_{23}	$Ho_6Fe_{23}H_{12}$	Cubic, Th_6Mn_{23}-type
AB	TiFe, ZrNi	$TiFeH_2$	Cubic, CsCl- or CrB-type
A_2B	Mg_2Ni, Ti_2Ni	Mg_2NiH_4	Cubic, $MoSi_2$- or Ti_2Ni-type

[a] A is an element with a high affinity to hydrogen and B is an element with a low affinity to hydrogen.

Far from the metal surface the potentials of a hydrogen molecule and of two hydrogen atoms are separated by the dissociation energy ($H_2 \rightarrow 2H$, $E_D = 435.99$ kJ mol^{-1}). The first attractive interaction of the hydrogen molecule approaching the metal surface is the Van der Waals force, leading to the physisorbed state ($E_{Phys} \approx 10$ kJ mol^{-1} H_2) approximately one hydrogen molecule radius (≈ 0.2 nm) from the metal surface. Closer to the surface the hydrogen has

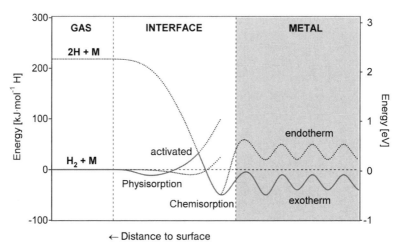

Fig. 6.14 Lennard-Jones potential of hydrogen approaching a metallic surface. Far from the metal surface the potentials of a hydrogen molecule and of two hydrogen atoms are separated by the dissociation energy. The first attractive interaction of the hydrogen molecule is the Van der Waals force leading to the physisorbed state. Closer to the surface the hydrogen has to overcome an activation barrier for dissociation and formation of the hydrogen metal bond. Hydrogen atoms sharing their electron with the metal atoms at the surface are then in the chemisorbed state. In the next step the chemisorbed hydrogen atom can jump in the subsurface layer and finally diffuse on the interstitial sites through the host metal lattice.

to overcome an activation barrier for dissociation and formation of the hydrogen metal bond. The height of the activation barrier depends on the surface elements involved. Hydrogen atoms sharing their electron with the metal atoms at the surface are then in the chemisorbed state ($E_{Chem} \approx 50$ kJ mol^{-1} H$_2$). The chemisorbed hydrogen atoms may have a high surface mobility, interact with each other and form surface phases at sufficiently high coverage. In the next step the chemisorbed hydrogen atom can jump in the subsurface layer and finally diffuse on the interstitial sites through the host metal lattice. The hydrogen atoms contribute with their electron to the band structure of the metal.

The hydrogen is, at a small hydrogen to metal ratio (H/M < 0.1), exothermically dissolved (solid-solution, α-phase) in the metal. The metal lattice expands proportionally to the hydrogen concentration by approximately 2 to 3 Å^3 per hydrogen atom [62]. At greater hydrogen concentrations in the host metal (H/M > 0.1) a strong H—H interaction due to the lattice expansion becomes important and the hydride phase (β-phase) nucleates and grows. The hydrogen concentration in the hydride phase is often found to be H/M = 1. The volume expansion between the coexisting α- and the β-phase corresponds in many cases to 10 to 20 % of the metal lattice. Therefore, at the phase boundary, a large stress is built up and often leads to decrepitation of brittle host metals such as intermetallic compounds. The final hydride is a powder with a typical particle size of 10 to 100 µm.

The thermodynamic aspects of hydride formation from gaseous hydrogen are described by means of pressure–composition isotherms (Figure 6.15). While the solid solution and hydride phase coexist, the isotherms show a flat plateau, the length of which determines the amount of H$_2$ stored. In the pure β-phase, the H$_2$ pressure rises steeply with the concentration. The two-phase region ends in a

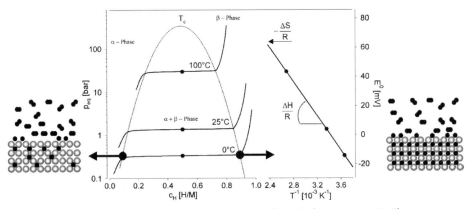

Fig. 6.15 Pressure composition isotherms for hydrogen absorption in a typical intermetallic compound on the left-hand side. The solid solution (α-phase), the hydride phase (β-phase) and the region of the coexistence of the two phases. The coexistence region is characterized by the flat plateau and ends at the critical temperature T_c. The construction of the Van't Hoff plot is shown on the right-hand side. The slope of the line is equal to the enthalpy of formation divided by the gas constant and the intercept is equal to the entropy of formation divided by the gas constant.

critical point T_C, above which the transition from the α- to the β-phase is continuous. The equilibrium pressure p_{eq} as a function of temperature is related to the changes ΔH and ΔS of enthalpy and entropy, respectively, by the Van't Hoff equation:

$$\ln\left(p_{eq}/p_{eq}^\circ\right) = \frac{\Delta H^\circ}{R} \cdot \frac{1}{T} - \frac{\Delta S^\circ}{R} \tag{6.9}$$

As the entropy change corresponds mostly to the change from molecular hydrogen gas to dissolved solid hydrogen, it amounts approximately to the standard entropy of hydrogen ($S^\circ = 130$ J K^{-1} mol^{-1}) and is therefore, $\Delta S_f \approx -130$ J K^{-1} mol^{-1} H$_2$ for all metal–hydrogen systems. The enthalpy term characterizes the stability of the metal hydrogen bond. To reach an equilibrium pressure of 1 bar at 300 K ΔH should amount to 39.2 kJ mol^{-1} H$_2$. The entropy of formation term of metal hydrides leads to a significant heat evolution $\Delta Q = T\Delta S$ (exothermic reaction) during the hydrogen absorption. The same heat has to be provided to the metal hydride to desorb the hydrogen (endothermic reaction). If the hydrogen desorbs below room temperature this heat can be delivered by the environment. However, if the desorption is carried out above room temperature the necessary heat has to be delivered at the necessary temperature from an external source which may be the combustion of the hydrogen. For a stable hydride like MgH$_2$ the heat necessary for the desorption of hydrogen at 300 °C and 1 bar is approximately 25 % of the higher heating value of hydrogen.

6.3.3
Empirical Models for Metallic Hydrides

Several empirical models allow the estimation of the stability and the concentration of hydrogen in an intermetallic hydride. The maximum amount of hydrogen in the hydride phase is given by the number of interstitial sites in the intermetallic compound for which the following two criteria apply: the distance between two hydrogen atoms on interstitial sites is at least 2.1 Å [64] and the radius of the largest sphere on an interstitial site touching all the neighboring metallic atoms is at least 0.37 Å (Westlake criterion) [65]. The theoretical maximum volumetric density of hydrogen in a metal hydride, assuming close packing of the hydrogen, is therefore 253 kg m^{-3}, which is 3.6 times the density of liquid hydrogen. As a general rule it can be stated that all elements with an electronegativity in the range 1.35 to 1.82 do not form stable hydrides [56]. Exceptions are vanadium (1.45) and chromium (1.56), which form hydrides and molybdenum (1.30) and technetium (1.36) where hydride formation would be expected but does not occur. The adsorption enthalpy can be estimated from the local environment of the hydrogen atom on the interstitial site. According to the rule of imaginary binary hydrides the stability of hydrogen on an interstitial site is the weighted average of the stability of the corresponding binary hydrides of the neighboring metallic atoms [66]. More general is the rule of reversed

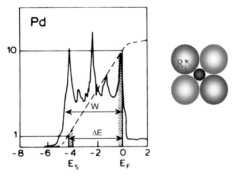

Fig. 6.16 The local band-structure model [63]. The band energy parameter $\Delta E = E_F - E_s$, the width of the d-band W and the distance R_j between the hydrogen atom and the next neighboring atoms.

stability (Miedema model): The more stable an intermetallic compound the less stable is the corresponding hydride and vice versa [67].

$$\Delta H(AB_n H_{x+y}) = \Delta H(AH_x) + \Delta H(B_n H_y) - (1-F) \cdot \Delta H(AB_n) \quad (6.10)$$

This model is based on the fact that hydrogen can only participate in a bond with a neighboring metal atom if the bonds between the metal atoms are at least partially broken.

Hydrogen absorption is, electronically, an incorporation of electrons and protons into the electronic structure of the host lattice. The electrons have to fill empty states at the Fermi energy E_F while the protons lead to the hydrogen induced s-band, approximately 4 eV below the E_F (Fig. 6.16). The heat of solution $\Delta \bar{H}_\infty$ of binary hydrides MH_x is related linearly to the characteristic band energy parameter $\Delta E = E_F - E_s$, where E_F is the Fermi energy and E_s the center of the host metal electronic band with a strong s character at the interstitial sites occupied by hydrogen. For most metals E_s can be taken as the energy which corresponds to one electron per atom on the integrated density-of-states curve [68].

$$\Delta \bar{H}_\infty = a \cdot \Delta E \cdot \sqrt{W} \cdot \sum_j R_j^{-4} + b \quad (6.11)$$

where $a = 18.6$ [kJ (mol H_2)$^{-1}$ Å^4 eV$^{-3/2}$] and $b = -90$ kJ mol^{-1}. The semiempirical models mentioned above allow an estimation of the stability of binary hydrides as long as the rigid band theory can be applied. However, the interaction of hydrogen with the electronic structure of the host metal in some binary hydrides, and especially in the ternary hydrides, is often more complicated. In many cases the crystal structure of the host metal and, therefore, also the electronic structure changes upon the phase transition and the theoretical calculation of the stability of the hydride becomes very complicated, if not impossible.

The stability of metal hydrides is usually presented in the form of Van't Hoff plots according to Eq. (6.9) (Fig. 6.17). The most stable binary hydrides have enthalpies

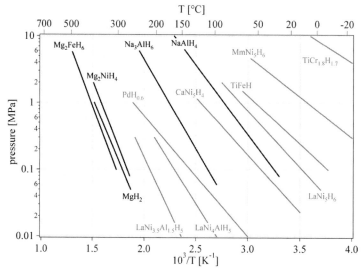

Fig. 6.17 Van't Hoff plots of some selected hydrides. The stabilization of the hydride of LaNi$_5$ by the partial substitution of nickel with aluminum in LaNi$_5$ is shown as well as the substitution of lanthanum with mischmetal (e.g. 51 % La, 33 % Ce, 12 % Nd, 4 % Pr).

of formation of $\Delta H_f = -226$ kJ (mol H$_2$)$^{-1}$, for example HoH$_2$. The least stable hydrides are FeH$_{0.5}$, NiH$_{0.5}$ and MoH$_{0.5}$ with enthalpies of formation of $\Delta H_f = +20$ kJ (mol H$_2$)$^{-1}$, $\Delta H_f = +20$ kJ (mol H$_2$)$^{-1}$ and $\Delta H_f = +92$ kJ (mol H$_2$)$^{-1}$, respectively [69].

Due to the phase transition upon hydrogen absorption, metal hydrides have the very useful property of absorbing large amounts of hydrogen at a constant pressure, that is the pressure does not increase with the amount of hydrogen absorbed as long as the phase transition takes place. The characteristics of the hydrogen absorption and desorption can be tailored by partial substitution of the constituent elements in the host lattice. Some metal hydrides absorb and desorb hydrogen at ambient temperature and close to atmospheric pressure. Several families of intermetallic compounds listed in Table 6.1 are interesting for hydrogen storage. They all consist of an element with a high affinity to hydrogen, the A-element, and an element with a low affinity to hydrogen, the B-element. The latter is often at least partially nickel, since nickel is an excellent catalyst for hydrogen dissociation.

One of the most interesting features of the metallic hydrides is the extremely high volumetric density of the hydrogen atoms present in the host lattice. The highest volumetric hydrogen density known today is 150 kg m^{-3} found in Mg$_2$FeH$_6$ and Al(BH$_4$)$_3$. Both hydrides belong to the complex hydrids and will be discussed in the next chapters. Metallic hydrides reach a volumetric hydrogen density of 115 kg m^{-3}, for example LaNi$_5$. Most metallic hydrides absorb hydrogen up to a hydrogen to metal ratio of H/M = 2. Greater ratios up to H/M = 4.5, for example BaReH$_9$ [70], have been found, however, all hydrides with a hydrogen to metal ratio of more than 2 are ionic or covalent compounds and belong to the class of complex hydrides.

Metal hydrides are very effective for storing large amounts of hydrogen in a safe and compact way. All the reversible hydrides working around ambient temperature and atmospheric pressure consist of transition metals; therefore the gravimetric hydrogen density is limited to less than 3 mass%. It is still a challenge to explore the properties of the lightweight metal hydrides.

6.3.4
Hydrogen Density in Metallic Hydrides

Hydrogen will be stored in various ways depending on the application, for example mobile or stationary. Today we know of several efficient and safe ways to store hydrogen, however, there are many other new potential materials and methods possible to increase the hydrogen density significantly. The material science challenge is to better understand the electronic behavior of the interaction of hydrogen with other elements and especially metals. Complex compounds like $Al(BH_4)_3$ have to be investigated and new compounds from the lightweight metals and hydrogen will be discovered.

6.4
Complex Transition Metal Hydrides
Klaus Yvon

6.4.1
Etymology

Complex transition metal hydrides derive their name from the presence of discrete transition (T) metal–hydrogen complexes in their solid-state structures. Typical examples are the members of the series Mg_2FeH_6–Mg_2CoH_5–Mg_2NiH_4. As shown in Fig. 6.18 they contain octahedral $[FeH_6]^{4-}$, square-pyramidal $[CoH_5]^{4-}$ and tetrahedral $[NiH_4]^{4-}$ anion complexes, respectively, that are surrounded by Mg^{2+} cations. The iron-based compound Mg_2FeH_6 is of particular interest for hydrogen storage

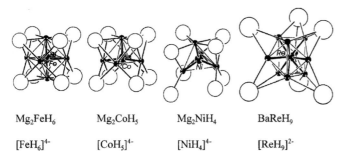

Mg_2FeH_6 Mg_2CoH_5 Mg_2NiH_4 $BaReH_9$

$[FeH_6]^{4-}$ $[CoH_5]^{4-}$ $[NiH_4]^{4-}$ $[ReH_9]^{2-}$

Fig. 6.18 Metal–hydrogen complexes and cation environments in complex transition metal hydrides. Small filled circles: hydrogen; large open circles: Mg^{2+} and Ba^{2+}.

applications because it contains 5.6 wt.% hydrogen and has a volume efficiency for hydrogen storage that is more than twice that of liquid hydrogen. On the other hand, BaReH$_9$ contains [ReH$_9$]$^{2-}$ anions and is relatively heavy and expensive, but its hydrogen-to-metal ratio (H/M = 4.5) is the highest known among metal hydrides. Since their discovery in the 1980s the number of complex T metal hydrides has continuously increased and now totals over 80 compounds containing some 30 different T metal hydride complexes [71].

6.4.2
Composition

There exist two broad families of complex T metal hydrides. The first displays mononuclear T metal hydrogen complexes in which the hydrogen atoms are only terminal ligands. In some of these hydrides, all the hydrogen is bonded to the T metal according to the general composition $M_m^{\delta+}[TH_n]^{\delta-}$ (T = 3d, 4d, 5d elements; M = alkali, alkaline-earth, rare-earth elements; $m, n = 1, 2, 3\ldots$), whereas in others some hydrogen is also surrounded by metal cations $M^{\delta+}$ only, corresponding to the general composition $M_m^{\delta+}[TH_n]^{\delta-} \cdot M_o^{\delta+}H_p^{\delta-}$ ($\delta, m, n, o, p = 1, 2, 3\ldots$).

The second family displays polynuclear complexes in which hydrogen forms not only terminal but also bridging ligands. This family is of particular interest because of its potentially rich crystal chemistry (see below).

6.4.3
Synthesis

Most hydrides are prepared by one-step solid-state reactions from the elements under hydrogen pressure (up to 200 MPa) and combine metals that do not form stable binary compounds between themselves, such as magnesium and iron (2Mg + Fe + 3H$_2$ → Mg$_2$FeH$_6$) or magnesium and manganese (3Mg + Mn + 7/2H$_2$ → Mg$_3$MnH$_7$). Only a few hydrides derive from stable binary metal compounds such as Mg$_2$NiH$_4$ (Mg$_2$Ni + 2H$_2$), LaMg$_2$NiH$_7$ (LaMg$_2$Ni + 7/2H$_2$) and La$_2$MgNi$_2$H$_8$ (La$_2$MgNi$_2$ + 4H$_2$), or from two-phase mixtures such as Ba$_7$Cu$_3$H$_{17}$ (Ba–Cu alloy with compositional ratio Ba/Cu~7/3). Some hydrides can also be prepared by inexpensive ball milling (Mg$_2$FeH$_6$) and combustion synthesis (Mg$_2$NiH$_4$). Solution methods are rarely used (BaReH$_9$) because the complexes are usually insoluble. Many hydrides must be handled with care because they are air sensitive and pyrophoric.

6.4.4
Mononuclear Complexes

These complexes have one T metal atom at their center and have terminal hydrogen ligands only. Those currently known are summarized in Table 6.2.

The T elements range from Group 7 (Mn) to closed d-shell elements of Group 12 (Zn) in the periodic system. No complexes have been reported for Groups 4 (Ti), 5 (V) and 6 (Cr) and none for Ag, Au and Hg. Of particular interest for hydrogen

Table 6.2 Mononuclear transition metal–hydrogen complexes in solid state metal hydrides as of May 2007; av = average due to ligand disorder

Mn	Fe	Co	Ni	Cu	Zn
$[MnH_4]^{2-}$ $[MnH_6]^{5-}$	$[FeH_6]^{4-}$	$[CoH_4]_{av}^{5-}$ $[CoH_5]^{4-}$	$[NiH_4]^{4-}$	$[CuH_4]^{3-}$	$[ZnH_4]^{2-}$

Tc	Ru	Rh	Pd	Ag	Cd
$[TcH_9]^{2-}$	$[RuH_4]^{4-}$ $[RuH_5]_{av}^{5-}$ $[RuH_6]^{4-}$ $[RuH_7]^{3-}$	$[RhH_4]^{3-}$ $[RhH_5]_{av}^{4-}$ $[RhH_6]^{3-}$ $[PdH_4]^{4-}$	$[PdH_2]^{2-}$ $[PdH_3]^{3-}$ $[PdH_4]^{2-}$		$[CdH_4]^{2-}$

Re	Os	Ir	Pt	Au	Hg
$[ReH_6]^{3-}$ $[ReH_6]^{5-}$ $[ReH_9]^{2-}$	$[OsH_6]^{4-}$ $[OsH_7]^{3-}$	$[IrH_4]^{5-}$ $[IrH_5]^{4-}$ $[IrH_6]^{3-}$	$[PtH_2]^{2-}$ $[PtH_4]^{2-}$ $[PtH_6]^{2-}$		

storage applications are complexes based on 3d elements. Their currently known hydride representatives are summarized in Table 6.3.

A given complex usually occurs in different hydride structures such as octahedral $[FeH_6]^{4-}$ in Mg_2FeH_6, $Ca_4Mg_4Fe_3H_{22}$ and $SrMg_2FeH_6$, and tetrahedral $[NiH_4]^{4-}$ in Mg_2NiH_4, $CaMgNiH_4$, $LaMg_2NiH_7$ [72], $La_2MgNi_2H_8$ [73] and $Mg_2Na_2NiH_6$ [74]. Some hydrides contain different complexes within the same structure, such as $Mg_6Co_2H_{11}$ (square-pyramidal $[CoH_4]^{5-}$ and saddle-like $[CoH_5]^{4-}$). Hydrides in which different complexes have different T elements at their centers have not yet been reported. Interestingly, most complexes contain T elements that do not form stable binary hydrides under ordinary conditions such as iron ($[FeH_6]^{4-}$) and cobalt ($[CoH_5]^{4-}$), or relatively unstable hydrides such as nickel ($[NiH_4]^{4-}$).

The ligand geometries include hydrogen-rich configurations such as tricapped trigonal prismatic ($[ReH_9]^{2-}$), pentagonal bipyramidal ($[RuH_7]^{3-}$), octahedral ($[FeH_6]^{4-}$), square-pyramidal ($[CoH_5]^{4-}$) and tetrahedral ($[NiH_4]^{4-}$), and hydrogen-poor configurations such as triangular ($[PdH_3]^{3-}$) and linear ($[PdH_2]^{2-}$). No trigonal bipyramidal and cubic antiprismatic configurations have been reported as yet, although indications for $[OsH_8]^{2-}$ complexes exist in osmium-based Cs_3OsH_9 and Rb_3OsH_9. The complexes are usually ordered at room temperature and tend to become disordered at high temperature, such as tetrahedral $[NiH_4]^{4-}$ in Mg_2NiH_4 ($T_1 \sim 240\,°C$). No di-hydrogen complexes have been reported as yet for this class of hydrides. Examples of "composite" hydride structures are the iron-based quaternary hydride $Ca_4Mg_4Fe_3H_{22}$ that derives from ternary Mg_2FeH_6 by an ordered substitution of Mg by Ca, and of $[FeH_6]^{4-}$ complexes by hydride anions H^-, and members

Table 6.3 Metal hydride complexes in ternary and quaternary 3d transition metal hydrides, as of May 2007

Manganese		Nickel	
$[MnH_4]^{2-}$ tet	M_3MnH_5, M = K, Rb, Cs	$[NiH_4]^{4-}$ tet	Mg_2NiH_4
$[MnH_6]^{5-}$ oct	Mg_3MnH_7		$MMgNiH_4$, M = Ca, Sr, Eu, Yb
			$LaMg_2NiH_7$, $La_2MgNi_2H_8$,
			$Mg_2Na_2NiH_6$

Iron		Copper	
	M_2FeH_6, M = Mg, Ca, Sr, Eu, Yb		
$[FeH_6]^{4-}$ oct	$M_4Mg_4Fe_3H_{22}$, M = Ca, Yb	$[CuH_4]^{3-}$ tet	$Ba_7Cu_3H_{17}$
	MMg_2FeH_6, M = Sr, Ba, Eu		

Cobalt		Zinc	
	$Mg_6Co_2H_{11}$		
$[CoH_4]^{5-}$ sad	$Mg_6Co_2H_{11}$	$[ZnH_4]^{2-}$ tet	M_2ZnH_4, M = K, Rb, Cs
$[CoH_5]^{4-}$ pyr	Mg_2CoH_5		M_3ZnH_5, M = K, Rb, Cs
	$M_4Mg_4Co_3H_{19}$, M = Ca, Yb		

^{a} oct = octahedral, tet = tetrahedral, sad = saddle-like, pyr = square pyramidal.
^{a} hydrogen ligands are partially disordered.

of the osmium-based series $LiH \cdot nMg_2OsH_4$ ($n = 1, 2, \infty$) that are built up by slabs of covalently bonded $[OsH_6]^{4-}$ complexes and sheets of ionically bonded Li^+H^- (Fig. 6.19). Finally, only relatively few 3d analogs exist for the more numerous 4d and 5d hydrides, that is only one manganese analog (Mg_3MnH_7) for the rhenium compounds and only one iron analog (Mg_2FeH_6) for the ruthenium compounds. In particular, no manganese analog for $BaReH_9$ has been found as yet, and no nickel analog for any of the platinum compounds, and no iron analogs for the above osmium series.

6.4.5
Polynuclear Complexes

Hydride complexes containing more than one T metal atom are less common but of considerable interest because of their possible links to the so-called "interstitial" hydrides. Those reported so far all display T—H—T bridges and some also T—T bonds.

As shown in Fig. 6.20, dimers and tetramers without T—T bonds occur in nonmetallic $Li_5Pt_2H_9$ and $La_2MgNi_2H_8$, dimers and polymers with T—T bonds in Mg_3RuH_3 and Mg_2RuH_4, respectively, and tetramers with some possible T—T interactions in metallic $MgRhH_{0.94}$ and $NdMgNi_4H_{\sim 4}$.

6.4 Complex Transition Metal Hydrides

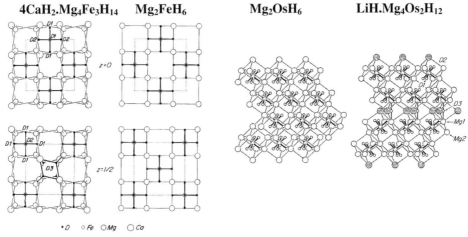

Fig. 6.19 Composite structure of cubic $Ca_4Mg_4Fe_3H_{22}$ as compared to cubic Mg_2FeH_6, and of trigonal $LiMg_4Os_2H_{13}$ as compared to cubic Mg_2OsH_6; large circles Ca (open) or Li (hashed), medium circles Mg, small circles H (D) bonded either to Fe or Os, or to Mg and Ca or Li only.

6.4.6
Complex Formation in "Interstitial" Hydrides

Until recently the only known case of hydrogen-induced complex formation in a metallic host structure was Mg_2NiH_4. The brownish colored hydride is non-metallic and was originally classified as an "interstitial" hydride. Only after its room-temperature structure was shown to contain tetrahedral $[NiH_4]^{4-}$ complexes was it classified as a "complex" metal hydride. Similar cases have now also been identified in other systems such as Mg_3Ir-H, $LaMg_2Ni$-H and La_2MgNi_2-H. Hydrogenation of metallic Mg_3Ir, for example, leads to the intensely red colored

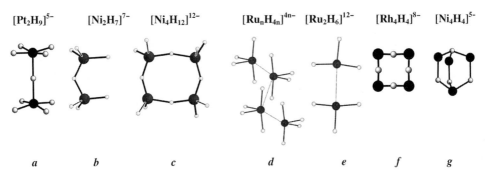

Fig. 6.20 Polynuclear complexes in (a) $Li_5Pt_2H_9$, (b) and (c) $La_2MgNi_2H_8$, (d) Mg_4RuH_4, (e) Mg_3RuH_3, (f) $Mg_4Rh_4H_{\sim 4}$ and (g) $NdMgNi_4H_{\sim 4}$. Large filled circles: T metals; small gray circles: hydrogen; dotted lines: Ru—Ru bonds.

nonmetallic hydride $Mg_6Ir_2H_{11}$, the structure of which can be rationalized in terms of square pyramidal $[IrH_5]^{4-}$ and saddle-like $[IrH_4]^{5-}$ 18-electron complexes and hydrogen anions H^-, in agreement with the limiting ionic formula $4Mg_6Ir_2H_{11} = 5MgH_2 \cdot 19Mg^{2+} \cdot 2[IrH_5]^{4-} \cdot 6[IrH_4]^{5-}$. Hydrogen-induced insulating states were also observed in the systems $LaMg_2Ni_2$-H [72] and La_2MgNi_2-H [73] at the compositions $LaMg_2NiH_7$ and $LaMg_2Ni_2H_8$, respectively. The structure of the former hydride can be rationalized in terms of tetrahedral $[NiH_4]^{4-}$ complexes and hydrogen anions H^- corresponding to the limiting ionic formula $La^{3+}Mg^{+2}{}_2 \cdot [NiH_4]^{4-} \cdot 3H^-$. The latter hydride displays two types of polynuclear hydrido complexes having novel geometries, dinuclear $[Ni_2H_7]^{7-}$ and tetranuclear $[Ni_4H_{12}]^{12-}$ (Fig. 6.20(b) and (c)). All these compounds are stoichiometric and must be considered as "complex" rather than "interstitial" T metal hydrides. On the other hand, hydrogenation of intermetallic compounds such as MgRh and $NdMgNi_4$ leads to nonstoichiometric hydrides that are metallic. Yet, their stereochemistry suggest a clear tendency for complex formation (tetramers $[Rh_4H_4]^{8-}$ in $MgRhH_{\sim 0.9}$ and $[Ni_4H_4]^{5-}$ in $NdMgNi_4H_{\sim 4}$, see Fig. 6.20(f) and (g). Interesting directional bonding effects have also been reported in other "interstitial" hydride systems such as RT_3H-H [75] and R_2Ni_7-H [76] (R = Rare earth, T = Co, Ni) in which nickel tends to adopt tetrahedral and cobalt octahedral (or square-pyramidal) H atom configurations. Taken together, these findings suggest that complex formation could be a general phenomenon in "interstitial" T metal hydrides, at least on a local level.

6.4.7
Bonding

The compositions and ligand geometries of T metal hydride complexes can be rationalized in terms of "magic" electron counts by taking into account all valence electrons and full charge transfer from the surrounding cation matrix. Those displaying terminal H ligands only (mononuclear complexes) can be described in terms of two-center–two-electron (2c–2e) bonds and various s-p-d hybridization schemes such as $(d^{10})sp^3$ for 18-electron tetrahedral $[NiH_4]^{4-}$ and $[PdH_4]^{4-}$, $(d^8)dsp^2$ for planar 16-electron $[PdH_4]^{2-}$, $(d^8)dsp^3$ for square-pyramidal 18-electron $[CoH_5]^{4-}$, $(d^6)d^2sp^3$ for octahedral 18-electron $[FeH_6]^{4-}$, and $(d^{10})sp$ for linear 14-electron $[PdH_2]^{2-}$. However, other electron counting rules based on partial charge transfer from the cation matrix and s–d hybridization schemes involving three-center–four-electron bonds are also possible [77]. The complexes displaying additional bridging H ligands (polynuclear complexes) can be described in terms of both 2c-2e and 3-centre-2-electron (3c-2e) bonds, such as dinuclear 34-electron $[Ni_2H_7]^{7-}$ ($2d^{10}$ + six 2c-2e + one 3c-2e bonds) and tetranuclear 64-electron $[Ni_4H_{12}]^{12-}$ ($4d^{10}$ + eight 2c-2e + four 3c-2e bonds). Within this picture each T atom conforms to the 18-electron rule.

Complexes having more than 18 electrons have not been reported as yet. As the H/M ratio decreases the systems tend to become metallic and electron counting becomes less obvious. Two interesting examples are Mg_2RuH_4 and Mg_3RuH_3 that contain saddle-like 16-electron $[RuH_4]^{4-}$ and T-shaped 17-electron $[RuH_3]^{6-}$

complexes, respectively. Both systems become 18-electron if one postulates Ru–Ru bonds that join the complexes to linear $[Ru_nH_{4n}]^{4n-}$ polymers and $[Ru_2H_6]^{12-}$ dimers, respectively. Clearly, limiting ionic formulas based on full charge transfer and "magic" electron counts are not suited to describing all bonding aspects of complex metal hydrides. However, they are useful for rationalizing (and predicting) hydrogen contents, which is generally not possible with their "interstitial" (metallic) counterparts. Finally, an important bonding feature of complex T metal hydrides is the interaction between the hydrogen ligands of the complexes and the surrounding cations. As can be seen from the 18-electron series Mg_2FeH_6–Mg_2CoH_5–Mg_2NiH_4 shown in Fig. 6.18 the cations adopt cubic – or nearly cubic – configurations that maximize the Mg^{2+}–H interactions. Clearly these interactions not only stabilize the complexes but also contribute greatly to the thermal stability of the overall structure (see below). As for the T–H interactions, they do not appear to contribute much to thermal stability, as shown by the relatively weak force constants measured for Mg_2FeH_6 (Fe–H stretching mode: $\sim$1.9 mdyn Å^{-1}). Theoretical band structure calculations for some systems are available (for recent examples see Mg_3MnH_7 [78], $BaReH_9$ [79] and $LaMg_2NiH_7$ [72]), but have not yet addressed the issue of thermal stability.

6.4.8
Properties and Possible Applications

In contrast to their "interstitial" counterparts complex metal hydrides are usually nonmetallic. They are often colored and sometimes transparent. Those deriving from intermetallic compounds show hydrogenation-induced transitions from metallic (Mg_2Ni, Mg_3Ir, $LaMg_2Ni$, La_2MgNi_2) to nonmetallic (brownish-red Mg_2NiH_4, red $Mg_6Ir_2H_{11}$, dark gray $LaMg_2NiH_7$ and $La_2MgNi_2H_8$) states. Such transitions are of both fundamental and technological interest as shown for the Mg_2Ni–H system that displays switchable optical properties [80]. Magnetic properties are known for hydrides containing magnetic ions that order, such as rose-colored K_3MnH_5 (Mn(II), $\mu_{eff}=4.5\mu_B$, $T_N=28$ K) and faint violet-colored Eu_2PdH_4 (Eu(II), $\mu_{eff}=8\mu_B$, $T_C=15$ K). The vibrational spectra indicate stretching and bending modes of the complexes in the expected ranges 1600–2000 cm^{-1} and 800–1000 cm^{-1}, respectively. As to properties of interest for hydrogen storage applications the hydrogen weight and volume efficiencies and desorption temperatures of a few compounds are summarized in Table 6.4. Clearly, some hydrides have outstanding properties, such as Mg_2FeH_6 that shows one of the highest known hydrogen storage volume efficiencies of all materials known (150 g l^{-1}), and $BaReH_9$ that has a H/M ratio that surpasses even the hydrogen-to-carbon ratio of methane (H/C = 4). The weight efficiencies are also remarkable as shown by Mg_2FeH_6 and Mg_3MnH_7 that can store up to 5.5 wt.% hydrogen, that is considerably more than currently used "interstitial" metal hydrides.

Unfortunately, regarding thermal stability complex metal hydrides perform less well than their interstitial counterparts. Only a few decompose near room temperature, such as $BaReH_9$ and $Ba_7Cu_3H_{17}$, that are, however, relatively heavy and expensive and not completely reversible. Mg_2NiH_4, which is the only

Table 6.4 Hydrogen storage properties of selected complex transition metal hydrides (found and/or characterized in Geneva)

Formula	Hydrogen anions	Hydrogen [wt.%]	Density [kg m^{-3}]	Desorption temperature at 1 bar H$_2$ [°C]
Mg$_3$MnH$_7$	[MnH$_6$]$^{5-}$, H$^-$	5.2	119	~280
BaReH$_9$	[ReH$_9$]$^{2-}$	2.7	134	<100
Mg$_2$FeH$_6$	[FeH$_6$]$^{4-}$	5.5	150	320
Ca$_4$Mg$_4$Fe$_3$H$_{22}$	[FeH$_6$]$^{4-}$, H$^-$	5.0	122	395
SrMg$_2$FeH$_8$	[FeH$_6$]$^{4-}$, H$^-$	4.0	115	440
Mg$_2$CoH$_5$	[CoH$_5$]$^{4-}$	4.5	126	280
Mg$_6$Co$_2$H$_{11}$	[CoH$_4$]$^{5-}$, [CoH$_5$]$^{4-}$, H$^-$	4.0	97	370
Ca$_4$Mg$_4$Co$_3$H$_{19}$	[CoH$_5$]$^{4-}$, H$^-$	4.2	106	>480
Mg$_2$NiH$_4$	[NiH$_4$]$^{4-}$	3.6	98	250
CaMgNiH$_4$	[NiH$_4$]$^{4-}$	3.2	87	405
Ba$_7$Cu$_3$H$_{17}$	[CuH$_4$]$^{3-}$, H$^-$	1.5	63	20
K$_2$ZnH$_4$	[ZnH$_4$]$^{2-}$	2.7	57	310
K$_3$ZnH$_5$	[ZnH$_4$]$^{2-}$, H$^-$	2.7	56	360
α-MgH$_2$	Mg$^+$, H$^-$	7.7	109	280
H$_2$(liquid)	H°	100	71	−253

commercialized hydride of this class, decomposes only above ~250 °C, corresponding to a desorption enthalpy of $\Delta H = 64$ kJ (mol H$_2$)$^{-1}$. Most other complex T metal hydrides are more stable (decomposition temperatures >300 °C, $\Delta H > 80$ kJ (mol H$_2$)$^{-1}$) and must be heated to recover hydrogen, which represents a penalty in energy. Such hydrides, however, are of interest for thermochemical thermal energy storage, for example Mg$_2$FeH$_6$ can be cycled relatively easily around 500 °C and 80 bar pressure [81]. Another drawback of complex T metal hydrides is their tendency to liberate hydrogen only stepwise, such as the quaternary hydride Ca$_4$Mg$_4$Fe$_3$H$_{22}$ that decomposes according to the double reaction

$$Ca_4Mg_4Fe_3H_{22} \rightarrow 2Ca_2FeH_6 + 4Mg + Fe + 5H_2$$

and

$$Ca_2FeH_6 \rightarrow 2CaH_2 + Fe + H_2.$$

Clearly, complex T metal hydrides share this undesirable feature with complex p-metal hydrides such as alanates (e.g. NaAlH$_4 \Rightarrow 1/3$Na$_3$AlH$_6$ + 2/3Al + H$_2$). As to the factors responsible for thermal stability the nature of the cation matrix plays a major role. As can be seen from Table 6.5 the calcium-containing hydrides Ca$_4$Mg$_4$Fe$_3$H$_{22}$ and CaMgNiH$_4$ are much more stable than the corresponding calcium-free hydrides Mg$_2$FeH$_6$ and Mg$_2$NiH$_4$, respectively, and this correlates with the thermal stability of the binary hydrides CaH$_2$ and MgH$_2$. This confirms

Table 6.5 Desorption enthalpies of Mg- and Ca-based complex transition metal hydrides

	ΔH [kJ mol^{-1} H$_2$]
Mg$_2$FeH$_6$	98
Ca$_4$Mg$_4$Fe$_3$H$_{22}$	122
Mg$_2$NiH$_4$	64
CaMgNiH$_4$	129
MgH$_2$	74
CaH$_2$	184[a]

[a] Model prediction.

that the interactions between hydrogen of the complexes and the surrounding metal cations govern to a large extent the thermal stability.

6.4.9
Conclusions and Outlook

The currently known complex T metal hydrides are capable of storing hydrogen at volume densities that exceed by far those of compressed hydrogen gas and liquid hydrogen. Their weight efficiencies reach 5.5 % and their hydrogen dissociation temperatures at 1 bar hydrogen pressure are in the range 100–400 °C. Some hydrides are relatively inexpensive to fabricate (Mg$_2$FeH$_6$) but thermally too stable for room temperature applications, while others are sufficiently unstable (BaReH$_9$) but too expensive for large scale applications. At present none of the complex T metal hydrides known combines both requirements, that is their use is limited to niche markets where the price of materials plays a minor role. Finally, some systems show hydrogen-induced transitions from delocalized (alloy) to localized (hydride) electron states, which are of interest for hydrogen detectors and optical applications. Thus, a challenge for future work is to synthesize new compounds based on light and inexpensive 3d elements and to reach a better understanding of the metal–hydrogen interactions that govern thermal stability. The prospects of finding such compounds and achieving that goal are good, since the possible element combinations are numerous and not yet fully explored.

6.5
Tetrahydroborates as a Non-transition Metal Hydrides
Shin-ichi Orimo and Andreas Züttel

6.5.1
p-Metal Hydrides

All the elements of Group 13 (boron group) form polymeric hydrides (MH$_3$)$_x$. The monomers MH$_3$ are strong Lewis acids and are unstable. Borane (BH$_3$) achieves electronic saturation by dimerization to form diborane (B_2H_6). All other hydrides in this group attain closed electron shells by polymerization. Aluminum hydride, alane, $(AlH_3)_x$ has been extensively investigated [82], the hydrides of gallium indium and thallium much less so [83].

The hydrogen in the p-element complex hydrides is often located at the corners of a tetrahedron with boron or aluminum in the center. The bonding character and the properties of the complexes $M^+[BH_4]^-$ and $M^+[AlH_4]^-$ are largely determined by the difference in electronegativity between the cation and the boron or aluminum atom, respectively. The IUPAC has recommended the names tetrahydroborate for $[BH_4]^-$ and tetrahydroaluminate for $[AlH_4]^-$. The alkali metal tetrahydroborates are ionic, white, crystalline, high melting solids that are sensitive to moisture but not to oxygen. Group 3 and transition metal tetrahydroborates are covalent bonded and are either liquids or sublimable solids. The alkaline earth tetrahydroborates are intermediate between ionic and covalent. The tetrahydroaluminates are very much less stable than the tetrahydroborates and therefore considerably more reactive. The difference between the stability of the tetrahydroaluminate and the tetrahydroborates is due to the different Pauling electronegativity of B and Al, 2.04 and 1.61, respectively. The properties of the complex hydrides can be varied by the partial substitution of the boron or aluminum atom.

In contrast to the interstitial hydrides, where the metal lattice hosts the hydrogen atoms on interstitial sites, the desorption of the hydrogen from the complex hydride leads to a complete decomposition of the complex hydride and a mixture of at least two phases is formed. For alkali metal tetrahydroborates and tetrahydroaluminates the decomposition reaction is described according to the following equations:

$$A(BH_4) \rightarrow \text{``}ABH_2 + H_2\text{''?} \rightarrow AH + B + \tfrac{3}{2}H_2$$

and

$$A(AlH_4) \rightarrow \tfrac{1}{3}A_3AlH_6 + H_2 \rightarrow AH + Al + \tfrac{3}{2}H_2$$

For alkaline earth metal tetrahydroborates and tetrahydroaluminates the decomposition reaction is described according to the following equations:

$$E(BH_4)_2 \rightarrow EH_2 + B$$

and

$$E(AlH_4)_2 \rightarrow ? \rightarrow EH_2 + 2Al + 3H_2$$

The physical properties of the tetrahydroborates and the tetrahydroalanates are to a large extent still not known.

The tetrahydroalanates are discussed extensively in Chapter 6.6. The following part will therefore focus on the tetrahydroborates.

6.5.1.1 Structure of Tetrahydroborates

Selected structural information on the alkali and alkaline earth alkali borates is summarized in Table 6.6. There are many controversial results for the structure of hydroborates because in lightweight borohydrides a significant fraction of the electrons participate in the bonds and therefore the atoms appear to be different when the structures are determined by X-ray and neutron diffraction.

Harris [84] suggested, in 1947 in a letter to the editor, the space group *Pcmn* for Li[BH]$_4$ at room temperature, which turned out to be wrong. The low- and high-temperature structures of Li[BH]$_4$ were investigated by means of synchrotron X-ray powder diffraction [85, 86]. An endothermic (4.18 kJ mol^{-1}) structural transition was observed at 391 K from the orthorhombic low-temperature structure to the hexagonal high-temperature structure (Fig. 6.21). The high temperature structure melts at 287 °C with a latent heat of 7.56 kJ mol^{-1}. The hexagonal structure is still under discussion [87] since the theoretical calculation of the phonon density of states [88] has shown strange features. Both transitions are reversible, although a small hysteresis in temperature was observed. While the hexagonal phase grows from the liquid as a single crystal the orthorhombic phase has not yet been found as a single crystal because of the structural transition which prevents the formation of a single crystal of Li[BH]$_4$.

According to powder neutron diffraction studies performed at room temperature by Davis and Kennard [89] on a Na[BD$_4$] sample, the compound crystallizes with a NaCl-type structure. Concerning the low-temperature phase only lattice parameters are known from X-ray investigations [90]. The authors mentioned that the sodium and boron atoms may form a body-centered tetragonal array and that the hydrogen atoms could lie on a primitive tetragonal lattice. The structure of Na[BD$_4$] was recently found [91] to belong to space group *P* –4 21 *c*. Heat capacity measurements on NaBH$_4$ samples [92] showed a lambda-type phase transition at 190 K, which

Table 6.6 Structural data of selected alkali and alkaline earth metal tetrahydroborates

Complex	Space group	a [nm]	b [nm]	c [nm]	Ref.
Li[BH$_4$]	*Pnma* (# 62) (<381 K)	0.7178	0.4437	0.683	[84, 85]
	*P*6$_3$*mc* (#186) (>391 K)	0.4276		0.6948	[84, 85]
Na[BH$_4$]	*P*4$_2$/*mnc* (# 137) (<190 K)	0.4332		0.5949	[90]
	F –43*m* (#216) (>190 K)	0.6148			[89, 91]
K[BH$_4$]	*P*4$_2$/*mnc* (# 137) (<197 K)	0.4684		0.6571	[95, 96]
	Fm-3*m* (>197 K)	0.6727			[95, 96]
Be[BH$_4$]$_2$					
Mg[BH$_4$]$_2$	*P*-3*m*$_1$ (# 164)	1.359		1.651	[97]
	Fm-3*m*	1.55			[97]
Ca[BH$_4$]$_2$	*Fddd* (# 70)	0.8791	1.3137	0.750	[106]
Al[BH$_4$]$_3$	*C*2/*c* (# 15)	2.2834	0.6176	2.2423	[99]
	*Pna*2$_1$ (# 33)	1.8021	0.6138	0.6199	[99]

LiBH$_4$ AT 293K (20°C)

Orthorhombic symmetry
space group: **Pnma (#62)**
a = 7.17858(4) Å
b = 4.43686(2) Å
c = 6.80321(4) Å
Vol: 216.685 Å^3, Z = 4

Atom	x	y	z
Li	0.1568	0.250	0.1015
B	0.3040	0.250	0.4305
H$_1$	0.900	0.250	0.956
H$_2$	0.404	0.250	0.280
H$_3$	0.172	0.054	0.428

LiBH$_4$ AT 408K (135°C)

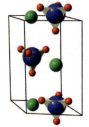

Hexagonal symmetry
space group: **P6$_3$mc (#186)**
a = 4.27631(5) Å
b = a
c = 6.94844(8) Å
Vol: 110.041 Å^3, Z = 2

Atom	x	y	z
Li	0.3333	0.6666	0.0000
B	0.3333	0.6666	0.5530
H$_1$	0.3333	0.6666	0.3700
H$_2$	0.1720	0.3440	0.6240

Fig. 6.21 Low- and high-temperature structure of Li[BH$_4$] determined from X-ray diffraction (Soulié, Renaudin, Cerny, and Yvon (2002), Ref. [85] and Züttel, Rentsch, Fischer, Wenger, Sudan, Mauron and Emmenegger (2003) [86]).

was interpreted as an order–disorder transition associated with reorientations of [BH$_4$]$^-$-tetrahedra [93]. According to recent theoretical work [87, 94], similar transitions could exist in LiBH$_4$.

Neutron diffraction of K[BD$_4$] shows a tetrahedral array of hydrogen atoms about the boron atom with a B–H distance of 0.1219 nm [95] and the crystal is face-centered cubic at room temperature [96].

Currently several groups are working on the preparation and structure determination of Mg[BH$_4$]$_2$ and Ca[BH$_4$]$_2$ [97, 98] and Al(BH$_4$)$_3$ [99].

6.5.1.2 Stability of Tetrahydroborates

The stability of metal tetrahydroborates has been discussed in relation to their percentage ionic character and those compounds with less ionic character than diborane are expected to be highly unstable [100]. Steric effects have also been suggested to be important in some compounds [101, 102]. The special feature exhibited by the covalent metal hydroborate is that the hydroborate group is bonded to the metal atom by bridging hydrogen atoms, similar to the bonding in diborane, which may be regarded as the simplest of the so-called "electron-deficient" molecules. Such molecules possess fewer electrons than those apparently required to fill all the bonding orbitals, based on the criterion that a normal bonding orbital involving two atoms contains two electrons.

The bonding character and the properties of the complexes M$^+$[BH$_4$]$^-$ and M$^+$[AlH$_4$]$^-$ are, therefore, largely determined by the difference in electronegativity between the cation and the boron or aluminum atom, respectively [56, 113]. The localization of the negative charge on the boron atom is crucial for the stability of the anion [BH$_4$]$^-$. Figure 6.22 shows the dependence of the decomposition temperature on the Pauling electronegativity of the cation for tetrahydroalanates and

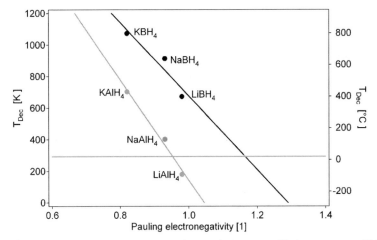

Fig. 6.22 Decomposition temperature for metal tetrahydroalanates and metal tetrahydroborates versus the Pauling electronegativity of the metal atom, which forms the cation. The horizontal line is at 300 K. A hydrogen equilibrium pressure of 1 bar at room temperature is achieved when the electronegativity of the metal is 0.9 and 1.15 for tetrahydroalanates and tetrahydroborates, respectively.

tetrahydroborates. The decomposition temperature is the temperature at which hydrogen gas with a pressure of 1 bar is in equilibrium with the solid hydride ($T_{dec} = \Delta H/\Delta S$). The partial substitution of the cation in an alkali metal tetrahydroborate leads to a change in the stability of the complex [103] as long as a phase with mixed cations exists. The tetrahydroaluminates are very much less stable than the tetrahydroborates and therefore considerably more reactive. The difference between the stability of the tetrahydroaluminate and the tetrahydroborates is due to the different Pauling electronegativity of B and Al, 2.04 and 1.61, respectively. The properties of the complex hydrides can also be varied by partial substitution of the boron or aluminum atom or by partial substitution of the cation [104].

The stability of tetrahydroborates can be considered from different viewpoints:

1. The stability as the heat of formation (ΔH_f), the enthalpy difference between the complex and its elements in the standard state according to the reaction $M + B + 2H_2 \rightarrow M[BH_4]$.
2. The stability as an enthalpy difference between the complex hydride and an intermediate product in the decomposition reaction $M[BH_4] \rightarrow MH + B$. All tetrahydroborates where the MH exhibits a heat of formation between the complex hydride and its elements go through the above-mentioned intermediate product in the thermal desorption of hydrogen.
3. The stability of the bond between the cation and the hydroborate M–[BH$_4$] and the stability of the hydrogen–boron bond M[B–H$_4$] is different. This has an influence on the possible path of the thermal decomposition reaction.

Orimo et al. [105] have found a correlation between the electronegativity of the cation and the frequency of the bending and stretching modes of hydrogen in the anion as well as the melting temperature of the complex.

The thermodynamic stabilities for the series of metal tetrahydroborates $M[BH_4]_n$ (M = Li, Na, K, Cu, Mg, Zn, Sc, Zr and Hf; $n = 1–4$) have been systematically investigated [106] by first-principles calculations [107, 108]. The results indicated that the bond between M^{n+} cations and $[BH_4]^-$ anions in $M[BH_4]_n$ is ionic and the charge transfer from $[BH_4]^-$ anions to M^{n+} cations is responsible for the stability of $M[BH_4]_n$.

6.5.1.3 Sorption Mechanism and Kinetics of Tetrahydroborates

The thermal hydrogen desorption from $Li[BH]_4$ liberates three of the four hydrogens in the compound upon melting at 280 °C and decomposes into LiH and boron. The thermal desorption spectrum exhibits four endothermic peaks. The peaks are attributed to a polymorphic transformation around 110 °C, melting at 280 °C, the hydrogen desorption (50 % of the hydrogen was desorbed ("$LiBH_2$") around 490 °C) and when 3 of the 4 hydrogen are desorbed at 680 °C. Only the third peak (hydrogen desorption) is pressure dependent, all other peak positions (temperature) do not vary with pressure. The calculated enthalpy $\Delta H = -177.4$ kJ mol^{-1} H$_2$ and entropy $\Delta S = 238.7$ J K^{-1} mol^{-1} H$_2$ of decomposition are not in agreement with the values deduced from indirect measurements of the stability [109]. Especially, the value of the entropy is by far too high and cannot be explained, because the standard entropy for hydrogen is 130 J K^{-1} mol^{-1} H$_2$. Additives to the tetrahydroborate, for example SiO_2 [86], lower the hydrogen desorption temperature for $Li[BH_4]$ by approximately 100 K, which is, according to recent studies of the isotherms, due to a catalytic effect and not a destabilization of the complex [110].

The thermal hydrogen desorption spectrum of $Li[BH_4]$ at very low heating rate (0.5 K min^{-1}) exhibits three distinct desorption peaks [111] (Fig. 6.23). This is an

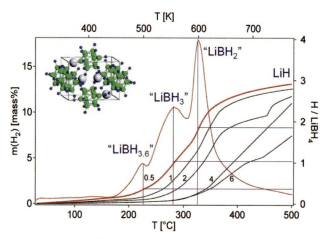

Fig. 6.23 Integrated thermal desorption spectra for $Li[BH_4]$ measured at various heating rates (values in figure in K min^{-1}). The differential curve for the lowest heating rate of 0.5 K min^{-1} is given together with the hypothetical compositions at the peak maxima. The structure of an intermediate phase is shown as an inset (from Chen, Xiong, Luo, Lin and Tan (2003), Ref. [123]).

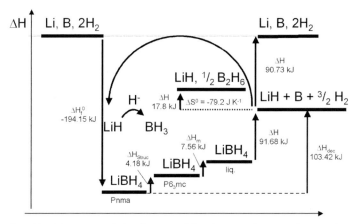

Fig. 6.24 Schematic enthalpy diagram of the phases and intermediate products of Li[BH$_4$].

indication that the desorption mechanism involves several intermediate steps. This was recently explained by the formation of a borohydride cluster as intermediate product [112] and Raman spectroscopy and XRD have delivered some evidence [112] for the existence of Li$_2$B$_{12}$H$_{12}$ as an intermediate in the hydrogen desorption reaction Li[BH$_4$].

Figure 6.24 summarizes the energy levels for the hydrogen desorption reaction of Li[BH$_4$] considering the enthalpy of the reactants. The most stable state is Li[BH$_4$] in the low temperature structure *Pnma*, which is transformed into the high temperature modification (*P6$_3$mc*) at 391 K (see also Fig. 6.21), followed by melting around 280 °C. Subsequently, desorption of hydrogen is observed, via the above discussed intermediate(s), resulting in LiH and solid boron. Due to the high stability of LiH ($\Delta H_f = -90.7$ kJ), its desorption proceeds at temperatures above 1000 K and is therefore usually not accessible in technical applications.

Alkali tetrahydroborates are usually prepared by chemical methods, for example by the reaction of the metal hydride with diborane in a solvent (2 LiH + B$_2$H$_6$ → Li[BH$_4$] in tetrahydrofuran). The hydrogen desorption reaction from Li[BH$_4$] is reversible and the end products lithium hydride (LiH) and boron absorb hydrogen (Fig. 6.25) at 690 °C and 200 bar to form Li[BH$_4$] [113, 114]. The absorption reaction requires a long time (>12 h) and does not necessarily complete. However, the result is in agreement with the claim in Goerrig's patent about the direct formation of Li[BH$_4$] from the elements. The mechanism of the absorption of hydrogen by LiH and B to form Li[BH$_4$] is under investigation and the two following reaction paths will be discussed:

1. LiH and B react with each other to form an intermediate state, which is "filled up" with hydrogen, resulting finally in Li[BH$_4$].
2. Boron and hydrogen react to form diborane, which is known to spontaneously react with alkali hydrides to give alkali tetrahydroborates [115] (e.g. in the case of Na[BH$_4$])

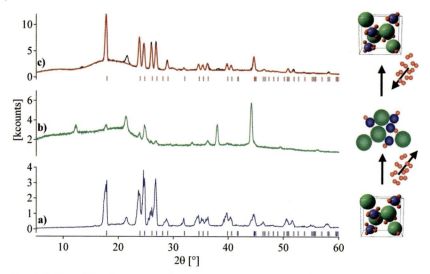

Fig. 6.25 X-ray diffraction pattern of Li[BH$_4$]: (a) the purchased sample under pure Ar, (b) after heating the sample to 1000 K in vacuum for 14 h and the desorption of 3 hydrogen atoms per formula unit (LiH + B), (c) after exposure of the desorbed sample at 1000 K to hydrogen at a pressure 15 MPa for 10 h and subsequently cooling the sample to room temperature.

The counterargument for the latter reaction path is that the reaction of hydrogen with boron is endothermic ($\Delta H° = 35.6$ kJ mol^{-1}, $\Delta S° = -79.2$ J K^{-1} mol^{-1}) [116] and will therefore only proceed at high temperatures and pressures (driven by entropy). The diborane in the gas phase reacts, upon cooling of the sample, with the solid LiH and the LiH transfers an H$^-$ to the BH$_3$ to form Li[BH$_4$] exothermically.

(a) H$_2$ + B → (BH$_3$)$_2$ at high temperature and high pressure
(b) 2 Li$^+$H$^-$ + (BH$_3$)$_2$ → 2 Li$^+$[BH$_4$]$^-$ upon cooling of the reactor

The release of diborane is observed during desorption of tetrahydroborates [117], probably as a concurrent reaction to the liberation of hydrogen. To form diborane during desorption, an H$^-$ atom is transferred from the borohydride complex to the positively charged Li leaving LiH and a neutral BH$_3$ molecule, which is the basic element of diborane (reaction (a)). Diborane evolution during desorption is thus likely to occur, but this does not provide us with evidence for its formation from the elements. Apart from the unfavorable thermodynamics for the formation of diborane, the inertness of boron in general is a kinetic barrier. It is not clear, whether this behavior originates from the particular modification of boron (passivation by oxides etc.) or from intrinsic properties of the material. Hints for the first point are given by the fact that the formation of Li[BH]$_4$ occurs if one starts from LiH and nanodispersed B obtained from desorption of the tetrahydroboride [113]. Recently, it was discovered that the kinetic barrier for the formation of tetrahydroborates is drastically reduced if metal–boron compounds, for example MgB_2, are used as starting materials instead of pure boron [118].

In the concurrent desorption reaction (second reaction), neutral H is removed from the complex leaving charged $(BH_x)^-$; 2H are recombined to H_2. However, $(BH_3)^-$ is highly unstable according to recent calculations [119]. Therefore, Ohba et al. [108] calculated the stability of complexes consisting of several Bs and Hs. This has the effect of weakening the influence of the charge by its 'distribution' over several atoms and the complex is stabilized. Figure 6.23 shows the proposed atomistic structure model of the most stable modification, monoclinic $(Li_2)^{2+}(B_{12}H_{12})^{2-}$. The recent progress of the study on the tetrahydroborates is summarized in the reviews [120, 121].

6.6
Complex Hydrides
Borislav Bogdanović, Michael Felderhoff, and Ferdi Schüth

6.6.1
Introduction

Complex hydrides are very interesting materials for hydrogen storage applications. They show a wide variety of compositions and structures and thus their properties with respect to hydrogen storage applications vary widely. Complex formation can be thought of conceptually as a means to stabilize unstable hydrides, such as the boron hydrides or aluminum hydride, by reaction with other binary hydrides. Most of these other binary hydrides are thermodynamically very stable, such as LiH, NaH or MgH_2. Thus not all the hydrogen stored in complex hydrides can be reversibly exchanged, upon decomposition of the complex hydrides part is retained in the stable binary light metal hydrides. From a thermodynamic point of view, the entropy of the decomposition reaction is dominated by the entropy contribution resulting from the formation of gaseous hydrogen. This amounts to 131 J K^{-1} mol^{-1}. At room temperature, the entropy contribution by release of gaseous hydrogen is therefore 39 kJ mol^{-1}. In order to be reversible, that is $\Delta G \cong 0$ kJ mol^{-1}, the enthalpy of decomposition also has to be 39 kJ mol_H^{-1}. Thus, the enthalpy of formation of the complex hydride gives a rough guideline as to whether a material may be reversible around room temperature. It should be around -40 kJ mol_H^{-1}.

The best known members of the class of complex hydrides are the aluminum hydrides and the boron hydrides, which have their original uses as mild and selective reducing agents in organic synthesis. Because they contain the light elements aluminum and boron, their gravimetric storage densities are rather attractive for mobile applications, especially if combined with a low atomic weight cation, such as lithium, beryllium, sodium, or magnesium. These materials will therefore be the focus of our attention in this chapter.

However, there are many more complex hydrides known. A continuously updated list can be found in the hydride database of Sandia National Laboratory [122]. Many entries in this list are complex platinum group metal hydrides or those containing rhenium, such as K_2ReH_9 [123] or Na_2PtH_4 [124]. These complex transition metal hydrides are well covered in a review by Bronger [125], who also described many of these compounds for the first time. Such complex hydrides, however, are for many reasons not suitable as hydrogen storage materials. Let us briefly recapitulate some of the main requirements, which need to be met by a commercially attractive hydride, keeping in mind that not all requirements have the same weight, depending on the field of application (mobile or stationary, small scale or large scale, cost-sensitive or non cost-sensitive product). Table 6.7 lists the properties that hydrides to be used in commercial applications should have. Disregarding all thermodynamic restrictions, especially the gravimetric storage capacity and the price, will exclude the complex platinum group metal hydrides from any applications. Even for the relatively "light" Li_3RhH_6 the gravimetric storage capacity is less than 5 wt.%, counting all six hydrogen atoms. However, one has to keep in mind that due to the stability of the alkali metal hydride for each alkali ion one hydrogen atom will not be useful for reversible hydrogen storage in the low or

Table 6.7 Requirements for complex metal hydrides as reversible hydrogen storage materials

Property	Target
Gravimetric storage density	>6.5 % for mobile high, but not as crucial for stationary
Volumetric storage density	>6.5 % for mobile high, but not as crucial for stationary
De-/rehydrogenation kinetics	Dehydrogenation <3 h, rehydrogenation <5 min
Equilibrium pressure	Around 1 bar near room temperature
Enthalpic effect	As low as possible (but basically fixed due to requirement of reversibility near room temperature)
Safety	As high as possible, that is no ignition on exposure to air or water
Cycle stability	>500 mobile
	>10 000 stationary
Memory effect (i.e. loss of storage capacity upon incomplete de- or rehydrogenation	Ideally absent
Price	As low as possible (rough guideline: 100 €/kg H_2)

Values are not strict thresholds and should be considered as approximate. Variations depend on target application, end-user philosophy, system requirements and state of the art already reached.

medium temperature range. Most other platinum group metal hydrides have even lower gravimetric storage capacities.

With respect to gravimetric storage density and price, the complex hydrides of the late elements of the first row of the transition metals seem to be more interesting. These compounds were extensively investigated by the group of Yvon. For instance, Mg_2FeH_6 [126] which crystallizes in the K_2PtCl_6 structure type, has a hydrogen storage capacity of about 5.5 wt.% and also a favorable volumetric storage capacity (density of $2.73\,g\,cm^{-3}$). However, the stability exceeds even the stability of MgH_2, which does not render it very promising for hydrogen storage applications. Similar stabilities were reported for other hydrides of this family, such as Mg_2CoH_5 [127]. Zn-based hydrides, such as K_2ZnH_4, appear to be less stable [128], but the gravimetric storage capacity is too low for practical applications. There is also a series of compounds based on complex magnesium hydride formed with the heavier alkali metals, which have interesting chemical or structural properties [129, 130]. However, all these compounds are again very stable and, since they only form with the heavier alkali metals, the gravimetric storage densities are too low.

This short survey shows, that the most promising classes of materials at present indeed appear to be the alanate- and boranate-based complex hydrides. There is a multitude of compounds known which fall into these groups (a selection is listed in Table 6.8), but unfortunately, many of them are so far only ill characterized and the thermochemistry of many of the compounds is either not known at all or the data are rather unreliable. The following discussion will therefore primarily focus on those materials which have been reasonably well characterized and show promise for reversible hydrogen storage in mobile and stationary applications.

Of the light metal hydrides, $NaAlH_4$ is the most thoroughly studied example. This is because this seems at present to be the most promising one for practical application in different areas. In addition to the favorable gravimetric storage density exceeding 5 %, this material was shown to be reversible, if titanium compounds are added to the alanate. Due to the high research activities and the promising properties of this system, the next (and main) section of this chapter will be devoted to titanium-doped $NaAlH_4$.

6.6.2
Titanium-Doped Sodium Aluminum Hydrides as Hydrogen Storage Materials

6.6.2.1 Preparation
The mixed ionic–covalent complex hydrides $NaAlH_4$ and Na_3AlH_6 have been known for many years and their crystal structures have been elucidated ($NaAlH_4$, Ref. [131]; Na_3AlH_6, Ref. [132]). $NaAlH_4$ has been synthesized by both indirect and direct methods and used as a chemical reagent (for a historical review of $NaAlH_4$ and Na_3AlH_6 synthesis, see Ref. [133]).

Table 6.8 Selected complex metal hydrides and their properties

Compound	Gravimetric storage capacity (total hydrogen^a)/wt.%	Enthalpy of formation/kJ mol$_H^{-1}$
LiAlH$_4$	10.5	−18
Be(AlH$_4$)$_2$	11.3	
NaAlH$_4$	7.4	−38
Mg(AlH$_4$)$_2$	9.3	
KAlH$_4$	6.1	
Ca(AlH$_4$)$_2$	7.8	
Ga(AlH$_4$)$_3$	7.4	
Ti(AlH$_4$)$_4$	9.3	
LiBH$_4$	18.4	−68
Be(BH$_4$)$_2$	20.7	
NaBH$_4$	10.6	−90
Mg(BH$_4$)$_2$	14.8	
Ti(BH$_4$)$_3$	13	decomposes at 25 °C
Zr(BH$_4$)$_4$	10.6	decomposes at 25 °C
Mg$_2$FeH$_6$	5.5	
K$_2$ZnH$_4$	2.7	

^{a} Due to the stability of the alkali metal hydrides, in most cases the reversible storage capacity close to room temperature is appreciably lower than this value.

For the preparation of *Ti-doped* sodium alanate three methods are, in principle, possible:

1. Doping of presynthesized NaAlH$_4$ by reaction with catalytic amounts of Ti-doping agents (Eq. (6.12));
2. The direct synthesis – consisting of reaction of NaH (or Na), Al powder with hydrogen in the presence of Ti-doping agents as catalysts (Eq. (6.13)) [134–137];
3. Hydrogenation of Al–Ti alloys together with NaH or Na (for the application of this method there exists little data [138, 139]).

$$\text{NaAlH}_4 + \text{Ti-doping agent} \rightarrow \text{Ti-doped NaAlH}_4 \quad (6.12)$$

$$\text{NaH(Na)} + \text{Al} + \text{Ti doping agent} + 3/2\text{H}_2(2\,\text{H}_2) \rightarrow \text{Ti-doped NaAlH}_4 \quad (6.13)$$

Reaction of NaAlH$_4$ with the Ti-doping agent (Eq. (6.12)) was originally carried out in organic solvents (toluene, ether, isopropanol) and, more recently, predominantly in the solid state via ball-milling. Titanium alkoxides, titanium tetra-n-butylate, Ti(OBu^n)$_4$ [133] and triisopropylate [140, 141] and, presently, especially α-titanium trichloride [142] and colloidal titanium nanoparticles [Ti*] [143, 144] are used as doping agents.

While the doping of NaAlH$_4$ (Eq. (6.12)) is still most convenient for research purposes, the direct synthesis (Eq. (6.13)) [134–137] appears to have some essential advantages:

- NaH or Na, and Al are cheap raw materials, anyway cheaper than NaAlH$_4$, so that the storage material has the potential for low cost.
- The Ti-doped storage material can be produced by ball-milling of NaH with Al in the presence of the doping agent and subsequent hydrogenation of the doped NaH + Al mixture (or even in one step?). This procedure seems to be economical and suitable for scale-up.
- There are indications [134, 136, 137] that the direct synthesis (Eq. (6.13)) delivers Ti-doped storage material with improved kinetics in comparison to that from the doping of NaAlH$_4$ (Eq. (6.12)).

6.6.2.2 Thermodynamics

The prime criterion for applicability of a metal hydride as a hydrogen storage material is its ready hydrogen dissociation and reassociation (reversibility) under the required range of temperature and pressure conditions. In thermodynamic terms (i.e. apart from kinetics), to achieve a dissociation pressure of 1–10 bar at 0–100 °C, suitable for many practical applications, the enthalpy of the dissociation must be in the range 15 to 24 kJ mol$_H^{-1}$ [145] ("thermodynamic window"). Due to this thermodynamic restriction, many of the (reversible) metal hydrides with a high hydrogen content are not applicable for hydrogen storage because of their too high or too low thermodynamic stability. So, for example, the reversible MgH$_2$/Mg system (7.6 wt.% H$_2$) does not seem to be practical for mobile hydrogen storage purposes [146, 147], since the dissociation pressure of 1 bar is achieved only somewhat below 300 °C [148]. However, because of the high dissociation enthalpy (74 kJ mol$_H^{-1}$) [149], the system has been investigated as a reversible heat storage system for the temperature range 300–450 °C [150]. Numerous attempts have been made, by alloying magnesium with a secondary metal(s), to increase the dissociation pressure of the resulting Mg-alloy hydrides ("thermodynamic tailoring") [145]. However, none of these systems appear to be applicable for instance as hydrogen stores for supplying PEM fuel cells with hydrogen. On the other, low thermodynamic stability side may, for instance, be mentioned LiAlH$_4$ (7.9 wt.% H$_2$, decomposition to LiH and Al) whose formation from LiH, Al and hydrogen is exothermic only to the extent of 9 kJ mol$_H^{-1}$ (Table 6.8) [151].

Already, many years before titanium-doped NaAlH$_4$ was recognized as a possible hydrogen storage medium [152], Dymova et al. [153, 154] showed that the two-step dissociation of NaAlH$_4$ into NaH, Al and hydrogen (Eq. (6.14)) is in fact reversible. However, due to the poor kinetics of the uncatalyzed system, the reverse reaction requires impractically severe reaction conditions (200–400 °C, that is, above the 183 °C melting point of the tetrahydride, and 100–400 bar).

$$NaAlH_4(l) \rightleftharpoons 1/3 Na_3AlH_6 + 2/3 Al + H_2 \rightleftharpoons NaH + Al + 3/2 H_2 \qquad (6.14)$$

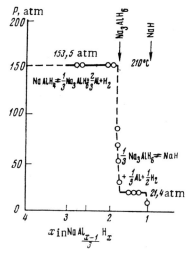

Fig. 6.26 Pressure–composition isotherms of the undoped NaAlH$_4$ and Na$_3$AlH$_6$ at 210 °C (from Dymova, Dergachev, Sokolov and Grechanaya (1975), Ref. [154]).

The PCI measurements carried out on the uncatalyzed system in desorption mode in the temperature range of liquid NaAlH$_4$ (Eq. (6.14)) (Fig. 6.26) resulted in values of 6.4 and 32.1 kJ mol$_H^{-1}$ for the first and the second dissociation step respectively [154]. These values are significantly higher than those based on measurements of PCIs on the Ti(OBu^n)$_4$ catalyzed solid NaAlH$_4$ (Eq. (6.15), Fig. 6.27) [133], 4.5^1 and 15.7 kJ mol$_H^{-1}$. For this reason, PCIs of metal-doped and undoped NaAlH$_4$ deserve to be re-examined (see also Section 6.6.2.7).

The thermodynamics of the Ti-catalyzed NaAlH$_4$ system (Eq. (6.15))

$$\text{NaAlH}_4(s) \underset{(a)}{\overset{\text{Ti}}{\rightleftharpoons}} 1/3\,\text{Na}_3\text{AlH}_6 + 2/3\,\text{Al} + \text{H}_2 \underset{(b)}{\overset{\text{Ti}}{\rightleftharpoons}} \text{NaH} + \text{Al} + 3/2\,\text{H}_2 \quad (6.15)$$
$$\hspace{6cm} 3.7\,\text{wt.\%} \hspace{3cm} 5.6\,\text{wt.\%}$$

$\rightarrow$ 1st dehydrogenation step
$\leftarrow$ 2nd hydrogenation step (6.15a)

$\rightarrow$ 2nd dehydrogenation step
$\leftarrow$ 1st hydrogenation step (6.15b)

are classic and comparable with many other metallic, ionic and covalent hydrides. As shown in Fig. 6.27, typical low-hysteresis, two-plateau absorption and desorption isotherms can be measured [133]. Temperature dependences of the plateau pressures P are given by the van't Hoff equation

$$\ln P = \Delta H/RT - \Delta S/R,$$

where ΔS is the entropy change, R is the gas constant and the enthalpy changes ΔH for the NaAlH$_4$ and Na$_3$AlH$_6$ decompositions are about 18.5 kJ mol$_H^{-1}$ and

---

^{1} Calculated value for liquid NaAlH$_4$ [12].

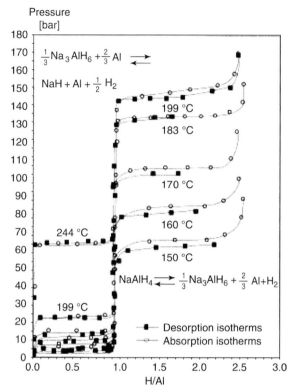

Fig. 6.27 Pressure–composition isotherms for the Ti-doped NaAlH$_4$ and Na$_3$AlH$_6$ (from Bogdanović, Brand, Marjanović, Schwickardi, and Tölle (2000), Ref. [133]).

15.7 kJ mol$_H^{-1}$, respectively. This allows direct comparison of the Na-alanate thermodynamic properties with those of other typical hydrides by means of the van't Hoff diagram of Fig. 6.28. We see that NaAlH$_4$ has thermodynamics comparable to those of the classic low-temperature (LT) hydrides LaNi$_5$H$_6$ and TiFeH – that is, in the range useful for a near-ambient-temperature hydrogen storage. Na$_3$AlH$_6$ requires about 110 °C for H$_2$ liberation at atmospheric pressure and is therefore less convenient for supplying a fuel cell that provides waste heat ($\approx$80 °C) The temperature level in such a system, which is discussed for applications in cars, is not sufficient for H$_2$ desorption. On the other hand, the Na$_3$AlH$_6$ van't Hoff line falls between the LT intermetallic hydrides and the high-temperature (HT) Mg and Mg-alloy hydrides, thus filling a moderate-temperature (MT) gap that might be useful for other applications than hydrogen storage, such as heat pumping and heat storage [133].

Figure 6.29 gives a general view of the possibilities for operation of these new hydride systems, neglecting their kinetic properties. The two curves in Fig. 6.29 divide the *linear P–T* map into three, or more precisely, four areas: in the stronger shaded area solid (s) NaAlH$_4$ is thermodynamically stable; above 183 °C is the area of stability of liquid (l) NaAlH$_4$. Na$_3$AlH$_6$ is stable inside the slightly shaded area

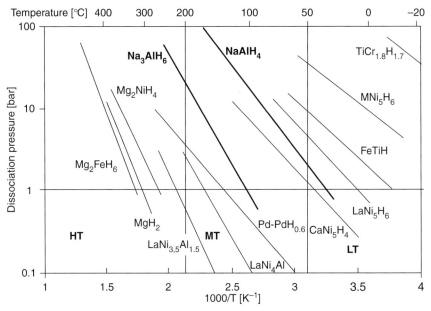

Fig. 6.28 A van't Hoff diagram comparing dissociation pressures of the Ti-doped NaAlH$_4$ and Na$_3$AlH$_6$ with those of other well-known hydrides (from Ref. [133]). HT, MT and LT stand for high-, medium- and low-temperature hydrides, respectively. Mm in MmNi$_5$H$_6$ is misch metal, a low-cost mixture of rare-earth elements.

and the blank area represents the region of stability of NaH/Al mixtures under hydrogen pressure. Crossing of the borders between the three (or four) areas in the P–T map from left to right is necessarily associated with absorption of heat and release of hydrogen and, in the reverse direction, with release of heat and absorption of hydrogen (marked in Fig. 6.29 by arrows). Thus, crossing the left curve from left to right results in the release of (maximum) 3.7 wt.% H and requires 18.5 kJ mol$_H^{-1}$ of heat. Starting from Na$_3$AlH$_6$ (i.e. in the absence of Al) and crossing the second border from left to right, is associated with evolution of a maximum of 3.0 wt.% H and absorption of 15.7 kJ mol$_H^{-1}$ of heat. If both borders are crossed in the same direction, a maximum of 5.6 wt.% of hydrogen can be released, while 17.7 kJ mol$_H^{-1}$ of heat have to be fed to the system. The doped NaAlH$_4$ system thus offers the possibility to be utilized over the full range of the hydrogen absorption, or the two dissociation steps can be applied separately for hydrogen or heat storage. In order to utilize only the first dissociation step for hydrogen storage (Eq. (6.15a)) the H$_2$-discharging temperature should be kept below or around 100 °C, at which temperatures the hydrogen dissociation pressure of Na$_3$AlH$_6$ (Fig. 6.29(a)) is still very low, while the dissociation of NaAlH$_4$ is already thermodynamically strongly favored.

As outlined above, utilization of the doped Na$_3$AlH$_6$ system (Eq. (6.15b)) for the purpose of reversible thermochemical heat storage, heat transformations and so

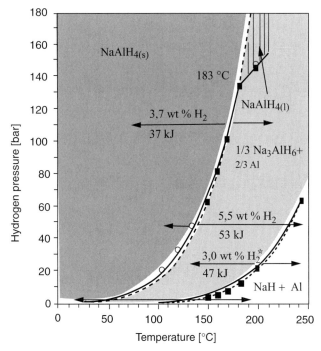

Fig. 6.29 Pressure–temperature relations for the Ti-doped (2 mol%) NaAlH$_4$ systems; (-■-) desorption pressure; (-○-) absorption pressure; (∗) hydrogen storage capacity of Na$_3$AlH$_6$ in absence of Al (from Bogdanović, Brand, Marjanović, Schwickardi, and Tölle (2000), Ref. [133]).

on, is of particular interest because the system could operate under moderate H$_2$ pressures in the *medium* temperature region (e.g. 150–250 °C; Fig. 6.29).

An obvious disadvantage of the doped NaAlH$_4$ (Eq. (6.15a, b)) as a *hydrogen storage* system is that pressure and temperature conditions for charging and discharging of hydrogen across both dissociation steps differ considerably from each other: in order to discharge hydrogen until the NaH + Al stage at normal pressure, the discharge temperature must be raised well above 100 °C; on the other hand, the high dissociation pressure of NaAlH$_4$ opposes charging of the system with hydrogen back to the NaAlH$_4$ stage, so that relatively high hydrogenation pressures must be applied. This lowers the energetic efficiency of the system. In principle it should be possible to tailor the *thermodynamic* properties of the system (Section 6.6.2.7) in such a way that the two curves in Fig. 6.29 come closer together, thus eliminating the problem. It has been shown [152] that the dissociation pressure of the doped Na$_3$AlH$_6$ can be lowered to the extent of 20 bar by substituting one of the Na atoms of the compound by Li, although one would rather have an increase in dissociation pressure. However, this demonstrates that the thermodynamic properties can be tuned to some extent.

Another possibility is perhaps the 'catalytic option': the difference between the dissociation pressures of NaAlH$_4$ and Na$_3$AlH$_6$ (Fig. 6.29) diminishes as the temperature is lowered, and thus conditions necessary for hydrogen discharging and recharging approach each other. The possibility to operate the system at low(er) temperatures depends primarily upon the activity of the applied catalysts (dopants). Improvement of the catalytic activity of the dopants would thus ameliorate a major disadvantage of this hydrogen storage system. With a perfect catalyst, it should be possible to discharge hydrogen from the system up to the NaH + Al stage under normal pressure at a temperature slightly above 100 °C and to recharge it with hydrogen at 30 °C and hydrogen pressures above 1 bar (marked by the arrow at the bottom of the diagram Fig. 6.29). Also for this reason further improvement of the catalyst's activity is of primary importance (see next section).

6.6.2.3 Influence of Catalysis and of Other Factors on NaAlH$_4$ Deabsorption and Reabsorption Rates

The practical use of Reaction (6.15)) requires catalysis to achieve sufficiently fast rehydrogenation and dehydrogenation rates. Our first work on catalyzed alanates at MPI-Mülheim was based on studies that used transition-metal catalysts for the preparation of MgH$_2$ [155]. We doped NaAlH$_4$ with Ti by solution-chemistry techniques whereby nonaqueous liquid solutions or suspensions of NaAlH$_4$ and either TiCl$_3$ or the alkoxide Ti(OBu^n)$_4$ catalyst precursors were decomposed to precipitate solid Ti-doped NaAlH$_4$ [133, 152]. An alternative approach was taken by Jensen *et al.* and Zidan *et al.* at the University of Hawaii, whereby the liquid Ti(OBu^n)$_4$ precursor was simply ball-milled with the solid NaAlH$_4$ [140, 141]. They also added Zr(OPr^i)$_4$ to help stimulate the kinetics of the second step (Na$_3$AlH$_6$ decomposition) of Reaction (6.15). Zaluska *et al.* [156] also obtained positive results by ball-milling with carbon as did Meisner *et al.* with diamond [157] (i.e. not using a transition-metal catalyst). Ball-milling with alkoxide catalyst precursors does result in *in situ* decomposition during at least the first several dehydriding/hydriding cycles, resulting in significant contamination of the H$_2$ with hydrocarbons; using the inorganic TiCl$_3$ catalyst precursor is clearly better if one wishes to use the ball-milling catalyst-doping approach [158].

Sandrock *et al.* have performed detailed desorption-kinetics studies (forward reactions, both steps, of Reaction (6.15)) as a function of temperature and catalyst level (Fig. 6.30) [142]. Both the NaAlH$_4$ and Na$_3$AlH$_6$ decomposition reactions obey classic, thermally activated behavior, consistent with the Arrhenius equation

$$\text{Rate} = k \, \exp(-Q/RT),$$

where k is the pre-exponential rate constant and Q is the activation energy. For example, the Arrhenius plots for undoped and 4 mol% Ti-doped Na alanate are shown in Fig. 6.31. Catalysis results in multiple order-of-magnitude increases in kinetics for both reactions. This important practical result of catalysis is a combined result of both changes in slopes (Q) and intercepts (k) of the Arrhenius lines. If one plots the activation energy Q as a function of catalyst loading, as shown in

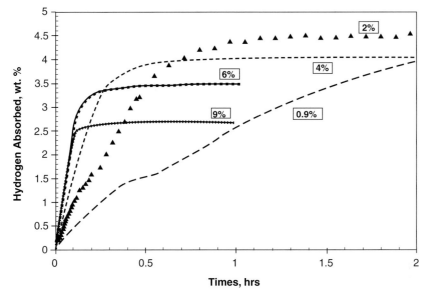

Fig. 6.30 H_2 absorption curves starting from NaH + Al (dehydrided Ti-doped NaAlH$_4$), as a function of added TiCl$_3$ catalyst precursor (expressed in mol %). Initial temperature $T_i = 125\,°C$. Applied hydrogen pressure $P_{H_2} = 81$–90 atm (from Sandrock, Gross, and Thomas (2002), Ref. [142]).

Fig. 6.32, it is very clear that the smallest Ti addition (0.9 mol%) has marked effects on lowering Q for both steps of Reaction (6.15). However, further increases in Ti level have no effects, at least not on Q. The dramatic decrease in Q indicates a major discontinuous change in the thermally activated mechanism. Further increases in desorption kinetics accrue with further increases in Ti level, but only through increases in the pre-exponential rate constant k [142].

Using the direct synthesis method [134–137], it has been shown that TiCl$_2$, TiF$_3$ and TiBr$_4$ as dopants also effectively improve NaAlH$_4$ desorption kinetics. Arrhenius data indicate that the catalyst precursors behave essentially in the same manner, with desorption rates in the order of highest to lowest corresponding to the dopants TiCl$_2$, TiCl$_3$, TiF$_3$, TiBr$_4$ [159, 160]. Furthermore, storage materials with comparable desorption rates can be directly prepared also by using TiH$_2$ or pre-reacted TiCl$_2$ with LiH as dopants [160].

The dramatic effects of Ti catalyst level, (using the dry TiCl$_3$–NaAlH$_4$ ball-milling/doping technique [142]) on $125\,°C$ *hydriding* kinetics and capacity are shown in the hydriding curves of Fig. 6.31. The kinetics are increased and capacity decreased with increasing catalyst level. Nearly identical catalyst effects are seen with desorption experiments, although the two steps of Reaction (6.15) are more clearly sequential in desorption Fig. 6.32 [142, 161]. It is thus obvious that catalysis affects both directions of Reaction (6.15) as well as both steps. For TiCl$_3$-doped NaAlH$_4$, dynamic *in situ*

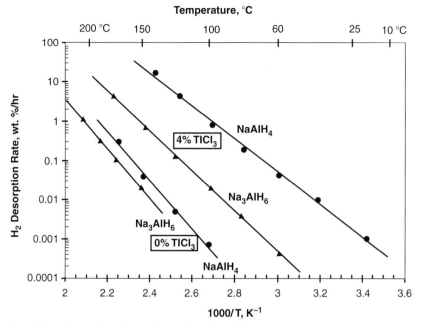

Fig. 6.31 Arrhenius lines for NaAlH$_4$ and Na$_3$AlH$_6$ decompositions for 0 and 4 mol% added catalyst precursor TiCl$_3$ (from Sandrock, Gross, and Thomas (2002), Ref. [142]).

X-ray diffraction allows one to determine hydrogen deabsorption and reabsorption rates under very low temperature and pressure conditions [161].

The highest rehydrogenation rates for Ti-doped NaAlH$_4$ recorded up to now were achieved by using colloidal titanium nanoparticles [162] (Ti*) as doping agents. At the beginning of cycle tests, under appropriate cycling conditions (102 °C/133–115 bar), full hydrogenation is achieved in less than 15 min [143], with a storage capacity of 4.6 wt.% H$_2$. Similarly high or higher (?) hydrogenation rates using the same dopant have been reported by Fichtner *et al.* [144]. It should, however, be added that the preparation of the Ti* dopant is complicated [162], and that the Ti*-doped NaAlH$_4$ exhibited low stability in cycle tests [143].

Most catalyst work thus far has concentrated on the transition metals Ti, Zr, Fe or on Ti/Fe [133] and Ti/Zr [140, 141] combinations, along with the nonmetals carbon [156] and diamond [157]. A screening of a large number of transition and non-transition metal chlorides as dopants for NaAlH$_4$ has been carried out with respect to the rate of first dehydrogenation of the doped materials. Titanium and zirconium have been identified as the most active catalysts [163], thus confirming the earlier assertions [133, 140, 141]. The influence of ball-milling time on the dehydrogenation rate of doped NaAlH$_4$ was examined using TiCl$_3$ as a dopant and the optimum was found to be ≈15 min [163]. It should, however, be added that with most ball-milling doping procedures thus far, the ball-milling times were

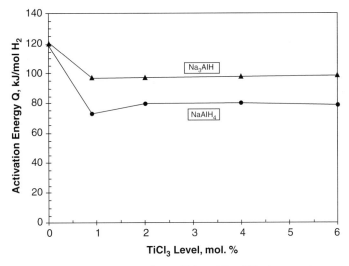

Fig. 6.32 Activation energies Q for NaAlH$_4$ and Na$_3$AlH$_6$ decompositions as function of added TiCl$_3$ (from Sandrock, Gross, and Thomas (2002), Ref. [142]).

considerably longer. In addition, one should keep in mind that the effect of ball-milling is highly dependent on various parameters, such as sample size, ball size, filling degree of the milling vessels and others, which are not easily transferable between mills of different types.

In an attempt to prepare a mixed Na–Li-alanate, a 1:1 NaH/LiAlH$_4$ mixture was ball-milled for a long period of time (30 h), resulting mainly in the formation of Na$_2$LiAlH$_6$. Upon addition of 10 wt.% of La$_2$O$_3$ as a further component during ball-milling, a significant increase in dehydrogenation rate and retention of hydrogen capacity in a four cycle test were observed [164].

6.6.2.4 Hydrogen Storage Capacity and Cyclic Stability

One of the main goals associated with the development of Ti- (or other metal) doped NaAlH$_4$ as hydrogen storage materials is to achieve the highest possible hydrogen storage capacity in a large number of cycles. The calculated hydrogen storage capacity of neat NaAlH$_4$ (Eq. (6.15)) is 5.6 wt.% H$_2$. However, in practice, the capacity is always lower, due to the added weight of the dopant, to the consumption of a fraction of NaAlH$_4$ in the doping reaction and incomplete rehydrogenation. In general, the lower the dopant concentration the higher the maximum storage capacity. If the stoichiometry of the doping reaction is known (see Section 6.6.2.5), the theoretically achievable storage capacity can be calculated.

A major disadvantage of the NaAlH$_4$ storage materials investigated previously is that storage capacities in cycle tests were always lower than the theoretically expected capacity. The causes for this detrimental effect were investigated by means of XRD analysis and solid state NMR spectroscopy and will be discussed in Section 6.6.2.5.

Reversible storage capacities of 4.9–5.1 wt.% H_2, close to the theoretical limit, were achieved in cycle tests using nanoparticulate titanium nitride [165] (TiN*) as a dopant [143]. For the second hydrogenation step (Eq. (6.15a)), a storage capacity of 3.3 wt.%, which is again close to the theoretical limit (3.7 wt.%), was achieved. Compared to Ti* (Section 6.6.2.3), however, hydrogenation rates with TiN* as a catalyst are more than 40 times lower [143].

Attaining *cycle stability*, both with respect to storage capacity and dehydrogenation and rehydrogenation rates, is still considered to be one of the most demanding tasks in bringing a hydride-hydrogen storage system to practical application.

A directly synthesized $NaAlH_4$ doped with 4 mol% of $TiCl_3$ exhibited reversible hydrogen absorption over 116 cycles. After an activation process, the particular sample showed very stable reversible hydrogen capacities and sorption rates [137].

Earlier, a 100 cycle test was carried out using Na_3AlH_6 doped with 2 mol% of $Ti(OBu^n)_4$; in the course of the test a minor decrease in storage capacity was observed [152]. Lately, cyclic stability with respect to storage capacity could be demonstrated in a 100 cycle test, using a commercial $NaAlH_4$ and doped with 2 mol% of $TiCl_3$ by ball-milling. For the test a self-designed automatic cycling facility was used. As can be seen in Fig. 6.33, under the specified hydrogenation conditions, the capacity of 4.0–4.2 wt.% H_2 is maintained throughout the test [166]. Also the dehydrogenation and rehydrogenation rates during the test were essentially constant. The observed almost complete hydrogen uptake in the second hydrogenation step (Eq. (6.15a)) is ascribed, among other factors, to the content of free Al in the used commercial $NaAlH_4$ (cf. Section 6.6.2.5).

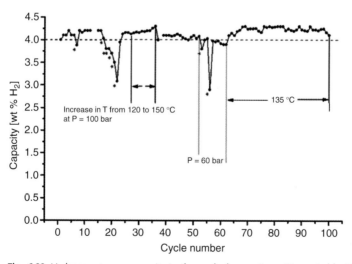

Fig. 6.33 Hydrogen storage capacity in the 100 cycles test. Hydrogenation conditions, unless otherwise noted, 120 °C/95–100 bar (isobaric). In cycles marked by asterisk*, insufficient time was allowed to complete hydrogenations (Presented by F. Schüth at the Gordon Research Conference on Hydrogen Metal Systems, July 13–17 2003, Colby College, Maine, USA, Ref. [166]).

6.6.2.5 Investigation of the Reversible Ti- or Zr-Doped NaAlH$_4$ System by Means of Physical Methods; Application of Computational Methods

Several authors have already investigated the doped NaAlH$_4$ system by means of X-ray diffraction [139, 157, 161], [167–169].

The thermal hydrogen desorption from both undoped and Ti/Zr-doped NaAlH$_4$ was studied by dynamic X-ray diffraction [167]. Enhanced desorption kinetics for the doped NaAlH$_4$ was clearly demonstrated. As a part of the dissociation reaction, formation of Al crystallites (>100 nm) was established. Hydrogen absorption and desorption from NaAlH$_4$ doped with TiCl$_3$ was measured by dynamic X-ray diffraction under conditions similar to those prevailing under fuel cell operation [161]: hydrogen absorption, 50–70 °C/10–15 bar; hydrogen desorption, 80–110 °C/5–100 mbar. Dynamic XRD enables direct measurement of individual reaction rates.

The processes occurring in the course of two sequential discharging and recharging cycles of Ti(OBu^n)$_4$-doped NaAlH$_4$ were investigated in parallel using X-ray diffraction and solid-state NMR spectroscopy. Both methods demonstrate that in hydrogen storage cycles (Eq. (6.15)) the main phases involved are NaAlH$_4$, Na$_3$AlH$_6$, Al and NaH (Fig. 6.34). Only traces of other, as yet unidentified phases are observed, one of which has been tentatively assigned to an Al–Ti alloy (but later to an amorphous Ti, see below) on the basis of X-ray diffraction analysis. The unsatisfactory hydrogen storage capacities heretofore observed in cycle tests are shown to be due

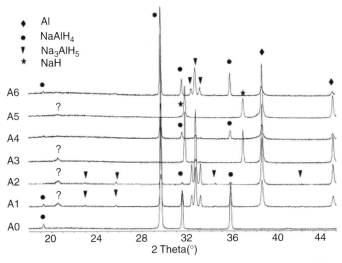

Fig. 6.34 Comparison of the powder pattern of the analyzed samples A0–A6, the reflections of the crystalline phases are marked by symbols (A0, NaAlH$_4$ doped with 2 mol% of Ti(OC$_4$H$_9$)$_4$ in toluene; A1–A6 denote samples after 50 % dehydrogenation of the first step, dehydrogenated first step, dehydrogenated second step, rehydrogenation, dehydrogenated second step and rehydrogenation, respectively). The reflections which could not be assigned to any known phase are marked by ? (from Bogdanović, Felderhoff, Germann, Härtel, Pommerin, Schüth, Weidenthaler, and Zibrowius (2003), Ref. [168]).

entirely to the second hydrogenation step (Eq. (6.15a)) being incomplete [168]. A possible scenario could be a mass transfer problem: during dehydrogenation rather large metallic aluminum particles are formed, as evidenced by the occurrence of eddy currents in the NMR experiment and by SEM investigations (see below). During rehydrogenation metallic aluminum particles would first react with NaH to Na_3AlH_6 and, in parallel, to Na_3AlH_6 and the residual aluminum would react to $NaAlH_4$. Those aluminum particles, which are rather large after dehydrogenation, would, after some rehydrogenation cycles, consist of a center of aluminum coated with $NaAlH_4$. The aluminum in the particle's core might then not be reached by the reactant Na_3AlH_6 formed in other parts of the sample, thus terminating the rehydrogenation reaction. This hypothesis could be confirmed by the XRD- and NMR-based finding that an excess of aluminum in the system results in complete rehydrogenation in the second step, so that the full storage capacity can be exploited [168].

X-ray diffraction and ^{27}Al-NMR spectroscopy measurements were used by Balema et al. [139] to study the ball-milling induced transformation of $LiAlH_4$ into Li_3AlH_6, Al and H_2, which is catalyzed by Ti and Fe.

One of the key questions relevant to the reversible Ti- (or Zr) doped $NaAlH_4$ system is the identity of the genuine catalyst and how it accelerates the forward and back reactions in such a dramatic manner [142].

Hydrogen dissociation by Ti-catalyzed $NaAlH_4$ was proven by static hydrogen isotope scrambling experiments [170].

Reduction of Ti-catalyst precursors (dopants) to the zerovalent (metallic) stage, both for wet chemical [133] and the ball-milling doping method [170], has been proven by measuring the amount of hydrogen evolved during the doping processes. For $Ti(OBu^n)_4$ [133] and $TiCl_3$ as dopants [142, 170], the reduction processes can be described by Eqs. (6.16) and (6.17) respectively:

$$xNaAlH_4 + Ti(OBu^n)_4 \xrightarrow{toluene} (x-1)NaAlH_4 + Ti + NaAl(OBu^n)_4 + 2H_2 \uparrow \quad (6.16)$$

$$xNaAlH_4 + TiCl_3 \xrightarrow[milling]{ball} (x-3)NaAlH_4 + Ti + 3Al + 3NaCl + 6H_2 \uparrow \quad (6.17)$$

It should, however, be mentioned that Eqs. (6.16) and (6.17) represent only the overall stoichiometry of the reduction processes, without referring to the possible postreduction changes of finely dispersed metals, for example formation of alloys [170], as shown below.

Haber et al. [171] described a "chemical synthesis" of Al–Ti alloys through reduction of $TiCl_3$ with $LiAlH_4$ in boiling mesitylene and subsequent heating of the reaction products to 550 °C. Depending on the molar ratio of the reactants, two different Al–Ti alloys, Al_3Ti and $AlTi$, could be prepared and identified by XRD. Based on XRD, Balema et al. report on the formation of Al_3Ti after ball-milling of a stoichiometric $LiAlH_4/TiCl_4$ mixture [139] and Majzoub and Gross observed via XRD formation of Al_3Ti upon ball-milling of stoichiometric $NaAlH_4/TiCl_3$ (3:1) mixture [159]. In contrast, no formation of alloys was observed by Sun et al. after ball-milling of NaAlH$_4$ with 10 mol% Ti- or Zr-alkoxides [173]. Instead, the authors report the incorporation of the catalyst into the crystal structure of NaAlH$_4$ via substitution of Na$^+$ by Ti$^{4+/3+/2+}$. This was inferred from observed changes of crystal lattice constants depending upon the level of doping. According to our own XRD measurements [169], however, the lattice constants of Ti- or Zr-doped NaAlH$_4$ determined for different samples by Rietveld refinement give no evidence for a change of lattice constants, and thus for incorporation of Ti or Zr ions into the crystal structure of NaAlH$_4$. In addition, replacement of Na$^+$ by a titanium ion would not be expected from a crystal chemical point of view.

On the other hand, the above-mentioned formation of the Al$_3$Ti alloy found in stoichiometric Na(Li)AlH$_4$/Ti-chloride reactions [139, 160], was not observed when NaAlH$_4$ was doped with less than 5 mol% titanium. The latter doping level is typical for preparation of samples for hydrogen storage experiments. Also no reflexions characteristic for metallic titanium could be observed [168, 169].

In a crucial experiment [169], after ball-milling NaAlH$_4$ with 4 mol% of TiCl$_3$ (Eq. (6.17)) and removal of excess NaAlH$_4$ by extraction with THF, the XRD of the solid residue (Fig. 6.35, RT) exhibits only reflexions for Al and NaCl. However, upon heating of the sample to 200, 400 and 830 °C (Fig. 6.35) the gradual formation

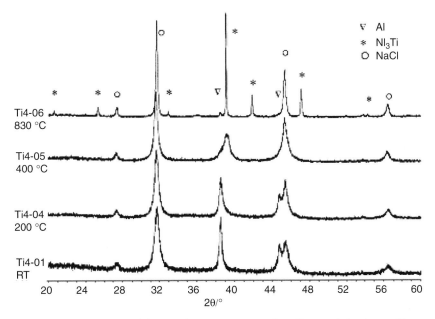

Fig. 6.35 Powder patterns of the sample Ti4-01, prepared by ball-milling NaAlH$_4$ with 4 mol% TiCl$_3$ for 3 h and removal of NaAlH$_4$ by extraction with THF. Ti4-04–06, powder patterns of materials obtained by thermal treatment of Ti4-01 at different temperatures. The reflections belonging to different phases are marked by different symbols (from Weidenthaler, Pommerin, Felderhoff, Bogdanović, and Schüth (2003), Ref. [169]).

of the Al$_3$Ti phase, with concomitant consumption of Al can be followed. (A similar reaction can be expected in the non-extracted doped alanate.) These results suggest that the doping reaction proceeds as given above (Eq. (6.17)), involving reduction of titanium to the zerovalent state, and that the titanium formed is present predominantly as an X-ray amorphous phase. Aluminum exists in the form of bulk metal. The simultaneous decrease in elemental Al and increase in Al$_3$Ti during heat treatment is an indication that initially titanium and aluminum form separate phases. Thermal treatment makes the atoms sufficiently mobile, so that the alloy can develop.

Interestingly, however, for the case of the Zr-doped NaAlH$_4$ system, according to XRD an Al–Zr alloy forms already during the doping procedure and remains present during dehydrogenation and rehydrogenation cycles [169].

The results of XRD investigations have been confirmed and extended by SEM- and TEM-EDX examinations of the materials. Ti(OBu^n)$_4$-doped NaAlH$_4$ in the dehydrogenated state was examined by the combined SEM-EDX method. According to Na and Al EDX mappings, in the dehydrogenated state the doped material is segregated into NaH and Al phases [133]. Also the microstructure of a Ti/Zr-doped NaAlH$_4$ was characterized by means of the SEM-EDX technique [174]. After five cycles, in the dehydrogenated state the Ti/Zr doped NaAlH$_4$ was found to be present in the form of smooth modules and porous surfaces. A correlation was observed between the Ti catalyst and Al-rich porous surfaces and the lack of Ti in the Na-rich modules. This finding suggested that the Ti catalyst may be chemically associated with Al from the alanate [174].

A recent TEM-EDX investigation [175] of dehydrogenated Ti-doped NaAlH$_4$ was carried out using Ti(OBu^n)$_4$ and Ti* [162] (cf. Section 6.6.2.3) as dopants. Accordingly, the Ti(OBu^n)$_4$-doped NaAlH$_4$ in the dehydrogenated state consists of a crystalline Al and an amorphous NaH phase. The striking result of EDX analyses is that the Ti-dopant is found to be present only in the Al phase. Similar, although not as clear-cut, results were obtained in the case of Ti* as a dopant, since the material lacks crystallinity and the boundaries between the Ti-containing Al phase and NaH phase are diffuse [175].

It is of particular relevance to discuss the present TEM-EDX result [175] in conjunction with the above-described XRD investigations [169]. From the results gained by these two methods, it follows that in both the hydrogenated (XRD) and the dehydrogenated state (TEM-EDX) the Ti-catalyst exists only in the Al phase and is absent in the NaAlH$_4$ and NaH phases. In the *hydrogenated* state there is only a small amount of the Al–Ti phase remaining and so the content of Ti in that phase is found to be high. In the *dehydrogenated* state the content of Ti in the Al phase depends on the doping level and is typically 2–4 at%. From this it follows that during hydrogen discharging and recharging cycles the content of Ti in the Al phase permanently "oscillates" between "Ti-poor" (dehydrogenated) and "Ti-rich" (hydrogenated) states. Further information concerning the state of titanium in the Al–Ti phase during the processes of hydrogen discharging and recharging has been obtained via XAFS investigations [175]. The XAFS and XANES spectra of Ti-doped NaAlH$_4$ after doping, after several and after 100 cycles are very close to each other. This reveals that there is little structural change in the material, and that the Ti species formed after doping remains almost unchanged in the course of the cycle test. The XAFS spectra indicate that Ti is coordinated predominantly by Al, that is, it is almost atomically dispersed in the Al phase.

On the basis of the considerations discussed above, it appears reasonable to assume that the location of reaction events (Eq. (6.15)) is at the phase boundary between Al/Ti and NaH, Na_3AlH_6 or $NaAlH_4$ phases (cf. discussion in the third paragraph of this section). This view seems to be supported by the clear analogy between the $NaAlH_4$ (Eq. (6.15)) and the Mg_2FeH_6 [126, 127] reversible system (Eq. (6.18)) [175–177]. Aside from the presence of the catalyst in the Al phase, in both cases formation of a *homogeneous ternary hydride phase* ($NaAlH_4$ or Mg_2FeH_6) takes place via a *heterogeneous* reaction of a metal (Al or Fe) with a binary metal hydride phase (NaH or MgH_2) and hydrogen. (In place of MgH_2, Mg can also be used.)

$$2\,MgH_2(Mg) + Fe + H_2(3\,H_2) \rightleftharpoons Mg_2FeH_6 \qquad (6.18)$$

According to TEM (Fig. 6.36a–c), the formation of Mg_2FeH_6 indeed takes place at the phase boundary between Fe particles and the growing Mg_2FeH_6 phase. The growth of the Mg_2FeH_6 phase takes place by insertion of newly formed Mg_2FeH_6 layers between the two phases (for details, see Ref. [176]).

Although at present only in the early stages, it can be expected that employment of computational methods, alone or, if necessary, in combination with experimental results, will in the future contribute to progress in the field of light metal hydrogen storage materials. So, for instance, a theoretical approach could be useful to find a way to selectively increase the Na_3AlH_6 plateau pressure without affecting that of $NaAlH_4$ [178].

Recent publications propose a reaction path where Ti (on Al-surfaces [179]) dissociates hydrogen [180] which reacts with the Al-surfaces and forms volatile AlH_x. Experimental support for this idea was provided by INS spectroscopy, which showed the presence of AlH_x during rehydriding of hydrogen-depleted sodium aluminum hydride [181].

Raman and IR spectra of mixed $Li_3AlH_nD_{6-n}$ compounds have been investigated in order to study the behavior of the corresponding Al–anions and Al–H interactions. A vibrational analysis *and force field calculations* have been performed using theoretical models and the experimental frequency assignments have been confirmed. The environment of aluminum atoms can be fully described in terms of strong covalent interactions. The alkali metal atoms (M = Li, Na) are purely ionic and there is no evidence for a M–H interaction. A model for the formation of $Li_2AlH_4D_2$ from $LiAlH_4$ and 2 LiD has been proposed [182].

The electronic structure and structural stability of $NaAlH_4$ have been studied with *ab initio* methods using the VASP code for different possible structure modifications. The study predicts that α-$NaAlH_4$ converts to β-$NaAlH_4$ at 6.43 GPa with a 4 % volume contraction [183]. It is suggested, in line with the findings for TiO_2 [184], to search for possibilities to stabilize the high-pressure phase β-$NaAlH_4$ at ambient pressure, perhaps by chemical means, that moreover may open up pathways for improved reversible hydrogen deabsorption and reabsorption.

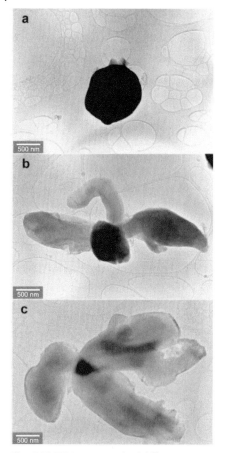

Fig. 6.36 TEM micrograph of different stages of the initial Mg_2FeH_6 formation are recorded. Dark regions of the particles are Fe regions and the lighter ones consist of Mg_2FeH_6. (a) Initial stage of the Mg_2FeH_6 formation; (b) vermicular excrescence of Mg_2FeH_6 out of the surface of an iron seed; (c) final stage of the Mg_2FeH_6 formation: only a small Fe seed, embedded in the Mg_2FeH_6 matrix, remains (from Bogdanović, Reiser, Schlichte, Spliethoff, and Tesche (2002), Ref. [176]).

Enthalpies of formation (ΔH_f) and disproportionation reaction enthalpies (ΔH_R) were computed for Na–Al–H compounds using density functional theory. For the optimized structures (calculated for a temperature of 0 K), the computed ΔH_R are in good agreement with experimental data. The calculations identified a ground state hydrogen position in all known Na–Al–H phases and two structural candidates for the lesser-known high temperature β-Na_3AlH_6 phase [185].

Using *ab initio* computations the stability of complex hydrides related to the alanates $NaAlH_4$ and Na_3AlH_6 was investigated [186]. It was found that Li substitution for Na reduces the hydrogen affinity of these materials (which is, however, opposite to the experimental finding [152]), while K increases it. Substitution on the

Al site by B or Ga markedly reduces the stability of the intermediate Na_3AlH_6 compound. The results indicate that both the overall stability and, as a consequence, the hydriding temperatures and rehydriding pressures, can be tailored by combined substitution on the Na and Al site of the complex hydride $NaAlH_4$. Thus a larger amount of hydrogen could be released in a narrower range of temperature and pressure [186].

6.6.2.6 Engineering and Safety Aspects; Encapsulation

There are only a few, very preliminary, data on the engineering properties of doped alanates [158]. Hydrogen desorption/absorption studies were done on a 100 g scale-up reactor, designed to simulate the heat transfer and gas impedance conditions of a larger engineering bed. The reactor was loaded with 100 g of $NaAlH_4$ doped with 2 mol% of $Ti(OBu^n)_4$ and of $Zr(OPr^i)_4$. (These dopants were later abandoned, with others, because of a high level of hydrocarbon impurities in the desorbed hydrogen, leading to a switch to the purely inorganic dopant $TiCl_3$ [158], see below). The work with the 100 g reactor gave an opportunity to observe an interesting thermal effect. The initial high hydrogen charging rate combined with limited heat transfer gave rise to a temperature increase from 155 to 234 °C in 1 min. The latter temperature is essentially the equilibrium temperature at the applied pressure of 172 bar H_2. This temperature is well above the melting point of $NaAlH_4$ (182 °C), so that $NaAlH_4$ was formed in the liquid state. Interestingly, however, the melting of the hydride bed had apparently "no or at least not much" detrimental effect on the subsequent performance of the alanate. After 4 cycles the bed was found to be sintered "into a porous, reasonably strong mass" (density $\approx 0.85\,\text{g}\,\text{cm}^{-3}$). In fact, the sintering of the hydride bed may be a benefit with respect to limiting particle migration, packing and expansion, maintaining stable internal gas impedance and safety.

Large volume changes of hydrides during dehydrogenation and rehydrogenation pose engineering problems. From published X-ray densities, the first step of hydrogen desorption (Eq. (6.15a)) would be expected to give a 30.3 % volume contraction and the second step (Eq. (6.15b)) an expansion of 13.8 %. The net volume change for complete dehydrogenation is expected to be $-16.5\,\%$. The volume change measured in the 100 g bed for dehydrogenation was found to be $-14.7\,\%$, only slightly lower than predicted [158].

Another scale-up reactor was charged with 72 g of $NaAlH_4$ doped with 4 mol% of $TiCl_3$ and the resulting storage material subjected to a 116 cycles test. The sample showed very stable hydrogen capacities and sorption rates. From internal thermocouple temperature measurement it could be concluded that the effective thermal conductivity is slightly lower, but typical of intermetallic hydride beds [137]. The above-mentioned sintering of the hydride bed into a rigid porous mass could be confirmed.

In this connection, it was found that aluminum is consumed in forming the alanates, independently of the aluminum origin (i.e. sorption materials or container vessel walls). However, the use of alanates that are not aluminum deficient did not appear to degrade the strength of aluminum-based alloys under long-term cycling conditions [137, 187].

The pyrophoric character and sensitivity to water of doped alanates represents a concern for any practical application. Up to now, no systematic investigation of the safety aspects of alanates has been carried out. The above-mentioned formation of a porous mass upon cyclization may to some extent help to moderate the safety problem.

A more thorough protection from air and humidity is found to be possible by encapsulation of doped alanates in highly porous matrices such as porous carbon or silica. Work in this direction is under way [188].

6.6.2.7 Outlook, Concepts, Thermodynamic Tailoring

There is still much research work needed if hydrogen storage via catalyzed $NaAlH_4$ is to reach practical utility, for example as a hydrogen storage material for PEM fuel cells (cf. Ref. [178]).

Since the production of the storage material is one of the main cost factors, development of the so-called direct synthesis ([134–137], Section 6.6.2.1) deserves special attention.

A great need still exists to increase hydrogenation rates, especially under low-pressure conditions. The lowering of hydrogen pressure necessary for charging a hydride storage tank has several beneficial effects, among others, upon safety, weight, material and production costs of the tank, increase in gravimetric energy density of the storage unit and lowering the energy required to fill the tank.

There are at present no kinetic measurements on the influence of hydrogen pressure and temperature on the *(re)hydrogenation* rate of Ti-doped $NaAlH_4$ (cf. Section 6.6.2.3). For the example of Ti*-doped $NaAlH_4$ [143], it has been shown, that the hydrogenation rate at a constant temperature decreases rapidly with decrease in hydrogen pressure. At constant *pressure* the increase in hydrogenation rate with increasing temperature is opposed by the approach to equilibrium conditions. Therefore, with increasing temperature at a constant pressure the hydrogenation rate will pass a maximum, which has been experimentally observed.

For $TiCl_3$ as a doping agent and ball-milling as a doping method, hydrogenation (Fig. 6.30) and dehydrogenation rates increase roughly in proportion to the increasing doping level [142]. About *specific* catalytic effects of metals other than titanium, there exists only some cursory data [133, 140, 141, 163]. We do not know yet if Ti, and perhaps Zr, are really the best catalysts for the doped $NaAlH_4$. It would, therefore, be desirable to perform a screening of other than Ti-catalysts in cycle tests, especially with respect to hydrogenation rates, preferably under some standard, say 100 and 50 bar hydrogen pressure conditions. Using recording of hydrogen evolution during ball-milling, it has been recently shown [171] that under the applied conditions reduction of some metallic compounds to the catalytically active elemental (metallic) stage is very incomplete. In such cases the catalytic potential of the respective metal is only partially utilized. Therefore, in screening of metallic compounds as dopants, their complete reduction during ball-milling should be controlled via measurement of hydrogen evolution [171]. In this regard also, the intriguing reports about the synergistic catalytic effect of Ti/Fe [133] and Ti/Zr [140, 141] combinations (or if possible other metals) deserves to be reinvestigated.

One of the main deficiencies for utilization of the doped $NaAlH_4$ system for hydrogen storage is the high temperature needed to discharge hydrogen in the second dehydrogenation step (Eq. (6.15b)). Lowering of the dissociation temperature (at a given pressure) is possible only by changing the thermodynamic properties of the alanates. Thermodynamic tailoring by means of partial substitution of the basic by secondary (or higher order) components is a well-known phenomenon in the field of intermetallic hydrides [189, 190]. As already mentioned (Section 6.6.2.2), the dissociation pressure of the Ti-doped Na_3AlH_6 can be *lowered* by substitution of one of the Na atoms of the compound by Li [152], which is of course an undesired effect. Nevertheless, partial or complete substitution of the sodium cation in $NaAlH_4$ or Na_3AlH_6 by other (light) metals is, in principle, a way to "tailor" plateau pressures of alanates.

More intriguing is the second possibility – tailoring on the side of dehydrogenated sodium or other metal alanate (or boranate) systems. In principle any stabilization of dehydrogenation products, NaH and/or Al (Eq. (6.15)), by complex or alloy formation should result in a change in the dissociation pressure of the respective reversible hydride system. Thus, it has been shown by Mössbauer spectroscopy that. upon doping of $NaAlH_4$ with Ti and Fe compounds, already after the first dehydrogenation step, an Al–Fe alloy is formed. In the course of the cycle test the alloy remained unchanged [133]. In a similar way, when $NaAlH_4$ is doped with $ZrCl_4$ by ball-milling, the resulting reduced Zr, as shown by XRD, forms with Al an Al–Zr alloy. Again, upon cyclization of Zr-doped $NaAlH_4$, the presence of the alloy in XRD persists [169]. In the case of doping of $NaAlH_4$ with $TiCl_3$ and subsequent removal of excess $NaAlH_4$ by extraction with THF, formation of an Al–Ti alloy could not be detected by XRD. However, after heating the sample to 400 °C, the XRD showed that the alloy Al_3Ti is formed [169].

Thus, it is probable that upon cyclization of Fe-, Zr- or Ti- (cf. XAFS results, Section 2.5) doped $NaAlH_4$, diluted alloys of the respective metals with Al are involved, not Al. Thermodynamic properties of such diluted alloys should differ from that of pure Al, and this should lead to a small increase in plateau pressures. Changes in the thermodynamics upon alloying small amounts (2–3 wt.%) of Ni and In with Mg–Al intermetallics have been decribed [191, 192]. This shows that by doping $NaAlH_4$ with catalytically active metals, simultaneously with improving the kinetics, the thermodynamic properties of the doped systems could also be changed, although maybe only to a small extent. This calls again (Section 6.6.2.2) for (re)investigation of the PCIs of doped alanate systems in comparison to undoped $NaAlH_4$.

6.6.3
Other Complex Hydrides

Of considerable interest is whether the concept of hydrogen storage via catalyzed sodium alanate can be extended to other complex hydrides. Among the large number of known or conceivable complex hydrides (see Introduction), as candidates for this purpose can be considered other than *sodium* alanates, boranates and transition metal complex hydrides, such as Mg_2FeH_6 [126, 127, 193]. Criteria for the selection of complex hydrides for hydrogen storage are availability, usable hydrogen storage capacity, response to catalysis and, especially, thermodynamic properties. Probably the most difficult task is to find hydrides which satisfy the rather stringent conditions of dissociation reaction enthalpy of 15–24 kJ mol_H^{-1} (Section 6.6.2.2) [145]. TG/DTA measurements can be applied as a quick method for determination of dissociation reaction enthalpies.

6.6.3.1 Alkali and Alkaline Earth Metal Alanates

Ti- and Fe-catalyzed solid state transformation of $LiAlH_4$ in Li_3AlH_6, Al and H_2 upon ball-milling has been reported by Balema et al. [139, 193, 194]. Reversible hydrogen storage via Ti-catalyzed Li_3AlH_6 (max. 1.8 wt.% H_2) has been reported by Chen et al. [195]. The result is based on DSC and TG analysis and has not yet been confirmed.

According to Morioka et al. [196], potassium alanate exhibits a rapid reversible dehydrogenation and rehydrogenation at 250–340 °C/<10 bar without the aid of a catalyst. Dehydrogenation of $KAlH_4$ (Eqs. (6.19) and (6.20)) proceeds stepwise in an analogous manner to that of $NaAlH_4$,

$$KAlH_4 \rightarrow \tfrac{1}{3} KAlH_6 + \tfrac{2}{3} Al + H_2 \qquad (6.19)$$

$$\tfrac{1}{3} K_3AlH_6 \rightarrow KH + \tfrac{1}{3} Al + \tfrac{1}{2} H_2 \qquad (6.20)$$

reaching a storage capacity of 3.5 wt.%. The dehydrogenation enthalpy of $KAlH_4$ is about 30 kJ mol_H^{-1} more endothermic than that of $NaAlH_4$. Such a situation makes $KAlH_4$ applicable as a typical high-temperature reversible hydrogen storage medium like MgH_2.

To date, the highest observed reversible hydrogen capacity of more than 8 wt.% was achieved by Zaluska et al. [197] with employment of lithium-beryllium hydrides. These can be prepared by ball-milling LiH and Be followed by hydrogenation at elevated temperatures. Reversible hydrogenation/dehydrogenation of lithium-beryllium hydrides proceeds according to Eq. (6.21), whereby the maximum capacity was obtained for $n:m = 3:2$.

$$n\,LiH + m\,Be + m\,H_2 \rightleftharpoons Li_nBe_mH_{n+2m} \qquad (6.21)$$

The pressure–composition isotherm for the Li–Be–H 3:2 composition shows a horizontal pressure plateau of ~8 wt.% H_2 with an equilibrium pressure of 1 bar at about 250 °C. The system can be discharged and recharged with hydrogen in the range 270–320 °C.

In recent years, a considerable amount of research has been done on the preparation and characterization of alkaline earth metal alanates. A detailed procedure for the preparation of solvent-free magnesium alanate has been described by Fichtner and Fuhr [198]. Lately, Fichtner and Ahlrichs et al. [199], starting from FTIR data of $Mg(AlH_4)_2$ and its solvent adducts, performed a quantum chemical calculation of Mg(AlH$_4$)$_2$ at the density functional theory (DFT) level. The calculated atomic positions were used to simulate an X-ray powder diffraction pattern, which was found to be congruent to experimental data. Accordingly, Mg(AlH$_4$)$_2$ possesses a CdI$_2$-analogous sheet structure. The dissociation enthalpy of Mg(AlH$_4$)$_2$ (Eq. (6.22)) was experimentally determined to be 5.48 kJ mol$_H^{-1}$ [200] or 0.91 kJ mol$_H^{-1}$.

$$Mg(AlH_4)_2 \rightarrow MgH_2 + 2\,Al + 3\,H_2 \tag{6.22}$$

Although more accurate measurements are still required, the thermodynamic stability of Mg(AlH$_4$)$_2$ appears to be too low for the purposes of practical hydrogen storage (Section 6.6.2.2).

A series of partially as yet unknown Mg, Ca and Sr alanates have been obtained by Dymova et al. [201–204] by application of the ball-milling method. Mg(AlH$_4$)$_2$ and Ca(AlH$_4$)$_2$ could be obtained in a mixture with alkaline earth metal chlorides from MgH$_2$ [204], CaH$_2$ [203] and AlCl$_3$ (Eqs. (6.23) and (6.24)) respectively. Sr(AlH$_4$)$_2$ [201, 202] was formed upon ball-milling of the two hydrides (Eq. (6.25)).

$$4\,MgH_2 + 2\,AlCl_3 \xrightarrow{b.m.} Mg(AlH_4)_2 + 3\,MgCl_2 \tag{6.23}$$

$$4\,CaH_2 + 2\,AlCl_3 \xrightarrow{b.m.} Ca(AlH_4)_2 + 3\,CaCl_2 \tag{6.24}$$

$$SrH_2 + 2\,AlH_3 \xrightarrow{b.m.} Sr(AlH_4)_2 \tag{6.25}$$

According to thermovolumetric measurements and IR and XRD evidence, the thermal dissociation of all three alanates follows the same hydrogen evolution pattern [201–204]: formation of the pentahydride (Eq. (6.26)), formation of the dihydride (Eq. (6.27)) and decomposition of the dihydride with formation of the corresponding alkaline earth metal-aluminum intermetallic compound (Eq. (6.28)). The pentahydrides can also be obtained by ball-milling M(AlH$_4$)$_2$ with dihydrides (Eq. (6.29)).

$$M(AlH_4)_2 \rightarrow MAlH_5 + Al + 1\tfrac{1}{2}H_2 \tag{6.26}$$

$$MAlH_5 + Al \rightarrow MH_2 + 2\,Al + 1\tfrac{1}{2}H_2 \tag{6.27}$$

$$MH_2 + 2\,Al \rightarrow MAl\text{-intermetallic} + H_2 \tag{6.28}$$

$$M(AlH_4)_2 + MH_2 \xrightarrow{b.m.} 2\,MAlH_5 \tag{6.29}$$

$$M = Mg,\ Ca,\ Sr \tag{6.30}$$

It should be noted that the thermal dissociation of alkaline earth metal alanates (Eqs. (6.26) and (6.27)) with formation of pentahydrides MAlH$_5$ follows a different pathway than that of alkali metal alanates (Eqs. (6.15), (6.19) and (6.20)), whose intermediates are hexahydrides of the type M'$_3$AlH$_6$, M' = Li, Na, K. Knowledge of the dissociation pathways of alkali and alkaline earth metal alanates and their structures (see below) and the relative stability of their intermediates can be useful in the search for new hydrogen storage systems based on alanates.

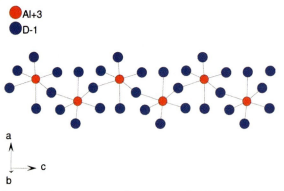

Fig. 6.37 Anion structure of the complex BaAlH$_5$ (from Zhang, Nakamura, Oikawa, Kamiyama, and Akiba (2002), Ref. [205]).

Novel barium and strontium alanates have been discovered and their structures elucidated by Akiba et al. [205, 206]. Both new alanates have been synthesized by hydrogenation of their corresponding intermetallic compounds. The Ba complex BaAlD$_5$ was found, by neutron diffraction, to possess a fascinating structure with one-dimensional zigzag chains of Al atoms and D(H) atoms as bridging atoms (Fig. 6.37). The strontium alanate Sr$_2$AlH$_7$, according to X-ray powder and neutron diffraction, is the first example that consists of isolated [AlH$_6$] units and infinite one-dimensional twisted chains of edge-sharing [HSr$_4$] tetrahedra along the crystallographic axis.

6.6.3.2 Reversible Hydrogenation/Dehydrogenation of Lithium Nitride and Imide as a Possible Means of Hydrogen Storage

A new type of a reversible light metal hydride system has been recently discovered by Chen et al. [207, 208]. The system is based on reversible lithium nitride (Eq. (6.31)) or lithium imide hydrogenation/dehydrogenation (Eq. (6.32)).

$$Li_3N + 2H_2 \rightleftharpoons Li_2NH + LiH + H_2 \rightleftharpoons LiNH_2 + 2LiH \quad (6.31)$$
$$\sim 10\,wt\%$$

$$Li_2NH + H_2 \rightleftharpoons LiNH_2 + LiH \quad (6.32)$$
$$\sim 6.5\,wt\%$$

The PCI diagram of the Li$_3$N/H$_2$ system exhibits two pressure plateaus corresponding to the first and second hydrogenation steps of Li$_3$N (Eq. (6.31)). Dehydrogenation of the Li$_2$NH + LiH mixture to Li$_3$N requires high temperatures. The second plateau has an equilibrium pressure of $\sim$0.5 bar at 230 °C. Starting from LiNH$_2$ and LiH (Eq. (6.32)) $\sim$6.5 wt.% H$_2$ can be reversibly stored at 255–285 °C. Co-evolution of NH$_3$ upon dehydrogenation of the LiNH$_2$–LiNH mixture (Eq. (6.32)) is reported to be completely suppressed using the 1:2 ratio of reactants [208]. Evolution of NH$_3$ can also be completely eliminated by utilization of TiCl$_3$ (1 mol%) as a catalyst, whereby at the same time the hydrogenation rate is considerably increased [209].

6.6.3.3 Hydrogen Storage Options for LiBH$_4$

LiBH$_4$ can deliver gaseous hydrogen by means of catalyzed hydrolysis and by thermal dissociation. Extraction of hydrogen from LiBH$_4$ via catalyzed hydrolysis is the subject of Chapter 4.7 (S. Suda) of this volume.

After Be(BH$_4$)$_2$, LiBH$_4$ has the second highest hydrogen content (18.4 wt.%) of all alanates and boranates (Table 6.8). The thermal hydrogen desorption of pure LiBH$_4$ starts at $\sim$320 °C and proceeds mainly in the temperature region 400–600 °C with a maximum around 500 °C. The total amount of desorbed hydrogen is 9 wt.%, which is half of the hydrogen present in the starting compound. Of interest in this connection is the observation that hydrogen desorption of LiBH$_4$ mixed with SiO$_2$ powder (25:75 wt.%) starts at lower temperature and 9 wt.% of H$_2$ can be liberated below 400 °C [210, 211].

6.6.3.4 Transition Metal Complex Hydrides

Numerous transition metal complex hydrides have been discovered and their structures elucidated by the groups of Yvon [212], Bronger [125] and Noreus [213]. Reasons for their inferior applicability for hydrogen storage have already been discussed in the Introduction. Remarkable among them is Mg$_2$FeH$_6$ [126, 127] (Section 6.6.2.5, Eq. (6.18)) which shows high storage capacity (5.5 wt.% H$_2$), an exceptionally high volumetric hydrogen density (150 kg/m^3) and excellent stability in cycle tests [176]. Although, due to its high thermodynamic stability, not suitable for hydrogen storage, the reversible system (Eq. (6.18) is a promising candidate for thermochemical high temperature (400–500 °C) thermal energy storage, as for instance required for solar thermal power plants.

6.6.3.5 Conclusion

Dissociation reaction enthalpies of complex hydrides other than sodium alanates, and thus their utility for hydrogen storage from a thermodynamic point of view (Section 6.6.2.2), are at present largely unknown. Experimental determination of these distinguishing features is therefore a task of high priority. Beyond that, the discovery of new practical hydrogen storage systems will, to a large extent, depend on our ability to find ways of modifying the relative thermodynamic stability of hydrogen-rich complex hydrides and their dissociation (dehydrogenation) products. Some ideas on that are given in Section 6.6.2.7. From a kinetic point of view, hints already exist that dehydrogenation of other than sodium alanates [139, 214], and even of Li$_3$N [209], is subject to catalysis by titanium.

6.7
Storage in Organic Hydrides
Andreas Borgschulte, Sandra Goetze, and Andreas Züttel

6.7.1
Introduction

In photosynthesis, water is split by sunlight to give gaseous oxygen and two protons and electrons, which can recombine releasing gaseous hydrogen. This indeed happens in plants (algae, etc.) – if one prevents the plant from using the protons and electrons for its own purpose it will produce hydrogen. The plant stores the energy of the hydrogen by chemically adding the protons to organic substances, which results, via a complex process, in the production of carbon hydrates. Thus, nature found a solution for the storage of hydrogen in a very effective way, and it is of course intriguing to have a closer look, how nature succeeded. From a chemical point of view, plants store hydrogen reversibly in carbon compounds. This idea is copied technically in hydrogen storage in carbon hydrides, for example methanol and cyclohexane, which is discussed in Section 6.7.3. These compounds have the advantage of being liquid fuels and can therefore be easily stored. A similar advantage applies for ammonia and ammonia-based compounds, which are discussed in Section 6.7.4.

6.7.2
Photosynthesis and Hydrogen Storage in Plants and Bacteria

Photosynthesis is the process of converting light energy to chemical energy and storing it eventually in carbon hydrates [215]. The overall chemical reaction involved in photosynthesis is:

$$6CO_2 + 6H_2O (+\text{light energy}) \rightarrow C_6H_{12}O_6 + 6O_2.$$

Photosynthesis occurs in two stages. In the first phase *light-dependent reactions* (also called *light reactions*) capture the energy of light and use it to make high-energy molecules. Of interest is that one of the involved high-energy molecules is actually *hydrogen* (proton plus electron) loosely bound to an organic molecule, which one could understand as *biological hydrogen storage* [216]. Furthermore, the hydrogen originates from water, previously split by the photochemical process. During the second phase, the *light-independent reactions* (also called the Calvin–Benson Cycle, and formerly known as the *dark reactions*) use the high-energy molecules to capture carbon dioxide (CO_2) and synthesize carbon hydrates – the long lasting chemical energy storage of plants. We will not discuss this so-called carbon fixation process, but focus on the water photolysis in the light reactions from the viewpoint of hydrogen storage.

The process of converting light energy into chemical energy is depicted by the mechanism of an electron transport chain starting from the splitting of water (see Fig. 6.38). Splitting water produces gaseous oxygen, protons and electrons on a low energy level. A series of proteins inside the chloroplasts transfers these electrons from one to another, enabling various biochemical reactions along the way as the electron drops in energy (red arrows).

The total process is rather complicated and thus only sketched schematically by the so-called z-scheme [217]. When sunlight strikes chlorophyll, specific molecules in it are excited and electrons are lifted to higher energy levels of the molecule. In detail, two so-called photosystems are involved in the electron chain. The two electrons from the water splitting are kept in photosystem II (P680). Then a photon

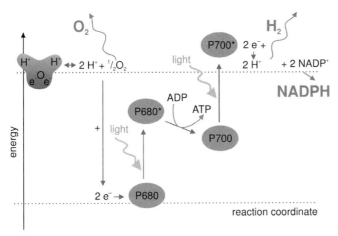

Fig. 6.38 The z-scheme of light reactions of photosynthesis. From (Hall and Rao (1972), Ref. [217]).

is absorbed by the chlorophyll core of P680, exciting the two electrons which are transferred to the acceptor molecule. The deficit of electrons is replenished by splitting another water molecule. The electrons are moved via several proteins, thereby producing the high-energy molecule ATP from ADP (for more detailed information see Refs. [215, 217]).

The still-excited electrons are transferred to the photosystem I complex (P700), which boosts their energy level to a higher level using a second solar photon. The highly excited electrons are transferred to the acceptor molecule, but this time they have reached an energy level high enough to recombine with H^+ and produce hydrogen. As gaseous hydrogen cannot be used by the plant, hydrogen is "stored" via bonding to the molecule $NADP^+$ (nicotinamide adenine dinucleotide phosphate) instead:

$$NADP^+ + 2H^+ + 2e^- \rightarrow NADPH + H^+$$

The reaction is catalyzed by an enzyme called Ferredoxin-NADP reductase FNR. If this enzyme is blocked, certain plants can produce gaseous hydrogen using the enzyme hydrogenase [218]. This is indeed used to produce hydrogen from algae and bacteria [219].

The NADPH and ATP are used later in the Calvin–Benson Cycle to synthesize carbon hydrates (glucose $C_6H_{12}O_6$) via:

$$6\,CO_2 + 12\,NADPH + 12\,H^+ + 18\,ATP \rightarrow C_6H_{12}O_6 + 6\,H_2O$$
$$+ 12\,NADP^+ + 18\,ADP$$

From the viewpoint of energy storage, eventually hydrogen is stored in carbon hydrates. Because this triggered the idea to use biofuels, that is carbon hydrates produced by plants, as an energy carrier of the future, it is worth estimating the total energy efficiency of the process. Interestingly, the quantum efficiency of the photomolecules is nearly 100 % [220]. However the cell needs more than one photon per reaction (see z-scheme). Measurements showed that the cell needs between 8 and 10 photons per oxygen molecule as the reduction potential of NADH is higher than the energy of the absorbable photons [221]. Furthermore, the cell wastes a certain amount of the high-energy molecules during the dark reactions (so-called photorespiration) [222]. In total, from 100 % light the losses are [223]:

- Only light in the range 400–700 nm can be used. This amounts to 50 % loss of total solar incident radiation.
- Reflection, absorption and transmission by leaves: 20 % loss.
- Limited light reaction efficiency: 10 (8) photons are needed to fixate one CO_2: 77 % loss (72 % loss).
- Respiration required for translocation and biosynthesis: 40 % loss.

The overall efficiency is then 5.5 % (6.6 %). In practice, the efficiency is lower: even very productive tropical grasses such as sugar cane do not yield crops at efficiencies of more then 0.6 % on an annual basis [224]. The water hyacinth, which is considered to be one of the most efficient solar energy converters, reaches 3–4 % maximum efficiency on a daily basis [225]. For technical use, additional losses due to processing and so on have to be considered. Consequences for the environment from intensive agriculture on a large scale are not foreseeable. A widespread introduction of biological energy conversion ("biofuel") is therefore questionable.

However, nature delivers us with interesting solutions for energy conversion/storage problems, which might be transformed into engineering solutions ("bionics"):

- Use of artificial chlorophyll membranes. ("biological solar cells")
- Direct formation of a hydride after photochemical/electrochemical water splitting instead of releasing gaseous hydrogen. The mandatory splitting of the hydrogen molecule during storage in hydrides is one of the major kinetic barriers and could thus be circumvented.
- Use of organic hydrides for hydrogen storage, in particular hydrogen storage by carbon hydrides. This option is discussed in the next section.

6.7.3
Hydrogen Storage in Carbon Hydrides

6.7.3.1 Carbon Hydrates (Methanol)
Hydrogen can be obtained from any hydrogen-containing compounds as long as its binding energy to the rest of the compound does not exceed that of the heating value of hydrogen itself. Thus any carbon hydrate ("biofuel") can be decomposed into hydrogen and carbon oxide (and water). We will discuss this alternative with

the representative example, methanol. It should be noted that these energy carriers release CO_2, which cannot be easily stored or removed. Furthermore, hydrogen is not reversibly stored on board. However, methanol can be chemically produced on a large scale from syngas at high temperatures and pressures [226]:

$$CO + 2H_2 \Rightarrow CH_3OH \qquad \Delta H = -91.7 \text{ kJ mol}^{-1}$$

Optimized catalysts (copper/zinc oxide) can reduce the process conditions to 50–100 bar and 250 °C. Methanol will then be burned as fuel in combustion engines or in fuel cells, according to [227]

$$CH_3OH + 2O_2 \Rightarrow CO_2 + H_2O \qquad \Delta H = -727 \text{ kJ mol}^{-1}$$

Direct methanol fuel cells (DMFC) are similar to PEMFCs (proton exchange membrane fuel cells) in that the electrolyte is a polymer and the charge carrier is the hydrogen ion (proton). The liquid methanol (CH_3OH) is oxidized in the presence of water at the anode generating CO_2, hydrogen ions and the electrons that travel through the external circuit as the electric output of the fuel cell. The hydrogen ions travel through the electrolyte and react with oxygen from the air and the electrons from the external circuit to form water at the anode, thus completing the circuit.

$$\text{Anode Reaction:} \quad CH_3OH + H_2O \Rightarrow CO_2 + 6H^+ + 6e^- \qquad (6.33)$$
$$\text{Cathode Reaction:} \quad 3/2\,O_2 + 6H^+ + 6e^- \Rightarrow 3H_2O \qquad (6.34)$$

One of the drawbacks of the DMFC is that the low-temperature oxidation of methanol to hydrogen ions and carbon dioxide requires a more active catalyst, which typically means that a larger quantity of expensive platinum catalyst is required than in conventional PEMFCs. In addition, the anode has a limited carbon monoxide tolerance. Further, the overall efficiency is smaller than for a PEMFC.

A disadvantage of the widespread use of methanol is its high toxicity to humans and the environment. Methanol's high solubility in water raises concerns that well-water contamination could arise from the widespread use of methanol as an automotive fuel. In general, CO_2-releasing fuels address the principle of an emission-free fuel. In the best case, the emitted CO_2 is reused for methanol production, for example by plants. There are other carbon hydrides, which can be used as on-board reversible hydrogen storage materials.

6.7.3.2 Reversible Hydrogen Storage in Carbon Hydrides

Hydrogen can be bonded to unsaturated carbon hydrides. This reaction takes place reversibly for several chemical systems at moderate temperatures with a relatively high storage capacity:

- decalin $C_{10}H_{18} \Leftrightarrow$ naphthalene $C_{10}H_8 + 5H_2$, $\Delta H = 297.3$ kJ mol^{-1}, 7.3 mass% [228]
- n-heptane $C_7H_{16} \Leftrightarrow$ toluene $C_7H_8 + 4H_2$, $\Delta H = 252$ kJ mol^{-1}, 8 mass% [229]

- methylcyclohexane $C_7H_{14} \Leftrightarrow$ toluene $C_7H_8 + 3\ H_2$, $\Delta H = 215.3\,\text{kJ mol}^{-1}$, 6 mass% [230]
- electrochemical hydrogen storage in Li-doped pentacene $C_{22}H_{14}$, 0.89 mass% [231].

In 1980 the idea of a seasonal hydrogen system in the form of methyl cyclohexane was presented by Taube and Taube [232, 233]. In 1984, a 16-ton demonstration truck was powered by hydrogen produced on board the truck and directly coupled to the combustion engine [234]. During 1985/86, an improved version with an on-board hydrogen production of approx. 3 $(gH_2)\,s^{-1}$, which corresponds to a thermal power of 360 kW, was constructed and experimentally tested [235].

Due to slow kinetics, the conventional heterogeneous catalysis of the dehydrogenation of decalin in the solid–gas phase is performed at temperatures of more than 400 °C, which might result in the formation of by-products or carbonaceous deposit on the catalyst in addition to thermal energy loss. In a recent study, an attempt was made to apply the so-called "liquid-film concept" to hydrogen evolution from decalin with carbon-supported platinum-based catalysts under reactive distillation conditions in order to obtain high electric power sufficient for PEMFC vehicle operations in the temperature range 200–300 °C [236].

Besides the chemical storage of hydrogen in carbon hydrides, hydrogen can also be electrochemically stored in organic polymers, for example in Li-doped pentacene. Here, the capacity is relatively small, though the discharge capacity of 239 mA h g^{-1} matches that of LaNi-based hydride batteries [237].

6.7.4
Hydrogen Storage in Ammonia and Ammonia-Based Compounds

Due to the low molecular weight of nitrogen, ammonia, NH_3, exhibits a high gravimetric density of 17.8 mass%. At ambient conditions (room temperature and 1 bar), ammonia is a gas and thus has to be condensed to achieve considerable volumetric densities (however, 8 bar is sufficient at room temperature). Liquid ammonia has a volumetric density of 1100 kg m^{-3} which is 50 % more hydrogen per volume than liquid hydrogen. Today ammonia is transported as such in large quantities as a fertilizer.

Ammonia is easily decomposed according to the reaction

$$2\,NH_3 \Leftrightarrow N_2 + 3\,H_2, \quad \Delta H = +91.86\,\text{kJ mol}^{-1}$$

The endothermic reaction is favored by high temperature and low pressure and is accelerated by the presence of nickel or iron catalysts. NH_3 can be burned directly in combustion engines or used in solid oxide fuel cells without preprocessing [238]. In alkaline and PEM fuel cells, the ammonia has first to be decomposed according to the above reaction. For the PEM cell, even trace amounts of ammonia left in the gas after decomposition must be removed [239].

Ammonia is one of the main basic compounds of industrial chemistry. Its production uses 1 % of the world's energy supply, as it is the preproduct of fertilizers used in the production of food for 40 % of the world's population [240]. Thus, handling, production and infrastructure on a large scale are well developed [241]. Disadvantages of ammonia are its harmful properties to humans and the environment. Ammonia is a strong irritant and excessive emissions to the environment must be avoided [242]. This puts strong demands on the storage container. One idea to ensure emission-free ammonia storage is its storage in metal ammine complexes.

6.7.4.1 Metal Ammine Complexes for Hydrogen Storage

An alternative way of storing hydrogen is storage of ammonia in the form of metal ammine complexes [243]. Most of the complexes of the form $M(NH_3)_n X_m$, where M is a metal cation such as Mg, Ca, Cr, Ni or Zn and X is an anion such as Cl^-, are solids and thus easy to handle (see Fig. 6.39). For the divalent metal ions, ammine complexes are generally formed easily by passing ammonia over the anhydrous MX_m salt. The heat of formation depends on M and X and varies between 40 and 80 kJ mol^{-1} NH$_3$ [244].

At elevated temperatures, ammine complexes decompose and release ammonia. By combining such complexes with an ammonia decomposition catalyst (see above) one obtains a very versatile hydrogen source (see Fig. 6.40). Christensen et al. [243] have shown that the kinetics of ammonia adsorption and desorption are reversible and fast, even at moderate temperatures.

(a)

(b)

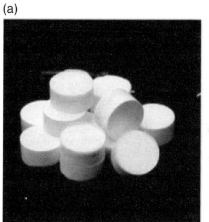
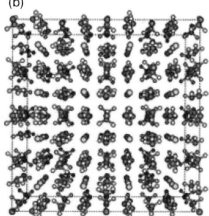

Fig. 6.39 Pellets of Mg(NH3)$_6$Cl$_2$ (left). The skeleton density is 1.25 g cm^{-3} and the pellets have a density within 5 % of this. It crystallizes in a cubic unit cell with a lattice constant of 10.19 Å [244]. The atomic structure of the crystalline complex is shown to the right, as it comes out of an energy minimization using density functional theory calculations. Mg^{2+} is shown in gray, Cl$^-$ in green, nitrogen in blue and hydrogen in white. From (Christensen, Sørensen, Johannessen, Quaade, Honkala, Elmøe, Køhler, and Nørskov (2005), Ref. [242]).

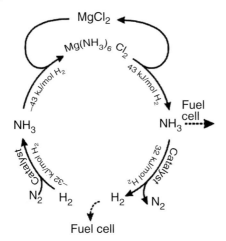

Fig. 6.40 Schematic illustration of the process whereby hydrogen can be transformed into ammonia, which is stored as Mg(NH$_3$)$_6$Cl$_2$. The Mg(NH$_3$)$_6$Cl$_2$ can be transported safely and when needed ammonia is released and decomposed into molecular hydrogen and nitrogen. The standard enthalpies of formation and decomposition are indicated [243]. If all the enthalpy needed for decomposition (43 + 32 = 75 kJ mol^{-1} H$_2$) has to be taken from the H$_2$ produced, about 30% of the heating value of the H$_2$ (242 kJ mol^{-1} H$_2$) would be lost. Taken from (Christensen, Sørensen, Johannessen, Quaade, Honkala, Elmøe, Køhler, and Nørskov (2005), Ref. [242]).

The Mg(NH$_3$)$_6$Cl$_2$ complex shows a considerably high hydrogen content of 9.1 mass% – in the form of ammonia. The desorbed ammonia could be used directly as a fuel for a solid oxide fuel cell without further reaction. With alkaline or low temperature (PEM) fuel cells, the ammonia must be decomposed (see above).

6.8
Indirect Hydrogen Storage via Metals and Complexes Using Exhaust Water
Seijirau Suda and Michael T. Kelly

6.8.1
Introduction

Over the last few years, various hydrogen–metal complex compounds have been investigated as high hydrogen-capacity materials. The following binary and ternary hydrides are well known for their high hydrogen capacity:

- binary saline hydrides: NaH, CaH$_2$, MgH$_2$
- aluminum-based complex hydrides: LiAlH$_4$ and NaAlH$_4$
- boron-based complex hydrides: LiBH$_4$, KBH$_4$ and NaBH$_4$.

Most of these hydrides react violently in air to generate hydrogen. Due to their highly sensitive nature, these materials, except MgH$_2$ and NaBH$_4$, must be treated with special care when they are used to generate hydrogen by thermal decomposition. In contrast, MgH_2 is rather stable in air and generates hydrogen moderately on hydrolysis. From the capacity viewpoint, $NaBH_4$ that contains 10.6 mass% (10.8 mass% by hydrolysis) is preferable as a hydrogen storage material to MgH_2 that has a capacity of 7.6 mass%.

Among various complex compounds, $NaBH_4$ is the only material that can be treated in the liquid state and generates hydrogen by catalytic hydrolysis. It can be stored stably as both powder or liquid under ambient conditions for a long time. The stoichiometric amount of hydrogen generated by hydrolysis, as shown in Eq. (6.35), is 10.8 mass%.

$$NaBH_4 + 2H_2O \rightarrow 4H_2 + NaBO_2 \quad (6.35)$$

The amount of hydrogen generated is of special significance because water takes part in the hydrolysis to generate 50 % of the hydrogen, as shown in Eq. (6.35). This is the most significant part of hydrogen generation by $NaBH_4$ hydrolysis.

The practical advantage of this compound is its safety when stored or delivered as a powder or liquid. The quality of water is not a serious issue and any water, such as rain, underground, river, waste, or seawater, is suitable when it is used to make liquid solutions for supplying hydrogen to actuate proton exchange membrane fuel cells (PEMFCs) in emergency and portable uses.

However, there are several drawbacks to the practical use of this compound. For various reasons, as pointed out below, its applicability as a hydrogen storage material is greatly reduced, even to impractical levels.

In this chapter, the properties and characteristics of $NaBH_4$ as a hydrogen storage material will be discussed from the practical application viewpoint.

6.8.2
Borohydride Complex Ions in Aqueous Alkaline Solution

Practically, the hydrolysis as given in Eq. (6.35) does not happen. $NaBO_2$ as "spent fuel" is unable to exist as anhydrous $NaBO_2$ in aqueous solutions. $NaBH_4$ gives an alkaline solution with a very high pH value, even diluted to a level of 1 % $NaBH_4$ in water (pH = 11) and it exists as a borohydride complex ion in the form of BH_4^- once it is diluted in aqueous solution:

$$BH_4^- + 2H_2O \rightarrow 4H_2 + BO_2^- \text{ (or } BH_4^- + 4H_2O \rightarrow 4H_2 + B(OH)_4^- \quad (6.36)$$

6.8.2.1 Protide as a State of Hydrogen
In the BH_4^- complex ion, B–H bonding is easily broken to generate hydrogen by catalytic hydrolysis. Protide is the state of hydrogen that exists as anionic hydrogen in binary hydrides such as NaH, CaH and MgH_2, and also Al- and B-based ternary hydrides such as $LiAlH_4$, $LiBH_4$, $NaAlH_4$, $NaBH_4$.

6.8.2.2 Water as the Hydrogen Source

Water supplies 50% of the hydrogen, as seen in Eq. (6.36). It is surprising that water decomposes to take part in hydrolysis under ambient temperature conditions whereas very high thermal energy levels, above 3500 K, are required to split hydrogen into protium when it is thermally decomposed.

The fact that the hydrolysis takes place under ambient conditions is worthy of special mention in the application of $NaBH_4$. The quality of water does not influence the generation of hydrogen in the hydrolysis although it is necessary to investigate in more detail the quality of water when it is required to generate hydrogen in such mobile and field applications as military and emergency uses.

6.8.3
Physico-Chemical Properties and Characteristics

6.8.3.1 Solubility Limit

The solubility of $NaBH_4$ in NaOH solutions is illustrated in Fig. 6.41 as a function of temperature. The solubility is also reduced significantly by the formation of sodium metaborate ($NaBO_2 \cdot 4H_2O$ or $NaBO_2 \cdot 2H_2O$) that extracts water from the aqueous solution of BH_4^- to form water of crystallization. The solubility of $NaBO_2$ is lower than that of $NaBH_4$, as shown in Fig. 6.42. The freezing point of $NaBH_4$ dissolved in NaOH solution is shown in Fig. 6.43 as a function of its concentration.

It should be remembered that the theoretical amount is reduced considerably by the solubility limit of $NaBH_4$ in aqueous NaOH solutions.

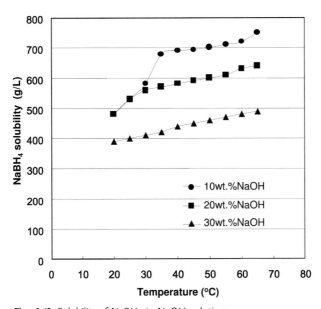

Fig. 6.41 Solubility of $NaBH_4$ in NaOH solutions.

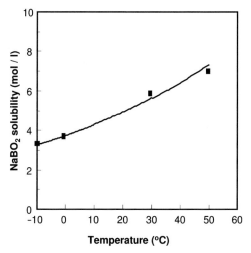

Fig. 6.42 Solubility of NaBO$_2$ in 10 wt.% NaOH solution.

6.8.3.2 Viscosity

The higher the concentration of NaOH solution, the higher the viscosity of the solution. The viscosity is influenced significantly by temperature and also by the formation of a metastable slurry of crystalline sodium metaborate, even at the beginning of hydrolysis. The crystalline growth increases the viscosity greatly and the liquid–solid phase separation reaches unfavorable levels from an engineering viewpoint.

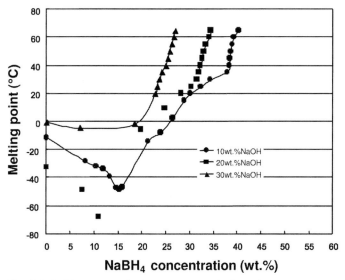

Fig. 6.43 Melting points of NaBH$_4$ in aqueous NaOH solutions.

6.8.3.3 Crystallization

Equation (6.37) represents the hydrolysis of sodium borohyride solutions to generate hydrogen and sodium metaborate;

$$Na^+ + BH_4^- + 2xH_2O \rightarrow 4H_2 + NaBO_2 \cdot 2(x-1)H_2O \quad (x = 1-3) \quad (6.37)$$

After initiating hydrolysis, BO_2^- ions tend to form crystalline sodium metaborate depending on the concentration and temperature conditions. The theoretical hydrogen amount is greatly reduced by the formation of sodium metaborates ($NaBO_2 \cdot 4H_2O$), down to 5.48 mass%, as can be seen in the following equations:

$$Na^+ + BH_4 + 2H_2O \rightarrow 4H_2 + NaBO_2 (10.8\,mass\%) \quad (6.38)$$

$$\Delta G^0 (298\,K) = -299\,kJ\,(mol\,BH_4^-)^{-1}, \quad \Delta H^0 (298\,K)$$
$$= -213\,kJ\,(mol\,BH_4^-)^{-1} \quad (6.39)$$

$$Na^+ + BH_4 + 4H_2O \rightarrow 4H_2 + NaB(OH)_4 (7.28\,mass\%) \quad (6.40)$$

$$\Delta G^0 (298\,K) = -315\,kJ\,(mol\,BH_4^-)^{-1}, \quad \Delta H^0 (298\,K)$$
$$= -247\,kJ\,(mol\,BH_4^-)^{-1} \quad (6.41)$$

$$Na^+ + BH_4^- + 6H_2O \rightarrow 4H_2 + NaB(OH)_4 \cdot 2H_2O \cdot (5.48\,mass\%) \quad (6.42)$$

$$\Delta G^0 (298\,K) = -319\,kJ\,(mol\,BH_4^-)^{-1}, \quad \Delta H^0 (298\,K)$$
$$= -213\,kJ\,(mol\,BH_4^-)^{-1} \quad (6.43)$$

The hydrolysis in Eq. (6.38) is an impossible reaction and, therefore, the theoretical value of 10.8 mass% is an unrealistic value, not practically available.

$NaB(OH)_4 \cdot 2H_2O$ is also known as $NaBO_2 \cdot 4H_2O$, which dissolves in its own water of crystallization at temperatures higher than the melting point of 53.5 °C, reducing the water content to $NaBO_2 \cdot 2H_2O$ (melting point; 57 °C). $NaBO_2 \cdot 2H_2O$ loses 1 mole of water to form $NaBO_2 \cdot H_2O$ at temperatures above 120 °C.

Crystallization during hydrogen generation may cause irreparable damage to hydrogen storage/generation devices due to the accumulation of sodium metaborate solidified in liquid flow lines such as valves, pipes and fittings. Crystalline growth must be avoided from any engineering viewpoint.

6.8.3.4 Exothermic Nature of Hydrolysis

As can be seen in Eqs. (6.38)–(6.43), the hydrolysis liberates a large amount of heat which contributes to maintaining the system temperature higher than the solubility limit of both $NaBH_4$ and $NaBO_2$. The rate of hydrogen generation is accelerated by increasing temperature but it becomes rather difficult to control the hydrogen generation kinetics. The heat generated during hydrolysis requires additional sink sources to release heat from a PEMFC, that is an additional heat exchanger is required other than that for the PEMFC.

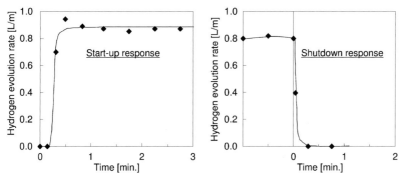

Fig. 6.44 Kinetics of hydrogen generation by NaBH$_4$ hydrolysis. Test equipment: hydrogen generator designed at 0.8 L H$_2$ min^{-1}. Fuel composition: NaBH$_4$/NaOH/H$_2$O = 158.5/76.5 (mass). Test conditions: 0.10 MPa, 295 K.

6.8.4
Hydrogen Generation by Catalytic Hydrolysis

For the generation of hydrogen, the alkaline stabilized borohydride solution requires a suitable catalyst such as Ru [245], fluorinated Mg$_2$NiH$_4$ [246, 247] or Raney-Co and Raney-Ni [248].

6.8.4.1 Rate of Hydrogen Generation
For PEMFC applications in automobiles, for example, the hydrogen generation rate must exceed at least 1.5–2.0 gH$_2$ s^{-1} during acceleration, even for a compact car of 50 kW power. In order to enhance the rate of hydrogen generation, it is necessary to increase the amount of catalyst or its surface area, or else NaBH$_4$ must be concentrated in aqueous solutions up to the solubility limit or much higher, although this is impractical in engineering devices.

Typical experimental data on the hydrogen generation kinetics as a function of the time elapsed are shown in Fig. 6.44.

6.8.4.2 Requirements of a Catalyst for Hydrolysis
The high rate of hydrogen generation is the special feature of this catalytic hydrolysis. It is important to know that all other hydrogen storage materials require thermal energy sources to liberate hydrogen at such a high rate as borohydride hydrolysis. Hydrogen storage materials such as metal hydrides and NaAlH$_4$ need to add heat for dehydrogenation and thermal decomposition because of their endothermic natures. Instead, NaBH$_4$ requires a heat sink to liberate heat due to the exothermic nature of the hydrolysis.

6.8.4.3 Hydrogen Generation at the Interface of the Liquid and Solid Phases
The essential requirement of the catalyst in hydrolysis is durability against the strong mechanical force that is developed at the outer surface of the catalyst. It

should be remembered that the hydrolysis occurs as an interfacial phenomenon between gas (hydrogen), liquid (BH_4^- solution) and solid (catalyst) phases.

Hydrogen as protide (H^-) in the BH_4^- solution is converted to gaseous hydrogen (H_2) within a few tenths or hundredths of a second at the surface of the catalyst. The growth of a gas bubble is accompanied by a great increase in the specific volume ($m^3\,kg^{-1}$) and brings the linear velocity ($m\,s^{-1}$) in the radial direction to the subsonic or supersonic regions, sufficient to develop a shock wave that increases significantly the stagnation enthalpy and leads to a change in the flow patterns [249]. The shock wave generated by the abrupt volume expansion may cause destruction of the surface between the catalyst and its substrate.

In a $NaBH_4$ system for hydrogen storage, the most serious problem is to create a powder-state catalyst that is highly durable against the shock wave. Such a catalyst bed has not yet been developed.

6.8.5
Hydrogen Generation Systems and Devices

The $NaBH_4$ hydrogen generation systems and devices are comparatively simple in their principles and components. They are basically composed of a catalytic reactor, fuel and fuel recovery tanks, mist and crystalline separators, condenser or heat exchanger, pump and pressure regulator. A hydrogen generation system for PEMFC is illustrated schematically in Fig. 6.45 and for an experimental set-up of 1 kW capacity in Fig. 6.46.

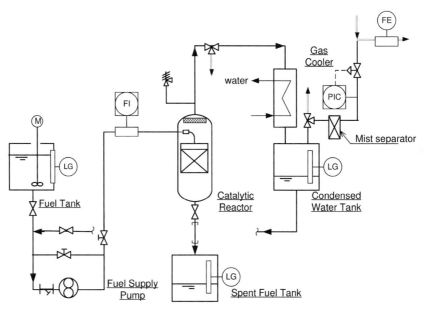

Fig. 6.45 Schematic diagram of a borohydride hydrogen storage device for PEMFC.

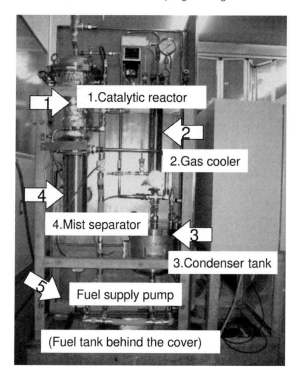

Fig. 6.46 Experimental (1 kW test unit) hydrogen storage set-up for PEMFC.

6.8.5.1 Operating Conditions
The temperature ranges widely between ambient and 100 °C and even to below the freezing point of $NaBH_4$ solution (usually less than $-20\,°C$). The pressure is controllable and is dependent on the type of application.

6.8.5.2 Compact Applications
Compact hydrogen storage devices have been developed for use with small PEMFC in the past few years. The cost effectiveness will be improved significantly by applying BH_4^- solutions as the hydrogen fuel source to handy or mobile electronic devices such as portable computers, PDA, CD-cameras and UPS. The fuel cost even now is estimated to be very reasonable for such compact devices with capacity ranges from a few tens to a few hundred watts. The borohydride fuel costs, estimated from today's market price are shown in Fig. 6.47 for the PEMFC and in Fig. 6.48 for the DBFC (direct borohydride fuel cell).

Very few auxiliary components are required in such compact devices. Hydrogen is supplied at a constant rate as required, depending on the electric capacity. Hydrogen reservoirs are of a disposable cartridge type that contain a cheap catalyst such as Raney-Ni or fluorinated-$Mg_2 NiH_4$. The future cost in 2020 is expected to be less than \$5/kg-$NaBH_4$ and \$0.1/kWh using an advanced production process introduced in the section below.

6 Hydrogen Storage

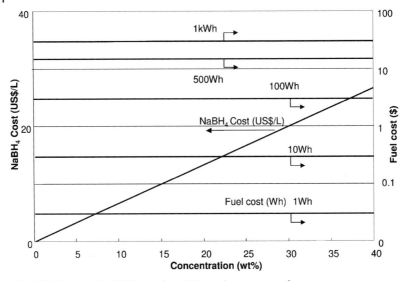

Fig. 6.47 Fuel cost for PEFC based on 2003 market price as a function of NaBH$_4$ concentration in NaOH solution.

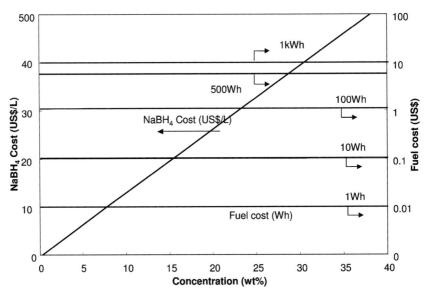

Fig. 6.48 Fuel cost for DBFC based on 2003 market price as a function of NaBH$_4$ concentration in NaOH solution.

6.8.6
Production and Reprocessing of NaBH$_4$

The development of a mass production process for NaBH$_4$ with reasonably cheap cost is the key issue for NaBH$_4$ application as a hydrogen storage material. Presently, there are two industrial processes available.

6.8.6.1 Conventional Process

The Rohm and Haas and the Bayer processes have been used commercially for the production of NaBH$_4$ and are briefly summarized in Eqs. (6.44) and (6.45).

Rohm and Haas process: [250]

$$4NaH + B(OCH_3)_3 \rightarrow NaBH_4 + 3NaOCH_3$$
$$\Delta G^0 (298\ K) = -129.5\ kJ\ (mol\ NaBH_4)^{-1} \qquad (6.44)$$

Bayer process: [251]

$$4Na + 2H_2 + 1/4\ Na_2B_4O_7 + 7/4SiO_2 \rightarrow NaBH_4 + 7/4Na_2SiO_3$$
$$\Delta G^0 (298\ K) = -411.3\ kJ\ (mol\ NaBH_4)^{-1} \qquad (6.45)$$

Those processes are less cost effective because only 25 % of the Na is used to produce 1 mole of NaBH$_4$ and the purification processes to separate NaBH$_4$ are rather complicated. The market price today is too expensive ($50–60/kg) to use NaBH$_4$ for storage applications. NaBH$_4$ is at present the most unsuitable and impractical hydrogen source for automotive applications.

6.8.6.2 New Processes

At least two different processes have been developed in the past few years [252–255]. In these processes, anhydrous sodium metaborate (NaBO$_2$) that is obtained easily from borax (Na$_2$B$_4$O$_7$ · 10H$_2$O) is used as the starting material to produce NaBH$_4$.

Mechano-Chemical Processs MgH$_2$ hydride plays an important role in the mechano-chemical process where it is used as an H-donor and O-acceptor. In the mechano-chemical process, which applies MgH$_2$ in a high-density ball mill and is operated under ambient conditions, the mechanical force is converted to chemical energy as shown in Eq. (6.46).

$$NaBO_2 + 2MgH_2 \rightarrow NaBH_4 + 2MgO$$
$$\Delta G^0 (298\ K) = -270\ [kJ\ (mol\ NaBH_4)^{-1}] \qquad (6.46)$$

Dynamic Hydriding/Dehydriding Process In the dynamic hydriding/dehydriding process, which makes use of the transitional state between hydriding and dehydriding states, thermal energy is rapidly applied to a mixture of NaBO$_2$ and Mg

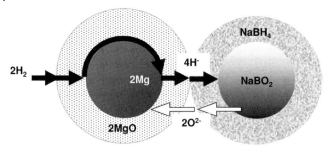

Fig. 6.49 Protide behavior in NaBH$_4$ synthesis showing the interfacial transition of protide and oxygen ion.

under a hydrogen atmosphere in order to generate a very reactive state of hydrogen, namely, "protide: H$^-$" that combines transitionally with Mg as Mg · 2(H$^-$).

$$NaBO_2 + 2H_2 + 2Mg \rightarrow NaBH_4 + 2MgO$$
$$\Delta G^0 \, (298 \, \text{K}) = -342 \, \text{kJ} \, (\text{mol NaBH}_4)^{-1} \qquad (6.47)$$

The reaction given in Eq. (6.47) is initiated at the outer surface of NaBO$_2$ and Mg particles and the reaction mechanism is shown schematically in Fig. 6.49 with regards to the surface transition of protide and oxygen ion (O$_2^-$). Hydrogen is first converted to protide at the outer surface of Mg under lower temperature conditions in hydriding regions and then protide reacts with NaBO$_2$ to form NaBH$_4$ under higher temperature conditions in the dehydriding regions where NaBO$_2$ releases O$_2^-$ which is transferred to the surface of Mg to form MgO.

The oxidation of Mg particles spreads from the surface toward the center of a particle, as can be seen from the SEM and EPMA shown in Fig. 6.50, and the rate-controlling factor is greatly dependent on the depth of the MgO layer and the

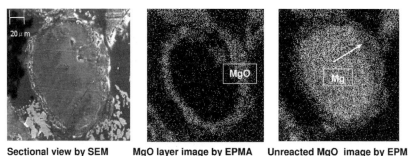

Sectional view by SEM   MgO layer image by EPMA   Unreacted MgO image by EPMA
Particle diameter: 90-106mm, Reaction rate: 30% (produced amount of NaBH$_4$)

Fig. 6.50 Sectional images of a large, nearly spherical Mg particle taken by SEM and EPMA showing growth of the MgO layer in a radial direction.

particle size. The smaller the particle size and the larger the specific surface area, the higher the reaction rate and the yield.

The conversion rate (recovery rate) of the mechano-chemical process reaches close to 100%, but that of the dynamic hydriding/dehydriding process reaches a maximum of 70%. The conversion rate can be improved considerably by reducing the size of the Mg particles.

6.8.7
Concluding Remarks

In this chapter, sodium borohydride ($NaBH_4$), a typical hydrogen–metal complex was introduced. It is the only complex compound that can generate hydrogen in a controlled manner and was chosen to illustrate the significant features of catalytic hydrolysis. The hydrolysis is shown to be an important means to utilize water as the source of hydrogen, even under ambient conditions.

Several technological difficulties encountered in the hydrolysis are based mostly on the physical-chemical properties and characteristics of $NaBH_4$ and $NaBO_2$ that coexist in alkaline solutions. High viscosity, solubility limits and crystal formation are the major sources of trouble in practical applications of $NaBH_4$ systems as a hydrogen source. In addition, it should be noted that the shock wave generated during hydrolysis requires a catalytic bed with a rigid physical structure.

Finally, the importance of developing an innovative production process for $NaBH_4$ was stated. The cost-effectiveness, that is, the generating cost per kg or per mole of hydrogen (\$/kg-H_2 or \$/mol-H_2), must be greatly reduced to be on a par with natural fuel sources available today. A possible production process under development, that applies a transitional state of protide (H^-), was introduced briefly.

We are still searching for the best hydrogen storage material and storage system. However, almost all materials and systems to be investigated in the near future will face various difficulties in their technological and engineering steps with regard to mass production, cost effectiveness and safety and handling measures.

6.8.8
Nomenclature

π	$\pi = 3.141592654$
l	length of the pipe
v	velocity
η	viscosity
n	number of mols
R	gas constant $R = 8.314 \, J \, K^{-1} \, mol^{-1}$
T	temperature in K
V	volume
p	gas pressure
p_0	standard gas pressure $p_0 = 1.013 \times 10^5$ Pa

a	repulsion constant $a = 2.476 \times 10^{-2}$ m^6 Pa mol^{-2}
b	volume constant $b = 2.661 \times 10^{-5}$ m^3 mol^{-1}
d_w	wall thickness of cylinder
d_o	outer diameter of cylinder
Δp	pressure difference
σ_V	tensile strength of the material
N_A	Avogadro constant $N_A = 6.022 \times 10^{23}$ mol^{-1}
S_{ml}	minimum surface area for one mol of adsorbate in a monolayer
M_{ads}	molecular mass of the adsorbate
ρ_{liq}	density of the liqud
p_{eq}	equilibrium gas pressure
$p°_{eq}$	standard equilibrium gas pressure
ΔH	enhalpy change
ΔS	entropy change
ΔG	change in Gibbs energy
ΔQ	heat change
ΔW	work change

References

1 Weast, R.C. (ed.) (1976) *Handbook of Chemistry and Physics*, vol. 57, CRC Press, Cleveland, Ohio.
2 Leung, W.B., March, N.H., Motz, H. (1976) *Phys. Lett.*, **56A** (6), 425–426.
3 Vollrath, F., Knight, D. P. (2001) *Nature*, **410**, 541.
4 Huston, E.L. (1984) *Proceedings of the 5th World Hydrogen Energy Conference*, vol. 3, July 15–20, 1984, Toronto, Canada.
5 Flynn, T.M. (1992) A liquefaction of gases. *McGraw-Hill Encyclopedia of Science & Technology*, 7th edn, vol. **10**. McGraw-Hill, New York, pp. 106–109.
6 Van Ardenne, M., Musiol, G., Reball, S. (1990) *Verlag Harri Deutsch*, pp. 712–715.
7 IUPAC Recommendations (1985) *Pure Appl. Chem.*, **57**, 603.
8 Rzepka, M., Lamp, P., Casa-Lillo, M.A. (1998) *J. Phys. Chem. B*, **102**, 10849; Dakrim, F., Levesque, D. (1998) *J. Chem. Phys*, **109**, 4981; Darkrim, F., Levesque, D. (2000) *J. Phys. Chem B*, **104**, 6773.
9 Wang, Q., Johnson, J.K. (1999) *J. Chem. Phys.*, **110**, 577.
10 Lee, S.M., Lee, Y.H. (2000) *Appl. Phys. Lett.*, **76**, 2877; Lee, S.M., Park, K.S., Choi, Y.C., Park, Y.S., Bok, J.M., Bae, D.J., Nahm, K.S., Choi, Y.G., Yu, S.C., Kim, N., Frauenheim, T., Lee, Y.H. (2000) *Synth. Met.*, **113**, 209, Baulischer, C.W. (2000) *Chem. Phys. Lett.*, **322**, 237; Baulischer, C.W. (2001) *Nano Lett.*, **1** (5), 223.
11 Malbrunot, P., Vidal, D., Vermesse, J., Chahine, R., Bose, T. (1997) *Langmuir*, **13**, 539.
12 Benham, M.J., Ross, D.K. (1989) *Z. Phys. Chem. NF*, **163**, 25.
13 Iijima, S. (1991) *Nature*, **354**, 56.
14 Dillon, A.C., Jones, K.M., Bekkeddahl, A., Kiang, C.H., Bethune, D.S., Heben, M.J. (1997) *Nature*, **386**, 377.
15 Hirscher, M., Becher, M., Haluska, M., Dettlaff-Weglikowska, U., Quintel, A., Duesberg, G.S., Choi, Y.-M., Downes, P., Hulman, M., Roth, S., Stepanek, I., Bernier, P. (2001) *Appl. Phys. A*, **72**, 129.
16 Liu, C., Fan, Y.Y., Liu, M., Cong, H.T., Cheng, H.M., Dresselhaus, M.S. (1999) *Science*, **286**, 1127.
17 Kajiaura, H., Tsusui, S., Kadono, K., Kakuta, M., Ata, M., Murakami,

Y. (2003) *Appl. Phys. Lett.*, **82**, 1105.
18 Chambers, A., Park, C., Baker, R., Rodriguez, N.M. (1998) *J. Phys. Chem. B*, **102**, 4253; Park, C., Anderson, P.E., Chambers, A., Tan, C.D., Hidalgo, R., Rodriguez, N.M. (1999) *J. Phys. Chem.*, **103**, 10572.
19 Ahn, C.C., Ye, Y., Ratnakumar, B.V., Witham, C., Bowman, R.C., Fultz, B. (1998) *Appl. Phys. Lett.*, **73**, 3378.
20 Lueking, A.D., Yang, R.T., Rodriguez, N.M., Baker, R.T.K. (2004) *Langmuir*, **20** (3), 714–21.
21 Fan, Y.Y., Liao, B., Liu, M., Wie, Y.L., Lu, M.Q., Cheng, H.M. (1999) *Carbon*, **37**, 1649.
22 Cheng, H.M., Liu, C., Fan, Y.Y., Li, F., Su, G., He, L.L., Liu, M. (2000) *Z. Metallkunde*, **91**, 306.
23 Poirier, E., Chahine, R., Bose, T.K. (2001) *Int. J. Hydrogen Energy*, **26**, 831.
24 Hirscher, M., Becher, M., Haluska, M., Quintel, A., Skakalova, V., Choï, Y.-M., Dettlaff-Weglikowska, U., Roth, S., Stepanek, I., Bernier, P., Leonhardt, A., Fink, J. (2002) *J. Alloys Compd.*, **330–332**, 654.
25 Ritschel, M., Uhlemann, M., Gutfleisch, O., Leonhardt, A., Graff, A., Täschner, C., Fink, J. (2002) *Appl. Phys. Lett.*, **80**, 2985.
26 Tibbetts, G.G., Meisner, G.P., Olk, C.H. (2001) *Carbon*, **39**, 2291.
27 Ströbel, R., Jörissen, L., Schliermann, T., Trapp, V., Schütz, W., Bohmhammel, K., Wolf, G., Garche, J. (1999) *J. Power Sources*, **84**, 221.
28 Harutyunyan, A.R., Pradhan, B.K., Tokune, T., Fujiwara, Y., Eklund, P.C. (2001) in Proceedings of the 25th International Conference on Carbon, CARBON '01, American Carbon Society, July 14–19 Lexington, KY.
29 De la Casa-Lillo, M.A., Lamari-Darkrim, F., Cazorla-Amorós, D., Linares-Solano, A. (2002) *J. Phys. Chem. B*, **106**, 10930.
30 Chahine, R., Bose, T. (1994) *Int. J. Hydrogen Energy*, **19**, 161.
31 Panella, B., Hirscher, M., Roth, S. (2005) *Carbon*, **43**, 2209.
32 Ye, Y., Ahn, C.C., Wutham, C., Fultz, B., Liu, J., Rinzler, A.G., Colbert, D., Smith, K.A., Smalley, R.E. (1999) *Appl. Phys. Lett.*, **74**, 2307.
33 Pradhan, B.K., Harutyunyan, A.R., Stojkovic, D., Grossman, J.C., Zhang, P., Cole, M.W., Crespi, V., Goto, H., Fujiara, J., Eklund, P.C. (2002) *J. Mater. Res.*, **17** (9), 2209.
34 Haluska, M., Hirscher, M., Becher, M., Dettlaff-Weglikowska, U., Chen, X., Roth, S. (2004) *Mater. Sci. Eng. B.*, **108**, 130.
35 Schimmel, H.G., Kearley, G.J., Nijkamp, M.G., Visser, C.T., de Jong, K.P., Mulder, F.M. (2003) *Chem. Eur. J.*, **9**, 4764.
36 Harutyunyan, A.R., Pradhan, B.K., Tokune, T., Fujiwara, Y., Eklund, P.C. (2001) in Proceedings of the 25th International Conference on Carbon, CARBON '01, American Carbon Society, July 14–19 Lexington, KY.
37 Züttel (2003) *Mater. Today*, **9**, 24.
38 Nijkamp, M.G., Raaymakers, J.E.M.J., van Dillen, A.J., de Jong, K.P. (2001) *Appl. Phys. A*, **72**, 619.
39 Evard, E.A., Voit, A.P., Gordeev, S.K., Gabis, I.E. (2000) *Mater. Sci.*, **36**, 499.
40 Orimo, S., Majer, G., Fukunaga, T., Zuettel, A., Schlapbach, L., Fujii, H. (1999) *Appl. Phys. Lett.*, **75**, 3093.
41 Orimo, S., Matsushima, T., Fujii, H., Fukunaga, T., Majer, G. (2001) *J. Appl. Phys.*, **90**, 1545.
42 Hirscher, M., Becher, M., Haluska, M., von Zeppelin, F., Chen, X., Dettlaff-Wegliskowska, U., Roth, S. (2003) *J. Alloys Compd.*, **356–357**, 433.
43 Atsumi, H. (2003) *J. Nucl. Mater.*, **313–316**, 543 [Atsumi, H., Tauchi, K. (2003) *J. Alloys Compd.*, **356–357**, 705].
44 Brosha, E.L., Davey, J., Garzon, F.H., Gottesfeld, S. (1999) *J. Mater. Res.*, **14**, 2138.
45 Fraenkel, D., Shabtai, J. (1977) *J. Am. Chem. Soc.*, **99**, 7074 [Weitkamp, J., Fritz, M., Ernst, S. (1995) *Int. J. Hydrogen Energy*, **20**, 967.
46 Chahine, R., Bose, T. (1994) *Int. J. Hydrogen Energy*, **19**, 161.
47 Langmi, H.W., Walton, A., Al-Mamouri, M.M., Johnson, S.R.,

Book, D., Speight, J.D., Edwards, P.P., Gameson, I., Anderson, P.A., Harris, I.R. (2003) *J. Alloys Compd.*, **356–357**, 710.
48. Otero Areán, C., Manoilova, O.V., Bonelli, B., Rodríguez Delgado, M., Turnes Palomino, G., Garrone, E. (2003) *Chem. Phys. Lett.*, **370**, 631.
49. Kazansky, V.B., Borovkov, V.Y., Serich, A., Karge, H.G. (1998) *Microporous Mesoporous Mater.*, **22**, 251.
50. Li, H., Eddaoudi, M., O'Keeffe, M., Yaghi, O.M. (1999) *Nature*, **402**, 276.
51. Yaghi, O.M., O'Keeffe, M., Ocking, N.W., Chae, H.K., Eddaoudi, M., Kim, J. (2003) *Nature*, **423**, 705.
52. James, S.L. (2003) *Chem. Soc. Rev.*, **32**, 276.
53. Rosi, N.L., Eckert, J., Eddaoudi, M., Vodak, D.T., O'Keefe, M., Yaghi, O.M. (2003) *Science*, **300**, 1127.
54. Férey, G., Latroche, M., Serre, C., Millange, F., Loiseau, T., Percheron-Guégan, A. (2003) *Chem. Comm.*, **24**, 2976.
55. Pan, L., Sandler, M.B., Huang, X., Li, J., Smith, M., Bitther, E., Bockrath, B., Johnson, J.K. (2004) *J. Am. Chem. Soc.*, **126** (5), 1308–9.
56. Rittmeyer, P., Wietelmann, U., Hydrides, in *Ullmann's Encyclopedia of Industrial Chemistry*, 5th, Completely Revised Edition, vol. **A13**: High-Performance Fibers to Imidazole and Derivatives, VCH, pp. 199–226.
57. Pearson, G.R. (1985) *Chem. Rev.*, **85**, 41–49.
58. Huheey, J.E. (1983) *Inorganic Chemistry*, Harper & Row, New York.
59. Allred, A.L., Rochow, E.G. (1958) *J. Inorg. Nucl. Chem.*, **5**, 264.
60. Mueller, W.M., Blackledge, I.R., Libowitz, G.G. (eds) (1968) *Metal Hydrides*, Academic Press, New York.
61. Lennard-Jones, J.E. (1932) *Trans. Faraday Soc.*, **28**, 333.
62. Fukai, Y. (1989) *Z. Phys. Chem.*, **164**, 165.
63. Griessen, R. (1988) *Phys. Rev. B*, **38**, 3690–3698.
64. Switendick, A.C. (1979) *Z. Phys. Chem. N.F.*, **117**, 89.
65. Westlake, D.J. (1983) *J. Less-Common Met.*, **91**, 275–292.
66. Miedema, A.R. (1973) *J. Less-Common Met.*, **32**, 117.
67. Van Mal, H.H., Buschow, K.H.J., Miedema, A.R. (1974) *J. Less-Common Met.*, **35**, 65.
68. Griessen, R., Driessen, A. (1984) *Phys. Rev. B*, **30** (8), 4372–4381.
69. Griessen, R., Riesterer, T. (1988) Heat of formation models, in *Hydrogen in Intermetallic Compounds I Electronic, Thermodynamic, and Crystallographic Properties, Preparation* (ed. L. Schlapbach), Springer Series Topics in Applied Physics, vol. **63**, pp. 219–284.
70. Yvon, K. (1998) *Chimia* **52** (10), 613–619.
71. Yvon, K., Renaudin, G. (2005) *Encyclopedia of Inorganic Chemistry* (ed. R.B. King), pp. 1814–1846, and references therein, ISBN 0-470-86078-2, John Wiley & Sons Ltd.
72. Yvon, K., Renaudin, G., Wei, C.M., Chou, M.Y. (2005) *Phys. Rev. Lett.*, **94** (1–4), 66403.
73. Chotard, J.-N., Filinchuk, Y., Revaz, B., Yvon, K. (2006) *Angew. Chem. Int. Ed.*, **45**, 7770–7773.
74. Kadir, K., Noréus, D. (2007) *Inorg. Chem.*, **46**, 2220–2223.
75. Filinchuk, Y.E., Yvon, K. (2006) *J. Solid State Chem.*, **179**, 1041–1052.
76. Filinchuk, Y.E., Yvon, K., Emerich, H. (2007) *Inorg. Chem.*, **46**, 2914–2920; see also Yvon, K., Filinchuk, Y.E. (2007) *J. Alloys Compd.* **446–447**, 3–5.
77. Firman, T.K., Landis, C.R. (1998) *J. Am. Chem. Soc.*, **120**, 12650–12656.
78. Orgaz, E., Gupta, M. (2002) *J. Alloys Compd.* **330–332**, 323–327.
79. Singh, D.J., Gupta, M., Gupta, R. (2007) *Phys. Rev. B*, **75** (1–6), 035103.
80. Richardson, T.J., Slack, J.L., Farangis, B., Rubin, M.D. (2002) *Appl. Phys. Lett.*, **80**, 1349–1351; see also Lohstroh, W., Westerwaal, R.J., van Mechelen, J.L.M., Chacon, C., Johansson, E., Dam, B., Griessen, R. (2004) *Phys. Rev. B*, **70** (1–11), 165411, and references therein.

81 Bogdanovic, B., Reiser, A., Schlichte, K., Spliethoff, B., Tesche, B. (2002) *J. Alloys Compd.* **345**, 77–89.
82 Fauroux, J.C., Teichner, Stanislas, J. (1966) *Bull. Soc. Chim. Fr.*, **9**, 3014–3016.
83 Wiberg, E., Amberger, E. (1971) *Hydrides of the Elements of Main Groups I-IV*, Elsevier, Amsterdam.
84 Harris, P.M., Meibohm, E.P. (1947) *J. Am. Chem. Soc.*, **69**, 1231–1232.
85 Soulié, J.-P., Renaudin, G., Cerny, R., Yvon, K. (2002) *J. Alloys and Comp.*, **346**, 200–205.
86 Züttel, A., Rentsch, S., Fischer, P., Wenger, P., Sudan, P., Mauron, P., Emmenegger, C. (2003) *J. Alloys Compd.*, **356–357**, 515–520.
87 Lodziana, Z., Vegge, T. (2004) *Phys. Rev. Lett.*, **93**, 145501.
88 Miwa, K., Ohba, N., Towata, Shin-ichi, Nakamori, Y., Orimo, Shin-ichi (2004) *Phys. Rev. B*, **69**, 245120.
89 Davis, R.L., Kennard, C.L. (1985) *J. Solid State Chem.*, **59**, 393.
90 Abrahams, S.C., Kalnajs, J. (1954) *J. Chem. Phys.*, **22**, 434.
91 Fischer, P., Züttel, A. (2002) Order-Disorder Phase Transition in Na[BD4], Trans. Tech. Publication Ltd. (2002) proceedings of EPDIC-8.
92 Johnston, H.L., Hallet, N.C. (1953) *J. Am. Chem. Soc.*, **75**, 1467.
93 Stockmeyer, W.H., Stephenson, C.C. (1953) *J. Chem. Phys.*, **21**, 1311.
94 Zarkevich, N.A., Johnson, D.D. (2006) *Phys. Rev. Lett.*, **97**, 119601.
95 Peterson, E. R. (1965) *Dissert. Abs., Washington State Univ.*, **25**, 5588.
96 Ford, P.T., Powell, H.M. (1954) *Acta Crystallogr.*, **7**, 604.
97 Konoplev, V.N., Bakulina, V.M. (1971) *Russ. Chem. Bull.*, **20** (1), 136–138.
98 Fichtner, M., Frommen, Ch., Züttel, A. (2006) Proceedings of International Symposium on Metal Hydrogen Systems Fundamentals & Applications, 1–6 October 2006 Lahaina, Maui, Hawaii, USA.
99 Aldridge, S., Blake, A. J., Downs, A.J., Gould, R.O., Parsons, S., Pulham, C.R. (1997) *J. Chem. Soc, Dalton Trans.*, 1007.
100 Schrauzer, G.N. (1955) *Naturwissenschaften*, **42**, 438.
101 Lippard, S.J., Ucko, D.A. (1968) *Inorg. Chem.*, **7**, 1051.
102 Lipscomb, W.N. (1963) *Boron Hydrides*, Benjamin, New York.
103 Nakamori, Y., Orimo, Shin-ichi (2004) *J. Alloys Compd.*, **370**, 271–275.
104 Miwa, K., Ohba, N., Towata, S., Nakamori, Y., Orimo, S. (2005) *J. Alloys Compd.*, **404–406**, 140–143.
105 Orimo, S., Nakamori, Y., Züttel, A. (2004) *Mater. Sci. Eng. B*, **108**, 51–53.
106 Nakamori, Y., Miwa, K., Ninomiya, A., Li, H., Ohba, N., Towata, Shin-ichi, Züttel, A., Orimo, Shin-ichi (2006) *Phys. Rev. B*, **74**, 045126.
107 Frankcombe T.J., Kroes G.-J., Züttel A. (2005) *Chem. Phys. Lett.*, **405**, 73–78.
108 Ohba, N., Miwa, K., Aoki, M., Noritake, T., Towata, Shin-ichi, Nakamori, Y., Orimo, Shin-ichi, Züttel, A. (2006) *Phys. Rev. B*, **74**, 075110.
109 Davis, William D., Mason, L.S., Stegeman, G. (1949) *J. Am. Chem. Soc.*, **71**, 2775–2781.
110 Mauron, P., Empa switzerland, personal communication.
111 Züttel, A., Wenger, P., Rentsch, S., Sudan, P., Mauron, Ph., Emmenegger, Ch. (2003) *J. Power Sources*, **118**, 1–7.
112 Orimo, Shin-ichi, Nakamori, Y., Ohba, N., Miwa, K., Aoki, M., Towata, Shin-ichi, Züttel, A. (2006) *Appl. Phys. Lett.*, **89**, 021920.
113 Orimo, S., Nakamori, Y., Kitahara, G., Miwa, K., Ohba, N., Towata, S., Züttel, A. (2005) *J. Alloys Compd.*, **404–406**, 427–430.
114 Sudan, P., Züttel, A. (2006) *Nature* submitted.
115 Soldate, A.M. (1947) *J. Am. Chem. Soc.*, **69**, 987.
116 Gunn, Stuart R., Green, LeRoy G. (1962) *J. Chem. Phys.* **36** (4), 1118 and Erratum: Stuart, R. Gunn, et al. (1962) *J. Chem. Phys.*, **37**, 2724.
117 Bielmann, M., Empa, Switzerland; personal communication.
118 Barkhordarian, G., Klassen, T., Bormann, R. (2006) *J. Alloys Compd.*, **407**, 249.

119 Kiran, B., Kandalam, A.K., Jena, P. (2006) *J. Chem. Phys.*, **124**, Art. No. 224703.
120 Züttel, A., Borgschulte, A., Orimo, S. (2007) *Scr. Mater.*, **56**, 823–828.
121 Orimo, S., Nakamori, Y., Eliseo, J.R., Züttel, A., Jensen, C.M. (2007) *Chem. Rev.*, **107**, 4111–4132.
122 http://hydpark.ca.sandia.gov/HydrideMaterialListing.html
123 Knox, K., Ginsberg, A.P. (1964) *Inorg. Chem.*, **3**, 555–558.
124 Bronger, W., Müller, P., Schmidt, D., Spittank, H. (1984) *Z. Anorg. Allg. Chem.*, **516**, 35–41.
125 Bronger, W., Auffermann, G. (1998) *Chem. Mater.*, **10**, 2723–2732.
126 Didisheim, J.-J., Zölliker, P., Yvon, K., Fischer, P., Schefer, J., Gubelmann, M. (1984) *Inorg. Chem.*, **23**, 1953–1957.
127 Zölliker, P., Yvon, K., Fischer, P., Schefer, J. (1985) *Inorg. Chem.*, **24**, 4177–4180.
128 Bortz, M., Yvon, K., Fischer, P. (1994) *J. Alloys Compd.*, **216**, 39–42.
129 Bertheville, B., Fischer, P., Yvon, K. (2002) *J. Alloys Compd.*, **330**, 152–156.
130 Gingl, F., Yvon, K., Fischer, P. (1994) *J. Alloys Compd.*, **206**, 73–75.
131 Lauher, J.W., Dougherty, D., Herley, P.J. (1979) *Acta Crystallogr., Sect. B*, **35**, 1454–1456.
132 Rönnebro, E., Noreus, D., Kadir, K., Reiser, A., Bogdanović, B. (2000) *J. Alloys Compd.*, **299**, 101–106.
133 Bogdanović, B., Brand, R.A., Marjanović, A., Schwickardi, M., Tölle, J. (2000) *J. Alloys Compd.*, **302**, 36–58.
134 Bogdanović, B., Schwickardi, M. (2001) *Appl. Phys A*, **72**, 221–223.
135 Zaluska, A., Zaluski, L. (2001) *Appl. Phys. A*, **72**, 157–165.
136 Gross, K.J., Thomas, G.J., Majzoub, E., Sandrock, G. (2001) Proceedings of the 2001 DOE Hydrogen Program Review, NREL/CP-570-30535.
137 Gross, K.J., Majzoub, E., Thomas, G.J., Sandrock, G. (2002) Proceedings of the 2002 U.S. DOE Hydrogen Program Review, NREL/CP-610-32405.
138 Maeland, A.J., Hauback, B.C., Fjellvåg, H., Sorby, M. (1999) *Int. J. Hydrogen Energy*, **24**, 163–168.
139 Balema, W.P., Wiench, J.W., Dennis, K.W.M., Pruski, M., Pecharsky, V.K. (2001) *J. Alloys Compd.*, **329**, 108–114.
140 Zidan, R.A., Takara, S., Hee, A.G., Jensen, C.M. (1999) *J. Alloys Compd.*, **285**, 119–122.
141 Jensen, C.M., Zidan, R.A., Mariels, N., Hee, A.G., Hagen, C. (1999) *Int. J. Hydrogen Energy*, **24**, 461–465.
142 Sandrock, G., Gross, K.J., Thomas, G.J. (2002) *J. Alloys Compd.*, **339**, 299–308.
143 Bogdanović, B., Felderhoff, M., Kaskel, S., Pommerin, A., Schlichte, K., Schüth, F. (2003) *Adv. Mater.*, **15**, 1012–1015.
144 Fichtner, M., Fuhr, O., Kircher, O., Röthe, J. (2003) *Nanotechnology*, **14**, 778–85.
145 Schlapbach, L., Züttel, A. (2001) *Nature*, **414**, 353–358.
146 Marinescu-Pasoi, L., Behrens, U., Langer, G., Gramatte, W., Rastogi, A.K., Schmitt, R.E. (1991) *Int. J. Hydrogen Energy*, **16**, 407–412.
147 Ritter, A. (1992) *VGB Kraftwerkstechnik (Engl.)*, **72**, 311.
148 Bogdanović, B., Ritter, A., Spliethoff, B. (1990) *Angew. Chem. Int. Ed. Engl.*, **29**, 223–234.
149 Bogdanović, B., Bohmhammel, K., Christ, B., Reiser, A., Schlichte, K., Vehlen, R., Wolf, U. (1999) *J. Alloys Compd.*, **282**, 84–92.
150 Bogdanović, B., Ritter, A., Spliethoff, B., StraSSburger, K. (1995) *Int. J. Hydrogen Energy*, **20**, 811–822.
151 Clasen, H. (1961) *Angew. Chem.*, **73**, 322–331.
152 Bogdanović, B., Schwickardi, M. (1997) *J. Alloys Compd.*, **253–254**, 1–9.
153 Dymova, T.N., Eliseeva, N.G., Bakum, S.I., Dergachev, Y.M. (1974) *Dokl. Akad. Nauk SSSR*, **215**, 1369, Engl. 256.
154 Dymova, T.N., Dergachev, Y.M., Sokolov, V.A., Grechanaya, N.A. (1975) *Dokl. Akad. Nauk SSSR*, **224**, 591, Engl. 556.
155 Bogdanović, B., Liao, S., Schwickardi, M., Sikorsky, P., Spliethoff, B. (1980) *Angew. Chem.*, **92**, 845–846; *Angew. Chem. Int. Ed. Engl.*, **19**, 818–819.

156 Zaluska, A., Zaluski, L., Ström-Olsen, J.O. (2000) *J. Alloys Compd.*, **298**, 125–134.
157 Meisner, G.P., Tibbetts, G.G., Pinkerton, F.E., Olk, C.H., Balogh, M.P. (2002) *J. Alloys Compd.*, **337**, 254–263.
158 Sandrock, G., Gross, K.J., Thomas, G., Jensen, C., Meeker, D., Takara, S. (2002) *J. Alloys Compd.*, **330–332**, 696–701.
159 Majzoub, E.H., Gross, K.J. (2003) *J. Alloys Compd.*, **356**, 363–367.
160 Gross, K.J., Majzoub, E.H., Spangler, S.W. (2003) *J. Alloys Comd.*, **356**, 423–428.
161 Gross, K.J., Sandrock, G., Thomas, G. (2002) *J. Alloys Compd.*, **330–332**, 691–695.
162 Franke, R., Rothe, J., Pollman, J., Hormes, J., Bönnemann, H., Brijoux, W., Hindenburg, T. (1996) *J. Am. Chem. Soc.*, **118**, 12090–12097.
163 Anton, D.L. (2003) *J. Alloys Compd.*, **356**, 400–404.
164 Genma, R., Uchida, H.H., Okada, N., Nishi, Y. (2003) *J. Alloys Compd.*, **356**, 358–362.
165 Kaskel, S., Schlichte, K., Chaplais, G., Khanna, M. (2002) *J. Mater. Chem.*, **13**, 1496–1499.
166 Schüth, F. (2003) Gordon Research Conference on Hydrogen Metal Systems, July 13–17 2003, Colby College, Maine, USA.
167 Gross, K.J., Guthrie, S., Takara, S., Thomas, G. (2000) *J. Alloys Compd.*, **297**, 270–281.
168 Bogdanović, B., Felderhoff, M., Germann, M., Härtel, M., Pommerin, A., Schüth, F., Weidenthaler, C., Zibrowius, B. (2003) *J. Alloys Compd.*, **350**, 246–255.
169 Weidenthaler, C., Pommerin, A., Felderhoff, M., Bogdanović, B., Schüth, F. (2003) *Phys. Chem. Chem. Phys.*, **5**, 5149–5153.
170 Bellosta von Colbe, J.M., Schmidt, W., Felderhoff, M., Bogdanovic, B., Schüth, F. (2006) *Angew. Chem. Int. Ed. Engl.*, **45** (22), 3663–3665.
171 Bellosta von Colbe, J.M., Bogdanović, B., Felderhoff, M., Pommerin, A., Schüth, F. (2004) *J. Alloys Compd.*, **370**, 104–109.
172 Haber, J.A., Crane, J.L., Buhro, W.E., Frey, C.A., Sastry, S.M.L., Balbach, J.J., Conradi, M.S. (1996) *Adv. Mater.*, **8**, 163–166.
173 Sun, D., Kiyobayashi, T., Takeshita, H.T., Kuriyama, N., Jensen, C.M. (2002) *J. Alloys Compd.*, **337**, L8–11.
174 Thomas, G.J., Gross, K.J., Yang, N.Y.C., Jensen, C. (2002) *J. Alloys Compd.*, **330–332**, 702–7.
175 Felderhoff, M., Klementiev, K., Grünert, W., Spliethoff, B., Tesche, B., Bellosta von Colbe, J.M., Bogdanović, B., Härtel, M., Pommerin, A., Schüth, F., Weidenthaler, C. (2004) *Phys. Chem. Chem. Phys.*, **6**, 4369–4374.
176 Bogdanović, B., Reiser, A., Schlichte, K., Spliethoff, B., Tesche, B. (2002) *J. Alloys Compd.*, **345**, 77–89.
177 Gennari, F.C., Castro, F.J., Andrade Gamboa, J.J. (2002) *J. Alloys Compd.*, **339**, 261–267.
178 Bogdanović, B., Sandrock, G. (2002) *MRS Bull.*, **27**, 712–716.
179 Leon, A., Kircher, O., Rothe, J., Fichtner, M. (2004) *J. Phys. Chem. B*, **108**, 16372–16380.
180 Chaudhury, S., Muckerman, J. (2005) *J. Phys. Chem. B*, **109**, 6952.
181 Fu, Q.J., Ramirez-Cuesta, A.J., Tsang, S.C. (2006) *J. Phys. Chem. B*, **110**, 711–715.
182 Bureau, J.-C., Amri, Z., Claudy, P., Letoffé, J.-M. (1989) *MRS Bull.*, **24**, 267–273.
183 Vajeeston, P., Ravindran, P., Vidya, R., Fjellvåg, H., Kjekshus, A. (2003) *Appl. Phys. Lett.*, **82**, 2257–2259.
184 Dubrovinsky, L.S., Dubrovinskaia, N.A., Swamy, V., Muscat, J., Harrison, N.M., Ahuja, R., Holm, B., Johansson, B. (2001) *Nature*, **410**, 653–654.
185 Opalka, S.M., Anton, D.L. (2003) *J. Alloys Compd.*, **356**, 486–489.
186 Arroyo y de Dompablo, M.E., Ceder, G. (2004) *J. Alloys Compd.*, **364**, 6–12.
187 Majzoub, E.H., Somerday, P.P., Goods, S.H., Gross, K.J. (2002) International Symposium on Hydrogen Effects on

Materials Behavior, Jackson, Wyoming, USA, September 23–27.
188 Schüth, F., Bogdanović, B., Taguchi, A., (MPI Mühlheim, Germany), patent application WO 2005014469.
189 Bowman R.C. Jr., Fultz, B. (2002) *MRS Bull.*, **27**, 688–693.
190 Joubert, J.-M., Latroche, M., Percheron-Guégan, A. (2002) *MRS Bull.*, **27**, 694–698.
191 Mintz, M.H., Gavra, Z., Kimmel, G. (1980) *J. Less-Common Met.*, **74**, 263–270.
192 Gavra, Z., Hadari, Z., Mintz, M.H. (1981) *J. Inorg. Nucl. Chem.*, **43**, 1763–1768.
193 Balema, V.P., Dennis, K.W., Pecharsky, V.K. (2000) *Chem. Commun.*, 1665–1666.
194 Balema, V.P., Dennis, K.W., Pecharsky, V.K. (2000) *J. Alloys Compd.*, **313**, 69–74.
195 Chen, J., Kuriyama, N., Xu, Q., Takeshita, H., Sakai, T. (2001) *J. Phys. Chem.*, **105**, 11214–11220.
196 Morioka, H., Kakizaki, K., Chung, S.-C., Yamada, A. (2003) *J. Alloys Compd.*, **353**, 310–314.
197 Zaluska, A., Zaluski, L., Ström-Olsen, J.O. (2000) *J. Alloys Compd.*, **307**, 157–166.
198 Fichtner, M., Fuhr, O. (2002) *J. Alloys Compd.*, **345**, 286–296.
199 Fichtner, M., Engel, J., Fuhr, O., Glöss, A., Rubner, O., Ahlrichs, R. (2003) *Inorg. Chem.*, **42**, 7060–7066.
200 Claudy, P., Bonnetot, B., Létoffé, J.M. (1979) *J. Thermal Anal.*, **15**, 119–128.
201 Dymova, T.N., Konoplev, V.N., Sizareva, A.S., Aleksandrov, D.P., Kuznetsov, N.T. (1998) *Dokl. Chem.*, **359**, 200–204.
202 Dymova, T.N., Konoplev, V.N., Sizareva, A.S., Aleksandrov, D.P. (2000) *Russ. J. Coord. Chem.*, **26**, 531–537.
203 Mal'tseva, N.N., Golovanova, A.I., Dymova, T.N., Aleksandrov, D.P. (2001) *Russ. J. Inorg. Chem.*, **46**, 1793–1797.
204 Dymova, T.N., Mal'tseva, N.N., Konoplev, V.N., Golovanova, A.I., Aleksandrov, D.P. (2003) *Russ. J. Coord. Chem.*, **29**, 385–389.
205 Zhang, Q.-A., Nakamura, Y., Oikawa, K., Kamiyama, T., Akiba, E. (2002) *Inorg. Chem.*, **41**, 6941–6943.
206 Zhang, Q.-A., Nakamura, Y., Oikawa, K., Kamiyama, T., Akiba, E. (2002) *Inorg. Chem.*, **41**, 6547–6549.
207 Chen, P., Xiong, Z., Luo, J., Lin, J., Tan, K.-L. (2002) *Nature*, **420**, 302–304.
208 Chen, P., Xiong, Z., Luo, J., Lin, J., Tan, K.-L. (2003) *J. Phys. Chem. B*, **107**, 10967–10970.
209 Ichikawa, T., Isobe, S., Hanada, N., Fujii, H. (2003) *J. Alloys Compd.*, **365**, 271–276.
210 Züttel, A., Rentsch, S., Fisher, P., Wenger, P., Sudan, P., Mauron, P., Emmenegger, C. (2003) *J. Alloys Compd.*, **356**, 515–520.
211 Züttel, A., Wenger, P., Rentsch, S., Sudan, P., Mauron, P., Emmenegger, C. (2003) *J. Power Sources*, **118**, 1–7.
212 Yvon, K. (1998) *Chimia*, **52**, 613–619.
213 Olofsson-Mårtensson, M., Häussermann, U., Tomkinson, J., Noreus, D. (2000) *J. Amer. Chem. Soc.*, **122**, 6960–6970.
214 Unpublished results.
215 Blankenship, R.E. (2002) *Molecular Mechanisms of Photosynthesis*, Blackwell Science; Gregory, R.P.F. (1971) *Biochemistry of Photosynthesis*, Wiley-Interscience, London.
216 This is a very uncommon way of looking at photosynthesis. For didactical purposes, it is easier to describe the photosynthesis circle via the reduction potential of the involved electrons.
217 Hall, D.O., Rao, K.K. (1972) *Photosynthesis*, Edition Arnold, London.
218 Wünschiers, R., Senger, H., Schulz, R. (1998) Photohydrogen production – Hydrogenases in the green alga Scenedesmus obliquus, in *Photosynthesis: Mechanisms and Effects*, vol. III (ed. G. Garab), Kluwer Academic Publishers, Dordrecht, The Netherlands, pp. 1951–1954.
219 Grätzel, M. (1999) in CATTECH: the magazine of catalysis sciences. *Technology and Innovation Issue 5*, **3**(1), 4–17 (Baltzer Science Publishers); see also Koch-Schwessinger, G. (1996)

"Wasserstoffproduktion mit Purpurbakterien, Frommann-Holzboog".

220 Voet, D., Voet, J.G. (1994) *Biochemistry*, Weinheim, New York, VCH, p. 597 (German edition).

221 Emerson, R., Chalmers, R., Cederstrand, C. (1957) *Proc. Natl. Acad. Sci.*, **49**, 137.

222 Heber, U., Krause, G.H. (1980) *Trends Biochem. Sci.*, **5**, 32.

223 Hall, D.O. (1976) *FEBS Lett.*, **64** (1), 6–16, and references therein.

224 Schneider, T.R. (1972) *Energy Conv.*, **13**, 77.

225 Calvin, M. (1974) *Science*, **184**, 375.

226 Häussinger, P., Lohmüller, R., Watsin, A.M., Hydrogen, in *Ullmann's Encyclopedia of Industrial Chemistry*, 5th, Completely Revised Edition, vol. **A13**, High-Performance Fibers to Imidazole and Derivatives, VCH, Weinheim, p. 358.

227 *Handbook of Chemistry and Physics*, 57th edn. CRC Press, Cleveland (1976), p. D-277.

228 Hodoshima, S., Takaiwa, S., Shono, A., Satoh, S., Saito, Y. (2005) *Appl. Catal. A*, **283**, 235.

229 Newson, E., Haueter, T., Hottinger, P., Von Roth, F., Scherer, G.W.H., Schucan, T.H. (1998) *Int. J. Hydrogen Energy*, **23**, 905.

230 Fang, B., Zhou, H., Honma, I. (2006) *J. Chem. Phys.*, **124**, 204718.

231 Taube, M., Taube, P. (1980) Proceedings of the 3rd World Hydrogen Energy Conference, Tokyo.

232 Taube, M., Rippin, D.W.T., Cresswell, D.L., Knecht, W. (1983) *Int. J. Hydrogen Energy*, **8**, 213.

233 Taube, M., Rippin, D.W., Knecht, W., Hakimifard, D., Milisavljvic, M., Grünenfelder, N. (1985) *Int. J. Hydrogen Energy*, **10**, 595.

234 Grünenfelder, N.F., Schucan, T.H. (1989) *Int. J. Hydrogen Energy*, **14**, 579.

235 Hodoshima, S., Arai, H., Takaiwa, S., Saito, Y. (2003) *Int. J. Hydrogen Energy*, **28**, 1255–62, and references therein.

236 Fang, B., Zhou, H., Honma, I. (2006) *Appl. Phys. Lett.*, **89**, 023102.

237 Wojcik, A., Middleton, H., Damopoulos, I., van Herle, J. (2003) *J. Power Sources*, **118**, 342.

238 Uribe, F.A., Gottesfeld, S., Zawodzinski, T.A., Jr. (2002) *J. Electrochem. Soc.*, **149**, A293.

239 *Science* **297**, 1654 (2002).

240 Dybkjær (1995) *Ammonia* (ed. A. Nielsen), Springer Verlag, Berlin, p. 199.

241 Galloway, N., Aber, J.D., Erisman, J.W., Seitzinger, S.P., Howarth, R.W., Cowling, E.B., Cosby, B.J. (2003) *BioScience*, **53**, 341.

242 Christensen, C.H., Sørensen, R.Z., Johannessen, T., Quaade, U.J., Honkala, K., Elmøe, T.D., Køhler, R., Nørskov, J.K. (2005) *J. Mater. Chem.*, **15**, 4106.

243 Lepinasse, E., Spinner, B. (1994) *Rev. Int. Froid*, **17**, 309.

244 Hwang, C., Drews, T., Seppelt, K. (2000) *J. Am. Chem. Soc.*, **122**, 8486.

245 Amendola, S.C. et al. (2000) *J. Power Source*, **85**, 186.

246 Suda, S., Sun, Y.-M., Liu, B.-H., Zhou, Y., Morimitsu, S., Arai, K., Tsukamoto, N., Uchida, M., Candra, Y., Li, Z.-P. (2001) *J. Appl. Phys. A*, **72**, 209–212.

247 Suda, S., Sun, Y.-M., Uchida, M., Liu, B.-H., Morimitsu, S., Arai, K., Zhou, Y., Tsukamoto, N., Candra, Y., Li, Z.-P. (2001) *Met. Mater. Int.*, **7** (1), 73–75.

248 Suda, S. (2003) US Patent (Appl.No.), 10/721479 (11-25-03).

249 Cengel, Y.A., Boles, M.A. (2003) *Thermodynamics-An Engineering Approach*, 4th Int. Edn., McGraw Hill.

250 Fedor, W.S., Douglas, M., Ingalls, D.O. (1957) *IEC*, **49** (10), 1664–72.

251 Schubert, F., Lang, K., Shabacher, W., Burger, A. (1963) US Patent, 3077356.

252 Kojima, Y., Haga, T. (2003) *Int. J. Hydrogen Energy*, **28**, 989–93.

253 Li, Z.-P., Morigasaki, N., Liu, B.-H., Suda, S. (2003) *J. Alloys Compd.*, **354** (1/2), 243–247.

254 Li, Z.-P., Liu, B.-H., Morigasaki, N., Suda, S. (2003) *J. Alloys Compd.*, **349** (1), 232–236.

255 Suda, S., Iwase, Y., Morigasaki, N., Li, Z.-P. (2004) *Advanced Materials for Energy Storage II* (eds. D. Chandra, R. G. Bautista, L. Schlapbach), TMS, USA, pp. 123–133.

7
Hydrogen Functionalized Materials
Arndt Remhof, Björgvin Hjörvarsson, Ronald Griessen, Ingrid Anna Maria Elisabeth Giebels, and Bernard Dam

7.1
Magnetic Heterostructures – A Playground for Hydrogen
Arndt Remhof and Björgvin Hjörvarsson

7.1.1
Introduction

Modern life would be unthinkable without magnetic thin films and heterostructures. They form the basis of magnetic sensors, magnetic reading heads to read out hard disks and magnetic data storage. Magnetic heterostructures consist of combinations of different materials which are in contact through at least one interface. They combine different physical properties which do not exist in nature. Examples are semiconductors/ferromagnets, superconductors/ferromagnets, or ferromagnets/antiferromagnets. These combinations promise to display new physical properties different from those of any single one of them.

Temporary hydrogen alloying of artificial magnetic heterostructures, for example superlattices and trilayers, leads to changes in the structural and electronic properties of the host. As a consequence of the hydrogen-induced modified electronic properties, the magnetic properties, that is the collective behavior of the individual magnetic moments within the metal, alter as well. In particular, the magnitude of the magnetic moment per atom, the ordering temperature as well as the strength and the sign of the interlayer exchange coupling are sensitive to the hydrogen concentration within the material. While a number of the nonmagnetic transition metals absorb hydrogen exothermically, the magnetic transition metals (Cr, Mn, Fe, Co and Ni) only absorb hydrogen endothermically. Consequently, the hydrogen tunability of the magnetic properties within the heterostructures only takes place via nonmagnetic spacer layers.

This chapter covers the properties and the various effects occurring in magnetic thin films and heterostructures based on the itinerant magnetism of the 3d elements. A review of the influence of hydrogen on the magnetism of the rare earth metals is given in the following chapter by A. Wildes.

Hydrogen as a Future Energy Carrier. Edited by A. Züttel, A. Borgschulte, and L. Schlapbach
Copyright © 2008 WILEY-VCH Verlag GmbH & Co. KGaA, Weinheim
ISBN: 978-3-527-30817-0

7.1.2
Hydrogen Loading Technique

Reversible hydrogen loading can only be accomplished within a limited concentration range, above which irreversible effects, for example cracking, appear. If the adhesion to the sample substrate is strong enough to prevent cracking or bending the one-dimensional reduction of symmetry can, for some thin film or thin film multilayers, give a tetragonal crystal distortion that allows reversible hydrogen loading to higher concentrations than for bulk specimens.

Typically, hydrogen loading is accomplished by exposing a sample to hydrogen gas, operating at room temperature or higher or by electrochemical loading. Even though many metals dissolve hydrogen exothermically, it is necessary to apply a certain activation energy in order to dissociate the hydrogen molecule and to overcome the surface barrier or to use a catalyst.

Usually a Pd capping layer is used for that purpose. Pick et al. demonstrated that the sticking coefficient of hydrogen molecules incident on a surface was greatly enhanced by capping the surface with Pd [1]. It not only acts as a catalyst for hydrogen dissociation and facilitates the hydrogen uptake, it also protects the surface from corrosion. Below $\approx$250 K, absorption and desorption is effectively hindered and the hydrogen concentration is thereby fixed in the sample. Desorption is a much slower process than absorption, this allows the use of measurement systems that do not facilitate hydrogen loading, for example the superconducting quantum interference device (SQUID) measurements at different hydrogen concentrations described in Ref. [2] were conducted in this way (Fig. 7.1).

Samples intended for hydrogen alloying experiments are typically capped by a 20–50 Å Pd layer. For magnetic purposes, however, the addition of the strongly paramagnetic and highly polarizable Pd is something that needs to be considered. The hydrogen uptake can be monitored by measuring the sample resistivity or the lattice parameter during loading. Pressure–concentration and pressure–resistivity isotherms have been recorded for different combinations of Fe and V [3,4,5]. Concentration–pressure relationships have been measured in Nb thin films [6,7] and for W/Nb and Fe/Nb multilayers [8,9,10]. Changes in the magneto-optical properties of Fe/Pd bilayers upon hydrogen loading have also been detected [11]

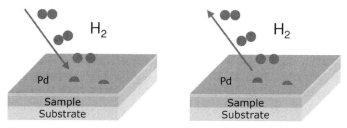

Fig. 7.1 Schematic sketch of the gas loading and unloading using the catalytically active Pd surface.

which, at least in principle, provides another method of monitoring the hydrogen uptake.

Hydrogen can be inserted into the magnetic elements that react endothermically with hydrogen by means of, for example implantation [12,13]. Implantation of highly energetic ions will, however, result in irreversible changes of structural quality in the host material. A surface layer of Pd increases the uptake rate of hydrogen [14].

7.1.3
Exchange Coupled Systems

Advances in thin film deposition techniques have promoted an unprecedented control over film thickness, interface quality and crystalline coherence. These factors have spurred a number of investigations in the area of magnetic thin films which could not be tackled before. Of particular interest are new possibilities to tune the magnetic anisotropy of thin films [17] and the interlayer exchange coupling (IEC) between magnetic layers [18, 19]. Usually these properties are controlled by the thickness and chemical composition of both the magnetic layers and the nonmagnetic spacer layer.

The Curie temperature of an individual ferromagnetic film decreases as one of the physical dimensions is reduced [20]. The finite size scaling for thin extended ferromagnetic films has been shown for a number of systems, such as Fe, Ni [21] and Gd [22]. However, since the ferromagnetic films of a few monolayer thickness are not self-supporting, the details of the finite size scaling depend on the elastic strain provided by the substrate, surface morphology, interface roughness and hybridization effects between the electronic structure of the film and the substrate. A well known example is Fe(110) on W(110) as compared to Fe(001) on Cu(001). In the former case Fe remains in the high spin bcc state down to a monolayer thickness [23], while in the latter case, below 10 monolayer thickness, Fe assumes the low spin fcc state [24] or takes on an incommensurate spin spiral structure [25]. Consequently the phase diagrams for both cases are drastically different. When investigating finite size scaling effects, great care has to be given not only to the precise determination of the thickness, but also to other perturbations, such as strain, magneto-elastic anisotropy, interface roughness, and so on.

Two thin ferromagnetic films F1 and F2, which are affected by finite size scaling, may interact via an interlayer exchange interaction mediated by a paramagnetic metal. Usually the *inter*layer interaction (J') is much weaker than the *intra*layer interaction (J), the ratio J'/J being of the order of 10^{-3} to 10^{-4} [26]. Furthermore, the sign of J' can be positive or negative, depending on the thickness of the nonmagnetic spacer layer. In various systems the IEC was found to be oscillatory and both Ruderman–Kittel–Kasuya–Yosida- (RKKY-)like models [27, 28] and a quantum interference model [29, 30] predict oscillations of the interlayer exchange coupling

J' with the Fermi wave vector k_F and the thickness of the nonmagnetic spacer layer, d_S, following

$$J' \propto \frac{\sqrt{2k_F d_s}}{\sin(2k_F d_s)} \tag{7.1}$$

The two ferromagnetic layers F1 and F2 involved may be of the same kind, or they may consist of different ferromagnetic layers. They may have the same thickness, or their thicknesses may be chosen to be different. Decreasing the thickness of the nonmagnetic interlayer to zero is not interesting if F1 and F2 are of the same kind, as this is equivalent to twice the thickness of the individual layers. However, if the layers F1 and F2 consist of different ferromagnetic materials with distinct Curie temperatures T_{C1} and T_{C2}, the situation is quite different. Extensive research in the past has shown that, in this case, the order parameter of the film with the lower T_C is increased while the order parameter of the film with the higher T_C is decreased [31]. A true phase transition is only observed for the higher Curie temperature. If we now separate the two different ferromagnetic layers F1 and F2 by a nonmagnetic spacer layer, individual phase transitions may exist. At least an oscillation of the ordering temperature of Ni by more than 40 K could be measured in Co/Cu/Ni trilayers as a function of interlayer exchange interaction by varying the Cu(001) spacer thickness [32] (Fig. 7.2).

To increase the amount of magnetic material (and thereby the magnetic signal), usually multilayers, that is systems composed of alternating magnetic and nonmagnetic layers, where the magnetic layers are individually ferromagnetic, are studied. Recently the tuning of magnetic properties in multilayers via controlled atomic scale interface roughness and intermixing has attracted much attention [33, 34]. Relatively short-range intermixing can have a long-range influence on the magnetic properties via change in the IEC. However, the modification of interface structures cannot be performed reversibly. Therefore there is considerable interest in finding a method to alter reversibly the chemical state of the spacer layer without modifying any other experimental parameters. This can, in particular, be achieved by interstitial loading of hydrogen. When hydrogen is dissolved in metals, the electronic states of the absorbing layer are strongly affected by, for example, the s–d hybridization. Within thin films and multilayers, the formation of stable hydride phases is suppressed, the bulk phase diagram does not apply. In thin metallic layers, H randomly occupies interstitial sites in the host metal. The stoichiometry is not fixed and the hydrogen concentration $c_H =$ (hydrogen atoms)/(metal atom) is a

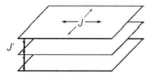

Fig. 7.2. Scheme of an exchange coupled system. Ferromagnetic films interact via interlayer coupling (J') across paramagnetic layers. J describes the interlayer coupling that is responsible for the ferromagnetic order within each layer.

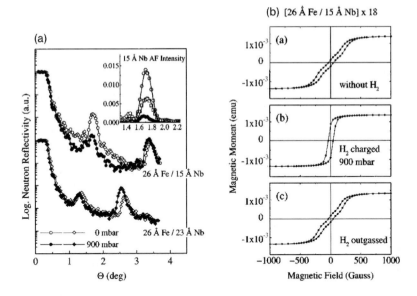

Fig. 7.3 H-tunability of the IEC demonstrated by neutron reflectivity measurements (A) and by magnetometric measurements (B) of a Fe/Nb superlattice. Images taken from Klose et al. [37].

continuous parameter, which can be adjusted via the applied hydrogen pressure or the temperature. Thus, the hydrogen concentration can be adjusted reversibly, and reversible tailoring of the electronic structure, in particular the density of states at the Fermi surface [35, 36] is conceivable. This should directly affect the interlayer exchange coupling J'. In fact, for Fe/Nb [37] (see Fig. 7.3) and Fe/V superlattices [38] it has been demonstrated that loading of hydrogen leads to a change in the IEC. Combining magnetization [39] and polarized neutron reflectivity measurements [38], it could be shown that a complete and reversible switching from ferromagnetic (F) to antiferromagnetic (AF) and vice versa can be achieved by changing the hydrogen concentration in the V interlayer. Changes in the IEC with hydrogen loading have also been demonstrated in heterostructures of for example Co/Nb [12] and Ho/Y [13].

Changes in the IEC through hydrogen loading are typically observed by the influence of, for example, the critical temperature or the disappearance or appearance of AFM scattering as in, for example polarized neutron scattering experiments.

Watts and Rodmacq found that while the bilinear IEC was highly sensitive to the hydrogen concentration, the biquadratic IEC was not significantly different for hydrogenated and nonhydrogenated samples of Co/Nb multilayers [12].

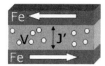

Fig. 7.4. H only enters the paramagnetic V spacer layers in a Fe/V superlattice, modifying the interlayer exchange coupling J'.

7.1.4
H in Fe/V

Due to their favorable properties, which will be discussed in the following, superlattices consisting of Fe and V became the model system to study the influence of H on exchange coupled superlattices and a variety of different phenomena have been discovered. In the following we will discuss this system in more detail (Fig. 7.4).

Both, Fe and V have bcc lattices with a lattice mismatch of less than 4%, allowing to grow them with a very high structural quality [40, 41]. Furthermore, V is a simple paramagnetic metal without magnetic structure in contrast to the Fe/Cr system [42, 43]. Although antiferromagnetic IEC in polycrystalline Fe/V multilayers was detected more than 10 years ago [38], the regions for F and AF coupling were identified only recently using epitaxial Fe/V(100) superlattices with sharp interfaces [39, 41]. According to these experiments, F coupling exists up to a V thickness of 12 monolayers (ML) and for V thicknesses larger than 14 ML, AF coupling only occurs between 13 and 14 ML of V with a relative precision for the AF range of ±1 ML.

V not only mediates the IEC in Fe/V multilayers, but it also displays an induced magnetic moment, aligned antiparallel to the magnetic moments of the neighboring Fe moments. At the same time a reduction in the Fe magnetic moment near the Fe/V interface has been observed [44] (Fig. 7.5).

Both effects lead to a strong reduction in the total magnetic moment with decreasing thickness of the Fe layers. Induced V moments of up to 1.1 µB/atom have been found [18]. Interface alloying plays an essential role in the formation of the magnetic structure of Fe/V interfaces. An increasing number of Fe (V) nearest neighbors to the V (Fe) atoms clearly enhances the effect [18]. Theoretically this

Fig. 7.5 Cross-section TEM of a Fe/V superlattice, showing well defined, smooth layers. Image taken from A. Broddefalk et al., Phys. Rev. B, **65** 214430 (2002).

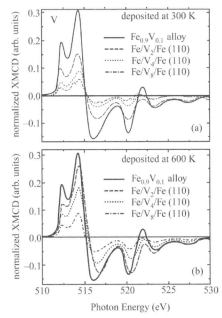

Fig. 7.6 XMCD measurements of different smooth Fe/V superlattices grown at 300 K compared with a homogeneous Fe/V alloy (A). The induced V moment obviously increases with the number of neighboring Fe atoms. Intermixing, achieved by higher growth temperatures (B) obviously enhances the effect. Image taken from A. Scherz et al., Phys. Rev. B, **68** 140401(R) (2003) [18].

induced moment is understood as a hybridization effect and has been calculated on the basis of self-consistent electronic structure calculations [45] (Fig. 7.6).

Unlike many other magnetic superlattices with interlayer exchange coupling, the mediating V interlayer provides a host for interstitial hydrogen atoms. Exposing Fe/V superlattices to a H_2-atmosphere at room temperature leads to the adsorption of hydrogen in the V spacer layers. The amount of hydrogen dissolved in the Fe matrix is negligible, as the solubility is endothermic [14,30]. Moreover, there is a hydrogen depletion zone of 1–2 ML in the V layers at the interface [15,16,25]. The hydrogen atoms reside on interstitial positions in the V layer. Due to the restoring strain of the substrate and the Fe layers, the expansion of the V spacer is restricted to the direction normal to the superlattice plane. The total thickness of the V layers can be changed reversibly by as much as 10% at moderate H pressures without any memory effects [39].

Hydrogen loading alters both the strength and the sign of the IEC as well as the magnetic moments. Considering the first effect, it has been observed that the interlayer ordering between ferromagnetic Fe layers in Fe/V(001) superlattices can be switched from parallel to antiparallel and vice versa upon introducing hydrogen to the V layers. This process is reversible upon removal of the hydrogen. One could view the introduction of hydrogen as a way of changing reversibly the thickness of the V spacer layer and thereby alter the coupling between the Fe

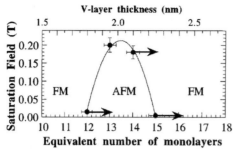

Fig. 7.7 Saturation field as a function of the V layer thickness for AF coupled Fe(3)/V(x) superlattices, obtained by magneto-optic Kerr effect and SQUID measurements. The thicknesses which are accessible by loading the samples with hydrogen are marked by arrows in the figure. Taken from Hjörvarsson et al. (1997), ref. [38].

layers. The tunable V thickness range which is accessible in a Fe/V superlattice, however, is too small to explain the observed effects. H would enhance the AFM exchange coupling in a Fe(3)/V(12) sample, however, the opposite was observed. [We will use the shortened notation, Fe(n)/V(m), where n and m refer to the number of monolayers. Fe(3)/V(12) denotes, therefore, 3 monolayers of Fe separated by 12 monolayers of V.] Obviously d_S is not responsible for the change in IEC. Equation (7.1) This then, from Eq. (7.1), only leaves the Fermi wave vector k_F of the vanadium to be affected by the hydrogen. To overcome the expansion, an even faster increase in the oscillatory exchange coupling period is required [38] (Fig. 7.7).

The fact, that the IEC can be tuned from parallel to antiparallel by H loading, suggest that there is a critical H concentration, at which the IEC vanishes and the Fe layers become uncoupled. Then the Fe layers in the superlattice would behave as independent quasi two-dimensional layers with spins in the plane and no in-plane anisotropy. Temperature-dependent *in situ* neutron reflectivity measurements were employed to investigate the magnetic J'–T phase diagram of a Fe(2)/V(13) superlattice with 300 repetitions [19]. The superlattice exhibits a critical concentration $x_c = 0.022$ at which the magnetic order changes from AF to F. In the vicinity of x_c the transition temperature depends strongly on x and decreases continuously as x_c is approached. The intuitive concept that J' vanishes at x_c is supported by an investigation of the AFM coupling field as a function of x. The resulting J'–T phase diagram is in agreement with a theoretical, schematic phase diagram proposed for an Ising model on a cubic lattice with fixed intralayer and variable IEC. Interlayer exchange coupled metallic superlattices in which the nonmagnetic spacer layers absorb hydrogen offer a new approach for studying systems in which magnetic order and dimensionality can be tuned continuously and reversibly (Fig. 7.8).

The modified electronic structure of the H-loaded V as compared to the pure metal is not only responsible for the change in the IEC in Fe/V superlattices, it also influences the magnetic moments. SQUID magnetization measurements of hydrogen-loaded Fe/V superlattices show a proportional increase in the saturation moments with the hydrogen concentration after hydrogen loading for all samples.

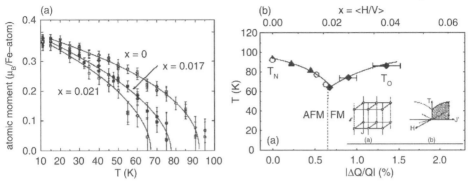

Fig. 7.8 (A) Magnetization of the AF coupled Fe(2)/VH$_x$(13) superlattice as a function of temperature for the H free state and for $x = 0.017$ H/V and 0.021 H/V, respectively. (B) Magnetic phase diagram for the superlattice determined by neutron reflectivity (open symbols) and SQUID (solid symbols). Lines are guides to the eye. Inset: phase diagram for a cubic Ising model with variable interlayer and fixed ferromagnetic intralayer coupling [51]. These images are taken from Leiner et al. (2003) [19].

A maximum increase, corresponding to a change in the atomic moments from 0.25 to 0.6 μ_B per Fe atom was determined for a Fe(3)/V(11) superlattice [39] (Fig. 7.9).

As the total measured moment consists of a ferromagnetic Fe moment and a proximity-induced antiferromagnetically aligned V moment, an element specific determination of the magnetic moments is required to distinguish between an increase in the Fe moment and/or a decrease in the V moment at the interface with increasing hydrogen concentration. Changes in the total magnetic moment under

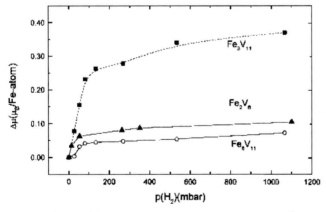

Fig. 7.9 Change of the atomic magnetic moment per Fe atom after saturating the samples with hydrogen at different H$_2$ pressures. The hydrogen loading was done at room temperature, the magnetization has been measured at 4.2 K. Taken from Labergerie et al. (2001), Reference [39].

hydrogenation were predicted by *ab initio* calculations [47], but the value for the calculated moment enhancement for superlattices with ideally smooth interfaces is quite small (4 %). Calculations within a model Hamiltonian approach taking into account interface alloying [45] have provided a larger effect (10 %), indicating that changes are dominated by the moments of the Fe layers, with only minute effects in the V layers. However, the mechanism of how changes in the electronic state of the spacer layer could remotely affect the Fe moment across the interface in the metallic system has not yet been clarified. Typical screening lengths are of the order of a few Å, which is significantly shorter than the presumed interface width. Experimentally, element specific X-ray resonant magnetic scattering (XRMS) was employed to investigate the interface magnetism in Fe/V superlattices [48] (Fig. 7.10).

To maximize the expected effect from the Fe interfaces, the minimal thickness of the Fe layers, two monolayers (2 ML), has been chosen, which still exhibits spontaneous magnetic ordering. The magnetic profile of the multilayer can be envisioned as subdivided into three parts: (i) two equivalent ferromagnetically ordered atomic layers of Fe, (ii) magnetically polarized V layers at the interfaces and (iii) the paramagnetic inner region of the V layers. The experiment clearly supports the idea of a stable V moment and an increasing Fe moment. While the magnetic asymmetry of the V remains constant, there is a strong increase in the Fe magnetic signal. The enhancement of the magnetic moment due to hydrogen absorption in the adjacent nonmagnetic vanadium layer is remarkable in two respects. First, although H definitively resides in the V layer thereby altering its electronic structure, there is no change in the magnetic asymmetry in the V signal. Secondly, even though

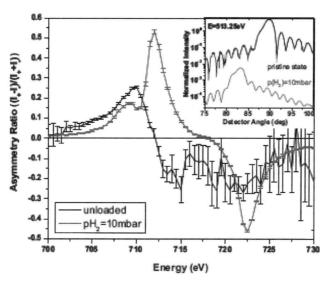

Fig. 7.10 Magnetic asymmetry ratios measured on a [Fe(2)/V(16)]$_{30}$ sample in the vicinity of the Fe L-edges in reflection at the second superlattice peak prior (bold line) and after (light line) hydrogen exposure. The inset shows the reflectivity curves in the vicinity of the second superlattice Bragg peak, recorded at a fixed photon energy of 513.25 eV. Image taken from Remhof et al., ref. [50].

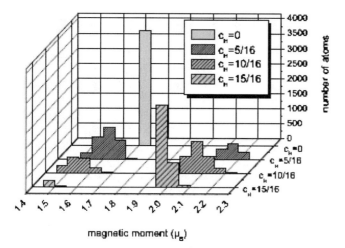

Fig. 7.11 Evolution of Fe magnetic moments with increasing hydrogen concentration c_H for a Fe[(2ML)/V(16ML)] superlattice with ideal interfaces. Image taken from Remhof et al., ref. [50].

hydrogen is confined to the central region of the V layer, its presence influences the magnetic moment of the Fe layer across the hydrogen depleted interface. Rephrasing, the Fe (V) wavefunctions have to be extended beyond the typical screening length in which all perturbations are usually screened within a metallic system. The reason for this long-range interaction is associated with a redistribution of the d- and s-electrons (Fig. 7.11).

Local electron neutrality is provided by the itinerant 4s-electrons and the number of d-electrons can be different depending on the atomic configuration. This is seen, for example, in the linear dependence of the chemical shift on the values of hyperfine fields measured in Mössbauer experiments [49], confirming the existence of a redistribution of d-and s- electrons as compared with the respective bulk states. Redistribution of d-electrons between Fe and V with hydrogen concentration leads to a shift of the d-band relative to the Fermi level which, in turn, explains the long-range global effect affecting the entire sample. The number of d-electrons on Fe atoms increases monotonically increases with increasing c_H. This results in an upward shift of the Fermi level relative to the bottom of the d-band. It is this shift which leads to a monotonic increase in the average Fe moment.

7.2
Optical Properties of Metal Hydrides: Switchable Mirrors
Ronald Griessen, Ingrid Anna Maria Elisabeth Giebels, and Bernard Dam

7.2.1
Introduction

For decades the great majority of experiments on metal–hydrogen systems were carried out without actually looking at the samples. In many cases there was no need to do so (e.g. for all pressure–composition isotherm, electrical resistivity, specific heat, neutron and X-ray scattering measurements). In other cases optical measurements could only be carried out for certain hydrides. For example, Weaver *et al.* [52, 53] measured in detail the reflectivity of dihydrides of Sc, Y and La but could not extend their interesting measurements to the trihydrides of Y and La. This is unfortunate since the pioneering work of Libowitz *et al.* [54] on Ce hydrides had demonstrated the existence of a metal–insulator transition between CeH_2 and CeH_3. In Y and La, however, hydrogenation from the dihydride to the trihydride leads irrevocably to the powdering of bulk samples. It was only much later, in a search for new high-T_c superconductors not based on copper oxide, that the Amsterdam group [55] discovered, in 1995, spectacular changes in the optical properties of metal hydride films of yttrium and lanthanum near their metal–insulator transition: the dihydrides are excellent metals and shiny while the trihydrides are semiconductors and transparent in the visible part of the optical spectrum (see Fig. 7.12). The transition from a shiny to a transparent state is reversible and simply induced at room temperature by changing the surrounding hydrogen gas pressure or electrolytic cell potential. Not only YH_x and LaH_x, but all the trivalent rare-earth hydrides and even some of their alloys exhibit switchable optical and electrical properties [56, 57]. In the transparent state they have characteristic colors: for example, YH_3 is yellowish, LaH_3 red, while some fully hydrogenated alloys containing magnesium are colorless. One of the most surprising results of these early measurements was, however, that the films retained their structural integrity even though they expanded by typically 15 % during hydrogenation of the pure parent metal to the trihydride. This meant that for the first time physical properties, such as electrical resistivity, Hall effect, optical transmission, reflection and absorption, were amenable to experimental investigation. This led to the discovery of new phenomena in the electrical, optical and mechanical properties of these materials. Furthermore, the possibility to fine-tune their properties by alloying and the ease of continuously changing their hydrogen content made them especially attractive for fundamental condensed matter physics. Soon after their discovery it became clear that switchable metal hydride films would pose intriguing questions. So far the nature of their insulating state is not understood and completely different mechanisms for their metal–insulator transition have been proposed. They also offer interesting possibilities to investigate continuous metal–insulator (quantum) phase transitions. Moreover, they can be used to tune magnetic interactions in, for example superlattices (see, e.g. Refs. [58, 59] and references therein).

The purpose of this article is to review the essential properties of switchable mirror materials with special emphasis on their optical properties. So far three generations of hydrogen-based switchable mirrors have been discovered:

- rare-earth switchable mirrors
- color neutral magnesium–rare-earth (Mg–RE) switchable mirrors
- magnesium–transition-metal (Mg–TM) switchable mirrors

Each of these generations has specific properties that are described separately in the next sections. The rare-earth switchable mirrors are historically important since they were the first found to exhibit optical switching as a result of hydrogen absorption. Their transparent appearance in the fully loaded state posed a serious problem to theorists. In particular, the yellow appearance of YH_3 seemed to be in contradiction with state-of-the-art band structure calculations that predicted a metallic state for this material. The physics of rare-earth switchable mirrors is thus closely related to that of electron correlation and metal–insulator transitions. The Mg–RE and the Mg–TM switchable mirrors both exhibit the three fundamental optical states of matter, that is shiny metallic, transparent and highly absorbing. The microscopic mechanisms responsible for their 'black state' are, however, different and need to be discussed separately. Furthermore, we show that alloys and multilayered samples can be used to fine-tune all the essential properties of switchable mirrors and exhibit a series of new phenomena. Finally, recent applications of switchable mirrors as smart coatings in electrochromic devices, as hydrogen indicators for catalytic and diffusion investigations, as the active layer in fiber optic hydrogen sensors and as hydrogen absorption detectors in a combinatorial search for new lightweight hydrogen storage materials are briefly described.

7.2.2
Preparation of Switchable Mirror Thin Films

The growth and microstructure of switchable mirror thin films involves specific aspects that need to be discussed before describing their physical properties. We focus first on Y, La and rare-earth films (for convenience all abbreviated as RE films), deposited in both their metallic and hydride forms. Some references to second-generation mirrors (Mg–RE) are included and third-generation switchable mirrors (Mg–TM) are shortly discussed separately.

7.2.2.1 Growth and Microstructure of Rare-Earth (Hydride) Thin Films
Metallic RE Films Although the deposition and electrical characterization of RE and RE hydrides was already reported in the seventies (see, e.g. Curzon and Singh [60]), these films were not yet phase-pure, due to the limited quality of the vacuum systems of those days. Typically, for pure RE films the background pressure has to be below 10^{-5} Pa [61, 62]. The method of deposition does not seem to be critical and both molecular beam deposition (MBD [61]), sputter deposition (SD [57, 63]) and pulsed laser deposition (PLD [64]) have been used. The purest metallic films are obtained by MBD, where the background pressure is lowest (typically 10^{-7} Pa). From the lattice expansion compared to pure Y one determines an as-deposited hydrogen content in YH_x of $x > 0.08$ [61].

Thermodynamically, one would expect a higher hydrogen content, since the equilibrium Y–YH_2 plateau pressure is $>10^{-24}$ Pa at room temperature [65]. At room temperature the polycrystalline films all tend to have a [0001]hcp or [111]fcc texture, that is the closed-packed layers are oriented parallel to the substrate plane.

From a fundamental point of view the underlying physical processes should preferably be investigated on single-crystalline or epitaxial films. Hayoz et al. [66]

were able to grow highly ordered yttrium on the metallic substrate W(110). Taking the recipe originally developed by Kwo et al. [67], Wildes et al. [62] and Remhof et al. [68, 69] managed to form pure, epitaxial yttrium films on sapphire substrates using a Nb template layer. These approaches have the disadvantage that they complicate optical transmission and electrical transport measurements. This problem was solved by Nagengast et al. [70, 71] who showed that epitaxial yttrium films can be grown directly on (111)-CaF_2 substrates. Shortly after, Jacob et al. [72] showed that good epitaxiality could even be obtained by using BaF_2 substrates. This is remarkable since there is a 20% lattice mismatch between Y and BaF_2. Due to the higher deposition temperature, the hydrogen impurity level of the epitaxial films is much lower than in polycrystalline films: $x = [H]/[Y] = 0.01-0.03$ [62, 73], as compared to $[H]/[Y] = 0.08$ in polycrystalline films. It is suggested that some fluorine, however, is absorbed by the RE film and acts as a surfactant facilitating growth [74]. The large lattice mismatch with respect to the substrate (>5% in the case of CaF_2) causes the coherent film to relax after a certain critical thickness. As shown by Borgschulte et al. [75] this relaxation process is responsible for the formation of so-called ridges [71, 76]: strips of material with their c-axis parallel instead of perpendicular to the substrate.

This stress-induced reorientation is due to (1012)-deformation twinning [77]. Indeed the as-grown density of ridges is much larger on films deposited on CaF_2 substrates as compared to Nb/sapphire substrates; moreover their density increases with film thickness. Formally, due to the ridges, the epitaxial films should be called bi-epitaxial since each of the twin orientations has a different substrate–film relation. The volume fraction of the ridges is however small and for most purposes except the Hall effect [73] their effect can be neglected. When the metallic epitaxial films are hydrogenated the number of ridges increases, and an extended self-organized ridge network is formed, which delineates micron-sized triangular domains [78, 79]. Although hydrogen absorption involves a large expansion of the lattice and a change in lattice symmetry, the epitaxiality of the film remains intact. Details about the switching process of epitaxial YH_x films are given in Section 7.2.3.5.

Cap Layers As-deposited RE films are very reactive. Exposing an uncovered film for several hours to air creates a 5 nm oxide layer while along grain boundaries oxygen is found at a depth of 150 nm [61, 80]. Therefore, palladium [61] and gold [62] cap layers are used, which at the same time act as catalysts for hydrogen dissociation and absorption [81]. To prevent the formation of a RE–Pd alloy (which blocks hydrogen transport) [82] these cap layers have to be deposited at room temperature. Alternatively, an intermediate (hydrogen transparent) thin oxide buffer layer is useful. In the case of La an intermediate AlO_x layer is essential to prevent oxidation of La [83]. After using a combinatorial technique (see also Section 7.2.6.2) to optimize the thickness of the Pd cap layer on Y, Van der Molen et al. [80] concluded that there is a well defined critical thickness of 4 nm, above which Pd can act as a catalyst for hydrogen absorption. This critical thickness is reduced to 2.7 nm, when an intermediate YO_x-layer is applied and to 0.5 nm with a 1.2 nm thick $AlOx$ layer [83]. Borgschulte et al. [84] concluded from photoemission and scanning tunneling microscopy studies (STM/STS) that this critical cap layer thickness is related to inactivation of Pd by encapsulation of the Pd islands by yttrium oxide or hydroxide strong metal–support interaction (SMSI effect). The substitution of Pd by cheaper, preferably transparent, catalytic cap layers is desirable, especially for smart window devices. In this respect, the catalytic properties of transition metal oxides [85] for storage materials could also prove to be useful for smart windows.

In Situ **Deposition of RE Hydride Films**   RE films are generally loaded with hydrogen *after* deposition. To circumvent the associated generation of large stress [86], it is advantageous to grow the RE films as hydrides. The conditions for growth depend strongly on the technique used.

1. *Molecular beam deposition:* The *in situ* formation of pure RE hydrides in MBD is difficult. The hydrogen source needs to be clean and the vacuum system needs to be conditioned by a steady flow of hydrogen. Hydrogen liberates adsorbed species (oxygen, water) from the UHV-chamber walls. As a result, in dirty systems, REO_x impurity phases are formed. Even when using up to 10^{-3} Pa molecular hydrogen and a substrate temperature of 800 K, a mixed oxide/hydride phase is obtained. Using an ultraclean atomic hydrogen source with a direct line of sight in the vicinity of the substrate, Hayoz et al. [87] were able to prepare high quality YH_2 films *in situ*. Epitaxial films form at a substrate temperature of 500 K and a background pressure of 5×10^{-4} Pa on W(110).
2. *Sputtering:* The *in situ* deposition of RE hydride films is comparatively easy in sputter deposition. As shown by Van der Sluis and Mercier [63] $GdMgH_5$ is formed when cosputtering the metals in a 5:1 hydrogen/argon atmosphere at room temperature and a 10^{-1} Pa total pressure. To prevent arcing (due to the formation of insulating deposits) RF sources are used. The *in situ* growth of the hydride phase probably benefits from the activated (atomic) hydrogen formed in the plasma.
3. *Pulsed laser deposition:* In PLD at elevated substrate temperatures, Lokhorst et al. [64, 88] found that RE hydrides can be made without adding hydrogen to the deposition chamber. In this case, the dihydride forms due to the fact that hydrogen dissolved in the RE target is preferentially liberated during the ablation process. Structurally these films range from nanocrystalline to epitaxial, depending on the deposition temperature and substrate used.

7.2.2.2   *Ex Situ* Hydrogen Loading

To transform the as-grown metallic films into hydrides they need to be exposed to hydrogen. This can be achieved by three methods:

1. *Gas loading* is the simplest and fastest procedure. It is well-suited for exploratory work but suffers from the fact that the amount of absorbed hydrogen cannot be

directly determined. Gas loading can be extremely fast. For example GdMgH$_2$ can be loaded to GdMgH$_5$ at room temperature within 40 ms by applying a pressure of a few times 10^5 Pa [79, 90]. The existence of kinetic barriers (e.g. stress, absorption or diffusion barriers) for hydrogen sorption is evident from the fact that a considerable overpressure (underpressure) is needed for the hydrogenation (dehydrogenation) reaction.

2. *Electrolytic loading* is the only technique that allows an easy and versatile *in situ* determination of the hydrogen concentration in a thin film. With an oxygen-free electrolyte it can be used to measure the hydrogen concentration in the films quantitatively and to determine pressure–composition isotherms [91]. At the solid/electrolyte interface the charge transfer reaction is represented by

$$H_2O + e^- \rightarrow H_{ad} + OH^-$$

Thus, there is a direct relation between the integrated current and the amount of hydrogen adsorbed. As the Pd cap layer is thin and as its plateau pressure at room temperature is over 1 mbar, the absorption by this layer is usually neglected. The potential of the Pd/RE electrode can be viewed as that of a Pd electrode in equilibrium with a low pressure hydrogen gas, of which the pressure is determined by the hydrogen concentration in the RE material underneath. Hence, the Pd/RE electrode potential is directly related to the equivalent hydrogen pressure of the underlying RE hydride. Electrolytic loading experiments were combined with optical transmission and reflection spectrometry as well as with electrical resistivity measurements by Kooij *et al.* [65]. They determined the hydrogen pressure–composition (*p–c*) isotherms of rare-earth hydrides over 30 orders of magnitude at room temperature. In Figure 7.12 the room temperature *p–c* isotherm of YH$_x$ for $0 < x < 3$ is given together with the optical transmission at 1.96 eV (635 nm) and the electrical resistivity. This shows that rare-earth metals such as Y have a strong hydrogen affinity. The dihydrides (β-phase) are already stable at very low pressures (10^{-24} Pa for Y films [65], 10^{-25} Pa for La films [92] and 10^{-28} Pa for Gd films [93] at room temperature). The trihydrides (γ-phase) are formed at a pressure of about 10^{-1} Pa. Therefore, the reversibility is limited to the transition between the dihydride and trihydride phase under practical conditions. Only at very high temperatures (1000 K) is it possible to go back to the metallic Y (α-phase) [66]. Note that compared to gas loading the total switching time increases to several minutes. Moreover, Notten *et al.* [91] report that the chemical stability of the Y-electrode is poor, especially when exposed to light.

3. *Chemical loading* of rare-earth films by immersion in an aqueous KOH solution containing NaBH$_4$ has been demonstrated by Van der Sluis [94] for Gd. The redox reaction forms BO$_2^-$ and GdH$_3$. The attractive feature of this technique is that, due to the high redox potential difference, it leads to almost stoichiometric trihydrides. The reaction is reversed by immersion in a 0.3 % H$_2$O$_2$ aqueous solution.

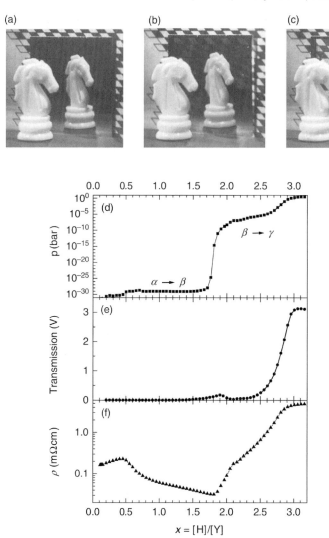

Fig. 7.12 (a)–(c) Photographs of a switchable mirror in the as-deposited, dihydride and trihydride states. (d) The pressure–composition isotherm (1 bar = 10^5 Pa), (e) the hydrogen concentration dependence of the optical transmission and (f) the electrical resistivity of a 300 nm thick yttrium film capped with 15 nm palladium. The optical transmission is measured at a photon energy $E = 1.96$ eV, corresponding to a wavelength $\lambda = 635$ nm. The weak transmission window of the β-phase around $x = 2$ and the transition to the transparent state in the β-phase are clearly visible. The metal–insulator transition is not of structural origin. It occurs within the hexagonal phase at a composition of approximately YH$_{2.86}$. The resistivity first increases slightly when hydrogen dissolves in Y (α-phase). The β-phase (dihydride) clearly has a lower resistivity and as soon as the γ-phase (trihydride) nucleates the resistivity increases to about 4 mΩ cm. This value is limited by the metallic Pd cap layer which shortcuts the semiconducting YH$_3$ film. (From Huiberts et al. (1996), Ref. [55] and Kooij et al. (1999) [65].)

7.2.2.3 Stress in Polycrystalline and Epitaxial Thin Films

Given the large lattice expansion, the fact that REH_2 films can be switched many times back and forth between the dihydride and trihydride phase is remarkable. To have reversible switching, plastic deformation has to be avoided. As shown by Pedersen et al. [95] Gd can be switched reversibly between well-defined stress states of the dihydride and trihydride phases, even though the stress involved is of the order of several GPa. Dornheim et al. [86] explained the relatively low net stress as the result of the fact that textured polycrystalline thin films benefit from a small in-plane tensile stress component. During the transformation from the dihydride to the trihydride phase the distance between the closed-packed planes mostly increases. Since the films have a texture such that these planes are parallel to the substrate most stress is relieved by an expansion of the film thickness. Clearly, to prevent hysteresis (see Section 7.2.3.3) and long-term degradation, a switching material with zero lattice expansion is preferred. Van Gogh et al. [96] found such a zero-expansion alloy, namely $La_{0.62}Y_{0.38}H_x$. It also exhibits the sharpest optical transition.

Epitaxial films also switch reversibly between the dihydride and trihydride state although some irreversible relaxation occurs during the first loading cycle (see Fig. 7.13). This irreversible relaxation is due to the ridges which develop in addition to those already formed during deposition. The process through which an epitaxial film can accommodate high reversible strains/stresses without deterioration of its crystallinity was recently described by Kerssemakers et al. [78, 79].

Using *in situ* atomic force microscopy (AFM), they showed that reversible relaxation takes place via the micron-sized triangular domains that are defined by the ridge network (see Section 7.2.3.5). The domains expand one by one,

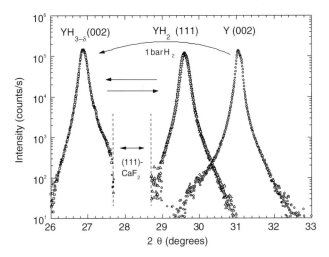

Fig. 7.13 X-ray $\theta-2\theta$-scan of a 300 nm thick epitaxial Y film on (111) CaF_2 capped with 8 nm Pd *in situ* showing the development of the close-packed lattice plane distance during reversible hydrogenation between the dihydride and trihydride state. Hydrogen loading is done in 1 bar H_2, unloading at 100 °C in air. (From Nagengast et al. (1999), Ref. [71].)

homogeneously and essentially independently, during hydrogen absorption (see Fig. 7.18(a)). From a transmission electron microscopy (TEM) study Kooi et al. [77] concluded that the reversible transition to and from the trihydride phase is greatly facilitated by numerous Shockley partial dislocations. These dislocations originate at the boundaries between two twin variants. These twins develop both within the triangles *and* within the ridges due to the phase transitions from the hcp phases Y and YH_3 to the fcc phase YH_2. This twin structure prevents an overall shape change associated with the symmetry change (see Fig. 7.21 of Ref. [77]).

7.2.2.4 Complex Metal Hydride Thin Films

The reversible switching of Mg–TM (TM = Ni, Co, Fe, Mn) films was discovered by Richardson et al. [97, 98]. While in bulk samples hydrogenation requires high temperatures (500 to 600 K) and pressures of 10^5 to 10^6 Pa [99, 100], for thin films it occurs readily at room temperature at low pressures when they are capped with a thin Pd layer. This class of materials does not involve rare-earth metals and might therefore be more resistant to oxidation than the earlier mirrors. This is especially important for applications.

As precursor for the complex metal hydride the metal alloy of interest is sputtered or evaporated and subsequently capped with a thin Pd layer. The grain size of these films is small ($\sim$30 nm). Complex hydrides of this type (Mg_2NiH_4, Mg_2CoH_5, Mg_2FeH_6) seem to form easily at room temperature at a few 10^2 Pa of hydrogen, albeit in an X-ray amorphous form [98, 101]. The reversibility is quite remarkable given the large lattice expansion involved ($\sim$32 %) and the large metal atom mobility involved. Note that, while Mg_2Ni forms an alloy, the corresponding Mg_2Fe or Mg_2Co compounds do not exist.

It is very remarkable that in these thin films the nucleation of the hydrogen-rich phase Mg_2NiH_4 starts preferentially near the film/substrate interface and not, as intuitively expected, close to the catalytic Pd layer at the surface of the sample [101, 102].

As we describe later complex metal hydride thin films can also be made *in situ* by MBD using an atomic hydrogen source. As a result of their different microstructure, these films nucleate randomly on rehydrogenation.

7.2.3
First-Generation Switchable Mirrors: Rare-Earth Metal Hydride Films

One of the most important aspects of the discovery of switchable mirrors is the fact that in the form of thin films rare-earth hydrides (REH_x) are amenable to a whole series of experiments, which were often impossible with bulk samples since hydrogen absorption resulted in a total disintegration of bulk samples.

In this section we show that the optical and electrical transition in REH_x is reversible, continuous, robust and that it occurs in the visible [89]. It also does not depend on the isotope mass, that is the material shows the same features when loaded with deuterium as with hydrogen [68, 103]. Moreover, we show that the rare-earth hydrides undergo a continuous quantum-phase transition [104, 105, 106].

Hysteretic effects are considered in Section 7.2.3.3. Although papers have appeared on, for example Pr [107], Sm [108], Gd [109, 110] and Dy hydrides [111] most investigations on switchable RE hydrides are dealing with Y. In this way Y has become the archetypal material for the first generation switchable mirrors. Scandium, although trivalent, does not form a trihydride under normal conditions (room temperature and 10^5 Pa) [112].

7.2.3.1 Optical Properties

In the Introduction we described how thin films of Y go from reflecting to transparent upon hydrogen absorption. From the photographs in Fig. 7.12 it is evident that this transition can be observed by the naked eye and that, consequently, it takes place in the visible part of the optical spectrum. The main optical changes from the reflecting to the transparent state occur smoothly and reversibly between the dihydride (YH_2) and the trihydride (YH_3) phase. The continuous evolution of reflection and transmission spectra during hydrogen loading between the dihydride ($x = 2$) and trihydride phase ($x = 3$) is displayed in Fig. 7.14 for a 300 nm thick YH_x layer capped with 15 nm Pd [65].

A typical feature of the dihydride phase is the weak transmission window in the red ($1.6 < E < 2.1$ eV for $YH_{2+\varepsilon}$ and $1.3 < E < 1.8$ eV for $LaH_{2+\varepsilon}$) at hydrogen concentrations $1.7 < x < 2.1$ [113] (see Fig. 7.14(a)). The existence of the dihydride transmission window is essential for the visualisation of hydrogen migration in switchable mirrors (see Section 7.2.6). In the transparent trihydride phase, the material has characteristic colors, for example yellow for $YH_{3-\delta}$ (absorption edge at 2.6 eV) and red for $LaH_{3-\delta}$ (absorption edge at 1.9 eV) [56]. Unfortunately, the optical contrast of the transition is reduced due to the Pd cap layer, which has a transparency of about 32 % in the visible for a 12 nm thick layer [114]. Thus, for technological applications it is important to reduce the amount of Pd [80, 83, 115] or to use another more transparent material as cap layer (see Section 7.2.2.1). Stoichiometric YH_3 cannot be obtained at 10^5 Pa H_2 pressure. At most $YH_{2.9}$ is formed during both gas loading and electrochemical hydrogenation. This means that there are many vacancies that act essentially as donors [116, 117] and $YH_{2.9}$ is consequently a heavily-doped semiconductor. Stoichiometric YH_3 is only reached above 4 GPa [118]. High pressure studies of optical transmission spectra show that the semiconducting gap of YH_3 decreases markedly with increasing pressure but remains open until at least 25 GPa. Extrapolating the pressure dependence of the gap, an insulator-to-metal transition is expected at 55.8 GPa.

Van Gogh et al. studied the optical properties of $La_{1-z}Y_zH_x$ with $0 < z < 1$ in great detail [56] and determined the dielectric function of these alloys. Figure 7.15 shows the development of the dielectric function of YH_x from the dihydride to the trihydride phase. For $1.9 < x < 2.1$ the film is in a single phase (β-phase) [119].[1] In

---

[1] Kooij et al. [120] stated erroneously that for $1.9 < x < 2.1$ thin films of YHx are in the coexistence region of the α-phase and β-phase. Remhof et al. [119] and Kerssemakers et al. [78] confirmed that YH_x follows the bulk phase diagram [121]. Thus, for $1.9 < x < 2.1$ the system is in the single β-phase and for $2.1 < x < 2.7$ in a mixed β-phase and γ-phase.

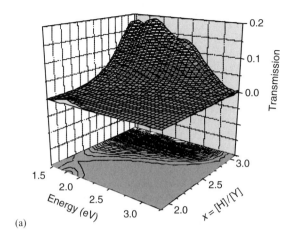

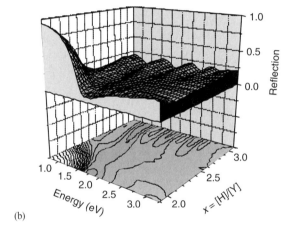

Fig. 7.14 Optical transmission (a) and reflection (b) of a 300 nm Y/15 nm Pd film as a function of hydrogen concentration x and photon energy during the second electrolytic loading from the dihydride to the trihydride state. In the lower part of the graphs contour plots are shown. The maximum transmission is limited by the Pd cap layer. (From Kooij et al. (1999), Ref. [65].)

this concentration range interband transitions exist ($\varepsilon_2 > 0$ around 2.5 eV) below the plasma energy, which is typically at 4 eV. These interband transitions arise from the flat d-band of YH_2. The weak dihydride transmission window appears at the photon energy where ε_1 crosses zero and ε_2 is low ($1.6 < E < 2.1$ eV). During loading to the trihydride (γ-phase) three features are observed (see Fig. 7.15): (I) The disappearance of the free electron optical response (the large negative ε_1 at low photon energies, that is typical for a metal is gradually reduced and replaced by positive ε_1) with increasing hydrogen concentration x. This means that our

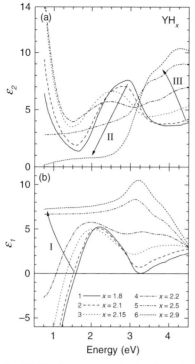

Fig. 7.15 Hydrogen concentration dependence of (a) the complex (ε_2) and (b) the real (ε_1) part of the dielectric function for YH$_x$. Similar dependences are observed in La$_{1-z}$Y$_z$H$_x$ alloys although the trihydrides La$_{1-z}$Y$_z$H$_3$ with $z < 0.67$ remain cubic while for $z > 0.86$ they are hexagonal. (From Van Gogh et al. (2001), Ref. [56].)

material changes from a metal into an insulator; (II) The suppression of interband absorption around 2.5 eV and, (III) a shift to lower energies of the second interband absorption peak in ε_2. The onset of interband absorption in the trihydride phase ($x = 2.9$) determines the final appearance of the film: YH$_3$ is yellow while LaH$_3$ is red. The photon energy at which ε_2 increases towards higher energy is the optical band gap of the material. For YH$_{2.9}$ this is 2.63 eV [56]. All La$_{1-z}$Y$_z$Hx alloys exhibit the same features as YH$_x$, irrespective of the crystal structure of the phases. In particular, it does not matter whether the fully loaded trihydride films are cubic or hexagonal: they are all transparent and semiconducting. Thus, the switching behavior is robust, in the sense that it is rather insensitive to chemical and structural disorder. Even the crystallographic phase appears irrelevant.

7.2.3.2 Electrical Properties and the Metal–Insulator Transition

The optical transition is always accompanied by an electrical transition from metallic to semiconducting when going from the dihydride to the trihydride phase (see Fig. 7.12(f)). This metal-insulator transition has also been investigated in detail. Van Gogh et al. [103] showed that the isotope mass is not important in the switching since YH$_x$ and YD$_x$ display the same electrical behavior. The metal-to-insulator (MI) transition takes place at a hydrogen concentration of about 2.86 where the resistivity measured at room temperature diverges [103].

A very important observation was that the MI transition in YH$_x$ occurs within a single structural phase and is not associated with a "trivial" first-order phase transition. This can be concluded from the fact that for $x > 2.7$ yttrium hydride is in a single phase (γ-phase) [119] and as we have seen the MI transition occurs at $x = 2.86$. This is consistent with the observation that there is also a MI transition in Mg-stabilized cubic YH$_{3-\delta}$ [122] and in LaH$_x$ which remains fcc at room temperature for $2 < x < 3$. Even La$_{1-z}$Y$_z$H$_x$ alloys show the same behavior independently of their crystallographic structure and their degree of structural disorder [56]. This is another demonstration of the robustness of the MI transition in rare-earth hydrides.

Intrigued by the possibility of a continuous quantum phase transition in YH$_x$, Hoekstra et al. investigated in great detail the temperature ($0.35 < T < 293$ K) and concentration dependences (for $0 < x < 3$) of the electrical conductivity σ, charge carrier density n and magnetoresistance of polycrystalline YH$_x$ switchable mirrors in magnetic fields up to 14 T [104–106]. In Fig. 7.16 the conductivity σ as a

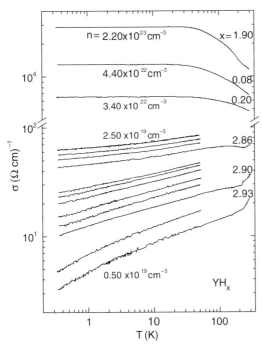

Fig. 7.16 Conductivity σ as a function of temperature T for a YH$_x$ film at a series of charge carrier densities n ($T = 0.35$ K). The metal–insulator transition is tuned by a combination of hydrogen gas loading (curves up to 300 K) and UV illumination (curves up to 50 K). Unlabeled curves have $n = (0.75, 0.89, 1.07, 1.18, 1.45, 1.51, 1.64, 1.88, 2.12$ and $2.32) \, 10^{19}$ cm^{-3}. The hydrogen concentration is indicated on the right. (From Hoekstra et al. (2001), Ref. [105].)

function of the temperature T is shown on a log–log scale for hydrogen concentrations between $x = 0.08$ and 2.93 [105]. The uppermost three curves with $n > 10^{22}$ cm^{-3} are for metallic YH$_x$ with $x < 2$. The curves relevant for a discussion of the MI transition are those with $n < 2.5 \times 10^{19}$ cm^{-3}. For x slightly smaller than 2.86 the curvature of the curves is positive which, according to scaling theories, implies that they correspond to a metallic state. For x larger than 2.90 their curvature is clearly negative. We conclude that somewhere between 2.86 and 2.90 the curvature vanishes and that, consequently, σ (T) is a power-law in temperature T. This is the signature of a genuine MI transition. An unusually small power of 1/6 can be estimated from the slope of the curves in the critical region. The characteristic response of thin YH$_x$ films is, according to Hoekstra et al., connected to a *continuous* quantum phase transition, a fundamental change of the ground state of the system as a function of a tuning parameter other than temperature T, namely the charge carrier density n in this case. For this quantum phase transition, the critical charge carrier density n_c is defined as the value at which the residual conductivity σ $(n, T=0) = \sigma_0(n)$ changes from a finite value in the metallic phase $(n > n_c)$ to zero in the insulating phase $(n < n_c)$. YH$_{3-\delta}$ is found to be a heavily doped semiconductor with a critical charge carrier density of $n_c(0.35 \text{ K}) = 1.39 \times 10^{-19}$ cm^{-3} at the MI transition. Moreover, the electrical conductivity data over a wide temperature and hydrogen concentration range in both the metallic and insulating regions can be collapsed onto a single curve with the same, unusually large critical exponents [104, 106]. These large critical exponents indicate the important role played by electron–electron interactions in the physics of switchable mirrors (see also Section 7.2.3.4 on page 292).

7.2.3.3 Hysteresis

Throughout this chapter it is repeatedly stated that the optical switching in YH$_x$ is reversible between $x = 2$ and 3. What is meant is that this system can be switched from metallic to semiconductor and back to the original metallic state. When this transition is followed in detail one observes, however, that there is a giant hysteresis. This can evidenced in two different ways: (i) in a gas loading experiment (which does not allow a determination of the actual hydrogen concentration in a film) by plotting the optical transmission as a function of the simultaneously measured electrical resistivity [96], or (ii) by measuring the optical transmission as a function of the hydrogen concentration loaded electrolytically into a film [56, 120]. A giant hysteresis is indeed found in YH$_x$ both in the pressure–composition isotherms and in transmission as a function of the hydrogen concentration x (see Fig. 7.17) [120]. LaH$_x$, however, switches without any hysteresis at all (see Fig. 7.17(e)) [56, 96]. Furthermore, in La$_{1-z}$Y$_z$H$_x$ alloys the magnitude of the hysteresis depends strongly on their composition. For $z = 0.67$ when the material stays fcc it vanishes altogether (see, e.g. the p–c isotherm of La$_{0.55}$Y$_{0.45}$ in Fig. 7.17(d)). As shown by Van Gogh et al. [96] these hysteretic effects in La–Y alloys are related to the presence of a phase transition from fcc to hcp accompanied by a large uniaxial expansion. This leads to different stress states between absorption and desorption and therefore to β- and γ-phases with different hydrogen concentrations and thus different optical properties

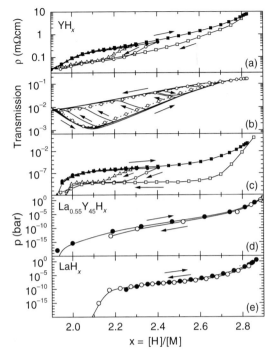

Fig. 7.17 Hysteresis in the p–c isotherms (c) and hydrogen concentration dependence of the electrical resistivity (a) and transmission (b) of YH_x during a hydrogen loading (filled circles) – unloading cycle (open circles). The giant hysteresis (factor 10^4) in the plateau pressure of the p–c isotherm of YH_x is absent in LaH_x and $La_{0.55}Y_{0.45}H_x$ alloys (lower two panels (d) and (e)). (From Van Gogh et al. (2001), Refs. [56] and Kooij et al. (2000) [120].)

during absorption and desorption [119]. Moreover, the optical transition is much sharper in LaH_x than in YH_x [96]. In single-phased La–Y alloys the sharpness of the optical transition seems to be related to the contraction or expansion of the material upon hydrogenation.

A closer look at the hysteretic effects in YH_x reveals that there is a four orders of magnitude difference between the absorption and desorption plateau pressures. This corresponds to a difference of $10 \text{ kJ (mol H)}^{-1}$ in the enthalpy of formation. The transmission as a function of x and ρ in combination with XRD and optical microscopy [119] shows that the intensity of the transmission window decreases within the single β-phase when going from $x = 1.9$ to 2.1. The same has been observed in epitaxial YH_x films [78]. Moreover, from $x = 2.7$ to 2.9 no hysteresis is observed in transmission because the system is in the single γ-phase. The difference between absorption and desorption is that during absorption the β-phase with $x = 2.1$ coexists with the γ-phase with $x = 2.7$, while during desorption the β-phase with $x = 1.9$ coexists with the γ-phase with $x = 2.65$. The fact that the phase coexistence is influenced by the hydrogen loading history, as correctly described by Remhof et al., is due to different stress states during loading and unloading [86, 119]. In single-phased materials such as LaH$_x$ there is no phase coexistence, but a continuous increase or decrease of the hydrogen content. Thus, no hysteresis will be observed. Recently, Lokhorst et al. were even able to make hysteresis-free YH$_x$ films by pulsed laser deposition [64, 88]. These films are nanocrystalline and stay fcc during hydrogenation. Apparently, the randomly oriented grains are too small to be transformed to the hcp phase. The same has been observed in sputtered Gd films by Dankert et al. [123].

7.2.3.4 Theoretical Models for the First-Generation Switchable Mirrors

Although the occurrence of a MI transition could have been expected on the basis of (i) the pioneering work of Libowitz and coworkers (see Refs. [54, 124] and references therein), (ii) the extensive work of Vajda and coworkers (see Ref. [121] and references therein) and Shinar et al. [125, 126] and (iii) the early bandstructure calculations of Switendick on Y, YH, YH$_2$ and YH$_3$ [127, 128], the transparency of YH$_{3-\delta}$ discovered in 1995 by Huiberts et al. [55, 129] came as a great surprise since, in 1993, Wang and Chou [130, 131] and Dekker et al. [132] had all concluded from self-consistent band-structure calculations, that YH$_3$ was a *semi-metal* with, in fact, a very large band *overlap* (1.5 eV). These difficulties stimulated theorists to reconsider the YH$_x$ and LaH$_x$ systems. Somewhat as for the high-T$_c$ superconductors two lines of thought have developed since 1995: (i) band-structure models including lattice (Peierls) distortions and (ii) strong electron-correlation models.

Peierls Distortion Models Kelly, Dekker and Stumpf [133] determined the ground state of YH$_3$ using density functional theory within the local density approximation and the Car–Parrinello method. They found a broken-symmetry structure with a direct electronic gap of 0.8 eV. The value of the gap depends strongly on the exact position of H in the Y lattice. As calculations based on the local density approximation systematically underestimate semiconducting energy gaps by about 1 eV, Kelly et al. considered their value as being consistent with the experimental optical data. Neutron scattering experiments on bulk (powders) and epitaxial YH$_{3-\delta}$ thin films by Udovic et al. [134, 135] and Remhof et al. [68] do not provide evidence for the occurrence of this broken-symmetry structure. Their data are compatible with the *P3c1* and *P63cm* structures. Kierey et al. [136] find with Raman scattering that the structure of YH$_3$ is either *P63cm* or *P63* (the broken symmetry structure) not *P3c1*. Zogał et al. [137] compare NMR data with electric-field gradient calculations and conclude that the *P63cm* shows very good agreement, better than *P3c1*. Van Gelderen et al. [138] have shown recently by parameter-free calculations that the *P63* structure is also compatible with the neutron diffraction data of Udovic. Thus, from all these experimental data we conclude that the space group of YH$_3$ is either *P63cm* or *P63*. However, the absence of an isotope effect [103, 139] in both electrical and optical properties of YH$_x$ indicates that the exact position of the interstitial hydrogen (or deuterium) has essentially no effect on the ground state of YH$_x$. Another indication of the robustness of the ground state of YH$_3$ is that all La–Y alloys [56] switch optically as a function of hydrogen concentration, irrespective of the chemical and structural disorder and also independently of the crystal symmetry of the trihydrides (fcc or hcp) [56, 122].

Electron Correlation Models Already in 1995 Sawatzky suggested that electron correlation effects might explain the insulating ground state of YH_3 and LaH_3. Since then several electron correlation models have been proposed. Ng, Zhang, Anisimov and Rice [117, 140] propose the following mechanism for the MI transition in LaH_x. The removal of a hydrogen atom from insulating LaH_3 leaves a vacancy which effectively donates an electron to the conduction band. This electron is, however, expected to be so strongly localized (the radius of the vacancy state is estimated to be about 0.3 nm in contrast to typically 10 nm in standard semiconductors) that an impurity band can only form at very high doping level (about 20 % impurities). This explains why the MI transition occurs at a large doping level, around $x = 2.8$. Eder, Pen and Sawatzky [141] introduce a further ingredient into the theory. They show that the strong dependence of the hydrogen 1s orbital radius on the occupation number (the radius of the 1s orbital is about 3 times larger for the negative H^- ion than for the neutral H atom), leads to the formation of localized singlet bound states involving one electron on the hydrogen and one on the neighboring metal orbitals. This breathing hydrogen orbital leads to a strong occupation number dependence of the electron hopping ("breathing Hubbard model") integrals. Both Ng et al. and Eder et al. consider the effect of correlation on the hopping integrals, an effect that was not taken into account by Wang and Chen [142].

The strong electron correlation models are similar to some of the models developed in the theory of high-temperature superconductors based on copper oxides. Instead of a linear combination of oxygen orbitals which couple to the copper 3d state in CuO, in YH_x it is a linear combination of yttrium 4d orbitals which couple to the 1s state of the hydrogen. Locally the structure then resembles strongly that of an isolated H^- ion, in agreement with the chemical view of a highly polarizable hydrogen ion in ionic hydrides. Electromigration experiments by Den Broeder et al. [143] and Van der Molen et al. [144] confirm that hydrogen in $YH_{3-\delta}$ behaves like a negative ion. Infrared transmission spectra by Rode et al. [145, 146] led also to the conclusion that significant charge transfer from yttrium to hydrogen is taking place. They find that hydrogen does not enter as a proton in the compound, but that it is negatively charged, with an effective charge of approximately $-0.5\,e$. Furthermore, angular resolved photoemission spectroscopy (ARPES) data by Hayoz et al. [147] bear the signature of strongly correlated electron systems. Hoekstra et al. [104–106] found unusually large critical exponents in scaling that hint to (strong) electron correlation (see Section 7.2.3.2).

Recent Calculations In the meantime new band-structure calculation methods beyond LDA have been used that treat correlation (many-body effects) such as GW and screened-exchange LDA. They all find gaps for both YH_3 and LaH_3, independent of the crystal structure (cubic or hexagonal) and the actual position of the hydrogen atoms (space groups $P3c1$, $P63cm$, $P63$, etc.) (see Fig. 7.18).

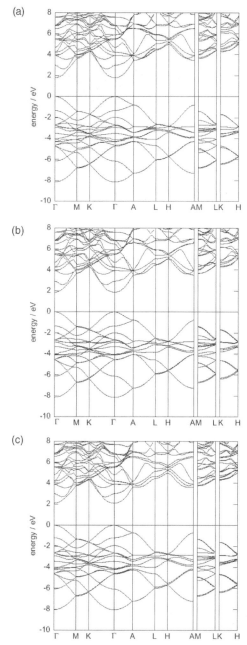

Fig. 7.18 Band structures calculated by Wolf and Herzig [150, 151] using screened-exchange LDA (sX-LDA) for hexagonal YH_3 with space group (a) $P3c1$ (b) $P6_3cm$ (c) $P6_3$. These band structure calculations demonstrate that the direct gap of YH_3 is insensitive to the chosen structure of YH_3. (From Wolf and Herzig (2002), Ref. [150] and Wolf and Herzig (2003) Ref. [151].)

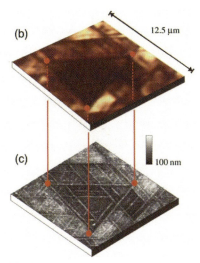

Fig. 7.19 (a) Optical domain switching. Real color optical micrograph in transmitted light of a 400 nm thick Y film capped with 7 nm of Pd. After loading to the trihydride phase YH$_{3-\delta}$, the hydrogen content is slowly lowered. The yellowish transparent YH$_{3-\delta}$ film switches back domain by domain to reddish opaque YH$_2$. The domains are bounded by a triangular network of reddish opaque lines. (b-c) One-to-one correlation between optical and structural texture. (b) Optical micrograph in transmitted light of a dihydride/trihydride mixed-phase region of a 150 nm thick Y/CaF$_2$ film capped with 20 nm of Pd. The image is 12.5 × 12.5 mm^2. The large triangle is almost entirely in the electrically conducting and opaque dihyride phase (YH$_2$). The bright yellow regions are in the transparent trihydride phase (YH$_3$). (c) AFM micrograph of the same area as in (b). A triangular network of ridges bounds micrometersized flat domains of various heights (dark is low, white is high). The surface topography closely matches the optical pattern in (b), indicating that domains are in different stages of switching from YH$_2$ (opaque, contracted domains) to YH$_3$ (transparent, expanded domains). (From Kerssemakers et al. (2000), Ref. [79].)

Miyake et al. [148] find that the GW approximation (GWA) leads to an insulating ground state for YH$_3$ and their calculated dielectric function agrees reasonably well with the experimental data obtained by Van Gogh et al. [56]. Chang, Blase and Louie [149] conclude, on the basis of their GW calculation, that LaH$_3$ is a band insulator with a direct dipole forbidden band gap of 1.1 eV and an optical band gap of 3.6 eV. Wolf and Herzig [150, 151] using screened-exchange LDA (sX-LDA) demonstrate that the direct gap of YH$_3$ is insensitive to the chosen structure of YH$_3$ (see Fig. 7.18). Van Gelderen et al. [152, 153] conclude from parameter-free quasi-particle calculations within the GW approximation that YH$_3$ is essentially a conventional semiconductor. The unusually large error in the band structure made by LDA is traced to its poor description of the electronic structure of the hydrogen atom. Their GW results predict a fundamental band gap of only 1 eV and an optical gap of 2.6 eV, in close agreement to optical measurements [56]. Alford et al. [154] have shown with GWA that both cubic YH$_3$ and LaH$_3$ are semiconducting. Finally, Wu et al. [155] recently studied hexagonal and cubic YH$_3$ and LaH$_3$ within the weighted-density approximation (WDA) and found a direct gap of 2.2 eV and a fundamental gap of 0.41 eV for hcp YH_3 and both cubic YH_3 and LaH_3 also show gaps. The authors conclude that the gap problem is not due to unusual correlations or quasiparticle corrections, but is a problem with the LDA and GGA exchange correlation functionals.

7.2.3.5 Specific Properties of Epitaxial Switchable Mirrors

Pixel-by-Pixel Switching As described earlier (see Section 7.2.2.3) *epitaxial* RE films deposited on (111)-CaF_2 substrates show a typical ridge structure [71]. The same was observed on other substrates, for example W [156] and Nb [77, 157]. The ridges mechanically decouple adjacent triangular domains [79]. Using *in situ* atomic force microscopy, combined with electrical resistivity and local optical transmission measurements, Kerssemakers *et al.* [78, 79, 158] discovered that micron-sized triangular domains switch one-by-one, homogeneously and essentially independently, during hydrogen absorption (see Fig. 7.19). These optically resolvable domains are delimited by an extended self-organized ridge network, created during the initial hydrogen loading [71, 76, 156]. The ridges have a higher effective hydrogen plateau pressure, that is they switch only after all domains have switched to the trihydride state. The ridges block lateral hydrogen diffusion and act as a sort of microscopic lubricant for the sequentially expanding and contracting domains. This block-wise switching results in a "Manhattan skyline" in which the optical state is directly related to the local, structural lattice expansion. The domain tunability is of technological relevance since it opens the way to a pixel-by-pixel switchable patterns with a minimal amount of inactive surface area.

In polycrystalline films *local* optical inhomogeneities are also observed in the two-phase region [119]. However, the ridges are a typical feature in epitaxial films only. Due to their unfavorable crystallographic orientation, their switching occurs at a significantly larger hydrogen pressure. However, on a *macroscopic* scale, both polycrystalline and epitaxial switchable mirrors exhibit the same optical properties [56]. This is in sharp contrast with their electrical transport properties, which are markedly different. Enache *et al.* [73] found that at all hydrogen concentrations the electrical resistivity ρ of the epitaxial films is substantially lower than that of polycrystalline films. The as-deposited epitaxial Y films have a residual resistivity ratio of typically 14 and a low residual resistivity of 5.2 $\mu\Omega$ cm at 4.2 K. The temperature dependence of ρ is essentially the same as for single-crystalline yttrium. The charge carrier density determined from Hall effect measurements is also similar to that of yttrium single crystals (i.e. 3.53×10^{22} cm^{-3} at 220 K). These features indicate that the quality of MBE deposited epitaxial Y films is excellent. The temperature dependence of the optical transmission at $E = 1.8$ eV in the dihydride state resembles that of the conductivity, that is $1/\rho$. These features are well described by a Drude model for free electrons, in which the only temperature-dependent parameter is the electron scattering time [73]. The values for the optical gap of the epitaxial and polycrystalline YH_3 films prepared under similar conditions (i.e. 1 bar H_2 at 300 K) are comparable [56, 73]. Due to the ridges, the electrical resistivities of these films are very different: for example at 220 K, 8 mΩ cm for epitaxial YH_3 and 40 mΩ cm for polycrystalline films. This is due to a difference in effective charge carrier density. The later switching of the ridges corresponds to a smaller average value of the H concentration in the epitaxial $YH_{3-\delta}$ films (i.e. under the same loading conditions the effective δ is larger in epitaxial than in polycrystalline films). The electrical transport properties of epitaxial YH_3 films are path-controlled by the ridges. This is clearly revealed by Hall effect measurements in magnetic fields up to 5 T: in sharp contrast to the isotropic electrical resistivity, the charge carrier density exhibits an in-plane anisotropy that is also temperature dependent.

7.2.4
Second-Generation Switchable Mirrors: Magnesium–Rare Earth Films

For technological applications it is desirable to switch from a metallic state to a transparent state that is color neutral instead of the colored rare-earth trihydrides, REH_3. To achieve this, alloying a RE with a metal for which the heat of hydride formation is similar to that of the REH_2–REH_3 transition and with a band gap large enough to have a fully transparent hydride, is an option. A group at Philips Research discovered in 1997 [57] that magnesium fulfils both demands [159]. This opened the way to the second-generation switchable mirrors. These alloys also turned out to be of fundamental importance since Mg-containing alloys exhibit a series of remarkable properties: (i) they disproportionate into REH_x and MgH_x and exhibit a microscopic shutter effect [160], (ii) MgH_2 stabilizes YH_3 in a cubic structure [122] and (iii) all $MgRE$–H_x exhibit a highly absorbing, black intermediate state [161]. In Fig. 7.20 pictures of the reflecting, absorbing and color neutral, transparent states are shown for a Gd–Mg alloy film. Further tailoring of the optical properties is achieved in multilayers of RE and Mg. As Mg plays a key role and, as we shall see later, can form Mg and MgH_2 inclusions in $MgRE$–H_x we describe first the physical properties of MgH_2.

7.2.4.1 Optical Properties of MgH_2

Mg is considered to be one of the most important candidates for the reversible storage of hydrogen due to its light weight, low cost and high hydrogen storage capacity (7.6 wt.% of hydrogen). However, only little is known about the intrinsic physical properties of the Mg–MgH_2 system. This triggered Isidorsson et al. [166] to study the optical properties of MgH_2 thin films in detail. MgH_2 turns out to be a color neutral, highly transparent insulator with an optical band gap of 5.6 eV, in close agreement with recent GW band structure calculations [162]. As shown by Krozer, Kasemo and others, the hydrogenation of Mg to MgH_2 is not straightforward [163, 164]. Palladium-capped Mg films exhibit unusual kinetics due to the formation of a blocking MgH_2 layer at the interface between Pd and Mg. This MgH_2 layer prevents hydrogen from diffusing into the underlying unreacted metallic Mg. Its formation can be circumvented by starting hydrogenation at relatively low (10^2 Pa) H_2 pressure at a temperature of 100 °C. Magnesium films with thicknesses up to 150 nm can be converted to MgH_2 in this way [164, 165]. To make sure that Mg is completely converted to MgH_2 up to 10 MPa H_2 was used in the optical studies

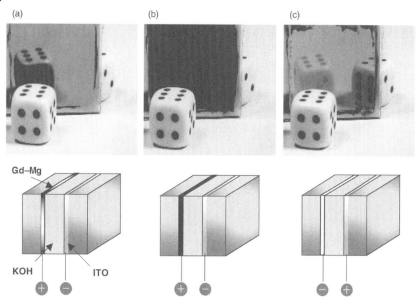

Fig. 7.20 The three optical states of a liquid electrolyte Gd–Mg switchable mirror: (a) shiny metallic at positive voltage (∼2 V), (b) strongly absorbing at ∼1 V and (c) transparent and insulating at negative voltages (−2 V). In this device the Gd–Mg layer deposited on a glass substrate is used as the negative electrode for hydrogen loading. The counter electrode is made of a transparent but conducting indium-tin-oxide (ITO) layer deposited on glass. The liquid electrolyte is a 1 M KOH solution in water. (From Griessen and Van Der Sluis (2001), Ref. [90].)

[166]. Using reactive MBE to make *in situ* MgH$_2$ films, Westerwaal [166] obtained a highly resistive thin hydride film, which however contained a 10 % fraction of metallic Mg. Therefore, we focus here on the properties of the *ex situ* hydrogenated films.

In Fig. 7.21(a) the total, specular and diffuse (scattered) transmission are shown for a 150 nm thick MgH$_2$ film capped with 12 nm Pd. As expected the transmission only drops at high energies, far beyond the visible part of the optical spectrum. The occurrence of diffuse transmission is due to surface roughness of the film. The diffuse transmission, T_d, has a strong wavelength dependence and is proportional to ω^4. It decreases strongly above the band gap as the film starts to absorb light. Therefore, it is easiest to determine the band gap E_g from this curve and one obtains $E_g = 5.6 \pm 0.1$ eV [166]. The dielectric function of MgH$_2$ is given in Fig. 7.21(b). In contrast to YH$_3$ and the other RE trihydrides, in MgH$_2$ there is no interband absorption below the band gap (i.e. $\varepsilon_2 = 0$). This means that the transmission of a MgH$_2$ film capped with Pd is only limited by the absorption in the metallic Pd film.

7.2.4.2 The Microscopic Shutter Effect

In contrast to the binary REH$_x$ switchable mirror films which have a weak red transparency window in their metallic dihydride phase (see Section 7.2.3.1),

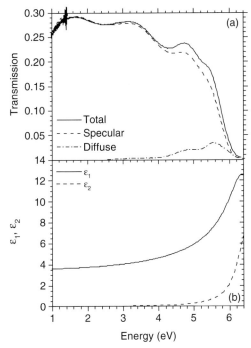

Fig. 7.21 (a) Total (solid line), specular (dashed line) and diffuse (dashed dotted line) transmission as a function of photon energy for a 150 nm thick MgH_2 film capped with 12 nm Pd and loaded at 100 °C in 10 MPa hydrogen. (b) Real ($\varepsilon 1$) and imaginary part ($\varepsilon 2$) of the dielectric function for MgH_2 determined from ellipsometry and transmission data. (From Isidorsson et al. (2003), Ref. [114].)

rare-earth alloys containing magnesium are remarkable for the large optical contrast [57, 160, 167] between their metallic dihydride and transparent trihydride phase. It has been shown by several techniques that this is due to a disproportionation of the Mg–RE alloy [93, 160, 168, 176]. This disproportionation is induced by the great hydrogen affinity of RE metals that form dihydrides already at very low hydrogen pressures (see Fig. 7.12(d) on page 283) Thus, while the RE-dihydride phase is formed, Mg separates out, remaining in its metallic, reflecting state. Upon further loading, insulating (transparent) MgH_2 is formed together with REH_3. In this way Mg acts essentially as a microscopic optical shutter [160]. It enhances the reflectivity of these switchable mirrors in their metallic state, suppresses the dihydride transmission window of their REH_2 phase (see Fig. 7.22) and increases their optical gap and transparency in their transparent state (see Fig. 7.25) [57]. The optical transmission ratio for thin films over the whole visible wavelength range can thus be increased to more than 10^3 (theoretically 10^9) [169] when at least 30 at.% Mg is added. This is the concentration where the dihydride transmission window disappears. YH_x and the other RE hydrides only have a contrast of 10 between the dihydride and trihydride. Di Vece et al. [93, 168] studied $Mg_yGd_{1-y}H_x$ alloys in detail. In Fig. 7.22 pressure–composition isotherms are shown together with the transmission at 1.85 eV (670 nm) [93]. GdH$_x$ behaves like YH$_x$ (see Fig. 7.12). The first plateau around 10^{-27} Pa corresponds to the transformation of α-Gd to β-GdH$_2$. A transparency window appears in transmission. The transition from β-GdH$_2$ to transparent γ-GdH$_3$ occurs at much higher pressures (around 1 Pa H$_2$).[2] For $y = 0.30$ and 0.62 a plateau at low hydrogen concentration is also observed with an enthalpy of formation close to that of GdH$_2$. The concentration range of this plateau suggests that all Gd is transformed to GdH$_2$. The disproportionation occurring in these Mg–Gd alloys thus results in gadolinium dihydride and pure magnesium clusters. The second (sloping) plateau corresponds to the formation of MgH$_2$ and GdH$_3$ since in thin films both have approximately the same enthalpy of formation. X-ray absorption fine structure (XAFS) studies indicate that upon hydrogen loading the coordination number of gadolinium by magnesium decreases markedly, thus confirming segregation [168]. Von Rottkay et al. [169] determined the index of refraction n and the coefficient of absorption k for several Mg–RE alloys (RE = Er, Gd, Sm). They found that k of Mg$_{0.5}$RE$_{0.5}$H$_{2.5}$ (in the transparent state) is quite low at energies below the band gap (typically $k = 10^{-3}$). This leads to the conclusion that the transparency of these films is mainly limited by the Pd cap layer just as in MgH$_2$. Van der Molen et al. [122] investigated the shift of the band gap in Mg$_z$Y$_{1-z}$H$_x$ films as a function of alloy composition. For fully hydrogenated Mg$_z$Y$_{1-z}$H$_{3-\delta}$ films with $0.10 < z < 0.50$ in which only fcc YH$_3$ and MgH$_2$ are present, the optical band gap E_g^{opt} increases almost linearly with increasing z ($\Delta E_g^{opt}/\Delta z = 1.1$ eV). Since the resistivity ρ increases almost exponentially with z from $z = 0$ to $z = 0.25$, they conclude that there is a change in some characterisitic energy parameter. With the simple relation

$$\rho(z) = \rho(z=0) \cdot \exp\left(E_g^\rho(z)/2k_B T\right)$$

they find that $\Delta E_g^\rho/\Delta z = 1.6$ eV. It is conceivable that the band structure of YH$_{3-\delta}$ changes due to quantum confinement (QC) effects since the grain size decreases from 18 nm (at $z=0$) to only 4 nm (at $z=0.3$) and the smaller the cluster, the higher the energy levels. The authors estimate theoretically that $E_g^{opt}/\Delta z = 1.0$ eV, in reasonable agreement with their optical and electrical data.

7.2.4.3 Cubic YH$_3$

Van der Molen et al. [122] found that cubic (fcc) YH$_3$ can be stabilized simply by addition of Mg. They studied Mg$_y$Y$_{1-y}$ thin films with $0 < y < 0.5$ in detail. X-ray diffraction showed that for $y = 0.10$ only fcc YH$_3$ is present in fully hydrogenated samples. Interestingly, the volume change upon loading from YH$_2$ to YH$_3$ is positive for all y, in contrast to the predictions by Sun et al. [170]. This is consistent

---

[2] The slope of the p–c isotherm in Fig. 7.22(a) for concentrations between 1.5 and 3 suggests that a fcc–hcp transition in GdH$_x$ might be absent in these sputtered films. This is in agreement with measurements by Dankert et al. [123] on sputtered Gd.

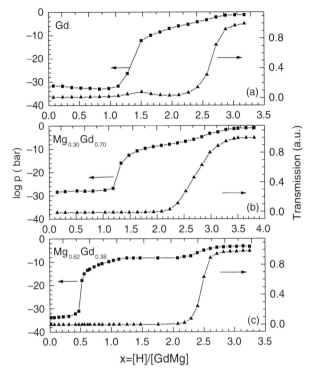

Fig. 7.22 Pressure–composition isotherms (squares) for (a) Gd, (b) $Mg_{0.3}Gd_{0.7}$ and (c) $Mg_{0.62}Gd_{0.38}$ alloys as determined electrochemically. The corresponding normalized transmission at 1.85 eV (triangles) is also shown. (1 bar = 10^5 Pa) (From Di Vece et al. (2002), Ref. [93].)

with the results obtained for $La_{1-z}Y_zH_x$ [56]. However, these samples show no segregation upon hydrogen loading, that is Y and La stay homogeneously mixed on an atomic scale. $La_{1-z}Y_zH_x$ expands for $z > 0.36$ and is cubic up to $y = 0.67$. This suggests strongly that if fcc YH_3 exists its volume should be larger than that of fcc YH_2. In the two-phase fcc-hcp YH_3 region ($y < 0.1$), the optical gap does not change significantly with y, although hcp YH_3 is increasingly substituted by fcc YH_3. Since MgH_2 is a large-gap insulator ($E_g = 5.6$ eV, see Section 7.2.4.1), this is only possible if the optical properties of fcc YH_3 are comparable to those of hcp YH_3. Therefore, this strongly suggests that fcc YH_3 is a large-gap semiconductor very similar to hcp YH_3. This is substantiated by electrical resistivity data [122]. That fcc YH_3 is semiconducting is another demonstration of the robustness of the metal–insulator transition in the rare-earth hydrides.

In Fig. 7.23 p–c isotherms together with the optical transmission at $E = 1.85$ eV ($\lambda = 670$ nm) are shown as a function of Δx, the amount of hydrogen per metal atom that is desorbed for both $YH_{3+\Delta x}$ and $Mg_{0.1}Y_{0.9}H_{2.9+\Delta x}$ [171]. The plateau in Fig. 7.23(a) is due to the coexistence of hcp YH_3 and fcc YH_2. On the other hand,

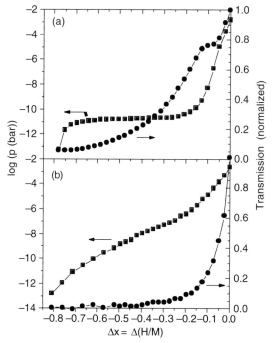

Fig. 7.23 Pressure–composition isotherms (squares) as determined by Giebels et al. [171] at 30 °C of (a) YH$_{3+\Delta x}$ and (b) Mg$_{0.1}$Y$_{0.9}$H$_{2.9+\Delta x}$ as a function of Δx, the amount of hydrogen per metal atom, that is desorbed. The measurements are performed electrochemically in the third desorption cycle. Also shown is the normalized transmission at 1.85 eV (l = 670 nm) (circles). At $\Delta x = 0.0$ the sample is fully loaded with hydrogen, around $\Delta x = -0.8$ the sample is unloaded. (1 bar = 10^5 Pa) (From Giebels et al. (2002), Ref. [171].)

there is no plateau in Fig. 7.23(b) since YH$_x$ remains cubic, as discussed above. Another consequence of the absence of the fcc–hcp phase transition is seen in Fig. 7.23(b): a steepening of the optical transition upon hydrogen desorption compared to pure YHx. This is in agreement with the observations in fcc La$_{1-z}$Y$_z$H$_x$ alloys [96]. As a consequence of this Mg$_{0.1}$Y$_{0.9}$H$_{2.9}$ shows a much larger thermochromic effect than YH$_3$ [171].

7.2.4.4 Black State Due to Coexistence of Mg and MgH$_2$

Mg-containing rare-earth hydrides exhibit, in addition to a transparent and a reflecting state, a state that has a low reflection and a very low transmission, that is it has a black appearance [57, 161, 172, 173]. These alloys can therefore be switched through the three fundamental optical states of matter by merely changing their hydrogen content. The key ingredient is the coexistence of nanograins of metallic Mg and insulating (dielectric) MgH$_2$ [161]. We discuss now the black state of MgY–H$_x$. Figures 7.24(a), (b) show the reflection and transmission of a 200 nm

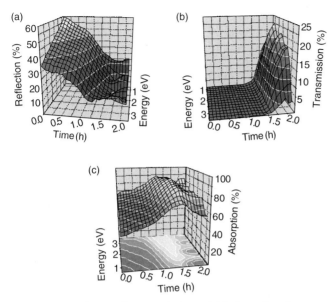

Fig. 7.24 (a) Reflection, (b) transmission and (c) absorption of a 200 nm thick $Mg_{0.5}Y_{0.5}$ film on quartz covered with 10 nm Pd during loading in 10^3 Pa H_2. The time is given in hours. At $t = 2.2$ h the sample is fully loaded with hydrogen. The overall composition is then approximately $Mg_{0.5}Y_{0.5}H_{2.5}$. (From Giebels et al. (2004), Ref. [161].)

thick $Mg_{0.5}Y_{0.5}$ film on quartz covered with 10 nm Pd during loading in 10^3 Pa H_2. From the measured reflection, R, and transmission, T, the absorption, A, is calculated, with $A = 1 - R - T$ (see Fig. 7.24(c)). Apart from a reflecting and a transparent state there also exists a highly absorbing, black state. The black state, which corresponds to the hill in A, occurs when the reflection is low and the film is just becoming transparent. Over the entire visible part of the spectrum the absorption is 85 to 90 %. After correction for the Pd cap layer and the quartz substrate 70–80 % absorption still remains. This is exceptionally high since we are looking (through the transparent substrate) at a smooth interface between the substrate and the film (and not at a rough surface such as in Si or Ni–P black [174, 175]). Moreover, it is not a narrow absorption line but it spans a wide energy range from the ultraviolet to the near-infrared. The same is observed in Mg–Gd [57, 172], Mg–La [161, 177] and Mg–Sc [177, 178] films and in Mg/Y and Mg/Ni multilayers [179]. Thus, it is a general feature of Mg–RE films with Mg playing a major role. Recently, Giebels et al. [161] observed that thin pure Mg films can also have a highly absorbing black state. For this, Mg films need to be hydrogenated carefully at elevated temperatures (50–100 °C) and to be rather thin ($\sim$100 nm). In order to observe the black state it is very important that the film is loading (or unloading) homogeneously, that is MgH_2 (or Mg) nucleates everywhere in the sample. This confirms early reports by Hjörvarsson [180] on a black state in MgH_x films. In the black state structural X-ray diffraction data indicate that Mg is transforming to MgH_2 both in Mg–RE films

[176] and in pure Mg. A sharp increase in the resistivity can be observed as soon as the reflection drops, that is when the absorption increases dramatically [161]. This hints to percolation phenomena. The optical and electrical behavior of these films can be reconstructed using Bruggeman's effective medium approximation [181] for a Mg–MgH$_2$ composite [161]. It shows that the coexistence of Mg and MgH$_2$ grains plays an essential role. The unique feature of the switchable mirror systems considered here is that they can switch from a metallic, reflecting film via a black, highly absorbing state near the percolation threshold to a transparent and insulating material, simply by changing the hydrogen concentration in a given sample. In all other metal–dielectric composites (such as Au–glass, Ag–glass and Co–Al$_2$O$_3$ [182]) a new sample has to be made for every desired volume percentage of metal in the dielectric.

7.2.4.5 Tailoring Optical Properties Using Multilayers

Giebels et al. [183] showed that multilayers offer much freedom to tailor the optical properties by playing with their periodicity. Multilayers have superior reflection in the low-hydrogen state (when the sample is unloaded). In the transparent fully loaded state the absorption edge of the mulilayers with the same overall composition is shifted in energy on increasing the individual layer thicknesses. Figure 7.25 shows the reflection and transmission spectra of the unloaded and fully loaded state, respectively, for multilayers with layer thicknesses ranging from 2 to 10 nm and an overall composition of Mg$_{0.6}$Y$_{0.4}$. For comparison, spectra of YH$_x$ and a disordered alloy of the same composition are added. The character of the two different materials, YH$_x$ and MgH$_x$, can clearly be distinguished. YH$_2$ exhibits a large dip in reflection around 1.75 eV (see Fig. 7.25(a)) and is even slightly transparent around this energy, while the Mg–Y alloy and multilayers have superior reflection due to the highly reflective Mg. In the fully loaded state (see Fig. 7.25(b)) Giebels et al. [183] observe a clear difference between the fully loaded disordered alloy and the multilayers, in contrast to observations by Van der Sluis [184]. With decreasing layer thickness the absorption edge of the multilayers shifts in energy from that observed in YH$_3$ to the edge of the alloy. This hints at changes in the band structure of the YH$_x$ layers when the layer thickness becomes thinner, in the same manner as in the alloys (see Section 7.2.4.2 and Ref. [122]). This may be the result of quantum confinement effects. Moreover, the optical contrast of multilayers is much higher than for pure YH$_x$ and their optical switching does not suffer from hysteretic effects. X-ray diffraction shows that YH$_x$ in multilayers still undergoes a fcc–hcp transition when going from the dihydride to the trihydride, in contrast to Mg–Y alloys of the same composition. In Fig. 7.26 the p–c isotherm between the low- (unloaded) and high-hydrogen (fully loaded) states for a Mg/y multilayer is shown. Only one plateau is observed. This plateau corresponds to the coexistence of YH$_2$ and YH$_3$ as well as Mg and MgH$_2$. Strikingly, there is still a hysteresis of two orders of magnitude in the pressure–composition isotherm, but there are no hysteretic effects in the optical transmission. By selecting appropriate multilayer periods, samples can be built that switch much faster than alloys with a comparable Mg content [184]. An optimal Mg layer thickness of 1.12 nm is found for a composition Mg$_{0.40}$Gd$_{0.60}$. The switching time is reduced to 0.08 s compared to

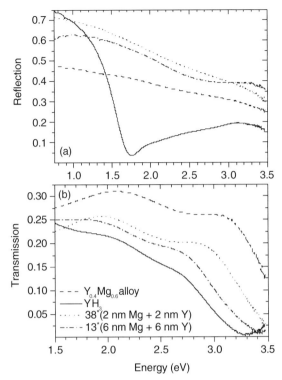

Fig. 7.25 (a) Reflection spectra after hydrogen desorption and (b) transmission spectra in the fully loaded state of a disordered $Mg_{0.6}Y_{0.4}$ alloy, Mg/Y multilayers with the same overall composition, and Y. The films all have a total thickness of 300 nm and are capped with 10 nm Pd. (From Giebels et al. (2002), Ref. [183].)

0.65 s for the alloy. However, when the Mg layer is increased to 4.45 nm the switching time (80 s) is considerably longer than in the corresponding alloy. The enhanced kinetics of multilayers is even more pronounced for higher Mg contents. For instance, a $Mg_{0.80}Gd_{0.20}$ alloy has a switching time (at room temperature) of about 20 min whereas a multilayer with the same average composition and 1.12 nm thick Mg layers has a switching time of only 70 s.

7.2.5
Third-Generation Switchable Mirrors: Magnesium–Transition-Metal Films

Richardson et al. [97, 185] reported in 2000 that alloys of Mg and Ni exhibit also optical switching as a function of H concentration (see Fig. 7.27). At first sight this does not seem to be of major importance since it has been known for decades that Mg_2Ni reacts readily with gaseous hydrogen to form Mg_2NiH_4 [100] that is considered to be an attractive material for storage purposes. The reaction occurs by formation of Mg_2NiH_4 from the hexagonal intermetallic compound Mg_2Ni upon

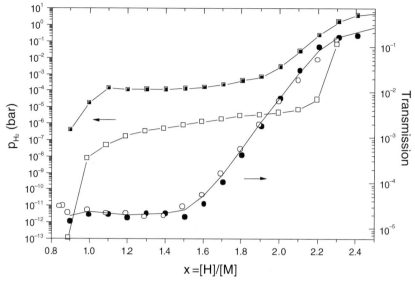

Fig. 7.26 Pressure–composition isotherm determined electrochemically (squares) of a 15 × (10 nm Mg + 10 nm Y) multilayer capped with 10 nm Pd between the unloaded (YH$_2$ and Mg) and fully loaded state (YH$_3$ and MgH$_2$) as a function of the hydrogen concentration per metal atom, $x = $ H/M. Also shown is the transmission (circles) at 635 nm (1.96 eV). The filled symbols refer to hydrogen absorption and the open ones to desorption. (1 bar = 10^5 Pa) (From Giebels et al. (2002), Ref. [183].)

hydrogen absorption. The hydrogen solubility range of the hexagonal Mg$_2$Ni is limited to 0.3, that is it extends up to a composition Mg$_2$NiH$_{0.3}$ [100, 186]. The hydride Mg$_2$NiH$_4$ is the only stable ternary compound known for the Mg–Ni–H system. This compound shows a well-defined stoichiometry, essentially independent of temperature and hydrogen partial pressure. During heating under a H$_2$ pressure of one atmosphere Mg$_2$NiH$_4$ undergoes a structural phase transition at 510 K to a cubic high-temperature form with the metal atoms in an antifluorite-type structure [187–189]. The low-temperature form of Mg$_2$NiH$_4$ is monoclinic, albeit with only slight distortions from the antifluorite structure [188, 190]. The discovery of Richardson et al., however, is important for two reasons. First, it demonstrated that Mg–Ni films absorb hydrogen very easily. While in bulk samples hydrogenation requires high temperatures (500 to 600 K) and pressures of 10^5 to 10^6 Pa [99, 100], for thin films it occurs readily at room temperature at low pressures when they are capped with a thin Pd layer. Secondly, it opened the way to the *third-generation* switchable mirrors. This class of materials does not involve rare-earth metals and might therefore be more resistant to oxidation than the earlier mirrors. This is especially important for applications. So far all alloys of Mg with the transition metals (TM) Ni, Co, Fe, Mn, V have been found to switch with hydrogen [98] (see Fig. 7.30). As for the first and second generation switchable mirrors, thin films provide us with a unique possibility to determine the intrinsic physical properties of Mg$_2$NiH$_x$ and other so-called complex metal hydrides such as Mg$_2$CoH$_5$ and Mg$_2$FeH$_6$. A first indication that the physics of Mg$_2$NiH$_4$ is not trivial is that, until very recently,

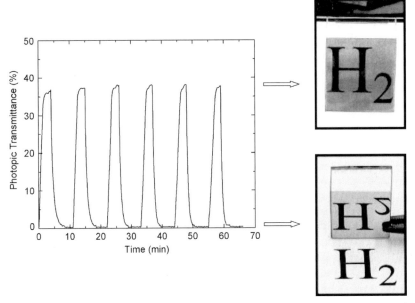

Fig. 7.27 Optical transmission of a 40 nm thick Mg–Ni–H$_x$ film evaporated on ITO/glass and capped with 5 nm Pd during electrochemical loading/unloading cycles in 8 M KOH/1 M LiOH. The upper photograph shows transparent Mg–Ni hydride, the lower photograph is Mg–Ni in the reflecting state. (From Richardson et al. (2000), Ref. [185].)

the position of H in Mg$_2$NiH$_4$ was not known precisely. By combining total energy calculations with experimental data Garcia et al. [191] conclude that in the HT phase the four H are arranged around a Ni atom in a tetrahedrally distorted square planar configuration. The negatively charged complex (NiH$_4$)$^{4-}$ is ionically bound to the Mg^{2+} ion. For the LT phase Myers et al. [192] found from an energy minimalization that hydrogen is in an almost perfect tetrahedron. As a whole every added hydrogen removes one electron [193] from the conduction band of Mg$_2$NiH$_x$, as expected on the basis of the so-called anionic model for hydrogen in Groups I, II and III of the periodic table [194].

7.2.5.1 Electrical Properties and the Metal–Insulator Transition

Enache et al. [193] have investigated the temperature (between 2 and 280 K) and concentration (for $0 < x < 4$) dependences of the electrical resistivity, charge carrier density and magnetoresistance of Mg$_2$NiH$_x$ switchable mirrors. Although the electrical resistivity increases by almost three orders of magnitude between metallic Mg$_2$Ni and transparent Mg$_2$NiH$_4$ (see Fig. 7.28) the charge carrier concentration n decreases only gradually from typically 10^{23} to 10^{21} cm^{-3}. These metallic-like features indicate that Mg$_2$NiH$_{4-\delta}$ prepared at room temperature under 1.3×10^5 Pa H$_2$ is a heavily-doped semiconductor [193, 195]. The estimated nonstoichiometry is $\delta = 0.05$. By combining Hall effect, electrical resistivity and electrochemical data, it is shown that n varies like $(4-x)$, which confirms that the role played by hydrogen in the host lattice is that of a negatively charged H$^-$ ion. In the limit $\delta \to 0$,

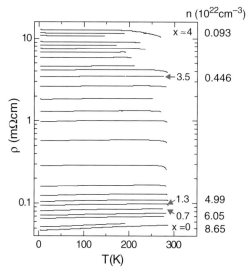

Fig. 7.28 Electrical resistivity ρ of a 232 nm thick Mg$_2$NiH$_x$ film (with $0 < x < 4$) covered with 2 nm Pd as a function of temperature. The curves corresponding to Mg$_2$NiH$_x$ samples with concentrations $x = 0$, 0.7, 1.3, 3.5 and ~4 are indicated explicitly. The decreasing electrical resistivity above 220 K is due to hydrogen loss during warming up of the film. (From Enache et al. (2004), Ref. [193].)

Mg$_2$NiH$_4$ is a semiconductor. This is in agreement with the electronic ground state of the low-temperature Mg$_2$NiH$_4$ phase calculated by Myers et al. [192] (see Fig. 7.29). For $0.3 < x < 4$ the hydrogen concentration dependence of the electrical resistivity of Mg$_2$NiH$_x$ can be modeled with an effective medium theory. From such an analysis Enache et al. concluded that the metallic Mg$_2$NiH$_{0.3}$ and Mg$_2$NiH$_4$ inclusions needed to be very flat ellipsoids with their small axes perpendicular to the plane of the film. This implies that a Mg$_2$NiH$_x$ film exhibits metallic behavior up to high hydrogen since the flat metallic inclusions short-circuit the sample.

7.2.5.2 Optical Properties

There have been several fragmentary reports on the optical properties of Mg$_2$NiH$_4$ in the past. Reilly and Wiswall noted already in 1968 [100] that bulk Mg$_2$NiH$_4$ was a dark red solid, Lupu et al. [196] determined a band gap of 1.68 eV for both the low-temperature (LT) and high temperature (HT) phases from resistivity data and reflection measurements on powder samples, Selvam et al. [197] measured two gaps near 2.0 and 2.4 eV, the direct and indirect gaps, respectively, of Mg$_2$NiH$_4$, and Fujita et al. [198] found a gap of $\approx$1.3 eV in a Mg$_2$NiH$_4$ film from resistivity data and optical absorption measurements for both the LT and HT phases. The large spread in these values is a direct indication of one major difficulty with these materials, that is the controlled deposition of films with well-defined composition and morphology (see also Section 7.2.6.2). Richardson et al. measured the optical transmission and reflection only for as-deposited films and fully hydrogenated films

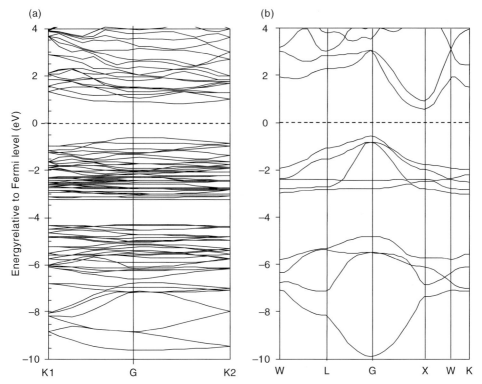

Fig. 7.29 Electronic band structures for Mg_2NiH_4 calculated with LDA in (a) the low-temperature monoclinic phase and (b) the high-temperature cubic phase with undistorted tetrahedral NiH_4 complexes. (From Myers et al. (2002), Ref. [192].)

(see Fig. 7.30). In many cases these films were very much Mg-rich [97, 98]. For films with a composition close to Mg_2NiH_4 Isidorsson et al. [245] estimated, from transmission edge measurements, a gap of 1.6 eV, in reasonable agreement with the band structure calculations of Myers et al. [192] and Haussermann et al. [199]. The first detailed study of hydrogen concentration dependence of the optical properties of Mg_2NiH_x was done by Isidorsson et al. [200]. They measured the reflection and transmission together with the electrical resistivity of a 232 nn thick $Mg_{2.1}NiH_x$ film capped with 2 nm Pd during slow loading with hydrogen. These measurements revealed a curious behavior of the reflection at low hydrogen concentration. Above $x = 0.3$ the reflection drops markedly and reaches a minimum at $x = 0.8$, that is at a hydrogen-to-metal ratio of less than 0.3.[3] The optical transmission remains very low (typically 0.001) until x approaches 4. The electrical resistivity at the

---

[3] The estimated hydrogen concentration in the original publication by Isidorsson et al. [200] was too low. At the time the authors were convinced that the drop in reflection occurred for $0.2 < x < 0.3$. Lohstroh et al. [101] have shown that the black state corresponds to an overall hydrogen concentration of $0.3 < x < 0.8$.

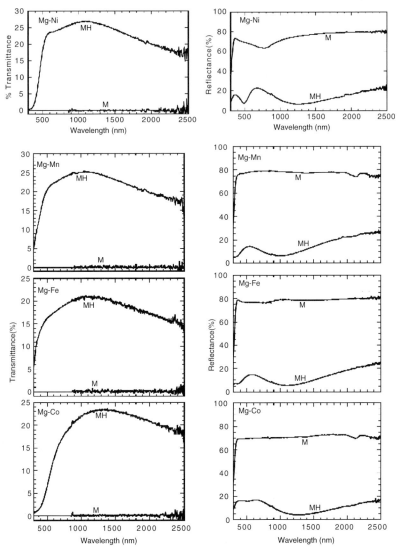

Fig. 7.30 Visible and near-infrared reflection and transmission spectra of 80 nm Mg–Ni, 60 nm Mg–Mn, 40 nm Mg–Fe and 40 nm Mg–Co films (Mg:TM 6:1) with 7 nm thick Pd overlayers on glass substrates in the metallic (M) and fully hydrogenated states (MH). Spectra were recorded from the substrate side. (From Richardson et al. (2001), Ref. [97] and Richardson et al. (2002), Ref. [98].)

minimum in reflection is still relatively low, typically below $100\,\mu\Omega\,cm$. The Hall effect data indicate that the charged carrier concentration is still rather high, typically $7 \times 10^{22}\,cm^{-3}$. This seems to be mutually incompatible since metals always have a large reflectance. This stimulated a large effort by Lohstroh et al. [101, 102] in order to understand the mechanism leading to this peculiar behavior.

7.2.5.3 Black State Due to Self-Organized Layering

An important clue about the origin of the black state in Mg_2NiH_x came from a visual observation from the front and backsides of the same sample during hydrogen loading [102]. Figure 7.31 displays photographs taken from two pieces of the same sample, (a) in the metallic state and (b) black $Mg_2NiH_{0.8}$. The sample on the left-hand side is viewed through the substrate and on the right-hand side the Pd cap layer faces the front. Without hydrogen, both sides are shiny reflecting and the small difference in appearance is mainly due to the different media for the incoming light (sapphire on the left-hand side and air on the right-hand side). After hydrogen is introduced at a pressure of 1600 Pa, it takes a couple of minutes for Mg_2NiH_x to become black when viewed through the substrate. Surprisingly, Mg_2NiH_x keeps its metallic luster when observed from the Pd side. Note, that the Pd cap layer (5 nm) has a rather high transparency ($T > 40\%$) and hence the Mg_2NiH_x layer underneath contributes significantly to the observed reflection. These photographs demonstrate vividly that the originally homogenous Mg_2Ni film starts to react with hydrogen at the substrate–film interface [101, 102]. Further H-uptake causes the hydrogen-rich layer to grow at the expense of the metallic part until eventually the entire film has switched to Mg_2NiH_4. This double layer model is compatible with the X-ray absorption spectroscopy data by Farangis et al. on Mg–Ni (see Fig. 7.13 in Ref. [201]) where they see that while the Mg–Ni peak gradually decreases the Mg–Ni hydride peak increases. This unusual loading sequence is also supported by the data in Fig. 7.32 that represents the evolution of the reflection from the substrate side as a function of hydrogen loading. Due to the transparency of Mg_2NiH_4 the reflection exhibits typical interference fringes when the optical path length of the light that is reflected between the two interfaces (substrate–Mg_2NiH_4 and Mg_2NiH_4–$Mg_2NiH_{0.3}$, respectively) is a multiple of the wavelength. With increasing hydrogen concentration the increasing Mg_2NiH_4

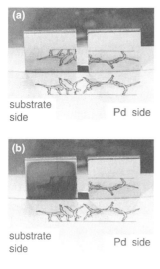

Fig. 7.31. Photographs of two identical films 200 nm Mg_2Ni/5 nm Pd on sapphire. On the left-hand side we look at the film through the substrate and on the right-hand side from the Pd layer side. (a) As-deposited, (b) in 1600 Pa H_2 at room temperature. Upon exposure to hydrogen the "substrate" side of the sample becomes black while the "Pd" side stays metallic. The difference in appearance is not due to the thin metallic Pd cap layer but due to the nucleation of the hydrogen-rich phase Mg_2NiH_4 at the film–substrate interface. (From Lohstroh et al. (2004), Ref. [103].)

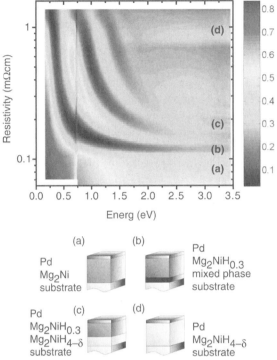

Fig. 7.32 Intensity map of the reflection R (measured through the substrate) during hydrogen uptake of a sample 250 nm Mg$_2$Ni/7 nm Pd. At low resistivity (i.e. in the metallic state) R is high. At around $r = 130$ mΩ cm ($x \sim 0.8$) R exhibits a deep minimum over the entire visible wavelength regime. A double layer system Mg$_2$NiH$_{0.3}$–Mg$_2$NiH$_4$ is formed and subsequently interference minima and maxima appear. As the transparent layer increases in thickness these interference fringes shift to lower energies. The hydrogen induced layering of the film is schematically indicated at four characteristic hydrogen concentrations (a)–(d). (From Lohstroh et al. (2004), Refs. [101, 102].)

thickness yields a shift of the interference fringes towards lower energies and a smaller difference between adjacent maxima until the entire sample has switched. In contrast, the reflection viewed from the Pd side is unchangingly high.

The increasing thickness of the evolving Mg$_2$NiH$_4$ layer can be directly determined from the experimental reflection data. During dehydrogenation the reflection, transmission and resistivity go through the same stages as during hydrogenation but, of course, in reversed order. The mechanism leading to the black state described above is not an exotic peculiarity of the Mg$_2$NiH$_x$ system. It is also observed in other Mg-based alloys (Mg–Co, Mg–Fe and Mg–Co–Ni) [202]. The black state is robust in the sense that it appears in sputtered films as well as in UHV evaporated films and the choice of the substrate is also not crucial, that is Al$_2$O$_3$, glass, ITO, SiO$_2$ and CaF$_2$ can be used. It is important to note here that the black state observed in Mg–TM hydrides is fundamentally different from that of Mg–RE hydrides described in Section 7.2.4.4. In the Mg–RE hydrides a spatial disproportionation into Mg and REH$_2$ is induced once and for all during the *first* hydrogen loading. These hydride nuclei are finely dispersed throughout the film. In the Mg–TM hydrides a macroscopic and reversible layering is the essential feature. The fact that the hydrogenation starts from the bottom of the film indicates some catalytic activity of the film/substrate interface. This suggests new strategies for the optimization of catalysts in, for example, nanostructured hydrogen storage materials. In this respect an appropriately catalyzed Mg$_2$FeH$_6$ is interesting since it has a relatively high hydrogen capacity: 5.7 wt.% and 171 kg H/m^3. As discussed in Section 7.2.8.2 the switching from a mirror to a black absorber also offers interesting possibilities for applications as smart coatings in solar collectors and antiglare rear-view mirrors or as a sensing layer in optical fiber hydrogen detectors. The fact that the switching takes place at a low hydrogen-to-metal ratio is a favorable factor for the cycleability of a device.

The "reversed loading" behavior is unexpected since the hydrogen enters through the top film surface. To explain the nucleation of Mg$_2$NiH$_4$ close to the substrate/film interface Westerwaal *et al.* characterized the microstructure of Mg$_2$Ni films with various thicknesses (20–150 nm). For films thinner than 50 nm, the film consists of small grains and clusters of small grains whereas on further growth the grain size increases and a columnar microstructure develops. They proposed, therefore, that close to the substrate, the relatively porous structure of the film with small Mg$_2$Ni grains locally reduces the nucleation barrier for Mg$_2$NiH$_4$ formation [203, 204]. This model was confirmed by the fact that the black state is not observed in films with a homogeneous microstructure. Such films can be made by growing complex Mg$_2$NiH$_4$ *in situ* by reactive MBD using an atomic hydrogen source in addition to the two metallic sources [205]. The hydride forms well below the decomposition pressure, showing on the one hand the hydriding power of the hydrogen source and, on the other, the barrier for decomposition of these compounds in vacuum. The microstructure of *in situ* grown Mg$_2$NiH$_4$ films is quasi-amorphous throughout the whole film thickness. When we first unload and then rehydrogenate this film, the nucleation of the hydride is random throughout the film, showing the importance of the microstructure for the occurrence of the black state.

The black state described here is a transient state. It may be stabilized by blocking the hydrogen absorption (stopping the gas flow or turning off the electrochemical potential). This is not very convenient for a device application. However, instead of relying on the self-organized double layering, we can obtain the same effect by depositing only a thin ~50 nm Mg$_2$NiH$_4$ layer and covering it with a suitable metallic cap layer. Indeed, a similar black state is found in this case after *fully* loading the sample [206].

7.2.6
Applications of Switchable Mirrors to Materials Science Problems

7.2.6.1 Switchable Mirrors as Indicator Layers
One of the striking properties of hydrogen in metals is its large mobility. Already at room temperature the H diffusion coefficient can be as high as 10^{-5} cm^2 s^{-1}, that is a value almost comparable to diffusion in liquids. Diffusion of hydrogen in

metals has also attracted considerable theoretical attention because it is fast even at low temperatures, involves only simple jumps between interstitial sites, and is ideal for the investigation of isotope effects when H is replaced by deuterium or tritium. A review of experimental data and techniques used so far to measure hydrogen diffusivity in bulk samples is given by Völkl and Alefeld [207]. Most of these methods are not applicable to thin films as they are either hampered by the influence of the substrate, for example in the Gorsky effect, or by the rather small volume of the film, for example in quasi-elastic neutron scattering.

Consequently, relatively little is known about hydrogen diffusion in thin metallic films and multilayers. Furthermore, very little is known about long-range diffusion of hydrogen in metals such as yttrium, lanthanum and rare-earths metals, for the simple reason that these materials are reduced to powders at high hydrogen concentrations. As mentioned in the Introduction, thin films of these materials do not suffer from this drawback. The understanding and manipulation of hydrogen transport through films is, however, important for the control and optimization of coatings and thin-film devices such as hydrogen detectors, metal hydride switchable mirrors or tunable magnetic elements. The fact that switchable mirror films have optical properties that depend strongly on hydrogen concentration offers the possibility to use them as two-dimensional hydrogen concentration indicators. The simplest application is to monitor hydrogen diffusion in switchable mirrors themselves. Another possibility is to deposit a thin switchable mirror on a sample of interest that might be opaque. In such a configuration hydrogen diffusion can be observed through a coloration change in the switchable mirror indicator. Several examples of indicator applications are described below.

Diffusion and Electromigration in Switchable Mirrors One can exploit the characteristic features of the transmission spectra of switchable mirrors to monitor hydrogen diffusion optically in a noninvasive way. The first experiments by Den Broeder *et al.* [143] were done on yttrium by using a film produced in the following way (see Fig. 7.33). First, an yttrium film is evaporated under ultrahigh vacuum conditions onto a transparent substrate (sapphire or quartz). Subsequently, a several nm thick palladium pattern (consisting, e.g. of a disk or one or more strips) is evaporated *in situ* on top of yttrium. When brought into contact with air the yttrium oxidizes, forming a thin Y_2O_3 layer that is impermeable to hydrogen atoms. However, areas covered with Pd do not oxidize and remain permeable to hydrogen. In a typical experiment, hydrogen gas (10^5 Pa) is introduced into the chamber containing the sample. The chamber that is equipped with optical windows and a temperature control system is placed onto the positioning table of an optical microscope. Using a white lamp behind the cell, optical transmission changes are monitored optically. In contact with H_2, the yttrium underneath the palladium pattern immediately starts absorbing hydrogen atoms and within a few seconds transparent $YH_{3-\delta}$ is formed under the Pd-covered region. Further hydrogen uptake can only take place if H diffuses out laterally into the Y underneath the transparent Y_2O_3 layer. For the disk geometry chosen in Fig. 7.33 hydrogen diffuses radially. As the hydrogen concentration x decreases from essentially 3 beneath the Pd disk to 0 far from

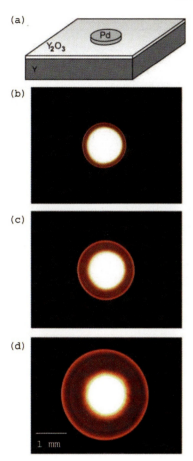

Fig. 7.33 Radial hydrogen diffusion in yttrium observed visually. Microscope photographs in transmitted light of (a) a 300 nm thick yttrium film on sapphire, covered by a 30 nm thick palladium disk of 1.1 mm diameter at various times t after introduction of H_2 (10^5 Pa). Within a few seconds transparent γ-YH$_{3-\delta}$ is formed beneath the Pd, which adsorbs and dissociates hydrogen. The remaining Y surface has been oxidized, making it impermeable to hydrogen. (b) The diffusion profile after 1 day, (c) after 5 days and (d) after 23 days at 60 °C. The red outer ring is due to the weak transparency window at $\lambda = 1.8$ eV, characteristic of the dihydride β-YH$_2$ phase (see Fig. 7.12e), while the dark surrounding region is nontransparent metallic α-YH$_x$. (From den Broeder et al. (1998), Ref. [143].)

it, several hydride phases are formed. As each phase has a characteristic optical transmission, H diffusion leads to the formation of typical rings that expand with time t as $\sqrt{t}$. The inner ring corresponds to the γ-YH$_{3-\delta}$ phase and the outer ring to the β-YH$_2$ phase, with its weak transparency window for red photons (1.8 eV). It is this characteristic optical feature that makes it possible to observe real-time hydrogen diffusion at room temperature.

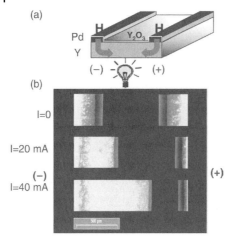

Fig. 7.34 Electromigration of hydrogen in a (a) 200 nm thick, 1.5 mm wide Y film on sapphire, covered on the left and right by 30 nm thick Pd strips. At $t=0$ H_2 gas (at 10^5 Pa) is introduced and hydrogen is allowed to migrate for 12:10 h at 110 °C. (b) In the absence of an electric current ($I=0$) the resulting diffusion profiles are symmetric. The lower two profiles show the hydrogen fronts when a DC current ($I=20$ and 40 mA, respectively) has been applied from right (+) to left (−). The asymmetry in the profiles is due to the force created by the electric field on the hydrogen atoms and proves the negative effective valence of H in $YH_{3-\delta}$. We note that the effect of electromigration on the β-phase growth is relatively small due to its lower resistivity. Therefore, it grows faster than $YH_{3-\delta}$ at the positive side of the sample, but shrinks to minimal size at the negative side, where it is almost overtaken by the accelerated $YH_{3-\delta}$ phase. The scale bar corresponds to 500 mm. (From Ref. [143].)

The same technique can be used to study electromigration of hydrogen in a switchable mirror [143, 144]. A nice example is shown in Fig. 7.34 for an yttrium film that is simultaneously loaded with hydrogen from the left and the right through Pd pads. In the absence of an electrical current (top panel) the diffusion pattern is symmetric. In the presence of a current a clear asymmetry is induced (middle panel). The asymmetry is increased by increasing the current (lower panel). These experiments show unambiguously that hydrogen behaves like a negative ion in YH_x. This is in agreement with the infrared data of Rode et al. [145] and the strong electron correlation models of Ng et al. [117, 140] and Eder et al. [141] (see Section 7.2.3.4).

Visualization of Hydrogen Diffusion in Opaque Materials The technique described above for YH_x is obviously not directly applicable to systems that are opaque. However, one should realize that a material such as YH_x also exhibits characteristic changes in its reflection (see Section 7.2.3.1). Remhof et al. [208] demonstrated that hydrogen diffusion in materials such as vanadium could be monitored optically in reflection by using samples as shown in Fig. 7.35a. A vanadium stripe of 10 mm length, 1 mm width and a thickness of typically 100 nm is covered with a thin layer of yttrium as an optical indicator for hydrogen diffusion. The indicator thickness

7.2 Optical Properties of Metal Hydrides: Switchable Mirrors

(a)

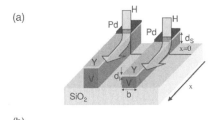

(b)

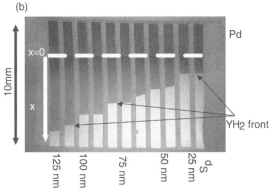

Fig. 7.35 (a) Schematic sample design used by Remhof et al. [208] to measure hydrogen diffusion in vanadium optically. Yttrium-covered vanadium stripes ($1 \times 10\,mm^2$) of thickness 25 to 125 nm are deposited onto a SiO$_2$ substrate ($15 \times 10 \times 0.53\,mm^3$). On one end the stripes are partially covered with a Pd cap layer to enable gas phase H loading. The actual sample shown in (b) consists of 11 stripes. The thickness of the vanadium stripes decreases from 125 to 25 nm from left to right. The sample is loaded in a hydrogen atmosphere ($p_{H2} = 100$ Pa, $T = 473$ K) for 10^4 s. The 11 composite V/Y stripes take up H via the Pd layer, which is located at the upper part of the photo. Lateral H migration occurs along the stripes away from the Pd-covered part. Within the Y-indicator layer, the presence of H leads to the formation of the YH$_2$ phase, which appears blue in reflection. Note the influence of the V thickness on the mobility of the YH$_2$ front in the indicator. The nonconstant thickness of the second, seventh and the ninth vanadium stripes leads to a deformation of the diffusion front. (From Remhof et al. (2002), Ref. [208].)

is typically 30 nm. One end of this V/Y sandwich is covered *in situ* with a 10 nm thick Pd cap layer. When brought into air a thin oxide layer forms on the uncovered yttrium. Loading with hydrogen proceeds via the Pd strip. The subsequent lateral hydrogen diffusion occurs mainly through vanadium, since H uptake cannot occur via the superficially oxidized Y and since hydrogen diffusion is several orders of magnitude faster in vanadium than in yttrium. The lateral migration of hydrogen in vanadium away from the Pd-covered region can easily be monitored optically as a change in the reflection of the Y indicator layer.

Figure 7.35(b) depicts a snapshot at $t = 10^4$ s after the Y/V sandwich was exposed to a hydrogen atmosphere of 100 Pa at $T = 473$ K. The Pd-covered part, where the hydrogen enters the samples and the lateral diffusion starts, is at the top of the photo. The optical discontinuity within the indicator corresponds to the boundary between the silvery gray low concentration α-YH$_{0.2}$ phase and the blue (in reflection) dihydride phase β-YH$_{1.9}$. This diffusion front has moved by approximately 0.9 cm within 10^4 s. This indicates that the diffusion coefficient is of the order of 10^{-4}–10^{-5} cm^2 s^{-1}, which is indeed very fast, in agreement with measurements on bulk samples. It should be pointed out here that reliable values for the diffusion coefficient can only be derived from simultaneous optical measurements on Y/V sandwiches with various thickness ratios in order to take into account the hydrogen leaking from the film under investigation (here V) to the indicator layer (here Y). With such an analysis Remhof et al. [208] determined $D = 1.2 \times 10^{-5}$ cm^2 s^{-1} at $T = 473$ K for a hydrogen concentration H/V = 0.62. Furthermore, the photographs in Fig. 7.35(b) indicate clearly that the overall diffusion rate of hydrogen in an Y/V sandwich depends on the thickness ratio of the two metallic layers. This means that the effective diffusion coefficient of a film can be tuned. In a patterned thin film this opens the opportunity to locally vary the hydrogen mobility and to investigate the behavior of hydrogen diffusion fronts when they are crossing the interface between two media with different diffusion coefficients [209, 210].

As an example we show the results obtained by Remhof et al. for a circular diffusion front generated in a "slow" medium (see Fig. 7.36). When it hits the boundary with a "fast" medium the front bulges out and a mushroom-shaped pattern is formed. The deformation of the diffusion front can be investigated in great detail and compared to theoretical calculations [211]. The results of this type of experiment are also relevant for the propagation of so-called diffusion waves (see, e.g. Refs. [212, 213]).

7.2.6.2 Combinatorial Research

Optimization of Catalytic Cap Layers/Buffer Layers The hydrogenation process of a metal hydride MH$_x$ capped by a catalyst layer is a complicated process [214]. Hydrogen molecules absorb at the catalyst surface (physisorption), split into chemisorbed hydrogen atoms, diffuse into the catalyst subsurface layers, diffuse towards the MHx boundary and cross the catalyst/MHx boundary.

To optimize catalytic layers one needs to study the interaction of two layers. To speed-up the optimization procedure of catalyst/metal-hydride layers Van der Molen et al. [80] introduced a combinatorial approach, using so-called matrix samples (see insert in Fig. 7.37). Using a shadow mask, they deposited two orthogonal stepped thickness gradients of both the switching metal (in this case yttrium or lanthanum) and the catalyst layer (palladium) and created in this way more than 200 different plaquettes on the same substrate. Monitoring the optical transmission during hydrogen loading, the behavior of these ~200 plaquettes is recorded. From a comparison the optimal layer thicknesses for the parameters of interest (i.e. for a short switching time, large optical contrast, reversibility, etc.) are easily obtained (see Fig. 7.37). Indeed a well-defined critical Pd thickness was found (see Section 7.2.2.1). The stepped matrix samples have the advantage that the optimal combination of layer thicknesses can easily be distinguished from the plaquette position on the matrix sample.

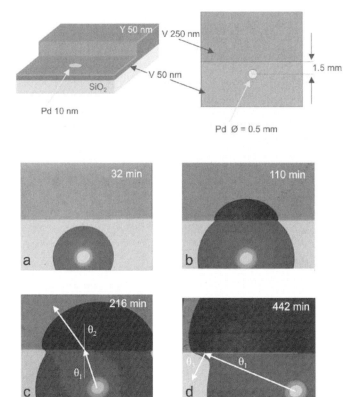

Fig. 7.36 Top panels: Schematic representation of the sample used to investigate hydrogen diffusion from a slow medium to a fast medium. Hydrogen enters the sample via the Pd dot in the lower half of the sample, 1.5 mm away from the interface. Photographs: Each image covers a 5.6 × 4.5 mm² area of the sample. They are recorded in reflection 32, 110, 216 and 442 min and after the sample has been brought into contact with a hydrogen atmosphere of 10^5 Pa at 373 K. The originally circular diffusion front is heavily deformed as soon as it hits the interface between the two media. On the basis of such investigations Remhof et al. [209] concluded that there were some analogies between diffusion and geometric optics (e.g. Snell's law is valid in (c)). (From Remhof et al. (2003), Ref. [209].)

With this technique it was also possible to look for appropriate buffer layers between the metal hydride and Pd to prevent interdiffusion and oxidation [83]. In the case of La switchable mirrors it was found that an AlO_x layer with a very well defined thickness (between 0.9 and 1.2 nm) is able to inhibit Pd/La interdiffusion and oxidation of La without significantly hampering hydrogen absorption.

The method can also be used to measure the catalytic activity of hydrogen uptake of various compounds. The metallic indicator layer has a high affinity for hydrogen but is not able to absorb it directly due to its nonactive oxide skin. This thin oxide layer does transport hydrogen to the optically active indicator layer, once the molecular hydrogen is dissociated by a catalyst (e.g. Pd-clusters [84]). On this

Fig. 7.37 Photograph of the sputtering chamber showing six magnetron sources. Five of them are covered with a shutter, one is uncovered (on the left) showing the magnets within the source. The sources are at various tilt angles.

thin film system (mostly Y/YO$_x$), various catalysts are deposited and their catalytic effect is studied as a function of composition, thickness, temperature and hydrogen pressure [215–217]. The technique is extended to a combinatorial screening method by the use of large area thin film matrix samples with controlled gradients in local chemical composition of two or more constituents [218]. The thin film approach opens the possibility to use well-established surface science techniques. As a demonstration of this technique Borgschulte *et al.* investigated the appropriate thermodynamic conditions for fast sorption kinetics of noble metal catalysts [215, 217]. Similarly, they studied the role of oxygen in the NbO$_x$-catalyzed sorption process on microscopic scale by X-ray photoelectron spectroscopy [216].

Search for New Lightweight Hydrogen Storage Materials In the standard approach followed so far, the exploratory search for new lightweight hydrogen storage materials is very time consuming since a bulk sample needs to be made for each chosen composition. However, metal hydrides and even complex metal hydrides are easily formed at room temperature from the elements, when a thin film of the corresponding metal composition is exposed to hydrogen. Thin films are therefore perfectly suited to study the reversible formation of complex hydrides from their elements. By making thin films with controlled gradients in the local chemical composition of three or more constituents, one may study the hydrogenation of typically 10^4 samples simultaneously, since the hydrogenation process can be monitored optically. Here one exploits the hydrogen concentration dependence of optical properties of metal-hydrides. Since essentially complex metal hydrides found so far have a band gap, the transparency is a good indicator for the presence of a (complex) hydride phase. For the fabrication of gradient samples Gremaud *et al.* [219] use a sputtering system with several off-axis sources, which can be tilted towards and away from the

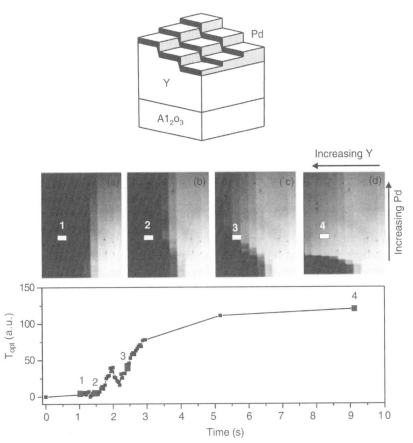

Fig. 7.38 Time dependence of the optical transmission of a Pd/YH$_x$ matrix sample during hydrogenation. The top panel is a schematic representation of a matrix sample. Four stages of the optical transmission of a matrix sample are shown in (a)–(d). The yttrium and Pd thicknesses increase from 0 to 150 and 1 to 25 nm respectively. In the lower panel the optical transmission versus time of one arbitrarily chosen plaquette (marked white, $d_{Pd} = 10$ nm, $d_Y = 100$ nm) is given. (From Van Der Molen et al. (1999), Ref. [80].)

substrate (see Fig. 7.37). Depending on the tilt angle, the thickness of a particular component increases up to sevenfold over the diameter of the 3 inch wafer.

Using Mg–Al and Mg–Ni as model systems, the well-known Mg$_2$NiH$_4$ and Mg(AlH$_4$)$_2$ hydrides indeed show up as transparent regions with a well-defined compositional width after hydrogenation. Reversible optical switching (and hence reversible hydrogen loading) was found in both systems over a very wide Mg-rich compositional range [219]. Another combinatorial approach is chosen by Olk et al. [220] who analyze the hydrogen sorption process by detecting the changes in the infrared emissivity during hydrogenation. This method basically detects both temperature changes (hence the enthalpy of formation) and changes in the optical properties of the material. They show that the hydrogen loading temperature of Mg–Ni–Fe alloys depends significantly on the composition. Not every lightweight complex metal hydride is suited for reversible hydrogen storage. The combination with PEM fuel cells dictates, for example a hydrogen gas pressure of $1-3 \times 10^5$ Pa at $T < 373$ K. In addition, the kinetics must be fast enough to fill a tank with 5 kg of hydrogen within 2 min. The goal is to find a hydrogen storage material (combination) with a hydrogen-mass density larger than 7.5 wt%, a volume density larger than 62 kg m^{-3} and suitable ab/desorption characteristics [221, 222]. To meet these objectives, one studies the effect of additives to stabilize or destabilize the crystal structure or to catalyze the sorption process. In a thin-film approach, the effect of substitutions on the loading behavior is easily studied by codepositing an additional gradient of a suitable element. To optimize the kinetic properties the gradient layers can be combined with various cap layers. Once the most promising spots have been identified, their local chemical composition, crystal structure and surface morphology are determined with Rutherford backscattering spectrometry, focused beam X-ray scattering and AFM, respectively. Thus the combinatorial thin-film-gradient technique is an efficient tool to optimize both the hydrogen content and the kinetics of the sorption reaction in both new and known complex metal hydrides.

Hydrogenography A breakthrough in combinatorial techniques for the search for new hydrogen storage materials occurred very recently when Gremaud et al. [223] demonstrated that "Hydrogenography" could be devised to measure optically the enthalpies (and entropies) of formation of thousands of alloy compositions simultaneously. Studying the quaternary metal hydride Mg–Ni–Ti–H they show that there is a composition region of Mg-rich Mg–Ti–Ni alloys that absorb hydrogen reversibly with enthalpies of formation between -40 and -37 kJ (mol H$_2$)$^{-1}$ and good hydrogenation kinetics.

Mg–Ni–Ti gradient films were sputtered on a 3 inch wafer by means of tilted Mg, Ni and Ti sputter guns positioned every $120°$ on a circle around the substrate. All the films with thickness 30–100 nm are covered *in situ* with a 20 nm cap layer to promote H$_2$ dissociation and to prevent oxidation of the underlying film. Pd is used as cap layer for temperatures up to 393 K, whereas Ni is preferred for higher temperatures, to avoid the formation of a Mg$_6$Pd alloy at the film–cap layer interface. After deposition, the metallic films are transferred into an optical cell to monitor their optical transmission during hydrogenation. The whole cell is placed in a furnace to control temperature up to 300 °C. A 150 W diffuse white light source illuminates the sample from the substrate side, and a CCD camera continuously monitors the transmitted light as a function of hydrogen pressure. According to Lambert–Beer's law, the logarithm of the optical transmission is a good measure of the local hydrogen concentration. By recording continuously the optical transmission as a function of increasing pressure one can essentially construct the pressure–composition isotherms for each pixel of the frame, that is for each composition of the ternary alloy Mg–Ti–Ni. The isotherm plateaus are observed as steep gradients in the optical transmission, which reflects the hydride formation at a well-defined pressure. By measuring the plateau pressure in this way at various temperatures, one easily constructs the Van't Hoff plots for all the compositions in the range. From this the corresponding enthalpy and entropy of hydrogenation are easily derived. The end product is an enthalpy map for hydrogenation in the ternary metal phase diagram. Here we easily find compositions with enthalpies of formation significantly less negative than that of both MgH_2 and Mg_2NiH_4. For example, for $Mg_{0.69}Ni_{0.26}Ti_{0.05}$ we find an enthalpy value $\Delta H = -40\,kJ\,(mol\,H_2)^{-1}$, close to the ideal value of $-39.2\,kJ$. By electrochemical analysis a gravimetric capacity of 3.5 wt.% is found. The corresponding entropy of formation is substantially reduced from the ideal gas entropy. Nevertheless, the plateau pressure of $Mg_{0.69}Ni_{0.26}Ti_{0.05}$ at 100 °C is still 2 orders of magnitude larger than for MgH_2.

7.2.7
Switchable Mirror-Based Devices

7.2.7.1 Electrochromic Devices

Electrochromism is the reversible change in optical properties that occurs when a material is electrochemically oxidized or reduced [224]. This working definition includes a change in optical properties anywhere in the solar (and even in some cases the microwave) range. In addition to the active electrochromic layer, a device consists of an electrolyte and a counter electrode, which may or may not be electrochromic. The electrolyte should be a good ionic conductor and electrically insulating in order to be nonvolatile.

At first sight the large change in optical transmission found in switchable mirrors seems comparable to that observed in other electrochromic materials such as hydrogen or lithium intercalated transition-metal oxide films [225]. However, the simultaneous occurrence of metallic reflectivity, high absorption over a large part of the solar spectrum, and colorless transparency in the visible part of the optical spectrum in the same material (e.g. in Mg–RE, see Section 7.2.4) is not found in other electrochromics. As far as we know, these are unique properties of metal hydrides. Well-known electrochromic materials such as WO_3, Prussian blue or viologen switch reversibly from a transparent to an absorbing state, which is active over a limited wavelength range only [225]. The intercalated blue bronzes, H_xWO_3, change, for example from a transparent colorless insulator at $x \sim 0$ to a transparent but deep blue semiconductor at $x \sim 0.3$. The high contrast of the metal hydrides between the metallic reflecting state and the semiconducting transparent state, indicates that gasochromic devices based on this materials could be much thinner (e.g. $\sim 100\,nm$) than the WO_3 layers used so far [226]. Interestingly, in the transparent trihydride state the switchable mirrors are in their oxidized state (Y^{3+}), which makes them complementary to WO_3 which is colored in its reduced state (W^{6+} mixed with W^{5+}) [227].

Device Demands Clearly, a feasible device needs to withstand a large number of switching cycles. To prevent mechanical damage to the device, lattice expansion during switching should be minimized. The expansion-free switching of the

$Y_{0.38}La_{0.62}$-alloy is in this respect very promising. Additionally, the absence of hysteresis in these alloys (see Section 7.2.3.3) is advantageous for a predictable operation of an electrochromic device. The power needed to switch the optical properties of a window determines to a large extent the applicability of an electrochromic device. The coloration efficiency of a device is given by the ratio of the change in optical density $\Delta(OD) = \log[T_1(\lambda)/T_2(\lambda)]$, with $T(\lambda)$ the transmission at a certain wavelength in the reduced and the oxidized state, respectively, divided by the amount of charge (or hydrogen in this case) ΔQ to be moved in/out of the active layer per unit area ($CE = \Delta(OD)/\Delta Q$ [225]). Clearly, to minimize the power needed for the switching of an electrochromic device, it is advantageous if the change in transmission is induced by a small amount of charge. Ouwerkerk [167] measured a $CE = -37 \, cm^2 \, C^{-1}$ in a 200 nm $Sm_{0.3}Mg_{0.7}H_x$ film. Values of $CE > 100$ are feasible since the minimum charge needed is only $\gg 40 \, mC \, cm^{-2}$ and the $\Delta(OD) = 4$ measured is limited by pinholes rather than intrinsic properties. The disadvantage of the metal hydrides is the fact that these materials need to be protected by a Pd cap layer, which reduces the maximum effective transparency to about 40%.

Towards an Electrochromic Device Compared to gasochromic switching, which can be as fast as 40 ms in $GdMgH_x$, the insertion of an electrolyte drastically increases the switching time. In the case of liquid electrolyte devices (see Fig. 7.20) the minimal switching time is of the order of minutes. Using a 2 M KOH solution, Ouwerkerk [167] showed reversible switching in a Pd-capped 200 nm $Sm_{0.3}Mg_{0.7}H_x$ thin film. Janner et al. [228] have studied the cycling durability of Gd–Mg switchable mirrors in 1 M KOH. After an initial decrease, the switching time increases again at the end of the lifetime. The increase in switching time is related to delamination and the formation of an oxide layer at the metal hydride–electrolyte interface. Additional protective layers have proved to be unsuccessful so far. Strain problems can possibly be reduced by depositing the metal hydride directly in the fully hydrogenated state.

The first all-solid-state metal hydride switchable mirror device [229] had switching times of up to 16 h and used GdMg as the active layer, hydrated ZrO_2 as the electrolyte and H_xWO_3 as the counter electrode and hydrogen storage layer. The transparent metal hydride is formed when 3 V is applied with respect to the WO_3 layer. The large currents involved led the authors to assume that the long switching times are due to electric shorts in the hydrated ZrO_2. In an inverted device (WO_3 as bottom electrode/H storage layer, see Fig. 7.39) Van der Sluis et al. [63] showed that the switching time can be reduced to 5 min, although the full switching cycle is still of the order of 1 h. The devices switch reversibly up to 400 consecutive cycles. Some degradation occurred, but this could be compensated by increasing the switching time. The self-discharge was negligible resulting in a retention time of over a year. Unfortunately, these results have not been reproduced yet. However, many improvements are possible. The Zr-based electrolyte [229] may not be ideal since it transports protons, while within metal hydrides such as YH_x or REH_x we

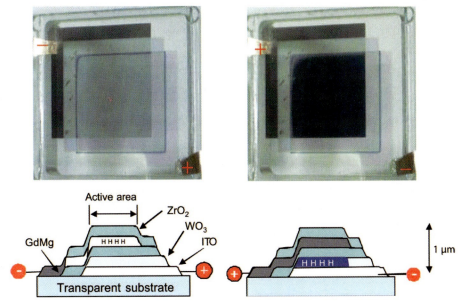

Fig. 7.39 Schematic cross-sections and photographs of an all-solid-state switchable mirror [63] in two optical states: (a) at negative applied voltage both the GdMgH$_x$ switchable mirror and the WO$_3$ storage electrode are in their transparent state; (b) at positive voltage GdMg is metallic and WO$_3$ dark blue.

are dealing with H$^-$. Alternatively, Y-doped CaF$_2$ might be used as a H$^-$-conducting electrolyte in conjunction with a suitable H$^-$-based counter electrode.

7.2.7.2 Thermochromic Devices

As shown by Giebels *et al.* [171] the temperature dependence of the equilibrium plateau pressures of metal hydrides can be used for an (*open*) thermochromic device. Y$_{0.9}$Mg$_{0.1}$ samples appear to be especially promising since Mg suppresses the fcc–hcp transition of the REH$_2$–REH$_3$ transition. Because of this more hydrogen is desorbed upon heating and thus the transparency change is much larger and varies almost linearly with temperature. An alternative scheme for an electrochromic device is proposed by Lokhorst *et al.* [230]. They use two *encapsulated* metal hydride layers with a slightly different temperature dependence of the hydrogen plateau pressure.

Any temperature change then results in a redistribution of the hydrogen over the two layers. The switchable metal hydrides are clearly different from other traditional (metal–insulator) thermochromic systems [225] such as the vanadium oxides (VO$_2$, V$_2$O$_5$). The change in contrast in the visible during the transition is much larger than in VO$_2$, which switches predominantly at wavelengths larger than 600 nm (2.1 eV).

7.2.8
Applications of Metal Hydride-Based Devices

7.2.8.1 Smart Windows

Smart coatings can play an important role in reducing the energy consumption of buildings and cars. In the US about 30 % of the primary energy consumption is used for heating/cooling/lighting of residential and office buildings [231]. By regulating the heat balance while maintaining outward visibility enormous energy savings are feasible. Several active and passive window coatings have been proposed [232]. In general one would like to regulate the solar power input, while maintaining visibility. Hence, the optical properties of the visible range should be different from those of the long wavelength range of the solar spectrum. Alternatively, heat dissipation is controlled by the emissivity around 0.1 eV (8 µm; 373 K black-body radiation). Windows with low thermal losses (e.g. based on SnO_2) should transmit radiation over the whole solar range and reflect energies below 0.5 eV (2.5 µm). Alternatively, to minimize heat input while maintaining outward visibility, one could use windows reflecting light between 0.5 and 1.65 eV (2.5 µm 750 nm). This would reduce the cooling requirement for buildings dramatically. The fact that switchable metal hydrides are able to reflect the incoming radiation over the whole solar spectrum make them of interest for use in cars, for example to reduce the thermal load of parked cars.

7.2.8.2 Switchable Absorbers

Devices based on Mg–Ni films have a potential as variable reflectance coatings. Recently Van Mechelen *et al.* [233] showed that these materials have an optical absorption contrast over the solar range of a factor 2, comparing the metallic Mg_2Ni to the "black" $Mg_2NiH_{0.8}$ state. As the blackbody emissivity at 100 °C is less than 15 % for this material, the "black state" is an interesting heat absorber. There is a present need for such variable reflectance metal hydride (VAREM) coatings particularly for use in combined photovoltaic cells/thermal solar collector devices. These devices consist essentially of a photovoltaic cell on top of a thermal solar collector [234]. Under certain conditions (high solar energy influx and low warm water consumption) the temperature of such devices can be well above 150 °C. This leads to a reduced total efficiency and (if plastic components are used) to an irreversible degradation of the photovoltaic component. The same heat management problem arises for solar collectors which are built from cheap plastics. VAREM coatings can be used to limit the temperature of such devices. When the thermal load of the system becomes too high, the VAREM is driven into its metallic reflecting state. In the opposite situation it is switched to its black absorbing state. In these devices VAREM coating can also be used to reduce heat radiation losses during night hours. The active control of the optical properties of a VAREM is also beneficial to heat collecting walls inside buildings: these so-called Trombé-walls [235] allow an optimal use of solar energy during all parts of the day in all seasons. A promising material to be used as a VAREM is, for example Mg–Ti–H [236].

7.2.8.3 Fiber Optic Hydrogen Sensors

Societal acceptance is a key ingredient for the implementation of a future hydrogen-based economy. As this acceptance depends essentially on safety of operation there is a great need for sensitive, selective, fast, reliable, stable and cheap hydrogen sensors. Currently, commercial hydrogen detectors are not useful for widespread use, particularly in transportation, because they are rather bulky, not stable over a long period of time and expensive. Furthermore, most of the sensors available commercially are based on electrical measurements at the sensing point. This might be undesirable in potentially explosive environments. These disadvantages can be circumvented by using optical detectors in which the end of an optic fiber is coated with a hydrogen-sensitive layer. The changes induced in the optical properties of this layer during absorption of hydrogen are detected optically at the other end of the fiber. Compared to other hydrogen sensors, optical fiber sensors have the advantage of being simple yet very sensitive, cheap, insensitive to electromagnetic noise, explosion safe and allowing multiple sensing with one central (remote) detector. This is the key advantage over conventional detectors that make optical sensors cost competitive. Furthermore, optic fibers are more resistant to corrosion than standard electrical wires.

To be competitive with other hydrogen sensors the coatings to be developed must satisfy at least the following performance targets:

1. Hydrogen concentration in air: 0–10 % (for safety detectors), 0–100 % (for sensors).
2. Operating temperature: −25 to 100 °C.
3. Response time: 1 s or less for 90 % full scale.
4. Long time stability: more than 1 year.
5. Selectivity: hydrogen specific.
6. Pollutant resistant: especially to hydrocarbons, H_2S.
7. Insensitivity to humidity.
8. Low price.

Several groups are presently developing fiber optic hydrogen sensors. For example at the National Renewable Energy Laboratory (NREL) researchers have obtained interesting results with fibers coated with a WO_3 film at the sensing end [237, 238]. This originally transparent material is colored blue when absorbing hydrogen. However, the presence of water molecules is essential for its switching [239, 240]. This limits its application range to temperatures above 0 °C. This problem can be avoided if metal hydride switchable mirrors are used as the sensing coating. In this context the Mg–TM hydrides are particularly interesting since their reflection decreases strongly already at low hydrogen concentrations, as shown in Fig. 7.31 for Mg_2NiH_x. In such a material coloration does not depend on the presence of water and, since hydrogen diffusion is sufficiently fast over a large temperature range (−40 to +90 °C), such sensors remain operational below 0 °C.

A breakthrough in sensor development was the introduction of triple layers: a catalytic cap layer, a buffer layer and the optically active Mg_2Ni layer [241]. The kinetics as well as the structural stability of the thin film devices could be improved by orders of magnitudes. The characteristics of such a device based on a Mg–Ti–H active layer is given by Slaman et al. [242].

7.2.8.4 Batteries

Notten et al. [243] discovered recently a very large (reversible) storage capacity of 5.6 wt.% (1500 mA h g^{-1}) in $Mg_{0.8}Sc_{0.2}Pd_{0.024}$ pellets in a 6 M KOH electrolyte at 25 °C. This is about 4 times more than in conventional Ni–MH batteries such as $LaNi_5H_6$. Their research was inspired by the switchable mirrors research on Mg–RE films done before at Philips Research Laboratories in Eindhoven [57, 167, 184]. To achieve such a large storage capacity Sc has to become ScH_3. This is rather surprising since pure Sc does not load to the trihydride phase at atmospheric pressure.

Only at 300 MPa is hcp ScH_3 formed [244]. Giebels et al. [179] confirmed the formation of ScH_3 recently in thin films of Mg–Sc. These films become very transparent when the Mg content is larger than 65 at.%. They behave as Mg–Y and Mg–La switchable mirrors, that is they show a black state and a shift of the band gap with excess Mg. As Sc is a very expensive material, successful attempts have been made to substitute it with Ti, V and Cr [246].

7.2.9
Conclusions and Outlook

In this chapter we have described the remarkable optical properties of switchable metal hydride films. So far three generations of switchable mirrors have been discovered. The first generation concerns the hydrides of the rare-earths and yttrium and lanthanum. The transparency of these materials posed intriguing questions to theorists about the origin of the insulating state. In the second and third generations magnesium plays an essential role. Alloyed with rare-earths it leads to color neutral films when fully hydrogenated. There is an intriguing highly absorbing state, the black state, which occurs at intermediate hydrogen concentrations. The coexistence of Mg and MgH_2 grains is essential for high absorption. Alloyed with the transition metals Ni, Co, Fe and Mn, magnesium forms complex metal hydrides that also exhibit the three fundamental optical states of matter, that is reflecting, absorbing and transparent. In these third-generation switchable mirrors, the mechanism leading to the black state involves a self-organized double layering of Mg_2NiH_x. Surprisingly, during hydrogenation Mg_2NiH_4 nucleates first in a thin layer at the interface to the substrate. Further hydrogen uptake yields growth of that layer at the expense of the remaining metallic layer on top.

The results presented in this chapter also show the tremendous advantage of thin films compared to bulk metal hydrides. During hydrogen absorption the films retain their structural integrity. In contrast to bulk samples they are not reduced to powder when they react with hydrogen, although the volume changes can be considerable (e.g. 32 % in the case of Mg_2NiH_4 and MgH_2). This makes it possible to investigate in detail the intrinsic physical properties (electrical resistivity, charge carrier density, optical properties, etc.) of these materials. The thin-film geometry also makes it possible to investigate and optimize catalytic layers. In many cases Pd works fine and makes it possible, for example, to hydrogenate Mg_2Ni at room temperature with hydrogen at pressures of a few hundred Pa. The possibility to study in detail the effect of catalytic cap layers is very important for all applications where reaction kinetics is important: energy storage, electrochromic devices, and so on. The formation of Mg_2NiH_4 shows that the chemistry of complex metal hydrides can also be studied. By using matrix samples or films with large compositional gradients it is possible to monitor optically the reaction of hydrogen with a great number of different samples evaporated on the same substrate. This allows a highly efficient combinatorial exploration of new lightweight hydrogen storage materials. Thin films of lightweight metals (e.g. alanates) will soon be investigated optically.

Although various electrochromic devices have been demonstrated, their performance still needs to be drastically improved. This will require a major research and development effort. On the other hand, the fiber optic metal hydride hydrogen sensor already shows that metal hydride applications may provide a clear advantage over competing systems.

References

1 Andersson, G., Hjörvarsson, B., Zabel, H. (1997) *Phys. Rev. B*, **55**, 15905–15911.
2 Andersson, G., Andersson, P.H., Hjörvarsson, B. (1999) *J. Phys. Cond. Mat.*, **11**, 6669.
3 Song, G., Geitz, M., Abromeit, A., Zabel, H. (1996) *Phys. Rev. B*, **54**, 14093.
4 Allain, M.M.C., Heuser, B.J. (2005) *Phys. Rev. B*, **72**, 054102.
5 Klose, F., Rehm, C., Fieber-Erdmann, M., Holub-Krappe, E., Bleif, H.J., Sowers, H., Goyette, R., Troger, L., Maletta, H. (2000) *Physica B*, **283**, 184.
6 Klose, F., Rehm, C., Nagengast, D., Maletta, H., Weidinger, A. (1997) *Physica B*, **234**, 486.
7 Klose, F., Rehm, C., Nagengast, D., Maletta, H., Weidinger, A. (1997) *Phys. Rev. Lett.*, **78**, 1150.
8 Pärnaste, M., Marcellini, M., Holmström, E., Bock, N., Fransson, J., Eriksson, O., Hjörvarsson, B. (2007) *J. Phys. Cond. Mat.*, **19**, 256213.
9 Hayashi, Y., Masuda, M., Tonomyo, K., Matsumoto, S., Mukai, N. (1999) *J. Alloys Comp.*, **193**, 463.
10 Kandasamy, K., Masuda, M., Hayashi, Y. (1999) *J. Alloys Comp.*, **288**, 13.
11 Strongin, M., Elbatanouny, M., Pick, M.A. (1980) *Phys. Rev. B*, **22**, 3126.
12 Watts, R., Rodmarq, B, (1997) *J. Mag. Magn. Mat.*, **174**, 70.
13 Leiner, V., Ay, M., Zabel, H. (2004) *Phys. Rev. B*, **70**, 104429.
14 Griessen, R., Riesterer, T. (1988) *Hydrogen in Intermetallic Compounds I*, Schlapbach L. (ed.), Springer, Berlin.
15 Andersson, G., Hjörvarsson, B., Isberg, P. (1997) *Phys. Rev. B*, **55**, 1774.
16 Burkert, T., Miniotas, A., Hjörvarsson, B. (2001) *Phys. Rev. B*, **63**, 125424.
17 Kuch, W., Chelaru, L.I., Fukumoto, K., Porrati, F., Offi, F., Kotsugi, M., Kirschner, J. (2003) *Phys. Rev. B*, **67**, 214403.
18 Scherz, A., Poulopoulos, P., Nünthel, R., Lindner, J., Wende, H., Wilhelm, F., Baberschke, K. (2003) *Phys. Rev. B*, **68**, 140401(R).
19 Leiner, V., Westerholt, K., Blixt, A.M., Zabel, H., Hjörvarsson, B. (2003) *Phys. Rev. Lett.*, **91**, 037202.
20 Allen, G.A.T. (1970) *Phys. Rev. B*, **1**, 352; Domb, C. (1973) *J. Phys. A*, **6**, 1296.

21 Li, Y., Baberschke, K. (1992) *Phys. Rev. Lett.*, **68**, 1208.
22 Srivastava, P., Wilhelm, F., Ney, A., Farle, M., Wende, H., Haack, N., Ceballos, G., Baberschke, K. (1998) *Phys. Rev. B*, **58**, 5701.
23 Kohlhepp, W.J., Elmer, H.J., Cordes, S., Gradmann, U. (1992) *Phys. Rev. B*, **45**, 12287.
24 Li, D., Freitag, M., Pearson, J., Qiu, Z.Q., Bader, S.D. (1994) *Phys. Rev. Lett.*, **72**, 3112.
25 Qian, D., Jin, X.F., Barthel, J., Klaua, M., Kirschner, J. (2001) *Phys. Rev. Lett.*, **87**, 227204.
26 Bürgler, D.E., Grünberg, P., Demokritov, S.O., Johnson, M.T. (2001) Interlayer exchange coupling in layered magnetic structures, in *Handbook of Magnetic Materials*, Vol. 13 (ed. K.H.J. Buschow), Elsevier Science.
27 Baltensperger, S., Helman, J.S. (1990) *Appl. Phys. Lett.*, **57**, 2954.
28 Bruno, P., Chappert, C. (1992) *Phys. Rev. B*, **46**, 261.
29 Bruno, P. (1993) *Europhys. Lett.*, **23**, 615.
30 Bounouh, A. et al. (1996) *Europhys. Lett.*, **33**, 315.
31 Ney, A., Wilhelm, F., Farle, M., Poulopoulos, P., Srivastava, P., Baberschke, K. (1999) *Phys. Rev. B*, **59**, R3938.
32 Srivastava, P., Wilhelm, F., Ney, A., Farle, M., Wende, H., Haack, N., Ceballos, G., Baberschke, K. (1998) *Phys. Rev. B*, **58**, 5701.
33 Kuch, W. et al. (2006) *Nature Materials*, **5**, 128.
34 Skubic, B. et al. (2006) *Phys. Rev. Lett.*, **96**, 57205.
35 Andersson, P.H., Fast, L., Nordström, L., Johansson, B., Eriksson, O. (1998) *Phys. Rev. B*, **58**, 5230.
36 Pierce, D.T., Unguris, J., Celotta, R.J., Stiles, M. (1999) *J. Magn. Magn. Mater.*, **200**, 290.
37 Klose, F., Rehm, C., Nagengast, D., Maletta, H., Weidinger, A. (1997) *Phys. Rev. Lett.*, **78**, 1150.
38 Hjörvarsson, B., Dura, J.A., Isberg, P., Watanabe, T., Udovic, T.J., Andersson, G., Majkrzak, G.F. (1997) *Phys. Rev. Lett.*, **79**, 901.
39 Labergerie, D., Westerholt, K., Zabel, H., Hjörvarsson, B. (2001) *J. Magn. Magn. Mater.*, **225**, 373.
40 Andersson, G., Hjörvarsson, B., Isberg, P. (1997) *Phys. Rev. B*, **55**, 1774.
41 Fullerton, E.E., Bader, S.D., Robertson, J.L. (1996) *Phys. Rev. Lett.*, **77**, 1382.
42 Schreyer, A., Ankner, J.F., Zeidler, T., Zabel, H., Schäfer, M., Wolf, J.A., Grünberg, P., Majkrzak, C.F. (1995) *Phys. Rev. B*, **52**, 16066.
43 Parkin, S.S.P. (1991) *Phys. Rev. Lett.*, **67**, 3598.
44 Harp, G.R., Parkin, S.S., O'Brian, W.L., Tonner, B.P. (1995) *Phys. Rev. B*, **51**, 3293.
45 Uzdin, V., Westerholt, K., Zabel, H., Hjörvarsson, B. (2003) *Phys. Rev. B*, **68**, 214407.
46 Manchester, F.D. (ed.) (2000) *Phase Diagrams of Binary Hydrogen Alloys*, ASM International, Materials Park.
47 Ostanin, S., Uzdin, V.M., Demangeat, C., Wills, J.M., Alouani, M., Dreysse, H. (2000) *Phys. Rev. B*, **61**, 4870.
48 Remhof, A., Nowak, G., Nefedov, A., Zabel, H., Björck, M., Pärnaste, M., Hjörvarsson, B. (2007) *Superlatt. Microstructures*, **41**, 127.
49 Uzdin, V.M., Häggström, L. (2005) *Phys. Rev. B*, **72**, 024407.
50 Remhof, A., Nowak, G., Zabel, H., Björck, M., Pärnaste, M., Hjörvarsson, B., Uzdin, V. (2007) *Europhys. Lett.*, **79**, 37003.
51 Griffiths, R.B. (1970) *Phys. Rev. Lett.*, **24**, 1479.
52 Weaver, J.H., Rosei, R., Peterson, D.T. (1978) *Solid State Commun.*, **25**, 201–20.
53 Weaver, J.H., Rosei, R., Peterson, D.T. (1979) *Phys. Rev. B*, **19** (10), 4855–4866.
54 Libowitz, G.G., Pack, J.G. (1969) *J. Chem. Phys.*, **50** (8), 3557–3560.
55 Huiberts, J.N., Griessen, R., Rector, J.H., Wijngaarden, R.J., Dekker, J.P., de Groot, D.G., Koeman, N.J. (1996) *Nature*, **380**, 231–234.
56 Van Gogh, A.T.M., Nagengast, D.G., Kooij, E.S., Koeman, N.J., Rector, J.H., Griessen, R., Flipse, C.F.J., Smeets,

R.J.J.G.A.M. (2001) *Phys. Rev. B*, **63**, 195105-1–21.

57 Van Der Sluis, P., Ouwerkerk, M., Duine, P.A. (1997) *Appl. Phys. Lett.*, **70** (25), 3356–3358.

58 Hjörvarsson, B., Chacon, C., Zabel, H., Leiner, V. (2003) *J. Alloys Compd.*, **356–357**, 160–168.

59 Lohstroh, W., Leuenberger, F., Felsch, W., Fritzsche, H., Maletta, H. (2001) *J. Magn. Magn. Mater.*, **237**, 77–89.

60 Curzon, A.E., Singh, O. (1978) *J. Phys. F.: Metal Phys.*, **8**, 1619–1625.

61 Huiberts, J.N., Rector, J.H., Wijngaarden, R.J., Jetten, S., de Groot, D., Dam, B., Koeman, N.J., Griessen, R., Hjörvarsson, B., Olafsson, S., Cho, Y.S. (1996) *J. Alloys. Compd.*, **239**, 158–171.

62 Wildes, A.R., Ward, R.C.C., Wells, M.R., Hjörvarsson, B. (1996) *J. Alloys Compd.*, **242**, 49–57.

63 Van Der Sluis, P., Mercier, V.M.M. (2001) *Electrochim. Acta*, **46**, 2167–2171.

64 Dam, B., Lokhorst, A.C., Remhof, A., Heijna, M.C.R., Rector, J.H., Borsa, D., Kerssemakers, J.W.J. (2003) *J. Alloys Compd.*, **356–357**, 526–529.

65 Kooij, E.S., van Gogh, A.T.M., Griessen, R. (1999) *J. Electrochem. Soc.*, **146**, 2990–2994.

66 Hayoz, J., Sarbach, S., Pillo, T., Boschung, E., Naumović, D., Aebi, P., Schlapbach, L. (1998) *Phys. Rev. B*, **58** (8), 4270–4273(R).

67 Kwo, J., Hong, M., Nakahara, S. (1986) *Appl. Phys. Lett.*, **49** (6), 319–321.

68 Remhof, A., Song, G., Sutter, C., Schreyer, A., Siebrecht, R., Zabel, H. (1999) *Phys. Rev. B*, **59** (10), 6689–6699.

69 Remhof, A., Song, G., Theis-Bröhl, K., Zabel, H. (1997) *Phys. Rev. B*, **56** (6), R2897–2899.

70 Kooij, E.S., Rector, J.H., Nagengast, D.G., Kerssemakers, J.W.J., Dam, B., Griessen, R., Remhof, A., Zabel, H. (2002) *Thin Solid Films*, **402**, 131–142.

71 Nagengast, D.G., Kerssemakers, J., van Gogh, A.T.M., Dam, B., Griessen, R. (1999) *Appl. Phys. Lett.*, **75**, 1724–1726.

72 Jacob, A., Borgschulte, A., Schoenes, J. (2002) *Thin Solid Films*, **414**, 39–42.

73 Enache, S., Leeuwerink, T., Hoekstra, A.F.T., Remhof, A., Koeman, N.J., Dam, B., Griessen, R. (2005) *J. Alloys Comp.*, **397**, 9–16.

74 Hayoz, J., Schoenes, J., Schlapbach, L., Aebi, P. (2001) *J. Appl. Phys.*, **90** (8), 3925–3933.

75 Borgschulte, A., Weber, S., Schoenes, J. (2003) *Appl. Phys. Lett.*, **82** (17), 2898–2900.

76 Grier, E.J., Kolosov, O., Petford-Long, A.K., Ward, R.C.C., Wells, M.R., Hjörvarsson, B. (2000) *J. Phys. D: Appl. Phys.*, **33**, 894–900.

77 Kooi, B.J., Zoestbergen, E., De Hosson, J.T.M., Kerssemakers, J.W.J., Dam, B., Ward, R.C.C. (2002) *J. Appl. Phys.*, **91** (4), 1901–1909.

78 Kerssemakers, J.W.J., Van Der Molen, S.J., Günther, R., Dam, B., Griessen, R. (2002) *Phys. Rev. B*, **65**, 075417.

79 Kerssemakers, J.W.J., Van Der Molen, S.J., Koeman, N.J., Günther, R., Griessen, R. (2000) *Nature*, **406**, 489–491.

80 Van Der Molen, S.J., Kerssemakers, J.W.J., Rector, J.H., Koeman, N.J., Dam, B., Griessen, R. (1999) *J. Appl. Phys.*, **86** (11), 6107–6119.

81 Pick, M.A., Davenport, J.W., Strongin, M., Dienes, G.J. (1979) *Phys. Rev. Lett.*, **43** (4), 286–289.

82 Borgschulte, A., Rode, M., Jacob, A., Schoenes, J. (2001) *J. Appl. Phys.*, **90** (3), 1147–1154.

83 Van Gogh, A.T.M., van der Molen, S.J., Kerssemakers, J.W.J., Koeman, N.J., Griessen, R. (2000) *Appl. Phys. Lett.*, **77** (6), 815–817.

84 Borgschulte, A., Westerwaal, R.J., Rector, J.H., Dam, B., Griessen, R., Schoenes, J. (2004) *Phys. Rev. B*, **70**, 155414.

85 Oelerich, W., Klassen, T., Bormann, R. (2001) *J. Alloys Compd.*, **315**, 237–342.

86 Dornheim, M., Pundt, A., Kirchheim, R., Van Der Molen, S.J., Kooij, E.S., Kerssemakers, J., Griessen, R., Harms, H., Geyer, U. (2003) *J. Appl. Phys.*, **93** (11), 8958–8965.

87 Hayoz, J., Pillo, T., Bovet, M., Züttel, A., Guthrie, S., Schlapbach, L., Aebi,

P. (2000) *J. Vac. Sci. Technol. A*, **18** (5), 2417–2431.
88 Lokhorst, A.C., Heijna, M.C.R., Rector, J.H., Giebels, I.A.M.E., Koeman, N.J., Dam, B. (2003) *J. Alloys Compd.*, **356–357**, 536–540.
89 Griessen, R. (2001) *Europhys. News*, **32** (2), 42.
90 Griessen, R., Van Der Sluis, P. (2001) *Physik in unserer Zeit*, **32**, 76–83.
91 Notten, P.H.L., Kremers, M., Griessen, R. (1996) *J. Electrochem. Soc.*, **143** (10), 3348–3353.
92 Kooij, E.S., Isidorsson, J., Giebels, I.A.M.E. (unpublished).
93 Di Vece, M., Zevenhuizen, S.J.M., Kelly, J.J. (2002) *Appl. Phys. Lett.*, **81** (7), 1213–1215.
94 Van Der Sluis, P. (1999) *Electrochim. Acta*, **44**, 3063–3066.
95 Pedersen, T.P.L., Salinga, C., Weis, H., Wuttig, M. (2003) *J. Appl. Phys.*, **93** (10), 6034–6038.
96 Van Gogh, A.T.M., Nagengast, D.G., Kooij, E.S., Koeman, N.J., Griessen, R. (2000) *Phys. Rev. Lett.*, **85** (10), 2156–2159.
97 Richardson, T.J., Slack, J.L., Armitage, R.D., Kostecki, R., Farangis, B., Rubin, M.D. (2001) *Appl. Phys. Lett.*, **78** (20), 3047–3049.
98 Richardson, T.J., Slack, J.L., Farangis, B., Rubin, M.D. (2002) *Appl. Phys. Lett.*, **80** (8), 1349–1351.
99 Blomqvist, H. (2003) Magnesium ions stabilizing solid-state transition metal hydrides. Ph.D. thesis, Stockholm University.
100 Reilly, J.J., Wiswall, R.H., Jr. (1968) *Inorg. Chem.*, **7**, 2254–2256.
101 Lohstroh, W., Westerwaal, R.J., van Mechelen, J.L.M., Chacon, C., Johansson, E., Dam, B., Griessen, R. (2004) *Phys. Rev. B*, **70**, 165411.
102 Lohstroh, W., Westerwaal, R.J., Noheda, B., Enache, S., Giebels, I.A.M.E., Dam, B., Griessen, R. (2004) *Phys. Rev. Lett.*, **93**, 197404.
103 Van Gogh, A.T.M., Kooij, E.S., Griessen, R. (1999) *Phys. Rev. Lett.*, **83** (22), 4614–4617.
104 Hoekstra, A.F.T., Roy, A.S., Rosenbaum, T.F. (2003) *J. Phys.: Condens. Matter*, **15** (9), 1405–1413.
105 Hoekstra, A.F.T., Roy, A.S., Rosenbaum, T.F., Griessen, R., Wijngaarden, R.J., Koeman, N.J. (2001) *Phys. Rev. Lett.*, **86** (23), 5349–5352.
106 Roy, A.S., Hoekstra, A.F.T., Rosenbaum, T.F., Griessen, R. (2002) *Phys. Rev. Lett.*, **89** (27), 276402.
107 Mor, G.K., Malhotra, L.K. (2000) *Thin Solid Films*, **359**, 28–32.
108 Kumar, P., Philip, R., Mor, G.K., Malhotra, L.K. (2002) *Jpn. J. Appl. Phys.*, **41**, 6023–6027.
109 Aruna, I., Mehta, B.R., Malhotra, L.K., Shivaprasad, S.M. (2004) *Adv. Mater.*, **16** (2), 169–173.
110 Lee, M.W., Lin, C.H. (2000) *J. Appl. Phys.*, **87** (11), 7798–7801.
111 Azofeifa, D.E., Clark, N. (2000) *J. Alloys Compd.*, **305**, 32–34.
112 McGuire, J.C., Kempter, C.P. (1960) *J. Chem. Phys.*, **33**, 1584–1585.
113 van Gogh, A.T.M., Griessen, R. (2002) *J. Alloys Compd.*, **330–332**, 338–341.
114 Isidorsson, J., Giebels, I.A.M.E., Arwin, H., Griessen, R. (2003) *Phys. Rev. B*, **68**, 115112.
115 Armitage, R., Cich, M., Rubin, M., Weber, E.R. (2000) *Appl. Phys. A.*, **71**, 647–650.
116 Huiberts, J.N., Griessen, R., Wijngaarden, R.J., Kremers, M., van Haesendonck, C. (1997) *Phys. Rev. Lett.*, **79** (19), 3724–3727.
117 Ng, K.K., Zhang, F.C., Anisimov, V.I., Rice, T.M. (1997) *Phys. Rev. Lett.*, **78** (7), 1311–1314.
118 Wijngaarden, R.J., Huiberts, J.N., Nagengast, D., Rector, J.H., Griessen, R., Hanfland, M., Zontone, F. (2000) *J. Alloys Compd.*, **308**, 44–48.
119 Remhof, A., Kerssemakers, J.W.J., Van Der Molen, S.J., Griessen, R., Kooij, E.S. (2002) *Phys. Rev. B*, **65**, 054110-1–8.
120 Kooij, E.S., van Gogh, A.T.M., Nagengast, D.G., Koeman, N.J., Griessen, R. (2000) *Phys. Rev. B*, **62** (15), 10088–10100.
121 Vajda, P. (1995) *Hydrogen in Rare-Earth Metals, Including RH_{2+x} Phases*, Elsevier Science, Amsterdam, Vol. **20** of

Handbook on the Physics and Chemistry of Rare Earths, Ch. 137.

122 Van Der Molen, S.J., Nagengast, D.G., van Gogh, A.T.M., Kalkman, J., Kooij, E.S., Rector, J.H., Griessen, R. (2001) *Phys. Rev. B*, **63**, 235116.

123 Dankert, O. et al. (unpublished).

124 Libowitz, G.G. (1972) *Ber. Bunsen-Ges. Phys. Chem.*, **76** (8), 837–845.

125 Shinar, J., Dehner, B., Barnes, R.G., Beaudry, B.J. (1990) *Phys. Rev. Lett.*, **64** (5), 563–566.

126 Shinar, J., Dehner, B., Beaudry, B.J., Peterson, D.T. (1988) *Phys. Rev. B*, **37** (4), 2066–2073.

127 Switendick, A.C. (1970) *Solid State Commun.*, **8**, 1463–1467.

128 Switendick, A.C. (1971) *Int. J. Quantum Chem.*, **5**, 459–470.

129 Huiberts, J.N. (1995) On the road to dirty metallic atomic hydrogen. Ph.D. thesis, Vrije Universiteit Amsterdam.

130 Wang, Y., Chou, M.Y. (1993) *Phys. Rev. Lett.*, **71** (8), 1226–1229.

131 Wang, Y., Chou, M.Y. (1995) *Phys. Rev. B*, **51** (12), 7500–7507.

132 Dekker, J.P., van Ek, J., Lodder, A., Huiberts, J.N. (1993) *J. Phys.: Condens. Matter*, **5**, 4805–4816.

133 Kelly, P.J., Dekker, J.P., Stumpf, R. (1997) *Phys. Rev. Lett.*, **78** (7), 1315–1318.

134 Udovic, T.J., Huang, Q., Erwin, R.W., Hjörvarsson, B., Ward, R.C.C. (2000) *Phys. Rev. B*, **61** (19), 12701–12704.

135 Udovic, T.J., Huang, Q., Rush, J.J. (1997) *Phys. Rev. Lett.*, **79** (15), 2920.

136 Kierey, H., Rode, M., Jacob, A., Borgschulte, A., Schoenes, J. (2001) *Phys. Rev. B*, **63**, 134109.

137 Zogał, O.J., Wolf, W., Herzig, P., Vuorimäki, A.H., Ylinen, E.E., Vajda, P. (2001) *Phys. Rev. B*, **64**, 214110.

138 van Gelderen, P., Kelly, P.J., Brocks, G. (2001) *Phys. Rev. B*, (63), 100301(R).

139 Kelly, P.J., Dekker, J.P., Stumpf, R. (1997) *Phys. Rev. Lett.*, **79** (15), 2921.

140 Ng, K.K., Zhang, F.C., Anisimov, V.I., Rice, T.M. (1999) *Phys. Rev. B*, **59** (8), 5398–5413.

141 Eder, R., Pen, H.F., Sawatzky, G.A. (1997) *Phys. Rev. B*, **56** (16), 10115–10120.

142 Wang, X.W., Chen, C. (1997) *Phys. Rev. B*, **56** (12), 7049–7052(R).

143 Den Broeder, F.J.A., Van der Molen, S.J., Kremers, M., Huiberts, J.N., Nagengast, D.G., van Gogh, A.T.M., Huisman, W.H., Koeman, N.J., Dam, B., Rector, J.H., Plota, S., Haaksma, M., Hanzen, R.M.N., Jungblut, R.M., Duine, P.A., Griessen, R. (1998) *Nature*, **394**, 656–658.

144 Van der Molen, S.J., Welling, M.S., Griessen, R. (2000) *Phys. Rev. Lett.*, **85** (18), 3882–3885.

145 Rode, M., Borgschulte, A., Jacob, A., Stellmach, C., Barkow, U., Schoenes, J. (2001) *Phys. Rev. Lett.*, **87** (23), 235502.

146 Schoenes, J., Borgschulte, A., Carsteanu, A.-M., Kierey, H., Rode, M. (2003) *J. Alloys Compd.*, **356–357**, 211–217.

147 Hayoz, J., Koitzsch, C., Bovet, M., Naumovic, D., Schlapbach, L., Aebi, P. (2003) *Phys. Rev. Lett.*, **90** (19), 196804.

148 Miyake, T., Aryasetiawan, F., Kino, H., Terakura, K. (2000) *Phys. Rev. B*, **61** (24), 16491–16496.

149 Chang, E.K., Blase, X., Louie, S.G. (2001) *Phys. Rev. B*, **64**, 155108-1–5.

150 Wolf, W., Herzig, P. (2002) *Phys. Rev. B*, **66**, 224112.

151 Wolf, W., Herzig, P. (2003) *J. Alloys Compd.*, **356–357**, 73–79.

152 Van Gelderen, P., Bobber, P.A., Kelly, P.J., Brocks, G. (2000) *Phys. Rev. Lett.*, **85** (14), 2989–2992.

153 Van Gelderen, P., Bobbert, P.A., Kelly, P.J., Brocks, G., Tolboom, R. (2002) *Phys. Rev. B*, **66**, 075104.

154 Alford, J.A., Chou, M.Y., Chang, E.K., Louie, S.G. (2003) *Phys. Rev. B*, **67**, 125110.

155 Wu, Z., Cohen, R.E., Singh, D.J., Gupta, R., Gupta, M. (2004) *Phys. Rev. B*, **69**, 085104.

156 Pundt, A., Getzlaff, M., Bode, M., Kirchheim, R., Wiesendanger, R. (2000) *Phys. Rev. B*, **61** (15), 9964–9967.

157 Grier, E.J., Jenkins, M.L., Petford-Long, A.K., Ward, R.C.C., Wells, M.R. (2000) *Thin Solid Films*, **358**, 94–98.

158 Kerssemakers, J.W.J., Van Der Molen, S.J., Günther, R., Dam, B., Griessen,

R. (2002) *J. Alloys Compd.*, **330–332**, 342–347.

159 Ellinger, F.H., Holley, C.E., Jr., McInteer, B.B., Pavone, D., Potter, R.M., Staritzky, E., Zachariasen, W.H. (1955) *J. Am. Chem. Soc.*, **77**, 2647–2648.

160 Nagengast, D.G., van Gogh, A.T.M., Kooij, E.S., Dam, B., Griessen, R. (1999) *Appl. Phys. Lett.*, **75** (14), 2050–2052.

161 Giebels, I.A.M.E., Isidorsson, J., Griessen, R. (2004) *Phys. Rev. B*, **69**, 205111.

162 Alford, J.A., Chou, M.Y. (2003) *Bull. Am. Phys. Soc.*, **48** (1), 300.

163 Krozer, A., Kasemo, B. (1990) *J. Less-Common Met.*, **160**, 323–342.

164 Rydén, J., Hjörvarsson, B., Ericsson, T., Karlsson, E., Krozer, A., Kasemo, B. (1989) *J. Less-Common Met.*, **152**, 295–309.

165 Westerwaal, R.J. (2002) Optimization of the Mg based switchable mirror. Master's thesis, Vrije Universiteit Amsterdam.

166 Westerwaal, R. (2007) Growth, microstructure and hydrogenation of Pd-catalyzed complex metal hydride thin films, PhD Thesis, Vrije Universiteit.

167 Ouwerkerk, M. (1998) *Solid State Ionics*, **115**, 431–437.

168 Di Vece, M., van der Eerden, A.M.J., van Bokhoven, J.A., Lemaux, S., Kelly, J.J., Koningsberger, D.C. (2003) *Phys. Rev. B*, **67**, 035430.

169 Von Rottkay, K., Rubin, M., Duine, P.A. (1999) *J. Appl. Phys.*, **85** (1), 408–413.

170 Sun, S.N., Wang, Y., Chou, M.Y. (1994) *Phys. Rev. B*, **49** (10), 6481–6489.

171 Giebels, I.A.M.E., Van der Molen, S.J., Griessen, R., Di Vece, M. (2002) *J. Alloys Compd.*, **80** (8), 1343–1345.

172 Armitage, R., Rubin, M., Richardson, T., O'Brien, N., Chen, Y. (1999) *Appl. Phys. Lett.*, **75** (13), 1863–1865.

173 Griessen, R. (1997) *Phys. Blätter*, **53** (12), 1207–1209.

174 Brown, R.J.C., Brewer, P.J., Milton, M.J.T. (2002) *J. Mater. Chem.*, **12**, 2749–2754.

175 Wu, C., Crouch, C.H., Zhao, L., Carey, J.E., Younkin, R., Levinson, J.A., Mazur, E., Farell, R.M., Gothoskar, P., Karger, A. (2001) *Appl. Phys. Lett.*, **78** (13), 1850–1852.

176 Isidorsson, J., Giebels, I.A.M.E., Kooij, E.S., Koeman, N.J., Rector, J.H., van Gogh, A.T.M., Griessen, R. (2001) *Electrochim. Acta*, **46**, 2179–2185.

177 Giebels, I.A.M.E., Griessen, R. (in preparation)

178 Ouwerkerk, M., Janner, A.-M., Van Der Sluis, P., Mercier, V.M.M. (XXXX) Optical switching device. United States patent application 20030227667.

179 Giebels, I.A.M.E., Lohstroh, W., Bjurman, M., Griessen, R. (unpublished)

180 Hjörvarsson, B. (private communication).

181 Bruggeman, D.A.G. (1935) *Ann. Phys. (Leipzig)*, **24**, 636–679.

182 Berthier, S. (1988) *Ann. Phys. Fr.*, **13**, 503–595.

183 Giebels, I.A.M.E., Isidorsson, J., Kooij, E.S., Remhof, A., Koeman, N.J., Rector, J.H., van Gogh, A.T.M., Griessen, R. (2002) *J. Alloys Compd.*, **330–332**, 875–881.

184 Van Der Sluis, P. (1998) *Appl. Phys. Lett.*, **73** (13), 1826–1828.

185 Richardson, T.J., Armitage, R.D., Slack, J.L., Rubin, M.D. (2000) Alternative materials for electrochromic mirror devices. Poster presentation at the Fourth International Meeting on Electrochromism (IME-4), August 21–23 2000, Uppsala, Sweden.

186 Soubeyroux, J.L., Fruchart, D., Mikou, A., Pezat, M., Darriet, B. (1984) *Mater. Res. Bull.*, **19**, 895–904.

187 Gavra, Z., Mintz, M.H., Kimmel, G., Hadari, Z. (1979) *Inorg. Chem.*, **18** (12), 3595–3597.

188 Genossar, J., Rudman, P.S. (1981) *J. Phys. Chem. Solids*, **42**, 611–616.

189 Schefer, J., Fischer, P., Hälg, W., Stucki, F., Schlapbach, L., Didisheim, J.J., Yvon, K., Andresen, A.F. (1980) *J. Less-Common Met.*, **74**, 65–73.

190 Zolliker, P., Yvon, K., Jorgensen, J.D., Rotella, F.J. (1986) *Inorg. Chem.*, **25** (20), 3590–3593.

191 Garcìa, G.N., Abriata, J.P., Sofo, J.O. (1999) *Phys. Rev. B*, **59** (18), 11746–11754.

192 Myers, W.R., Wang, L.-W., Richardson, T.J., Rubin, M.D. (2002) *J. Appl. Phys.*, **91** (8), 4879–4885.
193 Enache, S., Lohstroh, W., Griessen, R. (2004) *Phys. Rev. B*, **69**, 115326.
194 Mueller, W.M., Blackledge, J.P., Libowitz, G.G. (1968) *Metal Hydrides*, Academic Press, New York.
195 Blomqvist, H., Noréus, D. (2002) *J. Appl. Phys.*, **91** (8), 5141–5148.
196 Lupu, D., Sarbu, R., Biriş (1987) *Int. J. Hydrogen Energy*, **12** (6), 425–426.
197 Selvam, P., Viswanathan, B., Srinivasan, V. (1988) *J. Electron Spectrosc. Relat. Phenom.*, **46**, 357–361.
198 Fujita, Y., Yamaguchi, M., Yamamoto, I. (1989) *Z. Phys. Chem. Neue Folge*, **163**, 633–634.
199 Häussermann, U., Blomqvist, H., Noreus, D. (2002) *Inorg. Chem.*, **41**, 3684–3692.
200 Isidorsson, J., Giebels, I.A.M.E., Griessen, R., Di Vece, M. (2002) *Appl. Phys. Lett.*, **80** (13), 2305–2307.
201 Farangis, B., Nachimuthu, P., Richardson, T.J., Slack, J.L., Perera, R.C.C., Gullikson, E.M., Lindle, D.W., Rubin, M. (2003) *Phys. Rev. B*, **67**, 085106.
202 Lokhorst, A. (2006) Reflections on switchable mirror devices, PhD Thesis, Vrije Universiteit.
203 Westerwaal, R.J., Borgschulte, A., Lohstroh, W., Dam, B., Griessen, R. (2005) *J. Alloys Compd.*, **404–406**, 481–485.
204 Westerwaal, R.J., Borgschulte, A., Lohstroh, W., Dam, B., Kooi, B., ten Brink, G., Hopstaken, M.J.P., Notten, P.H.L. (2006) *J. Alloys Compd.*, **416**, 2–10.
205 Westerwaal, R.J., Slaman, M., Borgschulte, A., Broedersz, C.P., Borsa, D.M., Lohstroh, W., Kooi, B., ten Brink, G., Tschersich, K.G., Fleischhauer, H.P., Dam, B., Griessen, R. (2006) *J. Appl. Phys.*, **100**, 063518.
206 Pasturel, M., Slaman, M., Borsa, D.M., Schreuders, H., Dam, B., Griessen, R., Lohstroh, W.A. (2006) *Appl. Phys. Lett.*, **89**, 021913.
207 Völkl, J., Alefeld, G. (1978) *Diffusion of Hydrogen in Metals*, Springer-Verlag, Berlin, Vol. 28 of *Topics in Applied Physics – Hydrogen in Metals I*, Ch. 12.
208 Remhof, A., Van Der Molen, S.J., Antosik, A., Dobrowolska, A., Koeman, N.J., Griessen, R. (2002) *Phys. Rev. B*, **66**, 020101(R).
209 Remhof, A., Wijngaarden, R.J., Griessen, R. (2003) *Phys. Rev. Lett.*, **90** (14), 145502.
210 Remhof, A., Wijngaarden, R.J., Griessen, R. (2003) *J. Alloys Compd.*, **356–357**, 300–304.
211 Shendeleva, M.L. (2004) *J. Appl. Phys.*, **95** (5), 2839–2845.
212 Mandelis, A. (2000) *Phys. Today*, **53** (8), 29–34.
213 O'Leary, M.A., Boas, D.A., Chance, B., Yodh, A.G. (1992) *Phys. Rev. Lett.*, **69** (18), 2658–2661.
214 Schlapbach, L. (1992) *Surface Properties and Activation*, Springer, Berlin, **vol. 67** of *Topics in Applied Physics–Hydrogen in Intermetallic Compounds II*, Ch. 2.
215 Borgschulte, A., Westerwaal, R.J., Rector, J.H., Schreuders, H., Lohstroh, W., Dam, B., Griessen, R. (2005) *J. Alloys Compd.*, **404**, 699.
216 Borgschulte, A., Rector, J.H., Dam, B., Griessen, R., Züttel, A. (2005) *J. Catal.*, **235**, 353.
217 Borgschulte, A., Westerwaal, R.J., Rector, J.H., Dam, B., Griessen, R. (2006) *J. Catal.*, **239**, 263.
218 Borgschulte, A., Gremaud, R., de Man, S., Westerwaal, R.J., Rector, J.H., Dam, B., Griessen, R. (2006) *Appl. Surf. Sci.*, **253**, 1417–1423.
219 Gremaud, R., Borgschulte, A., Chacon, C., van Mechelen, J.M.L., Schreuders, H., Züttel, A., Hjörvarsson, B., Dam, B., Griessen, R. (2006) *Appl. Phys. A*, **A84**, 77–85.
220 Olk, C.H., Tibbetts, G.G., Simon, D., Moleski, J.J. (2003) *J. Appl. Phys.*, **94** (1), 720–725.
221 United States Department of Energy (DOE), *Hydrogen posture plan* (2004), URL http://www.eere.energy.gov/hydrogenandfuelcells/.
222 Schlapbach, L., Züttel, A. (2001) *Nature*, **414**, 353–358.
223 Gremaud, R., Broedersz, C., Borsa, D., Borgschulte, A., Mauron, P.A.,

Schreuders, H., Rector, J.H., Dam, B., Griessen, R. (2007) *Adv. Mat.* **19**, 2813–2817.

224 Rowley, N.M., Mortimer, R.J. (2002) *Science Progress*, **85**, 243–262.

225 Granqvist, C.G. (1995) *Handbook of Inorganic Electrochromic Materials*, Elsevier, Amsterdam.

226 Wittwer, V., Datz, M., Ell, J., Georg, A., Graf, W., Walze, G. (2004) *Solar Energy Mater. Solar Cells*, **84**, 305–314.

227 Bange, K. (1999) *Solar Energy Mater. Solar Cells*, **58** (1), 1–131.

228 Janner, A.-M., Van Der Sluis, P., Mercier, V. (2001) *Electrochim. Acta*, **46**, 2173–2178.

229 Mercier, V.M.M., Van der Sluis, P. (2001) *Solid State Ionics*, **145**, 17–24.

230 Lokhorst, A.C., Dam, B., Giebels, I.A.M.E., Welling, M.S., Lohstroh, W., Griessen, R. (2005) *J. Alloys Compd.*, **404–406**, 465–468.

231 Dresselhaus, M.S., Thomas, I.L. (2001) *Nature*, **414**, 332–337.

232 Granqvist, C.G., Wittwer, V. (1998) *Solar Energy Mater. Solar Cells*, **54**, 39–48.

233 van Mechelen, J.L.M., Noheda, B., Lohstroh, W., Westerwaal, R.J., Rector, J.H., Dam, B., Gricssen, R. (2004) *Appl. Phys. Lett.*, **84** (18), 3651–3653.

234 Stichting Energieonderzoek Centrum Nederland (ECN), Dutch patent application NL 1020281, International patent application PCT/NL03/00233.

235 URL: http://www.nrel.gov/buildings/highperformance/trombewalls.html

236 Borsa, D.M., Baldi, A., Pasturel, M., Schreuders, H., Dam, B., Griessen, R., Vermeulen, P., Notten, P.H.L. (2006) *Appl. Phys. Lett.*, **88**, 241910.

237 Benson, D.K., Tracy, C.E., Hishmeh, G., Ciszek, P., Lee, S.-H. (1998) Proceedings of the 1998 U.S. DOE Hydrogen Program Review, NREL/CP-570-25315.

238 Smith, R.D., Liu, P., Lee, S.-H., Tracy, E., Pitts, R. (2002) Proceedings of the 2002 U.S. DOE Hydrogen Program Review, NREL/CP-610-32405.

239 Georg, A., Graf, W., Neumann, R., Wittwer, V. (2000) *Solid State Ionics*, **127**, 319–328.

240 Georg, A., Graf, W., Neumann, R., Wittwer, V. (2001) *Thin Solid Films*, **384**, 269–275.

241 Pasturel, M., Slaman, M., Borsa, D.M., Schreuders, H., Dam, B., Griessen, R., Lohstroh, W., Borgschulte, A. (2006) *Appl. Phys. Lett.*, **89**, 021913.

242 Slaman, M., Dam, B., Pasturel, M., Borsa, D.M., Schreuders, H., Rector, J.H., Griessen, R. (2007) *Sens. Actuators B*, **123**, 538–545.

243 Notten, P.H.L., Ouwerkerk, M., van Hal, H., Beelen, D., Keur, W., Zhou, J., Feil, H. (2004) *J. Power Sources*, **129**, 45–54.

244 Bashkin, I.O., Ponyatovskii, E.G., Kost, M.E. (1978) *Phys. Status Solidi B*, **87**, 369–372.

245 Isidorsson, J., Giebels, I.A.M.E., Di Vece, M., Griessen, R. (2001) *Proc. SPIE*, **4458**, 128–137.

246 Niessen, R.A.H., Notten, P.H.L. (2005) *Electrochem. Solid-State Lett.*, **8**, A534.

8
Applications

K. Andreas Friedrich, Felix N. Büchi, Zhou Peng Li, Gerrit Kiesgen, Dirk Christian Leinhos, Hermann Sebastian Rottengruber, Robert C. Bowman Jr., and Bugga V. Ratnakumar

8.1
Fuel Cells Using Hydrogen
K. Andreas Friedrich and Felix N. Büchi

8.1.1
Introduction

It is often stated that fuel cells are a key technology of the twenty first century. Although this statement is debatable, it is certain that fuel cells are an enabling technology for a future hydrogen economy. Using pure hydrogen and air, fuel cells convert the chemical energy of the fuel directly into electricity with high efficiency, and they produce only water, thus eliminating all local emissions. The share of renewable energy from wind, water and sun will increase further, but these sources are not suited to cover the electrical base load due to their irregular availability. Therefore, power from renewable sources, in combination with production of hydrogen and fuel cells may well be an important option for future stationary and mobile power generation.

Fuel cells are not restricted to hydrogen as a fuel, but they can be operated with various other fuels (for a review on the fuel possibilities, see, for example Refs. [1–3]). However, the use of fuels containing carbon is associated with increased system complexity and carbon dioxide (CO_2) emissions. In contrast, the operation of fuel cells with pure hydrogen has the advantage of system simplicity and high system energy efficiency. In particular, the recently measured efficiency of hydrogen fuel cell powered cars is a factor of 1.5–2 higher than the hydrogen-powered internal combustion engine (ICE) vehicles (for more details see Section 8.1.7.2). Therefore, from a practical and theoretical point of view the energy conversion in fuel cells is ideally adapted for hydrogen as the fuel.

8.1.2
Brief History

The invention of fuel cells as an electrochemical energy conversion device, producing electrical energy from hydrogen and oxygen is attributed to William R. Grove, a British lawyer and scientist. The fuel cell principle however was discovered by Christian F. Schönbein, a Professor at the University of Basle from 1829 to 1868. After experimenting with hydrogen, oxygen and platinum Schönbein wrote the first publication on the fuel cell effect in December 1838 [4]. Grove, doing similar experiments with hydrogen, first mentioned his fuel cell experiments later in 1839 [5] and reported on a working fuel cell battery in 1842 [6] (Fig. 8.1). In the nineteenth century, fuel cells however did not reach any technical significance in electricity production. The lack of understanding of the basic electrochemical processes and the lack of appropriate materials prevented the development of powerful fuel cell devices. As a consequence electricity was produced more cheaply by generators, which were invented in the 1860s, a technology which soon developed into megawatt-sized plants for power production. For mobile supply of electricity, batteries were developed.

It was not until the early 1960s when fuel cells found their first application in space exploration. There hydrogen was used as the fuel for propulsion and was therefore available for electricity generation. Profound understanding of the electrochemical processes and newly developed materials allowed development of reliable and high performance fuel cell systems. As weight is one of the most critical issues in space craft, the superior gravimetric energy density of fuel cells, as compared to batteries, paved the way for this application. All manned spacecraft today generate their electricity by the use of fuel cells.

The interest in fuel cells for terrestrial applications increased in the 1980s when the ecological discussion on the use of energy and the associated emissions to

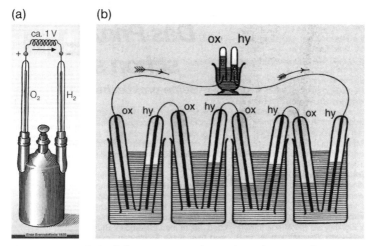

Fig. 8.1 Grove's fuel cell from 1839 (a) and "gas chain" from 1842 (b).

the atmosphere (for instance carbon dioxide) renewed interest in efficient energy conversion systems. The main drivers for the development of fuels cells are stationary and transportation applications. Decentralized stationary power generation based on fuel cells may improve overall fuel efficiency due to the possibility of the co-generation of electricity and heat. The use of fuel cells may even reduce the capital cost for the owners because distribution of heat on a small scale is easier and more efficient. However, for this application for the foreseeable future, natural gas will be the main fuel. Secondly, the increasing concern of the transportation sector over how to enable a future sustainable mobility has renewed interest in fuel cells as efficient energy converters in road transport. With hydrogen as the fuel, fuel cell-based power trains promise a considerable increase in the tank to wheel efficiency as compared to conventional internal combustion engines.

The broad renewed interest in fuel cell technology has triggered intensive research and development efforts for better materials and advanced process and production technologies. In a future hydrogen-based energy economy, fuel cells can play an important role as clean and efficient energy converters.

8.1.3
Principles and Thermodynamics

8.1.3.1 Principles
Batteries and fuel cells, both convert chemical energy directly into electrical energy and both are similar in their function. Whereas batteries store the chemical energy in the form of active electrode masses, for instance metallic lead and lead oxide in lead acid batteries, the chemical energy in fuel cells is provided in a continuous way, for example as hydrogen and air. Batteries are energy storage and conversion devices, fuel cells are only converters. This, at first glance, small difference between batteries and fuel cells leads to significant consequences for their technical application in energy conversion systems:

- Batteries are compact but tend to be heavy and are restricted in their energy density. Batteries do not allow a continuous long-term operation. Energy and power of the system are not independently scaleable.
- Fuel cells are lighter, need a storage tank for the fuel, and the size of the tank determines the operation time and the available energy. Energy and power can be scaled independently.

For smaller power requirements (e.g. flash lights, toys and so forth) or short-duration loads (for instance in car starter batteries) the inherent disadvantages of batteries are insignificant. With increasing performance and operation time, however, the small energy density of batteries is a considerable restraint and fuel cells may be better suited for the application since the size of the converter (fuel cell) and the size of the energy storage (tank) can be scaled independently.

The basic structure and principle of all fuel cells is similar: the cell consists of two electrodes which are separated by an electrolyte. The electrodes are connected through an external circuit. The electrodes are exposed to gas or liquid flows to supply fuel and oxidant (for instance hydrogen and oxygen). The electrodes have to be gas or liquid permeable and therefore possess a porous structure. The electrolyte should have a gas permeability as low as possible. For fuel cells with an acid electrolyte, hydrogen is oxidized at the negative electrode (the anode) according to the following equation. The protons formed enter the electrolyte and are transported to the cathode:

$$H_2 \rightarrow 2H^+ + 2e^- \tag{8.1}$$

At the positive electrode (cathode) oxygen reacts according to:

$$O_2 + 4e^- \rightarrow 2O^{2-} \tag{8.2}$$

Electrons flow in the external circuit during these reactions. The oxygen ions recombine with protons to form water:

$$O^{2-} + 2H^+ \rightarrow H_2O \tag{8.3}$$

The reaction product of this reaction is water, which is formed at the cathode in acidic fuel cells. It can be formed at the anode, if an oxygen ion (or carbonate) conducting electrolyte is used, as in the case of high temperature fuel cells or in the case of the liquid alkaline fuel cell (see Section 8.1.4). The reaction product water has to be removed from the cell.

An important advantage of fuel cells is the selectivity of the electrochemical reactions. Contrary to combustion processes where the reactions are controlled indirectly through the dependence of the rates on temperature and pressure, the electrochemical reactions are directly related to the cell voltage and are highly selective, that is no NO_x is produced when air is used as the oxidant at the cathode. The fuel cell itself has no moving parts and therefore it produces almost no noise. In some fuel cell systems a low noise level may arise from blowers but generally these systems are comparatively silent.

8.1.3.2 Thermodynamics

Fuel cells are galvanic cells, in which the free energy of a chemical reaction is converted into electrical energy (via an electrical current). The Gibbs energy change of a chemical reaction is related to the cell voltage via:

$$\Delta G = -nF\Delta U_0 \tag{8.4}$$

Where n is the number of electrons involved in the reaction, F is the Faraday constant and ΔU_0 is the voltage of the cell for thermodynamic equilibrium in the absence of a current flow. For the case of a hydrogen/oxygen fuel cell the overall reaction is:

$$H_2 + \frac{1}{2}O_2 \rightarrow H_2O \quad \text{with} \quad \Delta G = -237 \text{ kJ mol}^{-1} \tag{8.5}$$

For this reaction the equilibrium cell voltage ΔU_0 for standard conditions (standard temperature, pressure and activity) at 25 °C is determined by ΔG:

$$\Delta U_0 = -\frac{\Delta G}{nF} = 1.23 \text{ V} \tag{8.6}$$

In order to understand the performance of fuel cells under conditions deviating from ideal thermodynamic behavior it is necessary to discuss the relationship of cell voltage to the electrochemical potentials of the electrodes. The equilibrium cell voltage is the difference between the equilibrium electrode potentials of the cathode (positive electrode) $U_{0,c}$ and the anode (negative electrode) $U_{0,a}$, which are determined by the electrochemical reactions at the respective electrode:

$$\Delta U_0 = U_{0,c} - U_{0,a} \tag{8.7}$$

Similarly, the cell voltage under nonequilibrium conditions is always determined by the difference in electrode potential between the cathode and anode and reflects the losses which occur at both electrodes.

The open circuit voltage of a fuel cell varies with the partial pressure of the supplied reactants. The equilibrium condition is expressed by the Nernst equation which relates open circuit (OC) cell voltages and molar Gibbs energy to reactant pressure and activity (concentration) [2]. The dependence on reaction temperature is specified further below.

$$\Delta U_{OC} = \Delta U_0 + \frac{RT}{nF} \ln \left(\frac{\Pi(\text{react.\_activity})}{\Pi(\text{prod.\_activity})} \right) \tag{8.8}$$

With the activity expressed as $p' = p/p_0$, (partial pressure/standard pressure) the pressure dependence of the open circuit cell voltage can be written as:

$$\Delta U_{OC} = \Delta U_0 + \frac{RT}{nF} \ln \left(\frac{p'_{H_2} \cdot p'^{\frac{1}{2}}_{O_2}}{p'_{H_2O}} \right) \tag{8.9}$$

Therefore an increase in equilibrium cell voltage with the logarithm of the partial pressures is expected thermodynamically. However, the performance increase of fuel cells with pressure is normally much more pronounced because the kinetics of the reactions are strongly influenced by partial pressure increases (the reaction rate of the electrochemical reactions increases with higher concentrations).

8.1.3.3 Efficiency of Fuel Cells

In order to compare fuel cells with other energy conversion systems such as internal combustion engines an evaluation of the maximum thermodynamic efficiency is useful. An ICE converts chemical energy into mechanical energy, which can then be transformed into electrical energy by means of a rotating generator.

The combustion of a hydrocarbon or hydrogen (chemical energy) is accompanied by a rise in temperature as these reactions are exothermic and the reaction products are usually gases. The heat increase causes expansion of the formed gases which in their turn can produce mechanical work by causing the pistons in the ICE to move (or steam is generated to drive a steam cycle). The maximum efficiency of this system is given by the Carnot efficiency, $\varepsilon_r^{thermal}$, as follows:

$$\varepsilon_r^{thermal} = \frac{W_r}{(-\Delta H)} = 1 - \frac{T_2}{T_1} \qquad (8.10)$$

where W_r is the reversible work performed, ΔH is the enthalpy change of the reaction and T_1 and T_2 are the two absolute temperatures for the operation of the heat engine. In general, these efficiencies do not surpass 50 % for the most efficient engines (for instance steam turbines).

For a fuel cell the maximum thermodynamic efficiency can be calculated from the Gibbs energy (ΔG) and the enthalpy change (ΔH) of the electrochemical reaction. Ideally, the free energy of the reaction can be completely converted into electrical energy and the efficiency ε is given by:

$$\varepsilon_r^{cell} = \frac{W_e}{(-\Delta H)} = \frac{nF\Delta U_0}{(-\Delta H)} = \frac{\Delta G}{\Delta H} = 1 - \frac{T\Delta S}{\Delta H} \qquad (8.11)$$

where W_e is the electrical work performed, and (ΔS) is the isothermal entropy change of the reaction. $T\Delta S$ corresponds to the reversible heat exchanged with the external environment. For negative ΔS, as in the case of the H_2/O_2 fuel cell, an increase in temperature causes a decrease in thermodynamic efficiency of the fuel cell whilst for heat engines (e.g. ICE) the efficiency increases with higher temperatures (Fig. 8.2).

As can be seen from Fig. 8.3 the maximum efficiencies are higher for the electrochemical energy conversion than for heat engines at temperatures below 1150 K. However, this theoretical advantage has to be realized in practical systems and therefore the losses of fuel cells under realistic operation conditions have to be taken into consideration.

As discussed above, the electrochemical oxidation of a fuel can theoretically be accomplished at very high efficiencies (e.g. 96 % for gas-phase product water or 83 % for liquid product water for the H_2/O_2 reaction at 25 °C, see Fig. 8.3) as compared to heat engines utilizing the combustion of a fuel. However, in practice, fuel cells experience irreversible losses due to resistive and reaction kinetic losses (see Fig. 8.4), and efficiencies of fuel cell stacks rarely exceed 60 % at rated load. The irreversible losses appear as heat and, for example, a 1 kW fuel cell operating

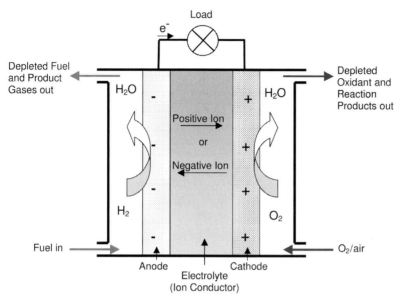

Fig. 8.2 Principle of fuel cells.

at 50% efficiency also has to dissipate 1 kW of heat. The efficiency losses due to sluggish electrode kinetics (overpotentials) and electrolyte resistance are expressed by the electrochemical efficiency which is also used to compare different fuel cells. Cells of different designs and different components can be compared by calculating the electrochemical efficiency given by:

$$\varepsilon_V = \frac{\Delta U_{cell}}{\Delta U_0} = 1 - \frac{\left(|\eta_a(j)| + |\eta_c(j)| + R_e j\right)}{\Delta U_0} \tag{8.12}$$

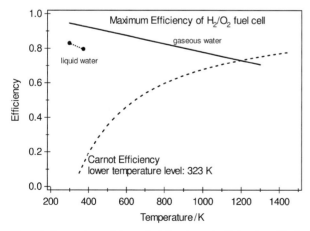

Fig. 8.3 Comparison of theoretical (maximum) efficiencies of fuel cells and heat engines.

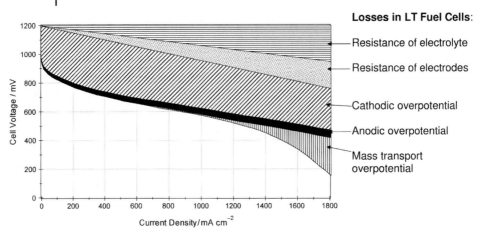

Fig. 8.4 Current–voltage characteristics of fuel cells and the associated losses.

The electrochemical efficiency gives more information about fuel cells than the thermodynamic efficiency as it is directly related to the performance of the cell. For a complete description of the fuel cell efficiency further factors associated with practical operation have to be considered. These include the faradaic efficiency ε_F which is defined as the ratio of the actual current I_{act} and the maximum possible current I_{max}. This faradaic efficiency considers the possibility of parallel reactions which can lead to a lower current yield than expected theoretically. Furthermore, the total efficiency of the cell requires the consideration of practical aspects concerned with the specific fuel used. In most cases fuel cells are not operated with 100 % fuel utilization in order to avoid fuel depletion in some areas of the electrodes. Therefore the fuel utilization should be included in the total efficiency, given as

$$\varepsilon_{fc} = \varepsilon_r^{cell} \cdot \varepsilon_V \cdot \varepsilon_F \cdot U \tag{8.13}$$

with ε_{fc} the fuel cell efficiency, ε_r^{cell} the thermodynamic efficiency, ε_V the electrochemical efficiency, ε_F the faradaic efficiency and U the utilization of the fuel.

For a fuel cell operating at 0.7 V, a power output of 0.5 W cm^{-2} requires a current density of at least 0.7 A cm^{-2}. If the open circuit voltage (OCV) is 1 V, then the allowed voltage loss of 0.3 V can be achieved only if the area specific resistance of combined ohmic and kinetic losses does not exceed $\approx$0.45 Ω cm^2.

Fuel cells, especially using low-temperature technologies, are sensitive to impurities in the gas streams. The electro-catalysts are prone to poisoning in the ppm range by adsorbates, like carbon monoxide (CO), sulfur compounds namely hydrogen sulfide (H_2S) and ammonia (NH_3). Therefore, precautions have to be taken to avoid these trace contaminants. Carbon dioxide and oxygen in the anode feed stream are responsible for an accelerated degradation of low-temperature fuel cells, but the effect is less dramatic as compared to the contaminants listed above and the power density is only affected after long-term operation. Water vapor, nitrogen and methane – apart from a dilution effect on hydrogen – do not influence significantly the performance of low-temperature fuel cells.

8.1.4
Types of Fuel Cells

Fuel cells are usually classified according to the electrolyte. Five types are currently being developed to a commercial level. All of these fuel cells operate well with hydrogen as the fuel. Depending on the nature of the electrolyte, the operating temperature ranges from 20 to 1000 °C. Dynamic operation and reliability with thermal cycling is superior for the fuel cells operated at lower temperatures (<200 °C). On the other hand the system complexity when operating with hydrocarbon feed stocks is reduced for high-temperature fuel cells. Low-temperature fuel cells are the alkaline fuel cell (AFC), the polymer electrolyte fuel cell (PEFC), the direct methanol fuel cell (DMFC) and the phosphoric acid fuel cell (PAFC). The high-temperature fuel cells operate at temperatures between 600 and 1000 °C and two different types are being developed, the molten carbonate fuel cell (MCFC) and the solid oxide fuel cell (SOFC). All types are described in this section in the order of increasing operating temperature. With pure hydrogen as the fuel, the low-temperature fuel cells exhibit major advantages regarding dynamic operation and durability, therefore these types will be discussed in more detail than the high-temperature fuel cells. Figure 8.5 shows a summary of all fuel cell types with an indication of reactants (pure hydrogen or carbon-containing fuels).

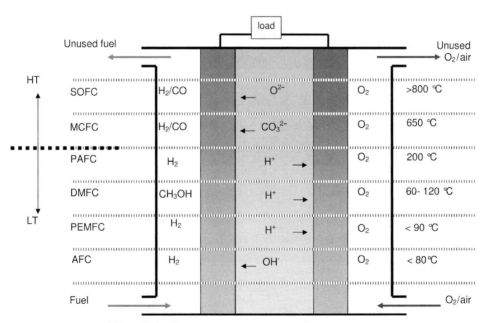

Fig. 8.5 Overview of different fuel cell types, temperature range and reactants.

8.1.4.1 Alkaline Fuel Cells

Alkaline fuel cells use an aqueous solution of potassium hydroxide as the electrolyte and operate at temperatures $\leq 80\,°C$. Although precious metal electrocatalysts are used for space applications, the AFC has the advantage that sufficient reactivity is also achieved with non-noble electrocatalysts like Ag and Ni. AFC are often operated with pure hydrogen and oxygen (or CO_2-free air) as fuel and oxidant since their electrolyte is sensitive to carbonate formation. However, the KOH electrolyte which is used in AFCs (usually in concentrations of 30–45 wt.%) has the advantage that the oxygen reduction kinetics are much faster in alkaline, than in acid electrolytes, making the AFC potentially a highly efficient system. AFCs are well suited for applications in CO_2-free environments such as space. AFCs were therefore used in the Apollo missions and the Space Shuttle program.

8.1.4.2 Polymer Electrolyte Fuel Cells

Polymer electrolyte fuel cells, also sometimes called SPEFC (solid polymer electrolyte fuel cells) or PEMFC (polymer electrolyte membrane fuel cell) use a proton exchange membrane as the electrolyte. PEFC are low-temperature fuel cells, generally operating between 40 and $90\,°C$ and therefore need noble metal electrocatalysts (platinum or platinum alloys on anode and cathode). Characteristics of PEFC are the high power density and fast dynamics. A prominent application area is therefore the power train of automobiles, where quick start-up is required.

The best-known electrolytes in PEFCs are perfluorinated sulfonic acid membranes. These membranes consist of a poly-tetrafluoroethylene- (PTFE)based backbone that is chemically inert in reducing and oxidizing environments and side chains with sulfonic acid groups. In contact with humidity, these membranes undergo a phase separation on the nm-scale into a hydrophilic (ionic aqueous) and hydrophobic (polymer backbone) phase. Proton transport takes place in the aqueous phase, similar to that in an aqueous solution. A dry membrane therefore has a low conductivity. Hence, water management in the membrane is one of the major issues in PEFC technology. Factors influencing the membrane water content are electro-osmotic water drag (for every proton part of its hydration shell is also transported through the membrane) and back transport of product water in the gradient between the anode and cathode.

8.1.4.3 Direct Methanol Fuel Cells

The direct methanol fuel cell is a special form of low-temperature fuel cells based on PE technology. In the DMFC, methanol is directly electro-oxidized at the anode without the intermediate step of reforming the alcohol into a hydrogen-rich gas. Since hydrogen is the focus of this contribution this type is not further discussed.

8.1.4.4 Phosphoric Acid Fuel Cells

The advantages of the phosphoric acid fuel cell is its relatively simple construction, its stability both thermally, chemically and electrochemically, and the low volatility of the electrolyte at operating temperatures (150–200 $°C$). These factors probably assisted their earlier deployment in commercial systems as compared to the other fuel cell types. The poly-phosphoric acid is usually stabilized in a SiC-based matrix. The high concentration of the acid increases the conductivity of the electrolyte and reduces the corrosion of the carbon-supported electrodes. PAFC need platinum-based noble metal electrocatalysts.

PAFC are mainly used in stationary power plants for distributed heat and power generation. Power plants, with outputs of single modules in the order of 200 kW, have been installed worldwide supplying, for example, shopping malls or hospitals with electricity, heat and hot water.

8.1.4.5 Molten Carbonate Fuel Cells

Molten carbonate fuel cells use a molten salt electrolyte of lithium and potassium carbonates and operate at about 650 °C. MCFCs promise high fuel-to-electricity efficiencies and the ability to consume coal-based fuels. A further advantage of the MCFC is the possibility of internal reforming due to the high operating temperatures (600–700 °C) and of using the waste heat in combined cycle power plants. The high temperature improves the oxygen reduction kinetics dramatically eliminating the need for precious metal catalysts. The molten carbonate (usually a Li–K or Li–Na carbonate) is stabilized in a matrix ($LiAlO_2$) that can be supported with Al_2O_3 fibers for mechanical strength.

Molten carbonate fuel cell systems can attain electrical efficiencies of at least 50 % or up to 70 % when combining the fuel cell with other power generators. MCFCs can operate on a wide range of different fuels and are not prone to CO or CO_2 contamination as is the case for low-temperature cells. For stationary power, MCFCs could play an important role in distributed power generation.

8.1.4.6 Solid Oxide Fuel Cells

In solid oxide fuel cells the most common material used as electrolyte is yttria-stabilized zirconia (YSZ). YSZ exhibits sufficient oxygen ion conductivity at temperatures >760 °C for the fuel cell application. The SOFC is a straightforward two-phase gas–solid system, so it has no problems with water management, flooding of the catalyst layer, or slow oxygen reduction kinetics. On the other hand it is difficult to find suitable materials which have the required thermal stability and matched thermal expansion coefficients for operating at such high temperatures. However, important advantages of SOFCs are the possibilities of internal reforming of carbonaceous fuels and the combination with other power generation systems due to the high temperature of the cell and the exhaust gases. Combined systems can reach high electric efficiencies. An efficiency of 60 % was realized by Siemens–Westinghouse by a combination of an SOFC and a gas microturbine.

Different concepts for SOFCs have been developed: Flat plates offer the possibility of easier stacking while tubular designs have smaller sealing problems. Monolithic plate and even single chamber designs have been considered and investigated.

The tubular design is probably the best-known design. It has been developed by Westinghouse (now Siemens Power generation) [8]. The first concept that was pursued by Westinghouse consisted of an air electrode supported fuel cell tube. In earlier days the tubes were made from calcium-stabilized zirconia on which the active cell components were sprayed. Nowadays this porous supported tube (PST) is replaced by a doped lanthanum manganite (LaMn) air electrode tube (AES) that increases the power density by about 35 %. The LaMn tubes are extruded and sintered and serve as the air electrode. The other cell components are deposited on this construction by plasma spraying.

A different type of SOFC design is under development by Hexis [9]. The HEXIS (heat exchanger integrated stack) stack concept can be used for small co-generation plants. The metallic interconnect in this case serves as a heat exchanger as well as a bipolar plate.

8.1.5
Active Components

8.1.5.1 Electrolytes

The principle of fuel cells is based on the spatial separation of the reaction between hydrogen and oxygen by an electrolyte. The electrolyte needs to conduct either positively charged hydrogen ions (protons) or negatively charged oxygen ions. For a technical realization the specific ionic conductivity of the electrolyte has to be in the range of 5×10^{-2}–1 S cm^{-1} and the electronic conduction of the electrolyte should be minimal. As already mentioned, the electrolyte should be gas impermeable in order to separate the reaction volumes effectively. Furthermore, a high chemical stability in oxidizing and reducing atmospheres is required. Often the electrolytes have a supporting function, leading to the necessity of good mechanical stability. Because of these stringent requirements only a few electrolyte systems are suitable for technical fuel cell applications.

The main requirement, a high specific conductivity of the electrolyte, is illustrated in Fig. 8.6 which shows the conductivity of selected electrolytes which are used in fuel cells. Materials with suitable properties are quite disparate, ranging from solid state ceramics to molten salts and aqueous electrolytes. As a consequence, the necessary ionic conductivity is obtained at very different temperatures and the specific conductivities also differ considerably, being higher for the liquids. It should be noted, however, that the important value for the fuel cell application is the area specific resistance with a target value, as rule of thumb, of <0.15 Ω cm^2. Therefore, although the specific conductivity of the oxygen ion conducting stabilized YSZ is lower than other electrolytes, this material can be integrated in a planar fuel cell with a thickness of only about 15 μm. In order to achieve a resistance of 0.15 Ω cm^2 the specific ionic conductivity should exceed 10^{-2} Ω cm^{-1}. Figure 8.6 indicates that this value is attained at about 700 °C for YSZ. The liquid electrolytes generally need a stabilizing matrix and therefore the resulting electrolyte layer is thicker. As a consequence, the specific conductivity has to be higher for these types. More details on the electrolytes are discussed in Section 8.1.4 with the different types of fuel cells.

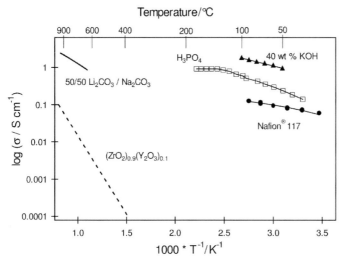

Fig. 8.6 Specific conductivities of electrolytes used in fuel cells as a function of temperature, data from Refs. [10–13].

8.1.5.2 Gas Diffusion Electrodes

The important electrochemical reactions take place at the interface between the catalyst and the electrolyte. Since reactants in fuel cells are mostly gaseous and poorly soluble in the electrolyte, a third phase, the gaseous one, has to be in contact with the interface leading to a three-phase boundary. A scheme of this zone in a PEFC is shown in Fig. 8.7. In order to optimize the kinetics in this area, the following features are required:

- highly active surface area of the catalyst
- good access of the reactants to the electrode–electrolyte interface (high area three-phase boundary)

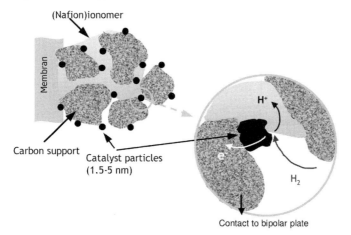

Fig. 8.7 Schematic of the active layer of a gas diffusion electrode PEFC.

- the water produced has to be removed effectively from the three-phase zone
- good electric and ionic contact of the reaction sites.

These requirements can be achieved best with a porous structure, but the realization depends strongly on the fuel cell type and on operating temperature. In particular, the solid oxide fuel cell utilizes cermet (ceramic-metal) anodes which consist of highly dispersed Ni metal mixed with YSZ particles. The YSZ component in the electrode introduces an ionic conductivity which increases the reaction zone in the electrode. In low-temperature fuel cells the same concept is used with modifications: Since costly precious metals (Pt and Pt alloys) are used as catalyst, a higher dispersion of this material is necessary in order to optimize the ratio of surface to mass. The metal is usually highly dispersed and the particle sizes are of the order of a few nanometers (platinum surface areas in the range of 20–60 $m^2\,g^{-1}$ are common). A distinction between unsupported and supported catalysts can be made. In supported catalysts the metal clusters are supported on larger carbon particles (usually about 50–1000 nm in diameter) in order to increase the active surface area, to provide electronic conductivity and to prevent agglomeration. These catalysts are introduced into the electrocatalyst layer (micropores) with proton conducting ionomer components. The ionomer content increases the catalyst utilization in the active layer and utilizations between 50 and 90 % have been reported [14, 15]. The catalyst layer has a thickness in the range 5–30 μm. The most common gas diffusion electrodes are partly hydrophobic due to the introduction of hydrophobic material (for instance PTFE) into the electrode. The hydrophobicity of the electrode helps in freeing pores for gas transport.

8.1.6
Fuel Cell Systems

8.1.6.1 Fuel Cell Stacks

The voltage of a single fuel cell at the rated point of operation is only 0.6–0.8 V. This voltage is too low for any technical application. In order to increase the voltage level to those required by the application, an appropriate number (2 to several 100) of cells are electrically connected in series.

Several concepts for this series connection have been developed. In cell designs optimized for high specific power, the series connection should exert only a minimal electric resistance to minimize losses and have low volume and weight. Generally, in these cases a bipolar type series connection is chosen, which interconnects individual cells over the entire active area and leads to filter-press type cell arrangements called stacks (see Fig. 8.8).

In the bipolar arrangement, the bipolar plate becomes the main construction element, of the fuel cell stack. Besides the performance of the electrochemical components, the characteristics of the bipolar plate determine the performance of the fuel cell stack with respect to specific power and volume. As the bipolar plate is also a significant cost contributor in the fuel cell stack, an optimized bipolar plate has to fulfil a series of requirements.

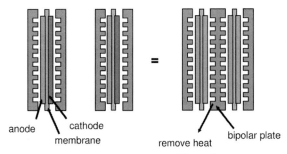

Fig. 8.8 Concept of bipolar stacking of fuel cells.

- Material requirements (for electrochemical performance):
  - electric conductivity $\geq 10\,\mathrm{S\,cm^{-1}}$
  - heat conductivity $\geq 20\,\mathrm{W\,m^{-1}\,K^{-1}}$, if the heat removed over the entire cross-section of the plate, $\gg 100\,\mathrm{W\,m^{-1}\,K^{-1}}$ if heat is removed from the edge of the plate
  - gas tightness: permeation of $H_2 < 10^{-7}\,\mathrm{mbar\,l\,s^{-1}\,cm^{-2}}$
  - corrosion resistance in contact with electrolyte, hydrogen, oxygen, heat and humidity (pure water).

    In addition to optimal bulk conductivity the contact resistance at the surface of the bipolar plate has also to be reduced as much as possible. Gas tightness is required to prevent leaks to the exterior as well as cross-leaks between the fluids in the stack.
- Construction requirements:
  - slim, for minimal volume
  - light, for minimal weight.
- Manufacturing requirements:
  - inexpensive material
  - short production cycle, for minimal manufacturing costs

Depending on operation temperature the material chosen for the bipolar plate is graphite (or graphite composites) or iron-based alloys (stainless steel) for low- and medium-temperature fuel cells (PEFC, PAFC, MCFC) and chromium or ceramic-based materials for the high-temperature systems (SOFC). In SOFC, for sealing reasons, tubular concepts have also been developed (see Section 8.1.4.6).

A polymer electrolyte fuel cell stack with 100 cells and graphite-based bipolar plates is shown in a partly expanded view in Fig. 8.9.

8.1.6.2 Fuel Cell System Designs

The fuel cell or fuel cells stack itself is only the electrochemical reactor, converting the chemical energy of the hydrogen fuel into electricity. This means that the reactants need to be supplied to the fuel cell and the products and the waste heat have to be removed from it. Therefore, the fuel cell itself needs to be complemented by auxiliary components which take care of supply and evacuation of the reactants,

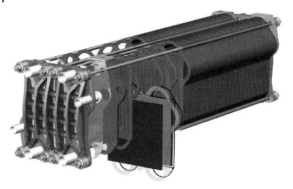

Fig. 8.9 Polymer electrolyte fuel cell stack with 100 cells and graphite-based bipolar plates. Partly expanded view for one cell with electrochemical components.

products and waste heat as well as control of all subsystems. The fuel cell together with these system components is called a "fuel cell system." Only a complete system is capable of delivering electric power to the user. The layout, design, arrangement and integration of such a system depend very much on the power and power density required by the user and the boundary conditions of the application.

Generally for low power fuel cell systems (0–10 W) the system is designed to be as simple as possible. The simplest possible fuel cell system consists of the hydrogen tank (or hydrogen generator), a valve, some tubing and the fuel cell. Figure 8.10 shows the concept of such a simple, low power system. No controls are needed as the system starts up by simply opening the valve. In this system no means for the transport of oxygen to the catalytic sites at the cathodes is present and the system relies on the passive diffusion of oxygen from the surrounding air. Removal of product water similarly occurs by diffusional transport out of the cell. Waste heat is also removed passively through radiation and natural convection on the surface of the stack.

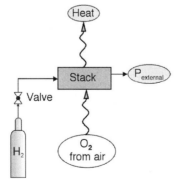

Fig. 8.10 Schematic of low power fuel cell system.

Simplicity is the advantage of such a system but in turn, because of the unforced oxygen transport, the achievable power density of such passive stacks is comparatively low. Electrochemical power densities of the order of 20–50 mW cm^{-2} active area allow for specific power densities of the stacks of the order of 20–100 W kg^{-1}. The lateral dimensions of the oxygen diffusion path are also restricted to a few centimeters and therefore cell areas are limited to a few tens of square centimeters. Consequently single cell power is limited to a few watts.

For the complete system, including the tank, the power density is even lower. However, with respect to specific energy, the systems can compete with batteries, a 20 W system (at 50 W kg^{-1}) including a metal hydride hydrogen tank of 1 kg (at 1 % H$_2$-storage) has a specific energy of more than 100 Wh kg^{-1} (at an efficiency of 57 % LHV). Using chemical hydrides with their high specific energy as the hydrogen storage, the specific energy of the system can be considerably increased.

Naturally, for more powerful fuel cell systems (1–100 kW) such as developed for the power train of cars, the passive approach of the low power systems is not feasible. Powerful systems require fuel cell stacks with power densities of the order of 1 kW kg^{-1} to be competitive with internal combustion engines. In such stacks active areas of several hundred square centimeters with current densities of over 1 A cm^{-2} are required. This leads to the need for forced air supply by a blower or compressor and also for very efficient heat removal as the heat load to be removed is also in the order of 1 W cm^{-2} of active area. Therefore, the complexity of such fuel cell systems increases considerably. Generally, at least four subsystems complementing the fuel cell stack are required:

- *Fuel subsystem*, including tank for hydrogen, piping, valves and controls to deliver the hydrogen fuel at the required pressure mass flow and humidity to the fuel cell stack.
- *Air supply subsystem*, including blower or compressor, air humidification, heat exchanger and pressure control devices to deliver process air at the required mass flow, pressure temperature and humidity to the fuel cell stack. Separation of water from the used air stream and, in some cases, an expander add more complexity to the air supply subsystem;
- *Cooling subsystem,* including pump, heat exchanger and conductivity control for the coolant. In mobile systems the heat is dumped unused into the ambient air, in the case of stationary systems, the waste heat of the fuel cell stack can be used or distributed.
- *Control subsystem*, including sensors for temperatures, pressures, relative humidities, a microprocessor-based control for the dynamic regulation of mass and heat flows, pressures and temperatures.

Figure 8.11 shows a (simplified) schematic of a mobile PEFC system with the three subsystems for fuel, oxidant and cooling.

In most applications, as described in the two examples above, oxygen from air is used as the oxidant in the fuel cell. However, especially in applications where no air is available, such as in submarines and for space applications, pure oxygen is used.

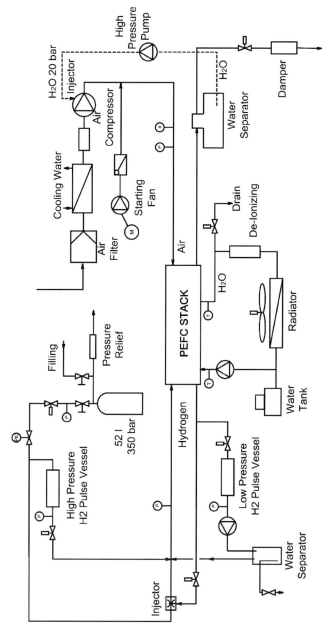

Fig. 8.11 Schematic of fuel cell system for automotive application with fuel (left), air (top right) and cooling (bottom right) subsystems.

In this case the oxidant supply subsystem becomes similar to the fuel subsystem, as oxygen is also fed from a tank to the stack and needs to be recycled to evacuate the product water from the stack.

The major applications for fuel cell systems are described in Section 8.1.7.

8.1.6.3 Efficiency Considerations

Hydrogen is a valuable energy carrier; therefore it should be used with the best possible efficiency.

The efficiency of the fuel cell or the fuel cell stack has been given in Section 8.1.3.3, as a function of cell voltage, faradaic efficiency and fuel utilization.

In small passive systems, without electrical consumers in the subsystems, the electrical efficiency of the fuel cell system equals that of the fuel cell stack, at rated load typically 50 to 60%.

However in most systems, the efficiency of the fuel cell system is lower than that of the stack itself, because of the power requirements of the peripheral consumers in the system (compressor, fans, valves, microprocessor). The parasitic power reduces the electrical output of the system, and therefore its efficiency is less than the one of the stack. Figure 8.12 shows a typical Sankey diagram for the power flows of a 25 kW PEFC system at rated load. Here the calculation is made from the chemical energy in the tank to the unregulated DC electrical outlet. If the user requires an AC voltage or a regulated DC voltage then the total efficiency is further reduced, depending on the efficiency of the electrical converter.

For the example in Fig. 8.12, the electrochemical efficiency is 57% (LHV) and the fuel utilization is 99%, therefore the fuel cell stack has an efficiency of 56.4% (see Eq. (8.13)). For an electric power of the stack of 29.6 kW, a hydrogen energy

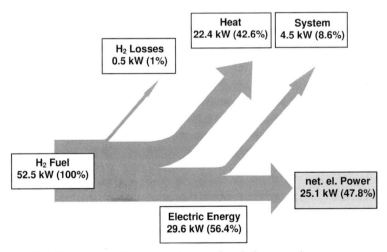

Fig. 8.12 Efficiency of a 25 kW fuel cell system from hydrogen in the tank to unregulated DC at full load.

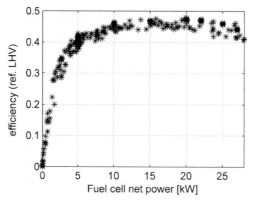

Fig. 8.13 Efficiency as function of power for the 25 kW fuel cell system.

inflow into the stack of 52.5 kW (or 0.44 g s^{-1}) is needed. The fuel-, air- cooling- and controlling-subsystems at this point of operation require 4.5 kW for air compression, cooling pumps, valves, sensors and controls. The system power output is therefore reduced to 25.1 kW and the total efficiency is 47.8 %.

For all fuel cell systems the efficiency is a function of the load. At low load the electrochemical efficiency of the fuel cell stack itself is highest, but except for systems with no parasitic load (as in Fig. 8.10), the overall system efficiency is low because of the base load of the subsystems. With increasing power the fraction of parasitic power decreases and the system efficiency is dominated by the electrochemical efficiency of the fuel cell stack.

In pressurized systems, as in Fig. 8.12, air compression is the main parasitic load. As the voltage at a given current increases with gas pressures, the efficiency and power density of the stack increases with increasing pressure. The gain, however, is reduced or offset by the power used by the compressor. The optimization of the fuel cell system with respect to efficiency is a complex task and has to be fit precisely to the requirements of the application (Fig. 8.13).

In the case of stationary systems, where 80 to 90 % of the heat is also extracted from the fuel cell system as valuable energy (and not considered waste) the total (electric and heat) efficiency of the combined heat and power generation can achieve >80 %.

8.1.7
Applications

8.1.7.1 **Portable**
The main advantage of a fuel cell system for portable applications is the flexible sizing of the energy conversion device and the energy storage device. Furthermore, the modular design of the fuel cell can be used to adjust the system to the power needs of the device. The energy density of such a system is high if the power demand is small and the energy demand is large. In an ideal situation the energy

density of the system is then dominated by the energy density of the fuel storage system.

Suitable applications, therefore, require long operation times at low power. Such applications are listed in the Table 8.1.

The portable application is attractive because the number of possible units is high and the cost limitations seem to be met more easily than for fuel cell systems in mobile or stationary applications [17–20]. This is especially true for small fuel cells for consumer applications.

For small power applications like laptops, camcorders and mobile phones the requirements of the fuel cell systems are quite specific: low temperatures are advantageous since the energy conversion system has to be integrated into a consumer electronic device and also an adequate load-following characteristic of the fuel cell is necessary. An option is the development of hybrid battery/fuel cell systems. There is now a growing pressure on battery manufacturers to further increase the energy density for the next generation (3G and 4G) of portable electronic equipment, which will require much higher power and energy densities to allow for new features like color displays and integrated computing. It is assumed that mobile phones and notebook computers will merge to provide users with broadband wireless and multifunctional portable computing capabilities. Due to potentially higher energy density, fuel cells appear to be the best technology for battery replacement, but several issues have to be resolved. The challenges of fuel cell development in micropower applications are associated with the complexity of the fuel cell system: fuel cell operation involves the management of electricity but also the management of the chemical media (fuel, water, air) as well as thermal management. As a consequence the fuel cell system is much more complex than a battery which is a sealed rechargeable power source without "interaction" with the ambient.

8.1.7.2 Automotive

Growing global concern about environmental problems caused by the intensive use of fossil fuels for road transport, as well as their limited availability, has triggered a large research and development effort for more efficient and cleaner power train technologies in order to reduce emissions from the transport sector, which account for about 26 % of the energy consumption and CO_2 emissions worldwide [21].

Conventional power trains for passenger cars based on internal combustion spark or compression ignition engines (ICE), today have a tank to wheel efficiency of only 19–21 % when benchmarked on standardized cycles, such as the New European Driving Cycle (NDEC) [22]. Therefore, there is a significant potential for CO_2 emission reduction through efficiency increase.

Electric power trains are locally emission free, however the energy storage, with today's battery technology, is limited in energy density to about 150 Wh kg^{-1} if advanced lithium batteries are used and refueling time takes a few hours. So the range and availability of the vehicle is lower than for cars with ICE power trains.

Fuel cell technology provides the potential for a true local zero emission power train if hydrogen is used as the fuel in combination with higher energy densities and fast refueling. The energy density is mainly limited by the energy density of

Table 8.1 Overview of the potential applications of portable fuel cells

Field of Application	Power Range	Devices	Solution Today	Critical Issues
Grid independent power	1–5 kW	E.g. small Uninterruptible power supply, UPS, small remote stationary systems (energy supply for signaling or weather stations)	Generator driven by ICE Lead acid battery	Noise, Emissions
Mobile communication	Up to 3 W	Cellular phone, cordless phones, pager	Secondary batteries	Time of operation, Volume
Entertainment	Up to 15 W	Discman, walkman, radio, tape recorders, CD-player, musical instruments	Primary batteries	Time of operation, volume
Computing	Up to 30 W	Notebook, PDA, organizer, calculator, dictating machine	Secondary batteries and primary batteries	Time of operation, volume
Power tools	Up to 1 kW	Drill, screwdriver, grinder, vacuum cleaner, angle sander, hedge clipper, pumps etc.	Secondary batteries	Time of operation, weight
Image processing	Up to 100 W	Consumer and professional Camcorder, digital camera	Secondary batteries	Time of operation, noise
Illumination	Up to 30 W	Flash light, emergency or warning light and signs, pocket lamp	Solar energy, primary and secondary batteries	Time of operation, noise
Toys	0.5–100 W	Model vehicles or aircrafts, game boy, etc.	Primary and secondary batteries	Time of operation, maintenance-free
Medical products	0.5–500 W	electric wheel chair, emergency medical equipment, portable devices	Secondary batteries	Time of operation, volume + weight
Scientific instruments	Up to 50 W	Oscilloscopes, multimeters, meteorological stations	Central electricity grid, primary batteries	Time of operation, volume + weight
Military applications	15–50 W	Central power source, sonar, radio set, noctoviser, radar, power supply for weapon systems	Secondary batteries	Time of operation, volume + weight, self discharge
Sensors	1 W	Glucose, gas, etc	Primary batteries	Volume, self discharge

the hydrogen storage to 300–500 Wh kg^{-1} which depends on the technology and size of the hydrogen tank. Tank to wheel efficiencies of the order of 40 % are expected [22].

However, the application of fuel cell systems in passenger cars is one of the most demanding applications, because of the stringent requirements with respect to limited space and weight. Power density of 500–1000 W kg^{-1} and gravimetric and volumetric energy density of 500–1000 Wh kg^{-1} and Wh l^{-1} are needed. Further, and probably most challenging, is the low cost (<100 € kW^{-1}) required to be competitive with established ICE technology.

With a possible huge market for automotive power trains (>50 billion € a^{-1}) all major automotive companies are developing fuel cell technology based on polymer electrolyte fuel cells for this application. While most technical goals such as power and energy densities can be met, the reduction of cost remains the major issue (see also Section 5.1.8.1). Prototype vehicles with fuel cell power trains have been presented by all major companies [23–25]. In particular, demonstration and concept cars with fuel cell propulsion systems have been presented by Daihatsu, DaimlerChrysler, Fiat, Ford Motor Co., GM/Opel, GM(Shanghai)/PATAC, Honda, Hyundai, Mazda, Mitsubishi, Nissan, PSA/Peugeot/Citroen, Renault, Suzuki, Toyota and Volkswagen.

Two different approaches are being pursued for the fuel cell-based power trains. First, a straight fuel cell power train is advocated and, secondly, an electric hybrid power train, combining a fuel cell system and an electric energy storage device such as a super capacitor or a battery is proposed (see Fig. 8.14). The advantage of the straight set-up is its simpler structure, however in this concept the fuel cell system

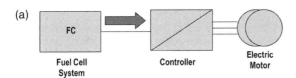

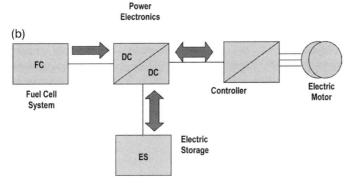

Fig. 8.14 (a) Straight fuel cell powertrain. (b) Hybrid electric power train with fuel cell and electric storage.

has to provide the fast dynamics and peak power for acceleration needed in a car. Further, the straight fuel cell concept does not allow recovery of braking energy.

Hybrid power trains allow one to combine the power of the fuel cell system and the energy storage device. Under conditions of high power (i.e. during acceleration), the power of the fuel cell system and the energy storage device can be combined. The battery or super capacitor energy storage provides the high dynamic power needs of the propulsion system and therefore does not require a highly tuned fuel cell system. If the storage efficiency is high (>90%), such as for a super capacitor, the breaking energy can be recovered in the hybrid concept. As a consequence, the efficiency of the power-train can be increased by 10–20% [26]. Hybrid systems however have a higher complexity because of the additional storage device and the controllers. Figure 8.15 illustrates the components of a straight fuel cell system concept (A) in comparison to a hybrid concept (B) with a super capacitor. Figure 8.15A shows the schematic of a concept car by DaimlerChrysler, the NECAR 4, which is powered by a PEFC system operated with liquid H_2. An efficiency of 37.7% (tank to wheel) in the New European Driving Cycle (NEDC) was measured for this prototype which corresponds to 53.46 mpg equiv. (4.0 l/100 km). Similarly the Opel HydroGen3 has been reported to exhibit an efficiency of 36% in the NEDC [27]. Figure 8.15B displays the arrangement of the components in an experimental hybrid fuel cell vehicle with a super capacitor.

In conclusion, many development challenges exist and have to be overcome to realize the potential of fuel cell-based power trains for automotive application under the conditions of daily vehicle operation. In particular, new materials have to be developed for improved component performance at lower cost. In addition the performance of advanced power trains based on internal combustion engines is a moving target, as here also further advancements with respect to efficiency are technically possible and probable.

8.1.7.3 Stationary

Today most electric power is produced in large centralized installations (>250 MW) and about 50% of the primary energy is converted into heat. Heat, however, is much more difficult to distribute than electricity. Therefore in the majority of installations heat is treated as waste or only a small fraction is used. Heat from low-power, decentralized power stations can be better utilized. But the combustion technologies have much lower electric efficiencies at lower power (i.e. below 100 kW), and therefore fuel cells in distributed combined heat and power generation (CHP) are an attractive alternative. The electric efficiency of fuel cell systems is sufficiently high (>30%) even at low power and also the heat can be utilized, so that the total efficiency can reach values >80% [29, 30].

With a hydrogen distribution infrastructure, decentralized power generation with fuel cells would have many advantageous attributes. At present and in the near future, however, hydrogen will not be available as fuel for distributed power, therefore systems for stationary co-generation are developed mainly for natural gas as the fuel. Natural gas cannot be converted directly in fuel cells but has to be reformed: hydrogen is extracted in a chemical reaction, which requires temperatures

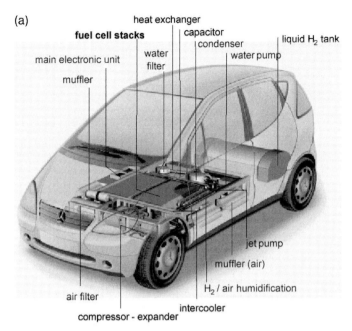

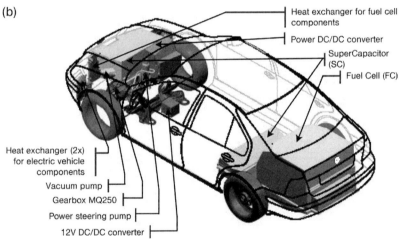

Fig. 8.15 (A) Schematic of concept car NECAR 4 representing a fuel cell vehicle with a PEM system operated with liquid H_2, ([22]); (B) Passenger car with hybrid fuel cell and super capacitor power train [28].

above about 700 °C. High-temperature fuel cell technologies (SOFC, MCFC) seem therefore better suited, because the reforming process can be integrated thermally.

In stationary applications the demanding requirement for fuel cell technology is life time. Minimum operation times of 40 000 h are required for combined heat and power applications.

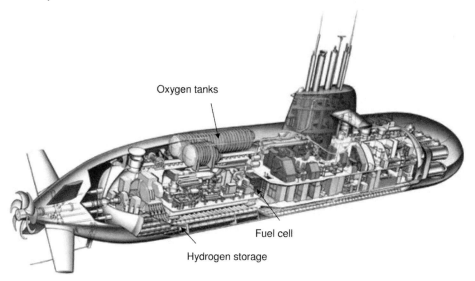

Fig. 8.16 Schematic of submarine U 31 (class 212A), a submarine from HDW with a hydrogen based fuel cell propulsion system, (Source HDW).

8.1.7.4 Special Applications

The applications described in Sections 8.1.7.1 to 8.1.7.3 are expected to have the potential to become high volume products. However, numerous niche applications with more relaxed cost requirements as compared to the mass markets are available. Electric power can be provided where there is no electric grid connection such as for light transportation (electric bikes, wheelchairs), as auxiliary power units (trucks, boats), remote area power supply (mountains, camping), for appliances (lawn mower, vacuum cleaner), for military and defence purposes, space travel or for electric grid back-up where availability is crucial (telecommunication, computing).

An example of such a grid-independent application is a hybrid system combining fuel cells and electrolysers as energy converters with renewable energy, namely wind or solar energy generators. Although the conversion of electricity into hydrogen and back into electrical energy by means of fuel cells is energy inefficient (efficiency is about 20 %), the combination of photovoltaic modules (PV) with independently operated fuel cells on the basis of gaseous hydrogen may have the advantage that the expensive PV system does not have to be scaled according to the month with the lowest solar yield. Furthermore, the reliability and availability is significantly improved because two parallel electricity generators are present.

While commercialization in the three high-volume application areas is not imminent, in some niche applications, such as back-up power or uninterruptible power supply, commercial products have recently become available. In these areas, the allowed cost of power is considerably higher and the required operation time is much less than in CHP applications. On the other hand, requirements with respect to power and energy density are lower than for mobile applications.

Fuel cells have the longest record of application in space exploration. Since the Gemini missions of NASA, fuel cells have been an indispensable technology in manned space travel. The fuel cell systems used in the NASA space shuttles are based on alkaline fuel cells from UTC. Each orbiter has three systems of 7 kW$_e$ each, which produce all electric power on board and provide potable water. These systems are hydrogen and oxygen fueled.

The military applications are diverse and numerous, ranging from soldier portable systems to air-independent propulsion systems for submarines, and cannot be explored in full detail here. In military submarines PEFC systems of several hundred kW are used as part of the electric drive trains for silent submerged cruising [31, 32]. After nearly two decades of development, a fuel cell powered submarine was presented by the Kiel shipyard Howaldtswerke Deutsche Werft AG (HDW) in 2002 (Fig. 8.16). The submarine's propulsion system operates in almost total silence and produces very little exhaust heat – two factors that make the submarine very difficult for enemies to detect. Because of the system's high efficiency, the submarine can stay under water for a considerably longer period than conventional submarines with diesel-electrical propulsion. The Siemens polymer electrolyte fuel cells in the HDW submarine are powered by hydrogen contained in a metal hydride tank on board the vessel whereas oxygen is stored as a liquid on board. The system consists of nine PEFC fuel cell stacks, providing 30 to 50 kW each.

8.1.8
Future Developments

Fuel cells have many potential benefits; however, only few commercial applications have been realized to date. For commercial success, fuel cell technology must compete with existing technologies, such as batteries in the field of portable devices or internal combustions engines in the automotive application. Besides technical superiority, fuel cell systems need to be competitive with existing technologies in cost and durability. These two important factors for future development are discussed in more detail below.

8.1.8.1 Cost
The costs which are accepted by the market for different applications are indicated in Fig. 8.17. The market introduction of fuel cells will be determined strongly by these costs requirements as fuel cells, similar to all new technologies, are cost intensive. Considering just these costs requirements, a market penetration of fuel cells as electrical power supply for consumer electronics (so-called 4-C area cellular phones, camcorders, computers, cordless tools) would be easiest. However, technological challenges still hamper fuel cells today in replacing batteries in these applications.

An earlier market introduction is expected in niche areas requiring larger portable systems, for instance, the independent electrical energy supply of sailing boats,

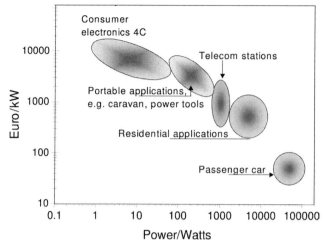

Fig. 8.17 Costs of energy supply devices as presently accepted by the market.

caravans or telecom stations. Later, co-generation applications in the residential sector are expected to become commercial. The last market introduction is now expected with the automotive sector, because there the cost and technological requirements are the most stringent.

In a detailed study on behalf of the US Department of Energy (DOE) Arthur D. Little Inc. (ADL) presented the cost structure of a gasoline-driven fuel cell system [33]. The analysis is based on the assumption of a 50 kW system with an annual production of 500 000 units. Under these conditions the total cost was estimated at $325/kW. The fuel cell subsystem was responsible for 67 % of this cost.

The price of the fuel cell stack is determined by the cost of the materials and the production technology. These costs are strongly determined by the number of units produced. For various components a strong decrease in the price can be expected as the demand increases. As a consequence, reliable cost estimations are difficult. However, the material expenses and some estimates of production costs can be compared with the actual market prices of the competing technologies, namely primary batteries, rechargeable batteries and the internal combustion engine. The comparison shows that fuel cell systems are still too expensive for most applications and a reduction of the cost by a factor of 10–50 is necessary to achieve competitiveness.

The present market situation is characterized by a small number of units and hand fabrication. Figure 8.18 compares the present situation of the stack production by a research institution with the prediction of ADL by plotting the cost distribution of PE fuel cell stacks. The present cost for the manufacture of a single unit is about a factor of 10–50 more expensive than the values of the ADL analysis. As can be seen, the importance of bipolar plates is presently about equal to that of the MEA and labor-intensive tasks like assembling and quality control contribute significantly to the cost of the stack system. These are the cost drivers that can be

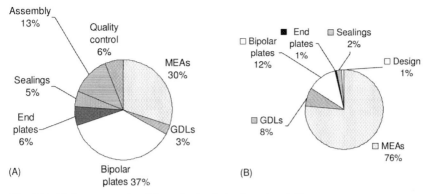

Fig. 8.18 (A) Present cost breakdown for the fuel cell stack and (B) the prediction of Arthur D. Little for 500 000 units/yr [33].

effectively reduced by a larger scale production; however, in this case the relative importance of the basic MEA components increases. Therefore, substantial efforts to lower the material costs of membranes and electrocatalysts through new material development are of paramount importance.

8.1.8.2 Lifetime

Daily use requires stringent reliability and lifetime characteristics for successful competition with existing power generation technologies. The durability requirements of a fuel cell stack in stationary combined heat and power applications is at least 40 000 h. The durability requirement in the transportation application for passenger cars is 5000 h over 10 years under severe climatic, on/off and transient cyclic conditions. The life/reliability characteristics of fuel cell technologies have not been verified satisfactorily, although 200 kW PAFC power plants have demonstrated acceptable reliability and availability [34] in extensive worldwide demonstrations. Furthermore, there is a clear lack of understanding of the fundamental degradation mechanisms in all fuel cell systems. Due to the length of time needed for testing and the tremendous importance of this data, this issue will receive much attention for a long period to come.

8.1.9
Concluding Remarks

Fuel cells can make a valuable contribution to a future energy economy and facilitate the establishment of a hydrogen-based energy system. They improve the flexibility of electrical power generation and increase the options for many applications, like distributed stationary power generation, vehicle propulsion and portable devices. Their main property is high electrical efficiency. The modularity of fuel cells makes this technology flexible as the required power can easily be met by changing the number of modules. Both the low- and high-temperature fuel cells have advantages and disadvantages, depending on the requirements of the application.

8.2
Borohydride Fuel Cells
Zhou Peng Li

8.2.1
Introduction

Hydrogen is a clean energy resource. It generates energy by chemical oxidation (combustion) or electrochemical oxidation (fuel cell). Fuel cells constitute an attractive electricity generation technology that directly converts chemical energy into electricity with high efficiency compared to the hydrogen combustion engine. They are being developed to run vehicles, portable and mobile devices. The polymer electrolyte membrane fuel cell (PEMFC), the alkaline fuel cell (AFC) and the phosphoric acid fuel cell (PAFC) require gaseous hydrogen as the fuel. However, hydrogen storage technologies still do not match practical needs. Compressed hydrogen is considered as a feasible solution to vehicle applications but is not suitable for portable devices due to its lower volumetric energy density. Using liquid fuel is considered to be a solution to improve the volumetric energy density, but it needs a fuel modification process that makes the fuel supply complicated. The direct methanol fuel cell (DMFC) using methanol as the fuel, is considered as a promising candidate for portable and mobile applications but its lower performance is a high hurdle to practical use. Besides carbon hydrogen fuels, boron hydrogen compounds (borohydrides) are considered to be potential fuels for fuel cells [35–41].

Borohydrides are a group of compounds with high hydrogen content. For example, $NaBH_4$ contains 10.6 wt% hydrogen which is much more than that of most hydrogen storage alloys. Although hydrogen gas as a fuel can be easily obtained from the hydrolysis reaction of borohydride ion, the direct anodic oxidation of borohydride ion provides more negative potentials than that of hydrogen gas. Some electrochemical oxidation mechanisms of borohydride ion have been reported [38, 39, 42–45]. Though these mechanisms were somewhat different because different electrode materials were used, the direct electricity generation from borohydride ion has been proven to be practical. The energy density can reach 9.3 Wh/g-$NaBH_4$ if using sodium borohydride as the fuel, which is higher than that of methanol (6.1 Wh/g-MeOH). In addition, borohydride solution can be used as a heat exchange medium to cool the fuel cell so that cooling plates are not needed. Furthermore, the electro-osmotic drag of water can be used as the cathode reactant so that the humidifier for hydrogen gas (PEMFC) or air (AFC) can be eliminated. These features are of benefit to the micro-fuel cell design. Nowadays, the direct borohydride fuel cell (DBFC) is considered as a potential candidate for portable and mobile applications.

8.2.2
Fundamental Aspects of the DBFC

8.2.2.1 Electrochemistry of the DBFC

There are several features common to the DBFC. The electrochemical reactions take place in an alkaline medium because borohydride ions are not chemically stable in acidic media. Thus, in principle, the electrode reactions and cell reaction for the DBFC can be described as follows:

$$\text{Anode}: BH_4^- + 8OH^- = BO_2^- + 6H_2O + 8e^- \quad E_a^0 = -1.24\,V \quad (8.14)$$
$$\text{Cathode}: 2O_2 + 4H_2O + 8e^- = 8OH^- \quad E_c^0 = 0.40\,V \quad (8.15)$$
$$\text{Cell reaction}: BH_4^- + 2O_2 = BO_2^- + 2H_2O \quad E^0 = 1.64\,V \quad (8.16)$$

where E_a^0 is the standard anode potential; E_c^0 the standard cathode potential and E^0 the electromotive force of the DBFC. It is very impressive that one ion of BH_4^- can generate 8 electrons and E^0 can reach 1.64 V, which is 1.33 times that of the PEMFC and 1.35 times that of the DMFC.

8.2.2.2 Thermodynamics and Electrode Kinetics of the DBFCs

Like other fuel cells, the DBFC is an electrochemical system in which the chemical energy of the borohydride is converted directly into electrical energy with the aid of oxygen or air. The conversion takes place with a reversible change in thermal energy ($T\Delta S$) where T is the absolute temperature in K and ΔS is the entropy change of the cell reaction. The energy conversion efficiency (η) can exceed the Carnot restriction but cannot reach 100%. The energy conversion efficiency is calculated from the Gibbs energy change (ΔG) and the enthalpy change (ΔH). The theoretical energy conversion efficiency of a fuel cell can obtained from the standard Gibbs energy change (ΔG^0) and the standard enthalpy change (ΔH^0) of the cell reaction. Figure 8.19 shows the thermodynamic comparison of the PEMFC and the DBFC. The thermodynamic calculation shows that the theoretical energy conversion efficiency of the DBFC (0.91) is larger than that of the PEMFC (0.83).

However, the cell reaction of the fuel cell is irreversible. This irreversibility is mainly caused by four major losses: activation loss related to electrode catalysts,

PEMFC	DBFC

• Cell reaction
  $H_2(g) + \tfrac{1}{2}O_2(g) = H_2O(l)$
  $\Delta H^0 = -285.83\,kJ\,mol^{-1}$
  $\Delta G^0 = -237.13\,kJ\,mol^{-1}$
• Electromotive force
  $E^0 = -\Delta G^0/nF = 1.23\,V$
• Theoretic energy conversion efficiency
  $\eta = \Delta G^0/\Delta H^0 = 0.83$
	• Cell reaction
	  $BH_4^-(l) + 2O_2(g) = BO_2^-(l) + 2H_2O(l)$
	  $\Delta H^0 = -1392.26\,kJ\,mol^{-1}$
	  $\Delta G^0 = -1267.56\,kJ\,mol^{-1}$
	• Electromotive force
	  $E^0 = -\Delta G^0/nF = 1.64\,V$
	• Theoretic energy conversion efficiency
	  $\eta = \Delta G^0/\Delta H^0 = 0.91$

Fig. 8.19 Thermodynamic comparison of the PEMFC and the DBFC.

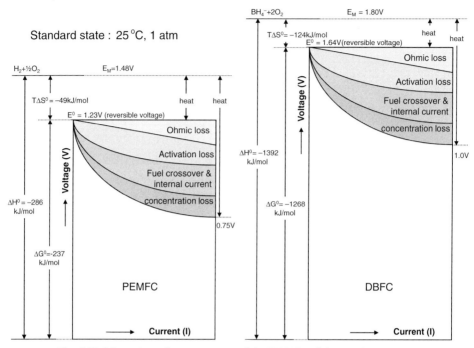

Fig. 8.20 Schematic performance curves of the PEMFC and the DBFC.

fuel crossover and internal current, Ohmic loss and concentration loss). Activation losses are caused by the slowness of the electrochemical reactions taking place on the catalyst surfaces. The migration of ions in the electrolyte and electron transportation between the catalyst particles and the electron collectors of the anode and cathode lead to ohmic losses. It is impossible to obtain the reversible cell voltage ($E^0 = \Delta G^0/nF$, where F is the Faraday constant and n is the number of electrons transferred according to the cell reaction) due to these losses during operation. Figure 8.20 shows the schematic performance curves of the PEMFC and the DBFC. E_M refers to the maximum cell voltage that is calculated from the standard enthalpy change (ΔH^0) of the cell reaction ($E_M = \Delta H^0/nF$).

If activation losses of the electrodes, ohmic losses and concentration losses of the DBFC were on the same level as the PEMFC with an operation voltage of 0.75 V, the DBFC would obtain 1 V of operation voltage under the similar operation conditions, as shown in Fig. 8.20. This high operation voltage will benefit the stack and the system design for fuel cell users because the number of cells can be reduced by 25 % compared with the PEMFC stacks.

8.2.2.3 Classification of the DBFCs

The DBFC can be classified into three models according to the electrolyte used, as shown in Table 8.2. Model 1 and 2 are very similar because the charge carrier and ion migration are the same. Model 3 is different from the models above in

Table 8.2 Classification and characteristics of the DBFCs

	Electrolyte	Charge Carrier (Ion Migration)	Product at Anode	Product at Cathode
Model 1	Alkali	OH^- (C→A)	$B(OH)_4^-$, H_2O	—
Model 2	Anion exchange membrane	OH^- (C→A)	$B(OH)_4^-$, H_2O	—
Model 3	Cation exchange membrane	Na^+ (A→C)	$B(OH)_4^-$, H_2O	NaOH

the charge carrier and ion migration direction. A schematic comparison of the cell components and electricity generation mechanism of the models is illustrated in Fig. 8.21.

The Model 1 type cell was suggested by Jasinski [44] who tested the KBH_4/O_2 cell with a Pt cathode, anodes made of Pt, Pd and Ni_2B and KOH as electrolyte. He used an asbestos membrane to separate the anode and the cathode compartments. The fuel cell using borohydride showed better performance than that using H_2 though there was a decrease in cathode performance due to the presence of fuel

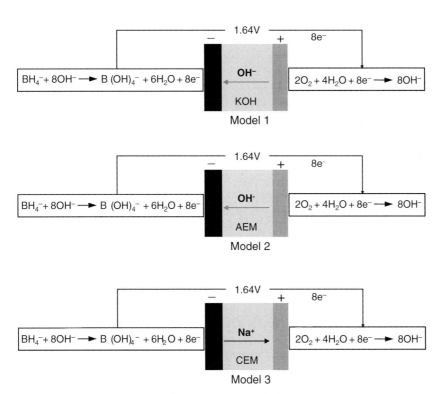

Fig. 8.21 Schematic comparison of cell components and electricity generation mechanism of the DBFC.

in the common electrolyte (asbestos membrane). He suggested this decrease could be offset by use of an electrolyte membrane.

Amendola et al. [38, 39] constructed Model 2 type cells with an air cathode and an anode made of highly dispersed Au/Pt particles supported on high-surface area carbon silk. An anion exchange membrane (AEM) was used as the electrolyte. The number of electrons utilized per molecule of BH_4^- oxidized (about 6.9 out of a possible 8) shows efficient utilization of the BH_4^- oxidation. Specific energy $>180\,Wh\,kg^{-1}$ and power densities $>20\,mW\,cm^{-2}$ at room temperature and $>60\,mW\,cm^{-2}$ at 70 °C have been reported.

Instead of an asbestos membrane and an anion exchange membrane, Li et al. [46] suggested a Model 3 type cell in which a Nafion membrane was employed as the electrolyte to keep borohydride from crossing to the cathode. It was verified that cation (Na^+) was the charge carrier in the Nafion membrane. Compared with anode polarization, cathode polarization was the main reason for the cell voltage drop. The power density of a Model 3 type cell can reach $190\,mW\,cm^{-2}$, as shown in Fig. 8.22.

Based on the Model 3, a 10-cell stack of microfuel cells was assembled. It was reported that an output of near 10 W was achieved, as shown in Fig. 8.23 [47]. Recently, through membrane electrode assembly (MEA) improvement, a power density $290\,mW\,cm^{-2}$ of the cell with an air cathode has been achieved. A 5-cell stack with effective area of $67\,cm^2$ demonstrated that the power reached 110 W when the operating temperature reached 60 °C, though the stack started at room temperature without humidification. The performances of single cell and 5-cell stacks are shown in Fig. 8.24.

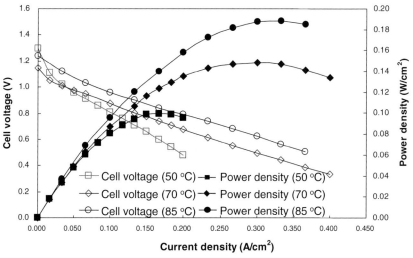

Fig. 8.22 Cell polarization and power density curves for the borohydride fuel cell at 50 °C, 70 °C and 85 °C. Anode: 0.2 g alloy cm^{-2}, 10 wt% NaBH4 in 20 wt% NaOH at a flow rate of $0.2\,L\,min^{-1}$. Cathode: 2 mg Pt cm^{-2}, humidified O_2 at $0.2\,l\,min^{-1}$ (1 atm).

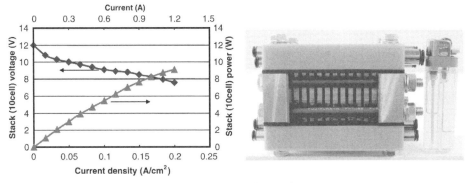

Fig. 8.23 A micro-fuel cell using borohydride as fuel and its performance.

8.2.2.4 Catalysts for the DBFCs

The noble metals are the only choice for the electrode catalyst in the FEMFC, the PAFC and the DMFC, because only noble metals can resist corrosion from the acidic electrolyte. Unlike these fuel cells working under an acidic electrolyte, the DBFC operates under an alkaline electrolyte. In an alkaline electrolyte, many non-noble metals such as Ni, Ag can resist corrosion by the electrolyte. The Ni catalyst was used successfully in the Apollo mission fuel cell in 1960s. Since then, the use

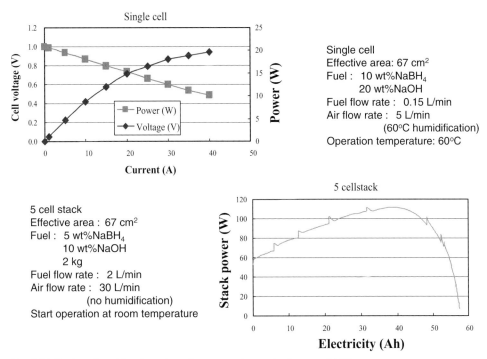

Fig. 8.24 Single cell and 5-cell stack performances when using improved MEA.

of Raney metals has been considered as an effective method of achieving a very active and porous form of electrodes for the AFC. Raney metals are prepared by mixing the active metal (e.g. Ni and Ag) with an inactive metal (usually Al and Si) by melting or ball milling. The mixture is then treated with a strong alkali that dissolves out the inactive metal to leave a porous material with a high catalytic activity. The Siemens AFC, using Raney Ni for the anode and Raney Ag for the cathode, was used successfully in submarines in the early 1990s [48].

In principle, in addition to the noble metal catalysts, the catalysts used in the AFC and Ni–MH battery are also suitable for the DBFC. Lee et al. [49] have reported the use of a $ZrCr_{0.8}Ni_{1.2}$ alloy (Laves phase alloy AB_2) as the anode catalyst. They proposed a stepwise electricity generation mechanism in which $ZrCr_{0.8}Ni_{1.2}$ takes up hydrogen from borohydride solution and then the hydrogen is electrochemically oxidized to generate electricity. Li et al. [46] reported a surface treated Zr–Ni Laves phase alloy AB_2 ($Zr_{0.9}Ti_{0.1}Mn_{0.6}V_{0.2}Co_{0.1}Ni_{1.1}$) as the anode catalyst in the DBFC. Liu et al. [50] reported a high power density achieved by using Ni powder (Inco type 210) as the anode material through electrochemical oxidation of borohydride.

8.2.3
Problem Areas in Research and Development

Theoretically, one ion of BH_4^- can generate 8 electrons, however, the number of electrons utilized per ion of BH_4^- oxidized is not limited to 8 electrons due to the anode catalysts used. It is reported that one ion of BH_4^- generates 6.9 electrons when using Au catalyst [39], 6 electrons when using Pd catalyst [51] and 4 electrons when using Ni catalyst [48]. In the case of Au as the anode catalyst, the anode reaction is considered to be an 8-electron reaction. With Pd as the anode catalyst, the anode reaction, cell reaction and standard electrode potentials (according to the thermodynamic calculation) are as follows:

$$6\text{-electron reaction}: BH_4^- + 6OH^- = BO_2^- + 4H_2O + H_2 + 6e^-$$
$$E_a^0 = -1.38\,\text{V} \tag{8.17}$$
$$\text{cathode reaction}: 1.5O_2 + 3H_2O + 6e^- = 6OH^- \quad E_c^0 = 0.40\,\text{V} \tag{8.18}$$
$$\text{cell reaction}: BH_4^- + 1.5O_2 = BO_2^- + H_2O + H_2 \quad E^0 = 1.78\,\text{V} \tag{8.19}$$

When using Ni as the anode catalyst, the anode reaction, cell reaction and standard electrode potentials are as follows:

$$4\text{-electron reaction}: BH_4^- + 4OH^- = BO_2^- + 2H_2O + 2H_2 + 4e^-$$
$$E_a^0 = -1.65\,\text{V} \tag{8.20}$$
$$\text{cathode reaction}: O_2 + 2H_2O + 4e^- = 4OH^- \quad E_c^0 = 0.40\,\text{V} \tag{8.21}$$
$$\text{cell reaction}: BH_4^- + O_2 = BO_2^- + 2H_2 \quad E^0 = 2.05\,\text{V} \tag{8.22}$$

When the electrons utilized per ion of BH_4^- decrease to 6 and 4, the electromotive force can reach 1.78 and 2.05 V respectively. However, the generated electricity will be decreased. For example, the energy density of sodium borohydride can reach 9.3 Wh/g-NaBH$_4$ in an 8-electron reaction, but only 7.6 Wh/g for a 6-electron reaction and 5.8 Wh/g for a 4-electron reaction. Moreover, the hydrolysis reaction of the borohydride is another problem leading to a decrease in the fuel utilization.

$$BH_4^- + 2H_2O = BO_2^- + 2H_2O + 2H_2 \tag{8.23}$$

Preventing the borohydride hydrolysis and obtaining an 8-electron reaction are the keys to improving the coulombic efficiency of the DBFCs. NaBH$_4$ concentration in the fuel is a factor in the hydrogen evolution. It was reported that when the NaBH$_4$ concentration was less than 1.5 M, BH$_4^-$ ion was electrochemically oxidized by an 8-electron reaction, but when the NaBH$_4$ concentration increased to 2 M, the anodic reaction was shifted to a 6-electron reaction if using Pt/C as the anode catalyst [52]. Using materials with high hydrogen overpotential and surface treatment technology are considered to be the ways to reduce the hydrogen evolution. Using the evolved hydrogen to supply the AFC or the PEMFC is another way to increase the fuel utilization from an engineering point of view.

Another problem is the cost of the fuel – sodium borohydride costs $55/kg. The fuel contribution to the cost of electricity (COE) for the DBFC will be $9.7/kWh, even if the operating voltage of the DBFC can reach 1.0 V. It is 100 times that of the hydrogen gas. At present it is very difficult to apply the DBFC to high power applications such as a car or a home power generator. However, it is very possible that DBFCs could be used in place of batteries in low power applications such as primary batteries or secondary batteries for notebook computers, cordless electric tools and so on because the COE for batteries is rather high. For example, the COE for an alkaline primary battery is up to $450/kWh.

Reduction in the cost of borohydride production is necessary in order to expand the application area for the DBFC. If the cost of borohydride production can be reduced to $0.55/kg, the DBFC will be a powerful competitor in vehicle applications of fuel cells.

8.3
Internal Combustion Engines
Gerrit Kiesgen, Dirk Christian Leinhos, and Hermann Sebastian Rottengruber

8.3.1
Hydrogen as a Fuel

Today's transportation sector, and hence large parts of the economy, depend almost completely on oil-based fuels with all the implications of geological and political availability. Since such resources are limited, the search for adequate alternative fuels has been going on for decades. The requirements imposed upon new fuels are basically derived from the knowledge and expertise gained by using gasoline and diesel fuels and are hard to meet for substitutes. For instance, customers have demanding expectations on the availability of fuel, ease of handling and safety of use. For evaluating the overall potential of an alternative fuel the economical aspects of production and distribution have to be considered, taking into account that a new fuel has also to be ecologically beneficial in terms of carbon dioxide emission, other greenhouse gases and emissions limited by legislation.

Hydrogen as an energy carrier has the potential to become this new fuel for the transportation sector. It can be produced in different ways using different prime energy sources, thus reducing the dependence on a single energy source. When using renewable sources of energy for its production, for example wind power for electrolysis or biomass, hydrogen is the preferred energy carrier for a sustainable economy with the absence of carbon making it the perfect fuel.

One way to make use of all these advantages in the transportation sector is to feed it into internal combustion engines and to power vehicles almost as with gasoline and diesel fuel. The standards set by today's gasoline and diesel engines are quite high, but there are ways to meet or even exceed them by developing the key technologies for the use of hydrogen in internal combustion engines.

8.3.2
The Hydrogen Internal Combustion Engine (ICE)

For a hydrogen-based power supply, as well as for mobile and stationary applications, an engine concept based on the reciprocating piston engine is an obvious choice. However, hydrogen with its low density and low ignition energy poses some new challenges for the internal combustion engine.

For an automotive propulsion system power density, instant power availability, the absolute power output, efficiency of the overall system and at the same time low environmental impact are of the highest importance.

For some time now, attempts have been made to utilize hydrogen in internal combustion engines. Beginning, for example with the works of Erren [53] and Oehmichen [54] in the 1930s and 1940s and the programs conducted at BMW [55] starting in the late 1970s and the ongoing efforts by other automotive companies, hydrogen engine development has made enormous progress.

Major distinctive features of hydrogen internal combustion engines are:

- the means of mixture formation (internal and external mixture formation)
- the ignition process (forced auto-ignition, ignition using a spark plug or fuel spray ignition)
- the basic thermodynamic process (Otto-cycle and Diesel-cycle)

Hydrogen internal combustion engines can be classified, like any other gaseous-fueled engines, into three basic engine concept categories [56]:

Otto-cycle gaseous-fuel engines are characterized by external or internal mixture formation and a timed ignition at a singular point, as, for example a spark plug. Most vehicle applications currently under development and testing fall into the category of 4-stroke Otto-cycle gaseous-fuel engines [57, 58]; but 2-stroke hydrogen engines [59] and rotary engines (Wankel-engine) [60] have been investigated in the past.

With Diesel-cycle gaseous-fuel engines, a so-called ignition spray ignites the air fuel mixture. Diesel fuel or MDF (Marine Diesel fuel) are the most common ignition fuels for these kinds of engines. Hence external mixture formation is most common for Diesel-cycle gaseous-fuel engines [61].

Gaseous-fuel Diesel-cycle engines are characterized by internal mixture formation immediately before ignition. Ignition is realized by pure auto-ignition or assisted by hot surface ignition. Therefore these engines usually run under lean burn conditions due to the late injection timing. Analogous to diesel engines, running on hydrocarbon fuel, size and bore limitations are not that strict compared to Otto-cycle engines [62, 63].

Gaseous-fuel Diesel engines as well as Diesel gaseous-fuel engines usually require a more complex infrastructure, for example a secondary fuel-system and possibly lean mixture exhaust gas after-treatment systems and are therefore mostly used for stationary power generation.

8.3.3
The Hydrogen Spark Ignition Engine

Considering the advantages and disadvantages of the above classified engine types, the Otto-cycle gaseous-fuel engine concept has the greatest potential, especially for automotive applications.

Its advantages are in mainly transient behavior and a less complex complete system. However, the discussed thermodynamic basics also apply to the other engine concepts which have not been discussed in depth here.

The power output and power density of hydrogen internal combustion engines are also a significant challenge, because the low density of gaseous hydrogen reduces the volumetric efficiency with external mixture formation [64]. Using pressure charging or direct injection these limitations can be overcome.

As shown in Fig. 8.25, the calorific value of the mixture within the cylinder, and therefore the engines power output, increases with internal mixture formation.

To maximize power output a stoichiometric mixture is demanded for engine combustion. The main limitation for stoichiometric operation in combination with external mixture formation is backfire into the intake manifold and uncontrolled preignition caused by the wide ignition limits of hydrogen–air mixtures and the low ignition energy demands.

To overcome these obstacles precise fuel metering, an accurately timed and optimized ignition system, variable cam phasing and optimized scavenging and cooling strategies are required over the entire operational range of the engine.

The only statutory limited emission component of hydrogen engines are oxides of nitrogen (NO_x). Formation of NO_x in hydrogen engines is entirely dependent on the combustion temperature [65].

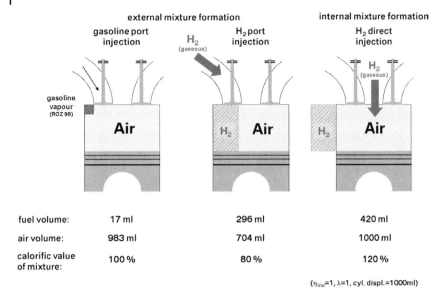

Fig. 8.25 Comparison of various engine concepts by means of the theoretical calorific value of mixture and power output.

Lean homogeneous hydrogen–air mixtures burn at very low temperatures causing very high efficiencies and minimal NO_x-emissions, but on the other hand cutting down the engines power density significantly.

To quantify the proportion of hydrogen and air in an engine, the combustion air/fuel ratio λ is commonly used:

$$\lambda = \frac{m_{Air}}{34.2 \cdot m_{Hydrogen}}$$

with λ being the relative air/fuel ratio; m_{Air}, the mass of air induced; $m_{Hydrogen}$, the mass of hydrogen induced.

The operation strategy for a hydrogen engine is therefore determined by the wide ignition limits of hydrogen as well as the NO_x formation limit. Homogeneous hydrogen–air mixtures can be burned very lean ($\lambda > 4$). This allows quality-based mixture control over a wide load range. Best engine efficiencies are reached in this mode of operation. On the other hand NO_x-formation is a strong function of the air/fuel ratio (Fig. 8.26). The NO_x generation limit can be observed for air/fuel ratio $\lambda = 2.0$–2.2.

For leaner mixtures NO_x-emissions are close to zero. For richer mixtures NO_x-emissions rise rapidly to significant levels.

A benefit of stoichiometric engine operation, besides the maximized power density, is in fact the possibility to reduce NO_x-emissions to negligible levels by using exhaust gas after-treatment systems [64]. With conventional automotive catalytic converters conversion ratios of more than 99.9 % percent can be reached.

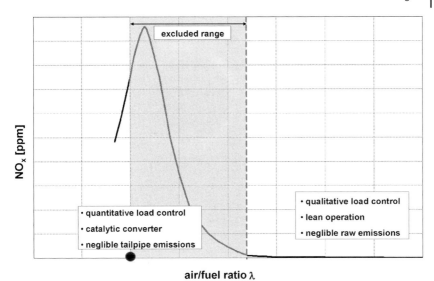

Fig. 8.26 NO_x-emissions as a function of air/fuel ratio λ and application strategy for hydrogen internal combustion engines.

But this also means that increasing engine load only by continuous enrichment of the hydrogen/air mixture (qualitative load control) will have undesired effects concerning engine emissions. So if an air/fuel ratio λ of approximately 2.0 to 2.2 is reached, engine control has to switch instantly to stoichiometric operation (quantitative load control). Mixture concentrations of $1.0 < \lambda < 2.0$ will be excluded by the engines governing systems. This ensures the lowest emissions over the engine's entire operational range.

Other emissions like CO and HC, resulting from lubricant oil, are negligible, but even these already low carbon-based emissions can be reduced to almost zero with the same conventional catalyst as required for NO_x reduction. Thus, an optimized operation strategy involves a combination of stoichiometric operation at high engine loads and lean operation at part load (Fig. 8.27).

As with conventional SI-engines, running on hydrocarbon liquid fuels, the means of fuel mixture, either internal or external, is important. As mentioned above, naturally aspirated hydrogen engines utilizing external mixture formation have a lower power output than comparable gasoline engines of the same cylinder capacity. This is mainly due to the lower calorific value of the mixture per cylinder volume.

The available fuel pressure for external mixture formation is usually significantly lower than with internal mixture formation. As a matter of fact the mixture of hydrogen and air takes place within the intake manifold and the available time for mixing will be comparatively long. This leads to low injection pressures in a range up to a maximum of 5 bar.

With external mixture formation the degrees of freedom for affecting the combustion process through injection and ignition timing are fairly limited. In addition

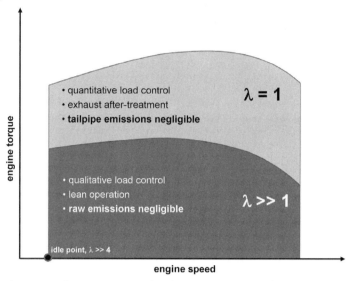

Fig. 8.27 Application strategy within the operational range of a hydrogen internal combustion engine.

to that, uncontrolled combustion phenomena such as backfire into the intake port and manifold, as well as preignition in the combustion chamber, pose major difficulties in designing an efficient combustion system.

Internal mixture formation on the other hand offers excellent potential for controlling combustion via injection rate, as well as increasing the engine's power output in general (Fig. 8.25). The key component for such a fuel system is an injector capable of handling high hydrogen pressures of 50 to 300 bar, high flow rates, accurate injection timing and thermal stresses due to combustion. Additional to that, a careful selection of the materials used for the nozzle and other parts of the injector in contact with hydrogen has to be made in order to minimize hydrogen embrittlement and ensure proper operation during the engine's lifespan.

Also with a combined mixture formation system it is possible to serve both the demands of high efficiency and high power density. At part load and lean operation external mixture formation has substantial benefits due to the demands of low-pressure hydrogen. On the other hand, at full load internal mixture formation helps to increase power density significantly. Combustion can be controlled by the injection rate and strategy. Figure 8.28 displays a possible cylinder head package of such an engine.

Figure 8.29(a) shows measured pressure curves of a conventional gasoline engine, a hydrogen Otto-cycle engine with external mixture formation, and a hydrogen engine with internal mixture formation at stoichiometric full load operation and an engine speed of 2000 rpm.

Due to the above-discussed effects of a more efficient gas exchange, the measured in-cylinder pressure during the compression stroke is significantly higher with the hydrogen engine utilizing internal mixture formation.

8.3 Internal Combustion Engines

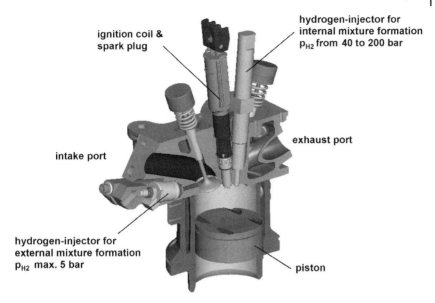

Fig. 8.28 Possible engine and cylinder head package.

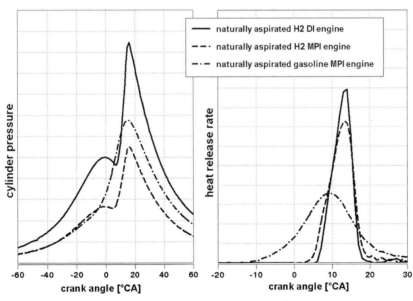

Fig. 8.29 Measured pressure curves (a) and heat release rates (b) at 2000 rpm and full load operation of different engine concepts.

Engine combustion and operation with hydrogen in general is characterized at part load by very stable combustion. Due to the wide ignition range of hydrogen–air mixtures, combustion losses from unburned fuel at lean operation are negligible.

For full load stoichiometric operation combustion velocity is significantly higher than with gasoline. Hence combustion duration is shorter than with gasoline according to the heat release rates shown in Fig. 8.29(b) This leads to high engine efficiency at full load operation. On the other hand these fast combustion rates induce high mechanical and thermal stress due to high pressure rise rates and high combustion temperatures.

The theoretical thermodynamic efficiency of an Otto-cycle engine is based on the efficiency of the constant volume process. Hence the engine's compression ratio and the specific heat ratio κ of the working gas are relevant for this basic consideration:

$$\eta_{th} = 1 - \frac{1}{\varepsilon^{\kappa-1}}$$

with η_{th} being the theoretical thermodynamic efficiency; ε the compression ratio; κ the specific heat ratio.

Higher compression ratios combined with a high specific heat ratio, provide higher thermal efficiencies. With a specific heat ratio of $\kappa = 1.4$ for hydrogen–air mixtures, hydrogen engines have certain advantages over engines running on hydrocarbon-based fuels ($\kappa = 1.35$ or lower). In Fig. 8.30 this effect is shown also in connection with the effect of a higher compression ratio.

According to these simplified considerations the achievable increase in theoretical thermal efficiency is highest with compression ratios up to $\varepsilon = 16$. But for real engines with compression ratios of $\varepsilon = 16$ and more, a gain in efficiency is hardly attained because of a superimposed increase in engine friction.

Looking towards real engine processes, hydrogen engines have a significantly higher efficiency compared to current technology gasoline and diesel engines over a widespread area of the operational range.

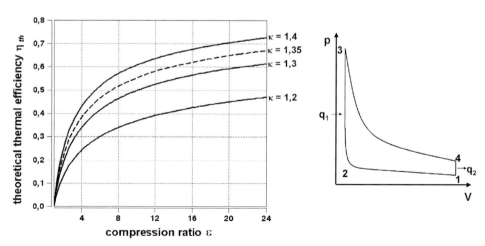

Fig. 8.30 Thermal efficiency of an idealized Otto-cycle process (constant volume process).

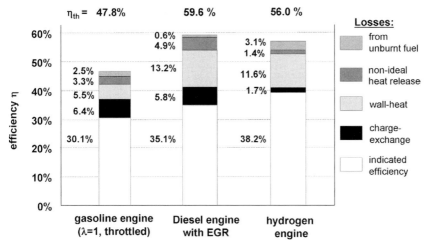

Fig. 8.31 Engine efficiency at 2000 rpm and part load operation of a hydrogen engine, compared to a current technology gasoline and diesel engine [64, 66].

The high theoretical thermodynamic efficiency results from the capability to run on very lean fuel–air mixtures, whereas the high theoretical thermodynamic efficiency of the diesel engine results mainly from a higher compression ratio (Fig. 8.31). Although the theoretical thermodynamic efficiency of the hydrogen engine is below that of a diesel engine, the hydrogen engine has a higher indicated efficiency. The main reasons are higher burning velocity resulting in lower losses due to a nonideal combustion process and significantly lower gas exchange losses due to unthrottled operation. Compared to a gasoline engine, the indicated efficiency of the hydrogen engine is approximately 8 % points higher, resulting in 25 % better fuel consumption [66].

8.3.4
The Hydrogen-ICE for Automotive Applications

The technological potentials of hydrogen S.I. engines are well-suited to the requirements of automotive applications and make them the preferred choice for the future with an established infrastructure for the production and distribution of hydrogen [67].

Besides working on technologies for the ultimate hydrogen internal combustion engine, there are efforts within the industry to bring dual-fuel hydrogen/gasoline engines into series production. For example the BMW Group has developed a production ready engine which will power the hydrogen version of the 7 Series the BMW Hydrogen 7 [68]. This car fulfills the same requirements and standards as all BMW products, on the road and in customer hands.

To allow for dual-fuel operation, some modifications on the engine itself and its subsystems have been carried out. For example the injection valves for the external mixture formation, the hydrogen supply rail and the ignition system. A concept

Fig. 8.32 Hydrogen 7 dual-fuel IC engine.

engine (Fig. 8.32) was built to prove the feasibility. It is based on a serial production gasoline V12 engine. It offers either an external hydrogen mixture formation or a direct-injection gasoline mixture formation. Power output is exceeding 191 kW with a torque of 390 Nm for the Hydrogen 7 application [69].

There are additional changes to the cooling and exhaust gas system required when integrating the powertrain into the vehicle body. A safety system is also installed to monitor all components and thus to ensure the highest possible safety standards.

The possibility of dual-fuel operation makes the hydrogen engine the ideal technology to close the gap between today's oil-based infrastructure of gasoline and diesel and tomorrow's infrastructure for hydrogen.

Having an additional gasoline tank on-board also provides a substantially increased range, which is of great importance for customers. Though the volumetric energy density of liquid hydrogen is about four times lower than for gasoline and diesel, cryogenic tank technology is the best choice in terms of energy density compared to compressed gaseous hydrogen. Nevertheless, an increased tank volume is required, which has a major effect on the package of hydrogen cars, while the influence of using an internal combustion engine on the geometry of the vehicle is negligible. The concept of a dual-fuel version of a BMW 7 Series, as realized with the Hydrogen 7, is shown in Fig. 8.33. The hydrogen ICE powertrain with its favorable power-to-weight ratio offers an attractive driving performance, a known driving characteristic and a high day-to-day usability. Even with the possibility of dual-fuel operation, the increase in weight is moderate compared to other alternative concepts of hydrogen powertrain.

Fig. 8.33 The dual-fuel BMW Hydrogen 7.

There are also some favorable economic aspects which emphasize the potential for industrialisation of the internal combustion engine fueled with hydrogen. First, the existing development and production resources of the automotive industry can be used, which provides a great financial benefit. Since the ICE is a well established, well understood and mature technology it minimizes the financial and technological risk to the industry when supporting a new fuel.

Additionally, the ICE still offers great potential in terms of improving efficiency and power output in the future. Furthermore, improvements and developments of conventional ICEs can also be applied to hydrogen ICEs (e.g. reducing friction, advanced valve trains, thermo-management, hybrid technologies, etc.), thus offering great synergies extending the effect of investments in today's engine technology into the future.

All the advantages and benefits of hydrogen (e.g. no CO_2-emissions, diversification of energy sources, sustainable energy supply when using regenerative sources, lowest emissions and low fuel consumption) can be economically exploited by using an internal combustion engine to power individual mobility, making it the ideal choice for the hydrogen society.

8.4
Hydrogen in Space Applications
Robert C. Bowman Jr. and Bugga W. Rathnakumar

8.4.1
Introduction

Although hydrogen is the most common chemical element in the universe, it is only the ninth most abundant element on Earth. Nevertheless, for nearly fifty years hydrogen has played a predominant role in the human exploration of space. This chapter will briefly review where and how hydrogen has been utilized in various space flight programs for launching the spacecraft (i.e. by serving as rocket fuel), for cooling scientific instruments and detectors to cryogenic temperatures, in the operation of electrical power systems such as nickel–hydrogen batteries and fuel cells, and other specialized applications. The demand for very large quantities of liquid hydrogen during the launching of spacecraft stimulated [70, 71] the commercial production of liquid hydrogen in the USA, starting in the early 1960s. To contain the costs of hydrogen liquefaction and minimize evaporative losses of this low boiling temperature (i.e. ~20 K) cryogenic liquid, extensive efforts were implemented to enhance the exothermic conversion rate of ortho-H_2 to para-H_2 to greater than 95+% during the liquefaction process [70, 71]. Complex facilities have been developed at launch locations in several countries to produce, store and transfer the enormous quantities of liquid hydrogen consumed during the launch of spacecraft for military, communications and space exploration missions.

In addition to all the current uses of hydrogen in space, there are a number of more novel concepts that are being developed or proposed for space technology over the next several decades. Some of these uses will be described at the end of this chapter. It should be very interesting to see whether any of these proposed roles for hydrogen will actually come to fruition in the foreseeable future.

8.4.2
Hydrogen as Rocket Fuel

Rocket propulsion moves a vehicle by ejecting mass out of the backside of the vehicle. Since there is no atmosphere in space, all of the reaction (i.e. propellant) mass must be carried on board and is usually released as a heated gas (i.e. the exhaust) expanding through a nozzle. All launches from earth into space have used chemical propulsion systems (i.e. reactions between a fuel and an oxidizer). The performance of a rocket vehicle is represented by the so-called "first rocket equation":

$$M_b/M_o = \exp(-\Delta V/g_c I_{sp}) \tag{8.24}$$

where M_o = initial total mass of the rocket and includes the propellant mass (M_p), M_b = the mass of the rocket after the propellant "burn" (i.e. reaction), ΔV = velocity change of the rocket, I_{sp} = the specific impulse (i.e. ratio of engine thrust in Newtons (N) divided by the propellant mass flow rate in kgs), and g_c = conversion factor between ΔV and I_{sp}, which equals one in the units used in this paper. A velocity of 11.2 km s^{-1} is needed to overcome the earth's gravity while ~8 km s^{-1} is required to obtain a low earth orbit (LEO). Larger I_{sp} values are desired to maximize rocket performance levels (i.e. to launch a larger payload mass [M_{PL}] and/or travel farther distances). Since I_{sp} is proportional to the square-root ratio of temperature (T) to molecular mass (M) for the exhaust gases, the very exothermic hydrogen reaction with pure oxygen gives the largest I_{sp} value for any chemical propulsion system. Furthermore, the high heat capacity of H_2 molecules reduces the exhaust temperature and protects the rocket engines from overheating. To illustrate the advantage of hydrogen as a rocket fuel, the delivered $I_{sp} = 4400\,\text{m s}^{-1}$ the liquid hydrogen (LH2) and liquid oxygen (LOX) engines in the USA NASA Space Shuttle compared to a calculated I_{sp} values of $\sim$2700 m s^{-1} for a rocket burning hydrazine and inhibited red fuming nitric acid. Hence, single or multiple LH2/LOX stages have been used to launch vehicles into space since the NASA Apollo program in the 1960s. Table 8.3 lists international examples where LH2/LOX has been used as rocket fuel. The information in this table has been compiled from a number of sources including three very useful web sites [72–74].

Although large I_{sp} values are important, rockets can increase their M_{PL}/M_o ratios by discarding empty tanks and engines at the beginning of their flights while using multiple stage launchers with ambient temperature liquid fuel (e.g. kerosene, hydrazine, N_2O_4) or solid fuel (e.g. aluminum and ammonium perchlorate) boosters in addition to the cryogenic LH2/LOX stages. In fact, all of the rockets in Table 8.3 are multiple stage launch vehicles. Staging also facilitates launches from the surface of the earth where the high-thrust and high-density solid rocket boosters (SRBs) lift the vehicles above the denser lower atmosphere where speeds must be kept low to minimize air friction and drag. Thus, the LH2/LOX engines are primarily used for acceleration into orbital velocities. The larger the payloads and further the travel distance (i.e. flights beyond earth orbit), the more critical becomes the overall optimization of design and performance improvement to reduce the propulsion system's "dry" mass M_{dry} (i.e. everything except the propellant). The refinements in component design and materials selection are the engineering issues that must be addressed in making "rocket science" work.

The two types of bipropellant liquid propulsion systems that have been used with LH2 and LOX rocket engines [75] are illustrated in Fig. 8.34. These cryogenic liquids are mixed in the rocket engine's combustion chamber with the aid of a pressure-regulated pressurization system (e.g. usually gaseous helium) that either directly forces the gases into the engines (i.e. pressure-fed) or into pumps to supply much higher pressures (i.e. >20 bar). The more complex and expensive pump fed systems have an advantage of providing high propellant flow rates for high-thrust launch vehicles such as the Space Shuttle and Ariane 5.

After several frustrating years and some spectacular failures, the first successful launch of a rocket with a liquid hydrogen fuel upper stage occurred on 27 November 1963 at the NASA Cape Canaveral Launch Complex in Florida. Since then over 450 space launches through the end of 2003 have involved single or multi liquid hydrogen fueled stages. The USA launches have been from either the Kennedy Florida complex or the Vandenburg Air Force Base in California. The LH2/LOX fueled Centaur stages have been integrated with various Atlas and Titan vehicles for launching unmanned scientific payloads including the Surveyor, Mariner 69 and Cassini instruments; numerous military satellites include those of the Defense Support Program (DSP) for intercontinental ballistic missile detection and the MILSTAR communications system; and commercial satellites for telecommunications and other applications. The massive Saturn vehicles used liquid hydrogen stages to

Table 8.3 Some properties of hydrogen powered space launch vehicles

Country	Launch Vehicle	Years Flown (First-Last)	Number LH2/LOX Stages	Number Engines	LH2/LOX Stage Burn Time (s)	Total LH2/LOX Propellant Mass (10³ kg)	Number[a] of Launches (Failures)	Examples of Missions Launched
USA	Atlas-Centaur	1963–1989	1	1	450	14	68 (12)	Surveyor, Mariner 69
USA	Atlas II-Centaur	1992–	1	2	377	16.9	49 (0)	GOES
USA	Atlas III-Centaur	2000–	1	1 or 2	928/464	20.8	4 (0)	
USA	Atlas V	2002–	1	1 or 2	926/463	20.8	3 (0)	
USA	Saturn IB	1966–1975	1	1	450	103	9 (0)	Apollo, Skylab
USA	Saturn V	1967–1973	2	5,1	390	451.7	13 (1)	Apollo, Skylab
USA	Titan IV-Centaur	1994–	1	2	600	21.0	16 (2)	DSP, Cassini
USA	Delta III	1998–2000	1	1	700	16.8	3 (3)	
USA	Delta IV	2002–	2	1	249/850	204/20.4	3 (0)	Spitzer, Eutelsat W5
USA	Space Shuttle	1981–	1	3	510	725	113 (2)	Galileo, ISS, Hubble
France	Ariane 1-3	1979–1989	1	1		8–10.8	22 (4)	Commercial satellites
France	Ariane 4	1988–2003	1	1	725	11.8	116 (3)	139 Telecom Satellites
France	Ariane 5	1996–	2	1,1	600	170/25	18 (4)	Rosetta (2004), HSO-Planck (2007)
Russia	Energia	1987–1988	1	4	470	797	2 (1)	Buran Space Shuttle
China	Chang Zheng (CZ-3)	1984–	1	2	470	18.2	26 (4)	Apstar, Asiasat
Japan	H-II	1994–1999	2	1,1	345/598	86/17	7 (2)	ISO
Japan	H-IIA	2001–	2	1,1	397/530	100/16.9	6 (1)	

[a]Total launches through end of 2003.

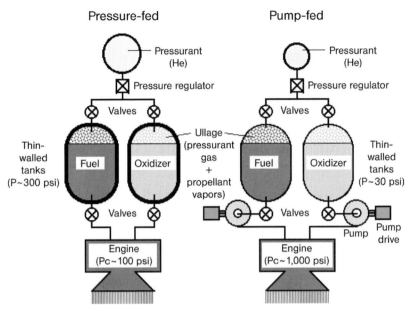

Fig. 8.34 The two generic types of bipropellant liquid propulsion systems used to launch spacecraft where liquid hydrogen is the fuel and liquid oxygen is the oxidizer.

launch men to the moon in the Apollo program and were also used for the Skylab missions in the 1970s. The more recent and current unmanned USA/NASA missions such as the Spitzer Space Observatory employed hydrogen upper stages on Delta III and IV launch vehicles.

In December 1979, France became the second country that successfully launched a vehicle (an Ariane 1) with a liquid hydrogen stage from its complex in Kourou, French Guiana. This effort was followed by the development of other Ariane launch vehicles that led to the workhorse Ariane 4 vehicle with its Viking booster rockets burning nitrogen tetroxide and unsymmetrical dimethylhydrazine propellants. The very versatile Ariane 4 has placed about 140 telecommunication satellites into orbit as well as supporting several space science missions. The larger Ariane 5 vehicle was subsequently developed with its Vulcain LH2/LOX engines to place dual 3–6 metric ton satellites into geosynchronous transfer orbits (GTO) at a time. Ariane 5 has also launched the Rosetta probe in March 2004 towards a rendezvous with the comet 67P/Churyumov-Gerasimenko in 2014 and will launch the Herschel–Planck instruments in 2009 to an orbit around the second Lagrangian Point L2, that is about 1.5 million km from earth [76].

Although the Soviet Union/Russia launched the first satellite into space in 1957 and has made more than 2800 launches since that time, liquid hydrogen was very rarely used except for the Energia launch vehicle developed for their Buran Space Shuttle. After the only successful launch of an unmanned Buran orbiter in 1988 using the Energia rocket, this program was canceled due to lack of funds.

Fig. 8.35 Launch of the Orbiter Endeavour as Space Shuttle Flight STS-77 on 19 May 1996 from the NASA Kennedy Space Center, FL, USA.

China used a liquid hydrogen stage in 1984 with its Chang-Zheng (CZ-3) or Long March vehicles and has had over 25 launches with these stages. Japan launched its first rockets with a hydrogen fueled stage (H-II) in 1994 and have developed a newer launch vehicle with H-IIA hydrogen stages in 2001. Both China and Japan continue to use LH2/LOX stages for some of their missions.

Between 1981 and 2003, liquid hydrogen has been used for 113 launches of the Shuttle orbiters of the NASA Space Transportation System (STS). The three engines on the Space Shuttle orbiters burned up to 230 000 kg of hydrogen during each launch. Figure 8.35 shows the launch of the Orbiter Endeavour as STS-77 mission on 19 May 1996 from the Kennedy Space Center. The three LH2/LOX engines aboard the orbiter along with the two solid-rocket boosters that burn aluminum metal powder (fuel) with ammonium perchlorate (oxidizer). During this typical Space Shuttle launch over 2.0 million liters of the cryogenic propellants are consumed in the 8.5 min of lift-off and ascent. The cryogenic hydrogen and oxygen fluids are stored in separate compartments of the external tank, which is the largest element of the Shuttle vehicles at 46.9 m tall and 8.0 m in diameter. Figure 8.36 provides a schematic of the external tank components along with its location to the orbiter and SRBs. The external tank is the only Space Shuttle system that cannot be reused. Three generations of external tanks have been used during the STS program. The first two versions were mostly fabricated with the weldable aluminum alloy 2219, which was replaced for flights starting in June 1998 with the stronger and lighter aluminum–lithium alloy Al–Li 2195. The combination of tank redesign and use of the Al–Li alloy has resulted in a total mass reduction of 23.0 % in the

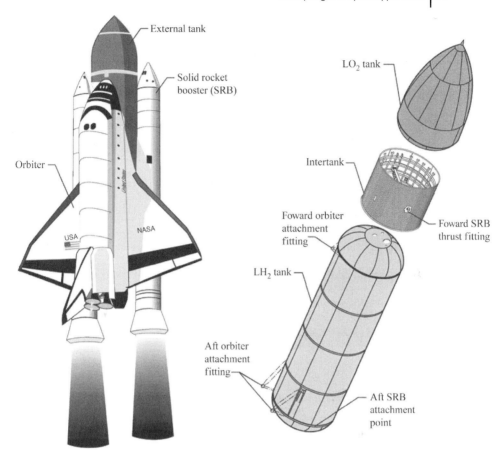

Fig. 8.36 Schematic diagram of Space Shuttle external tank that contains the liquid hydrogen and oxygen storage systems and supplies these propellants to the three main engines in the Orbiter.

mass of the external tanks, improving Shuttle performance by decreasing M_{dry} for heavier payloads or higher orbits.

The three Space Shuttle main engines are clustered at the aft end of the orbiter and were the world's first reusable rocket engines. Each main engine has a mass of ~3150 kg and is given postflight inspections and maintenance after each space flight. Figure 8.37 gives a schematic of the updated Block main engine, which is the fourth modification to its design and includes new high-pressure liquid hydrogen and oxygen turbopumps that can provide up to 440 kg s^{-1} of LOX and 73 kg s^{-1} of LH2 to the engine's main combustion chamber.

The Space Shuttle has been used for an extreme variety of missions from the deployment of scientific research satellites such as Galileo and the Hubble Space Telescope to numerous zero-gravity research experiments and the delivery of components and personnel to assemble the International Space Station (ISS). However,

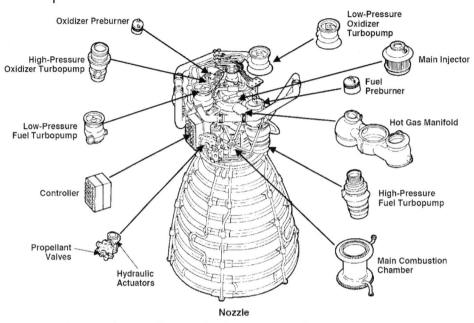

Fig. 8.37 Schematic of a Space Shuttle main engine with its components for the updated Block II versions that began flying missions in April 2002.

there have also been two tragic losses of human life during the Space Shuttle missions: (i) the destruction of the orbiter Challenger during its ascent on 28 January 1986 from failures of the O-rings in an SRB and (ii) the breakup of the orbiter Columbia during its re-entry on 1 February 2003 from damage to the carbon composite panels from pieces of foam insulation ejected off the external tank during launch. Flights of Space Shuttle orbiters were resumed with the July 2005 launch of the orbiter Discovery (i.e., STS-114 mission). There are significant risks with any space launch and about 5 % of all the hydrogen-fueled missions listed in Table 8.3 were failures.

As a final point in this section, the launching of spacecraft is a very expensive operation. For example, the Ariane 4 launches were in the range of US$ 60–100 million each. The Japanese H-II and H-IIA vehicles each cost about US$ 80–100 million to launch. The USA/NASA launches that use LH2/LOX stages cost from US$ 100–300+ million. These figures do not include the value of the payload, which usually ranges from $US100–1000+ million. Hence, sending satellites and people into space is a very expensive and risky activity.

8.4.3
Cryogenics Applications for Hydrogen

Although space is an intrinsically extremely cold environment with a background temperature of only ~2.7 K, it can also be considered as a gigantic vacuum Dewar where an object (i.e. spacecraft or satellite) in earth orbit will maintain temperatures between ~240 and 320+ K. These spacecraft temperatures are strongly dependent upon illumination of the radiated energy from the Sun and earth and internal heat generation from internal power sources (e.g. photovoltaic devices, fuel cells, etc.). While passive radiators on spacecraft can reject this heat to cool portions of the satellites to temperatures below ~100 K, cooling efficiency rapidly decreases for lower temperatures due to the T^4 dependence for radiant heat transfer, which can result in large and massive radiators [77]. Even with orbits far from earth (i.e. the L_2-Lagrangian for the Herschel–Planck missions [76]) and high-efficiency thermal shielding, it is extremely difficult for components to reach temperatures below ~40 K, even under the most favorable circumstances. In contrast, many high-performance detectors and their associated optics that operate in the middle (MWIR) and long (LWIR) wavelength infrared spectral regions require cooling to below 30 K for successful operation [77, 78]. To provide the very low temperatures for these sensors or other components (i.e. superconducting devices), spacecraft must contain either storage systems of cryogenic liquids/solids or cryogenic refrigeration machines (i.e. cryocoolers) [77–79]. As the substance with the second lowest boiling temperature, molecular hydrogen has a number of thermophysical properties that are useful for these cryogenic cooling applications. Some more importance properties for H_2 include: A critical temperature (T_c) of 33.2 K at 13 bar, a boiling point temperature (T_{bp}) of 20.4 K at 1.0 bar, a triple point temperature (T_{tp}) of 14.0 K at 0.07 bar, liquid density = 0.071 g cm^{-3} at T_{bp}, heat of vaporization = 448 J g^{-1} at T_{bp}, heat of fusion = 58.2 J g^{-1} at T_{tp} and heat of sublimation = 508 J g^{-1}. Hence, liquid hydrogen can provide cooling between 14 K and 30+ K while solid hydrogen can support operations from 14 K to below ~7 K. Only helium can produce colder temperatures [77].

The first use of hydrogen to cool scientific instruments during a space flight occurred with the Mariner 6 and 7 missions to the planet Mars that were launched in 1969. A two-stage open cycled Joule–Thomson (J–T) cryocooler [80] was flown on each mission for cooling Hg-doped germanium LWIR detectors to ~23 K for measurements of the infrared spectral energy emitted between 6 and 14.3 μm from the lower atmosphere and surface of Mars. Hydrogen gas at an initial pressure of ~380 bar was stored in a 1.3 l gas vessel fabricated from 2219 aluminum. Because the J–T inversion temperature for hydrogen is 204 K, the hydrogen gas needed to be cooled before reaching its J–T expander (i.e. a capillary tube with a 0.04 mm internal diameter). Nitrogen gas from a separate vessel filled to a 414 bar pressure was expanded in an upper J–T stage of the cryostat using a 0.05 mm internal diameter capillary tube to precool the H_2 gas below its inversion temperature. Both gases were directly vented to space vacuum [80]. These two stage H_2/N_2 J–T cryocoolers provided a stable 22 K temperature at the LWIR detector with a heat load of ~0.25 W for about 100 min while the spacecraft was in orbit about Mars. Although the J–T cryocooler in Mariner 6 failed due to apparent plugging at the N_2 J–T expander [80], the hydrogen cryocooler performed flawlessly with the Mariner 7 cameras and 130 spectra from Mars were obtained in the 1.0–14 μm region from the LWIR detector during the 27 min recording period.

For over thirty years, various organizations [81, 82] have been working on the development of closed-cycled J–T cryocoolers (i.e. "sorption coolers") that utilize metal hydrides to pressurize and circulate hydrogen for the formation of cryogenic liquid or solid hydrogen. These hydrogen sorption coolers offer the possibility of long-life (i.e. 2–10 years) operation without mechanically induced vibrations and electromagnetic fields from compressor motors for cooling highly sensitive LWIR sensors and other devices on spacecraft. The configuration for a generic hydrogen sorption cooler that generates a nominal 20 K liquid phase is shown in Fig. 8.38. The key element of the 20 K sorption cooler is the metal hydride compressor [81], an absorption machine that pumps hydrogen by thermally cycling several sorbent compressor elements (i.e. sorbent beds). The principle of operation of the sorption compressor is based on the unique properties of the hydride (e.g. $LaNi_{4.8}Sn_{0.2}$), which can absorb large amounts of hydrogen at relatively low pressures (i.e. ~1.0 bar at 280 K), and which will desorb to produce high-pressure hydrogen (i.e. 50–100 bar) when heated to ~500 K in a limited volume. Electrical resistance heaters accomplish heating of the hydride while the sorbent beds are cooled by thermally connecting the compressor elements to a radiator at ~275 K. As a sorption compressor element is taken through four steps (heat up, desorption, cool down, absorption) in a cycle – illustrated by the path A–B–C–D in Fig. 8.38(b), it will intake low-pressure hydrogen and output high-pressure hydrogen on an intermittent basis. In order to produce a continuous stream of liquid refrigerant several sorbent beds (i.e. usually 4 to

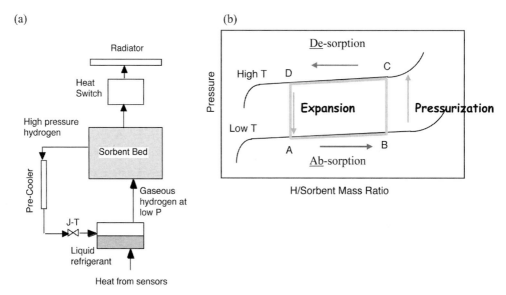

Fig. 8.38 (a) Generic schematic of a liquid hydrogen sorption cryocooler where the sorbent bed radiator rejects heat between 260 and 300 K and (b) schematic of the metal hydride isotherms where high $T \sim 470$ K and low $T \sim 280$ K and the box A–B–C–D shows the hydrogen pressurization changes during the complete J–T cycle.

Fig. 8.39 Photograph of the BETSCE sorption cryocooler in the laboratory that produced solid hydrogen at 11 K during the flight of Shuttle STS-77.

6) are needed to stagger their phases so that, at any given time, one is desorbing while the others are either heating, cooling, or reabsorbing low-pressure gas. In order not to lose excessive amounts of heat during the heating cycle, a heat switch can be provided to alternately isolate the sorbent bed from the radiator during the heating cycle, and to connect it to the radiator thermally during the cooling cycle [81, 82].

The first (and to date – only) operation of a hydrogen sorption cryocooler during a space flight was the BETSCE (Brilliant Eyes Ten-Kelvin Sorption Cryocooler Experiment) project that was managed by the Jet Propulsion Laboratory (JPL) for the former USA defense agency Ballistic Missile Defense Organization (BMDO). The BETSCE flight cooler system shown in Fig. 8.39 was launched in May 1996 on the Shuttle Mission STS-77 being mounted on the payload bay wall of the orbiter Endeavour. BETSCE had been developed for BMDO to support possible space-based missile detection and tracking by providing rapid, on-orbit periodic cooling of LWIR sensors in a surveillance satellite to ∼11 K from a stand-by temperature of ∼65 K in less than 120 s from activation. The BETSCE instrument characteristics and development are described thoroughly in Refs. [83, 84]. Briefly, a two-step process is used first to generate liquid hydrogen at ∼28 K and then to form solid hydrogen at ∼10 K, which is maintained for a minimum of 600 s when exposed to a simulated heat load of 100 mW. Prior to initiating the cooling process, hydrogen gas is stored at 100 bar pressure in a 4.0 l vessel. A sorbent bed of $LaNi_{4.8}Sn_{0.2}$ hydride collects the excess hydrogen at a pressure of ∼3.5 bar exiting from the J–T reservoir during the rapid cool-down phase. An independent $ZrNiH_x$ sorbent bed next reduces the pressure in the J–T reservoir to ∼1 mbar, which freezes the remaining hydrogen and produces temperatures as low as 9.44 K in the laboratory [83]. After all of the solid hydrogen had sublimed and the temperature started to rise in the J–T reservoir, a third hydride bed containing $LaNi_{4.8}Sn_{0.2}$ powder in an Al-foam matrix absorbed the hydrogen from the first two hydride beds and was heated to >500 K to transfer the hydrogen gas back into the storage vessel at a pressure of 100 bar to allow repeated periodic cooling. BETSCE was an adaptable test system for performance characterization of a sorption cryocooler within the highly constrained Space Shuttle safety and interface requirements. Microgravity was found to have no adverse effects on the ability of the J–T cryostat to form and retain both liquid and solid hydrogen and the hydride beds demonstrated similar heat and mass transfer parameters during ground testing before and after its space flight [83–85].

Two hydrogen sorption cryocoolers are currently being fabricated and assembled by JPL for delivery to the European Space Agency (ESA) Planck Mission, which has a planned launch in 2009. Each of these coolers is to provide continuous 1.0 W of cooling at $\sim$19 K for a minimum of 18 months of operation while the Planck spacecraft shown schematically in Fig. 8.40a is in the L_2 orbit about 1.5 million km from earth [76, 86, 87]. The objective of the Planck mission is to map the temperature anisotropy of the Cosmic Microwave Background (CMB) from the "Big Bang" formation of the universe over the entire sky with a sensitivity of $\Delta T/T \sim 2$ ppm and with an angular resolution of 10 arc-minutes in the frequency range 30 to 900 GHz. Through the use of special radiators and spacecraft orientation [76, 87], the Planck telescope and optics will be passively cooled below 60 K while the detectors on the two instruments must be actively cooled to much lower temperatures. The Low Frequency Instrument (LFI) will measure radio-frequency radiation between 30 and 100 GHz with high electronic mobility transistors cooled to $\sim$22–24 K. The High Frequency Instrument (HFI) will use infrared bolometers cooled below 0.1 K for signal detection in the 100–857 GHz range. The JPL sorption coolers will directly cool the LFI detectors and provide intermediate cooling for a mechanical helium J–T 4.2 K cooler and a ^{3}He/^{4}He dilution stage that produces temperatures below 0.1 K [76, 86, 87]. Extensive development was carried out over the past several years at JPL [86, 88] to enhance the performance and reliability of these hydrogen sorption cryocoolers and their critical components, including the fabrication and testing of a full-scale engineering model cooler, which achieved its goal of producing stable temperatures below 19 K in the J–T reservoirs. A schematic of the flight configuration for the Planck cooler is shown in Fig. 8.40(b) where the various components are identified. The six compressor elements and the low pressure storage bed contain $LaNi_{4.78}Sn_{0.22}$ hydride as the sorbent material and $ZrNiH_x$ is used to vary pressure within the gas gap heat switches between the compressor element sorbent beds and the radiators [81]. Additional information on the Planck cooler can be found in Refs. [81, 86, 88].

Although cryogenic liquid hydrogen has been stored in spacecraft to power fuel cells and for other uses (as will be described later in this chapter), liquid hydrogen Dewars/cryostats have not been used for cooling sensors or other devices [77–79]. However, solid hydrogen cryostats have been used on two past space missions [89] and are intended for a future NASA/ESA mission being planned for a launch around 2011.

The first use of solid hydrogen as a cryogen on a flight mission was for cooling the LWIR focal planes in the Spatial Infrared Imaging Telescope III (SPIRIT III)

8.4 Hydrogen in Space Applications | 393

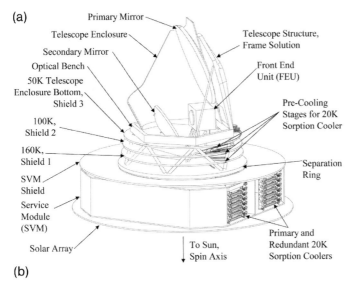

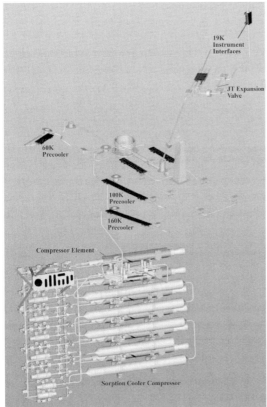

Fig. 8.40 (a) Schematic of 19 K hydrogen sorption coolers as to be mounted on Planck Spacecraft. (b) Layout of a Planck sorption cryocooler with its key components.

sensors systems [90] of the Mid-Course Space Experiment (MSX) sponsored by the BMDO agency. The SPIRIT III cryogen tank [89, 90] had a volume of 944 l and contained 80 kg of solid hydrogen at ~9 K that had been frozen within a 1.7% dense aluminum foam structure which provides heat conducting paths to the tank walls. The hydrogen was solidified *in situ* by circulating liquid helium though cooling coils surrounding the cryostat tank just prior to installing the MSX spacecraft onto a Delta II launch vehicle. The MSX instrument was launched into a near sun-synchronous LEO on 24 April 1996 from the Vandenberg AFB site. The MSX observatory tested a variety of multispectral imaging technologies to identify and track ballistic missiles during flight and studied the composition and dynamics of the Earth's atmosphere by observing the spectra of ozone, chlorofluorocarbons, carbon dioxide and methane. The SPIRIT III instruments were successfully controlled by the solid hydrogen cryogen to 10.5 K for nearly 11 months before all the hydrogen was depleted.

On 4 March 1999, the NASA funded Wide-field Infrared Explorer (WIRE) instrument [89] was launched from the Vandenberg AFB site by a Pegasus XL vehicle. The scientific objective of this mission was to study starburst galaxy evolution from a survey at two LWIR wavelengths near 12 μm and 25 μm using Si:As blocked-impurity-band detectors cooled to 7.5 K. A two-stage cryostat filled with 4.5 kg of solid hydrogen in a volume of ~50 l was to cool these LWIR focal planes to ~7 K and the optics to <13 K during a planned flight duration of four months while in LEO at an altitude of 540 km. Unfortunately, the WIRE mission failed due to the premature ejection of the instrument's telescope cover only 30 min after being launched. With the loss of the cover so early during the flight, the solid hydrogen cryostat had high solar and earth heat loadings that caused rapid venting of gaseous hydrogen. The resulting uncontrollable spacecraft spinning and venting led to complete depletion of the hydrogen within 48 h of the launch and no LWIR data could be obtained [78, 89]. Although this instrument was unusable for its original scientific objective, the WIRE's onboard star tracker has been used to measure oscillations in nearby stars, to probe their structure and to detect changes in brightness of stars caused by large planets passing between the star and WIRE.

As a partner for the future NASA James Webb Space Telescope (JWST) project [91] with a planned launch date in 2011 on an Ariane 5 vehicle, ESA is developing the conceptual design for a solid hydrogen cryostat that will hold ~1000 l just below 7 K. This system would cool the Mid-Infrared (MIRI) instrument for a minimum of five years to allow imaging and spectroscopy studies in the 5 μm to 27 μm infrared region. The JWST spacecraft would be sent to the same L_2 orbit as the Herschel–Planck missions where similar passive cooling of its telescope and optics would be to temperatures below 50 K. JWST is the planned successor [91] to the Hubble Space Telescope but extended into the infrared to observe highly red-shifted objects from the early universe that are obscured by dust at shorter wavelengths. This project will be in formulation phases through 2005 before completing detailed designs and the fabrication of flight hardware will actually start.

8.4.4
Nickel–Hydrogen Batteries for Space Application

Rechargeable (i.e. secondary) batteries are used on spacecraft in conjunction with a primary energy source, which may be a photovoltaic energy source or solar array for Earth satellites and planetary orbiters to the neighboring planets or nuclear power sources (i.e. a radioisotope combined with a thermoelectric generator (RTG) for outer planets). Requirements from these batteries typically include (i) providing power to the spacecraft, its equipment and instrumentation during solar eclipse periods, during nighttime and during mid-course attitude correction, (ii) supporting peak power demands during data transmissions, local mobility and/or subsurface operations and (iii) enabling pyrotechnic device firing. The batteries are charged back by the primary power source after each discharge to restore them to their original condition. The ability of the batteries to repeat multiple discharge and charge operations is defined in terms of their cycle life, which is a very important characteristic. For many space applications, such as planetary orbiters and satellites, the expected cycle life of the batteries may exceed thousands of cycles. In addition to long cycle life, the batteries are required to have (i) long shelf or operating life, (ii) high round trip efficiency in a charge and discharge cycle, (iii) good safety, (iv) no electrolyte spillage, (v) minimum self discharge, (vi) high energy density and above all (vii) a high degree of reliability for remote application, as in space. These constraints limit the number of rechargeable battery systems to a few, among which the nickel–hydrogen battery system stands out as the preferred choice for space missions, especially for multiple years of operation. Some of the notable NASA space missions powered by nickel hydrogen batteries include various Mars Orbiters, such as Mars Global Surveyor, Mars Odyssey and Mars Reconnaissance Orbiter, other long-life missions, such as Hubble Space Telescope (HST), International Space Station, Stardust, Genesis and several other LEO and Geosynchronous Earth Orbit (GEO) satellites. Indeed, the number of cumulative years nickel–hydrogen has thus far been in service far exceeds that of any battery system, even though it has been in use for less than two decades.

The first functional nickel–hydrogen (Ni–H_2) cell was developed in Russia. The first patent was granted in 1964. In 1970 INTELSAT (COMSAT & Tyco Laboratories) initiated the first aerospace Ni–H_2 cell development for GEO satellites [92, 93] In 1972, the US Air Force and Hughes Aircraft began cell development for LEO satellite applications. These developments were motivated by the fact that Ni–H_2 cells demonstrated a higher specific energy and cycle life than nickel–cadmium (Ni–Cd) cells. Reference [94] provides a good in-depth description of this battery technology as related to various NASA applications.

A Ni–H_2 cell may be viewed as a hybrid of the alkaline Ni–Cd cell with the alkaline hydrogen–oxygen fuel cell. Simply, the hydrogen electrode from the fuel cell is combined with the nickel oxide positive electrode from the Ni–Cd cell, thus forming a battery system with two of the most reversible electrodes. Overall, the reaction within the Ni–H_2 cell is:

$$NiOOH + (1/2)H_2 \Leftrightarrow Ni(OH)_2 \tag{8.25}$$

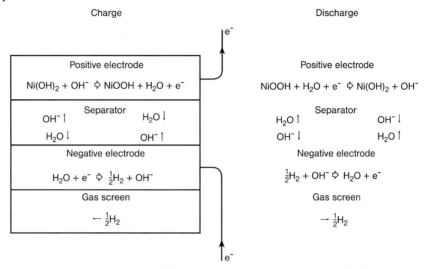

Fig. 8.41 Reactions in a Ni–H$_2$ cell during normal operation of charge and discharge.

The essential difference is that the fuel (i.e. hydrogen gas) is contained in a hermetically sealed vessel. The Ni–H$_2$ cell contains a sintered nickel oxyhydroxide (NiOOH) as the positive electrode and a porous platinum-catalyzed hydrogen electrode (similar to a fuel cell electrode) as the negative electrode in a 31 wt.% potassium hydroxide electrolyte solution with a Zircar (zirconium oxide) separator. The cell, therefore, contains hydrogen at a relatively high pressure ∼ 27–55 bar, which varies linearly with the state of charge of the cell. The cell is usually assembled in the discharged state, with nickel hydroxide as the positive active material. During the initial charge process or "formation" as the first charging is called, hydrogen is evolved at the negative electrode, while nickel hydroxide is converted to nickel oxyhydroxide. The reverse reactions take place during discharge, as illustrated in Fig. 8.41. During the latter portions of charge and overcharge, oxygen evolves at the positive electrode, as a parasitic process to the nickel electrode reaction. The oxygen thus evolved recombines electrochemically at the negative electrode, or chemically at the platinized wall wick, thus providing tolerance to overcharge. There is, thus, no net chemical change during overcharge. Likewise, the cell reversal also produces no net change in the cell chemistry. The nickel oxyhydroxide electrode can undergo reduction during open-circuit stand or discharge (termed as self-discharge) against either hydrogen oxidation or oxygen evolution. It may be noted that self-discharge against hydrogen, which contributes to the higher self-discharge rate of Ni–H$_2$ compared to Ni–Cd is similar to the cell discharge reaction, except that both the oxidation and reduction reactions occur on the same (positive) electrode. During self-discharge against oxygen, the evolved oxygen gas may combine chemically with hydrogen at the negative electrode, causing a proportionate decrease in the hydrogen pressure. The hydrogen pressure can thus be linearly correlated with the activity of the positive active material.

Fig. 8.42 Photograph of Ni–H$_2$ cell components prior to final assembly.

Since the hydrogen pressure is a linear function of the extent of charge, it is often used as a state of charge indicator for the cell. The H$_2$ pressure can build to ∼70 bar towards the end of charge. Therefore, the cell pack is to be contained in a pressure vessel with hemispherical caps (typically constructed of Inconel 718 alloy) on both ends to form a hermetically sealed unit, as shown in Fig. 8.42. The Inconel material is especially resistant to hydrogen embrittlement. There exist two design variations in terms of electrode ratios: (i) hydrogen precharge, that is excess hydrogen compared to the capacity of the positive electrode and (ii) positive precharge, where part of the hydrogen formed during initial charging is vented out. The positive precharge design seems to provide longer cycle life. The negative precharge design, on the other hand, provides complete protection against deep discharge, with the oxygen evolving from the positive electrode combining with the remaining hydrogen. Two levels of electrolyte concentrations have been widely tested, either 31 or 26 wt.% potassium hydroxide solution. Higher concentration electrolyte is believed to maximize the capacity, whereas the lower concentration electrolyte is reported to extend the life [95]. The current practice has been to use 31% solution for almost all of the space missions. Similarly, there exist three variations with respect to the pressure vessel: (i) individual pressure vessel (IPV), where each cell is encased in its own pressure vessel, (ii) common pressure vessel (CPV), where multiple cells, typically two, share the pressure vessel and (iii) single pressure vessel (SPV) where all the cells required for the battery are housed in a single pressure vessel.

Two basic types of electrode stack designs are used for aerospace cells: (i) The back-to-back design of Comsat [96] and (ii) the recirculating design developed by a USAF team working with the Hughes Company [97]. Figure 8.43(a) shows the

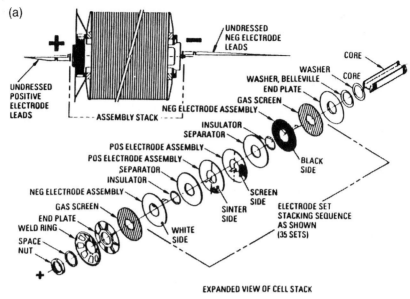

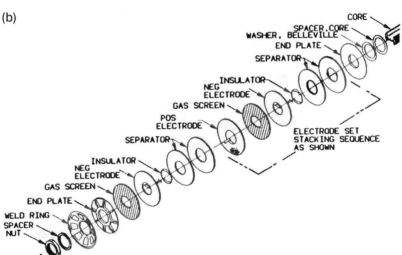

Fig. 8.43 Aerospace Ni–H$_2$ battery cell electrode stack designs: (a) back-to-back (Comsat version) and (b) recirculating (US Air Force version).

back-to-back design, where two positive electrodes are positioned back-to-back. On either side of the positive electrode are placed separators, followed by negative electrodes (with the platinum side facing the positive electrode), and gas screens [95]. These components constitute one module of the electrode stack and several such modules are housed in a pressure vessel, depending on the desired capacity. In the recirculating design shown in Fig. 8.43(b), a Ni–H$_2$ cell consists of several modules, each consisting of a gas screen, negative electrode, separator and positive electrode, with the last stage containing a hydrogen electrode and separator [96]. Again, there are two variations for the electrode configuration. In the first configuration, the bus bars are located on the outside of the electrode stack (Comsat design), located on opposite sides for the positive and negative electrodes. In the second configuration, the electrodes are in the shape of a pineapple slice (i.e. the Air Force/Hughes Design) with larger center hole for the tabs, which run in opposite electrodes for positive and negative electrodes. In recent years a modified design known as the "ManTech" design has combined the Comsat stacking arrangement with the pineapple-slice geometry and other features of the Air Force design. These systems provide higher energy density with improved gas electrolyte and thermal management [95]. After e-beam welding the pressure vessel shells to the weld ring, the pressure vessel is sealed by compression seals. The electrolyte fill tube is a part of the negative terminal and is welded after electrolyte filling. The cell pressure is monitored using a strain-gauge bridge mounted on the dome of the cell.

Multiple cells, typically 22 for a 30 V battery, are mounted on base plate either horizontally or vertically. In the latter design, which is more common, the cells are mounted using aluminum sleeves, which serve two purposes (i.e. to support the cells on the base plate under launch loads and to provide thermal management by mainly dissipating heat evolved from the battery during discharge). A flexible silicone rubber adhesive is used to provide the ability of the bond to sustain the shear stress between the cells and the sleeve that develops as a result of pressure vessel expansion during charge. For example, the Ni–H_2 battery [98] currently being used in the International Space Station has 76 cells, each with 81 Ah, and is packaged in two 38-cell assemblies, as illustrated in Fig. 8.44. The overall specific energy and energy density, with weight and volume of the assembly included are 24 Wh kg^{-1} and 10 Wh l^{-1}. These batteries are designed to provide a minimum of 6.5 years of mission life.

In the early stages of Ni–H_2 technology, there were many competitive manufacturers and alternative designs. Eventually, it became clear that the ManTech design was superior and that EaglePicher Technologies (Joplin, MO, USA) had become the dominant producer. A joint NASA/USAF LEO Life Test study verified

Fig. 8.44 Ni–H_2 battery module for the NASA International Space Station.

that the EaglePicher battery was clearly best for aerospace applications [94]. The NASA Glenn Research Center (Cleveland, OH, USA) made major contributions to the technologies of electrolyte concentration, wall-wicks and standardization of manufacturing processes. Problems with the baseline Ni–Cd batteries also eased the difficulties in infusing Ni–H_2 technology. After the 1986 Shuttle/Challenger disaster, a joint-services team dedicated itself to applying Ni–H_2 technology to the Hubble Space Telescope (HST). The original five batteries manufactured in 1990 are still supporting the mission. The impact of this government-wide acceptance was huge. Ni–H_2 technology has also made, and continues to make, a major contribution to USA national defense. The Ni–H_2 business in the USA peaked at approximately 6000 cells/year, (or $90 M/year) normalized to 50 Ah, with HST, ISS and Defense programs accounting for the bulk of this market. EaglePicher has supplied about 90 % of the Ni–H_2 cells and batteries for the various national and international space programs.

Over the years, several Ni–H_2 cell designs have evolved. The most common design is the individual pressure vessel (IPV) cell. This cell has a voltage under load between 1.20 and 1.27 V. Pressure vessel diameters range from 8.9 to 14.0 cm. Cell capacities range from 20 to 350 Ah. Because this cell requires a pressure vessel, the volume is large compared with the Ni–Cd and the energy densities are about 60 Wh l^{-1}. The specific energy, on the other hand, is higher than with Ni–Cd; the IPV Ni–H_2 cell has a demonstrated specific energy of 50 Wh kg^{-1}. Use of a common pressure vessel (CPV) or a single pressure vessel (SPV) battery can significantly reduce the volume. The CPV battery utilizes two cells in series sharing the same pressure vessel. The SPV battery contains as many as 22 cells sharing a common pressure vessel. SPV Ni–H_2 batteries have been manufactured and flown in the Clementine and Iridium programs. The capacities for these batteries are lower than the batteries using IPV cells. For example, cells for commercial IPV batteries can be as high as 350 Ah and the largest SPV battery flown (on the Iridium Program) to date is 50 Ah. Some of the notable NASA missions utilizing Ni–H_2 batteries are listed in Table 8.4 and the typical performance characteristics of these batteries are listed in Table 8.5. Apart from a higher specific energy compared with Ni–Cd cells, Ni–H_2 cells offer longer life at a higher depth of discharge (DOD). Cells designed for the ISS (for example) are designed and expected to reach 30 % DOD compared with Ni–Cd cells that are typically designed for 10–25 % DOD and often operate at lower DOD. Ni–H_2 cells cycled at 50–75 % DOD have far exceeded comparable performance by Ni–Cd cells. Cells in GEO orbit have been designed to meet a requirement of 75 % DOD at its maximum eclipse period compared with the 50–60 % for Ni–Cd cells. Additional benefits for Ni–H_2 cells include: (i) simple state of charge indication from their internal pressure, (ii) potential cell reversal without damage to the cell, (iii) greater tolerance to overcharge than Ni–Cd cells. The disadvantages, on the other hand, include: (i) heat generation on overcharge results in internal shorts caused by "popping" (i.e. the uncontrolled recombination of oxygen generated at the nickel electrode with hydrogen on the catalyzed surface of the hydrogen electrode [95]), "clam shelling" and faulty cell assembly, (ii) possible

Table 8.4 Some examples of NASA's space missions that use Ni–H$_2$ batteries (cell design types: IPV = individual pressure vessel, CPV = common pressure vessel, SPV = single pressure vessel)

Mission	Initial Launch Date (M/D/Y)	Cell Size (Ah)	Case Type and Diameter (cm)
Hubble Space Telescope, LEO	4/25/1990	80	IPV, 8.9
Landsat, LEO	ca. 1995	50	IPV, 8.9
GOES, GEO	ca. 1996	160	IPV, 11.4
MGS, low Mars orbit	11/7/1996	20	CPV, 8.9
Iridium, LEO	5/5/1996	50	SPV —
International Space Station, LEO	ca. 2000	87	IPV, 8.9
Clementine, lunar polar orbit	1/6/1998	30	SPV —
Stardust, comet sample return	2/7/1999	20	CPV, 6.35
EOS, LEO	12/18/1999	50	IPV, 8.9
Genesis, 5 loop halo orbit around L1	8/8/2001	20	CPV, 6.35
MAP, L2 Sun–Earth orbit	6/30/2001	23	CPV, 6.35
Odyssey, 1.98 hour Mars orbit	10/23/2001	16	CPV, 6.35

pressure vessel punctures/leaks/rupture, (iii) possible loss of capacity on storage and (iv) higher cell cost. Ni–H$_2$ batteries typically utilize cylindrical Ni–H$_2$ cell pressure vessels with hemispherical end-caps as shown in Fig. 8.42. The packing factor for the cylindrical IPV and CPV Ni–H$_2$ cells is less efficient than a Ni–Cd battery containing prismatic cells. The SPV design however, which consists of a 22 cells in a single case, is more efficient from a battery structure point of view and, therefore, exhibits a higher specific energy. Batteries using these cells typically contain between 16 and 76 cells. The buss voltage is usually between 28 and 123 volts, with the voltage from the battery sometimes being bucked from a higher voltage or boosted from a lower voltage to the required voltage using a DC-to-DC converter.

Ni–H$_2$ batteries have been used in several Earth and planetary orbiting missions since the first commercial flight by INTELSAT V in 1983. Because of its demonstrated long life in GEO, this battery has been the workhorse for commercial communications satellites over the past 20 years. NASA first utilized Ni–H$_2$ batteries in a LEO orbit when it launched the HST mission. These batteries, operating at a low 10% DOD, have performed flawlessly since 1990.

The life of a secondary battery depends on how it is operated. Depth of discharge, charge and discharge rate and temperature all influence the life and reliability. The most pronounced effect on secondary battery life is depth of discharge, and high temperature. Secondary batteries require a power system that is designed to control the charge limits of the battery. The thermal design of the battery structure must be capable of maintaining the temperature in the range 273–293 K to achieve long

Table 8.5 Performance characteristics of Ni–H$_2$ secondary batteries during NASA space missions where most operated in the temperature range 263 to 303 K

Technology	Use	#Batteries/ #cells	Ah Rated/ Actual	Voltage (V)	Specific Energy (Wh kg^{-1})	Energy Density (Wh l^{-1})	Design Life (year)	Cycle Life to Date (Cycles)
IPV Ni–H$_2$	IPV Cell	1	98/83	1.25	48	71		10
	Space Station	6/76	81/93	48	24	8.5	6.5	11 K
	HST	6/22	80/85	28	8	4	5	65 K
	Landsat 7	2/17	50/61.7	24			5	>50 K
CPV Ni–H$_2$	CPV Cell	2	16/17.5	2.50	43.4	77	10	50 K
	MIDEX MAP	1/11	16/17.5	28	36	21	5	1 K
	Odyssey	2/11	16/17.5	28	36	21	10–14	
	Mars 98	1/11	16/17.5	29	37	41	3	
	MGS	2/16	20/23	20	35	25	1 Mars year	50 K
	EOS Terra	2/54	50/	67		21	5	
	Stardust	1/11	16/17.5	28	36	21	7	1135
SPV Ni–H$_2$	SAR 10065	1/12	50/60	28	54.6	59.3	10	
	Clementine	1/22	15/18	28	54.8	78		200
	Iridium	1/22	60/70	28	53.4	67.7	3–5	50 K

life. The usual operating range for Ni–H$_2$ cells is 263 to 303 K. The temperature of the battery structure or base plate needs to be controlled in a spacecraft and may be controlled in either a passive or active manner.

One method to optimize battery life in a LEO application is to use a temperature-compensated voltage control (V_T), that is a temperature-dependent fixed voltage during charge when the cells reach a point at which gassing occurs. This method results in drawing a high current from the solar arrays until the battery reaches a preset voltage limit. The charge process is highly efficient during this period. When the battery reaches the preset voltage, the charge current tapers off until the end of the sunlight period. This reduces the overcharge and I^2R heat that is a major cause of cell degradation. Several voltage limit curves are available in the charge control system to account for unexpected high depths of discharge and/or imbalance between cells and/or batteries. This approach to charge control contributes to the long life achieved in many missions.

In a GEO orbit, the cells are in the Sun almost year round and, therefore, are trickle-charged to minimize the current and the heat generated during the long charge process during the full sun period. During the eclipse season, when the spacecraft is in the sun, the battery is charged at higher rates to return the energy used during the eclipse. Then the charge current is reduced using a V_T control, or in the Ni–H$_2$ case based on signals from pressure transducers on some cells in the battery indicating that the cells have reached a point of near full charge. The current is then reduced to minimize the overcharge current.

Another rechargeable battery chemistry that came into existence in the last decade is nickel–metal hydride (Ni–MH), which has been developed for commercial portable electronics applications [99, 100]. This system is very similar to Ni–H$_2$ technology, except hydrogen is stored in reversibly absorbing and desorbing solid metal hydrides at fairly low pressures (i.e. $\sim$1.0 bar). These cells can, therefore, be contained in thin metallic or even plastic cases, like the conventional Ni–Cd cells. Again, the Ni–MH cells are fabricated in the discharged state with sintered or plastic-bonded nickel hydroxide cathodes and plastic-bonded metal hydride anodes in a 31% KOH solution. Upon charging hydrogen generated at the metal hydride electrodes is absorbed within the solid phase hydride and is available for subsequent discharge. This chemistry has the same tolerance to overcharge and overdischarge as other nickel-based systems. Two classes of intermetallic alloys of rare earth and/or transition metals are typically used. The AB$_5$ alloys contain a naturally occurring mixture of lanthanides (La, Ce, Pr, Nd) known as mischmetal (Mm) with a typical alloy composition being MmNi$_{3.5}$Co$_{0.7}$Mn$_{0.4}$Al$_{0.3}$ [99]. The second common class of metal hydrides for Ni–MH [100] has the formula AB$_2$, similar to the Laves-type ZrV$_2$ alloy, with several substitutions for both V and Zr, such as Ti, Fe, Ni, Cr, Mn, and so on. AB$_5$ hydrides have low absorption pressures and flat isotherms and produce batteries with capacities in the range of 250 to 370 mAh g^{-1}. AB$_2$ hydrides, on the other hand, have higher pressures with sloping isotherms and have been claimed to give higher specific capacities [100]. Almost all of the commercial Ni–MH batteries use AB$_5$ compounds for reasons of durability and cost. Nickel metal hydride batteries had an impressive start [99] during the 1990s when they were introduced in the portable electronics market to replace conventional Ni–Cd batteries. They provide almost 100 % improvement in specific energy, compared to Ni–Cd, with almost identical performance characteristics otherwise. There are uncertainties relative to the cycle life in the Ni–MH cells, especially at partial depths of discharges, where Ni–Cd batteries have well-established cycle lives of over 30 000 cycles under these conditions. Ergenics (Ringwood, NJ, USA) has developed various segmented battery designs that incorporate moisture tolerant hydride beds with long-life Ni–H_2 battery cells [101] that might be attractive for aerospace applications. Before Ni–MH chemistry was thoroughly evaluated for space needs, the interest in developing Ni–MH batteries faded quickly, especially within the aerospace community, with the advent of lithium-ion batteries that provide almost a twofold benefit over Ni–MH batteries in specific energy. In fact, it is highly likely that lithium-ion batteries will gradually replace the Ni–H_2 batteries for many future space applications.

8.4.5
Other Applications in Past/Current Space Missions

The cryogenic tanks developed for NASA's Apollo program were the first space-qualified liquid hydrogen and oxygen storage vessels for operation in near-zero gravity environments [77]. These vessels held the reactants for the alkaline fuels that provided the electrical power in the Apollo Command/Service Modules (CSM) and also generated potable water (as a by-product) for drinking, personal hygiene and the reconstitution of dehydrated food by the astronauts as well as CSM cabin cooling [102, 103]. The Apollo fuel cells were located within the Service Module with H_2 and O_2 being supplied at 4.1 bar into chambers of concentrated (i.e. ~31 wt.%) potassium hydroxide separated by porous, sintered nickel electrodes that operated at nominally 483 K to form extremely pure water saturated with hydrogen gas. Each of the cryogenic hydrogen and oxygen liquids was distributed into three separate vessels where the total stored quantities were 36.5 and 432 kg, respectively, at lift-off. During a normal lunar mission [104], the three fuel cells would supply about 650 kWh energy at an average current of 77 A and 28.8 V. It should be noted that the rupture of an oxygen tank in the Service Module and not the hydrogen tank caused the famous accident during the Apollo 13 mission in April, 1970. The nominal generation rate for potable water was ~0.54 kg h^{-1} and this water was transferred via plumbing to the Command Module for consumption by the three astronauts. During the first Apollo flights, the excessive H_2 gas entrapped within this water produced major gastric discomfort for the Astronauts causing loud complaints to NASA Mission Control. This problem was solved for the Apollo 12 and later missions with the development of a filtering system using palladium–silver tubes through which only hydrogen gas would permeate and subsequently be vented into space. The degassed water was then conveyed to the water valve panel in the Command Module for consumption by much more comfortable astronauts [103].

Liquid hydrogen and oxygen are also stored in vacuum-insulated spherical tanks in the midfuselage under the payload bay liner of the Space Shuttle Orbiters to provide electrical power and potable water during these missions [105]. These cryogenic storage tanks are grouped into sets of one H_2 vessel and one O_2 vessel with up to five sets being installed, depending on the duration of a specific STS flight. The hydrogen vessels are constructed of aluminum alloy 2219 with an outside diameter of 1.16 m and total volume of 606 l. Each of these tanks has an empty mass of 98 kg and stores ~42 kg hydrogen at an initial temperature of 22 K. H_2 gas is provided to the alkaline fuel cell power plants at pressures between 13.8 and 15.4 bar. There are three fuel cell power plants located in the forward portion of each Orbiter's midfuselage. Each plant is capable of supplying 12 kW peak and 7 kW maximum continuous power in the voltage range 27.5–32.5 V(DC). A centrifugal water separator extracts liquid water from the fuel cell H_2-water exhaust stream and transfers the water under pressure to potable storage tanks in the lower deck of the crew cabin. This water is thus available for crew consumption and also for cooling the Freon-21 coolant loops in the orbiter. The residual circulating hydrogen is directed back to the fuel cell stacks for reaction.

Another occasional use of hydrogen during space missions is as a reference gas with specific scientific instruments. One example of this application involves the Cassini–Huygens Mission to Saturn and its moon Titan that was launched in October 1997 and arrived at its destination during the summer of 2004 [106]. Two metal hydride storage vessels constructed of Inconel alloy 750 and each filled with 58 g of cerium-free $MmNi_5$ hydride to store 0.86 g (9.6 sL) of hydrogen were fabricated by Ergenics. The hydride unit provided ultrahigh purity H_2 gas to the Gas Chromatograph Mass Spectrometer (GCMS) instrument of the Huygens probe that measured the composition of Titan's atmosphere during a descent to its surface on 14 January 2005. The GCMS successfully operated for 226 minutes and returned over 8000 mass scans of gas compositions [106] of the atmosphere during the Huygens descent onto the surface of Titan.

The planned use of metal hydride gas gap thermal switches in the compressor beds of the Planck sorption cryocoolers [81, 86, 88] was briefly mentioned in Section 8.4.3. These devices can provide alternating heat conduction and thermal isolation between two surfaces by varying the hydrogen pressure over the range ~0.5 to ~1000 Pa with a suitable metal hydride [108]. Laboratory tests at JPL on the compressor elements of the engineering version of Planck cooler have demonstrated that $ZrNiH_x$ controlled heat switches are very efficient and robust [81, 88]. Similar hydride-based thermal switches have been developed and evaluated for other types of cryogenic refrigeration systems [109, 110]. These devices also appear to have very good potential for terrestrial applications [108].

To conclude this section, two other metal hydride systems that have been proposed and partially evaluated for space technology but not yet implemented (to the best of the authors' knowledge) will be mentioned. In a study for the NASA Kennedy Space Center, Ergenics assessed the feasibility of using hydride beds to capture a portion of the extremely large quantities of hydrogen boil-off gas at the launch site to be converted back into the liquid [111]. A payback of about three years was projected. Another novel use suggested for metal hydrides was to exploit their endothermic behavior during the desorption process to cool astronauts' space suits during extravehicular activities in space [112]. While some initial studies on candidate alloys and sorbent bed designs were carried out, no complete prototypes of a space suit with a hydride cooling system have been developed.

8.4.6
Possible Future Applications in Space

During the next couple of decades or so, hydrogen may also play a dominant role in some novel technologies beyond those described in the previous sections. These opportunities generally involve improvement in efficiency, cost reduction, or both in propulsion technology.

Hydrogen may enable the collection of substantial quantities of samples from Mars to bring back to Earth or help facilitate manned exploration of Mars and beyond. The first idea is to transport sufficient quantities of hydrogen (presumably as cryogenic liquid/solid) to Mars for the *in situ* production of both fuel (i.e. methane) and oxygen from the carbon dioxide in the Martian atmosphere [113–115]. Only a fraction of the total mass of these products needs to be sent as hydrogen stored on the spacecraft. In principle, oxygen and water necessary for human existence on Mars could also be produced from this hydrogen, which would be recycled. Probably the most mature processes to accomplish these objectives would use hydrogen from Earth in the sequence of the catalyzed Sabatier reaction [75]

$$CO_2 + 4H_2 \rightarrow CH_4 + 2H_2O \tag{8.26}$$

after compression of the CO_2 from the Martian atmosphere and followed by the electrolysis process

$$2H_2O + \text{electricity} \rightarrow 2H_2 + O_2 \tag{8.27}$$

where the electricity comes from solar or RTG sources. The products from reactions (8.26) and (8.27) would be separated and stored in fuel tanks on the Martian landing vehicle. The Sabatier/electrolysis processes work with high efficiency in the laboratory but would require further development and optimization for use on Mars. However, the major challenge is to maximize the hydrogen storage density via either compact H_2 "slush" (i.e. mixtures of solid and liquid phases that maximize storage density) [113] or "zero-boil-off" cryogenic liquid H_2 Dewars [115] to eliminate hydrogen evaporative losses during and after flight.

Hypersonic air-breathing propulsion systems are promising to reduce launch costs into LEO from current levels of ~US$20 000/kg towards ~US$200/kg within the next few decades [116, 117]. Payload capacity of these reusable space launch vehicles can be increased by discarding the heavy oxygen tanks of rockets and using the oxygen in the earth's atmosphere as the vehicles fly at many times the speed of sound. The first successful hypersonic flight of an unmanned research aircraft with a supersonic combustion ramjet (scramjet) occurred in March 2004 when a NASA X-43A scramjet operating on pressurized hydrogen gas set a new record speed of Mach-7 (i.e. seven times the speed of sound or $\sim$2.3 km s^{-1}) for an air-breathing propulsion system during the 10 s duration of its gaseous hydrogen fuel supply. After igniting the hydrogen using injected hypergolic silane, the X-43A engine burned $\sim$1.4 kg of H$_2$ from bottles filled at initial pressures of $\sim$545 bar. To reach the speeds of about Mach-25 that are needed to go into orbit, supplemental rockets fueled with LH2/LOX will be used with the hypersonic engines that burn hydrogen from slush storage tanks [116, 117].

The energies (per unit mass) available from nuclear reactions (i.e. fission and fusion) are about 100 times larger than provided by chemical reactions. This results in the potential for enormous increases in the specific impulse I_{sp}. However, in practice there are many practical difficulties [75] in converting energy from the nuclear reactions into energy of the exhausted propellant "working fluid." For example, an O$_2$/H$_2$ chemical rocket engine operating at an oxidizer-to-fuel (O/F) ratio of 5 can transfer in excess of 80 % of the ideal O$_2$/H$_2$ chemical reaction energy (at a stoichiometric O/F of 8) to the exhaust gases, most nuclear propulsion concepts have much lower efficiencies [75]. Namely, fission reactor-based concepts such as nuclear thermal propulsion (NTP) and nuclear electric propulsion (NEP) have a much lower overall effective energy density because of the fundamentally low fraction of nuclear fuel "burn-up" in their reactors [70]. Figure 8.45 shows a "Bi-Modal" hybrid NTP/NEP system [75] where a nuclear-thermal rocket (NTR) provides high thrust-to-weight (T/W) propulsion to minimize gravity losses and trip times and then operates in the NEP mode for low-T/W, high-I_{sp} interplanetary transfer. As shown in Fig. 8.45, NTR uses hydrogen as the working fluid to give a moderate T/W > 0.1 at I_{sp} values in the range 8000–9000 m s^{-1}. In this engine the H$_2$ gas propellant is heated as it passes through a heat-generating solid fuel core (e.g. uranium oxide or carbide) [75, 118]. As described by Frisbee [75], an expander cycle drives turbopumps, and control drums located on the periphery of the core

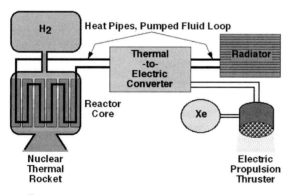

Fig. 8.45 A proposed dual mode nuclear propulsion system.

control the reactivity of the reactor. Material constraints are a limiting factor in the performance of solid core nuclear rockets. The maximum operating temperature of the hydrogen must be less than the melting point of the fuel, moderator and core structural materials. Electric power for the NEP system is obtained by operating the NTR reactor at a low thermal power level (so that no NTR H_2 propellant is required for reactor thermal control) with a closed-loop fluid loop (e.g. heat-pipes or pumped fluid loop) extracting heat from the reactor. This thermal energy is in turn used in a static or dynamic thermal-to-electric power conversion system for electric power production. This electric power is supplied to ion thrusters that exhaust a flux of heavy mass ions (e.g. xenon). While NEP has a very low T/W (e.g. vehicle T/W $< 10^{-3}$), it gives very high I_{sp} values (e.g. 20 000–50 000 m s^{-1}).

Approximately US$7 billion (1996 dollars) was invested in solid-core nuclear rocket development [70, 75] in the US prior to 1973. This work was mainly directed at the manned Mars mission and concentrated on the development of large, high-thrust engines. A series of engines based on hydrogen-cooled reactor technology was built and tested, starting in the1960s and through the early 1970s. The flight-rated graphite engine that was developed as a result of this program [75] was called NERVA (Nuclear Engine for Rocket Vehicle Application). This engine was designed to operate at 1500 MW, provide 333 kN of thrust at a specific impulse of 8100 m s^{-1} and have an engine weight of 10.4 t. It was engineered for a ten-hour life and 60 operating cycles [75]. The NERVA engine development program was very near completion when terminated in 1972. The next step would have involved a flight demonstration in Earth orbit. Since that time, there has been some limited work on fuels and materials; several "NERVA-derivative" engines have been proposed which would employ modern materials, turbopumps and turbopump cycles to take NERVA performance into the 8800 m s^{-1} I_{sp} range. Finally, solid-core nuclear thermal rocket propulsion technology has been under development in the former Soviet Union. Thus, solid-core nuclear thermal rocket propulsion represents a relatively mature advanced propulsion technology that may be suitable for exploration of the outer planets and their moons [118].

The more likely fusion technology to be used for space flights is Inertial Confinement Fusion (ICF) with high-power lasers or particle beams to compress and heat pellets of fusion fuel to fusion ignition conditions [70, 75]. A pellet of fusion fuel (i.e. mixtures of the hydrogen isotopes deuterium–tritium, D–T) is placed at the locus of several high-power laser beams or particle beams that simultaneously compress and heat the pellet, as described by Frisbee [75]. The pellet's own inertia is theoretically sufficient to confine the resulting plasma long enough so that a useful fusion reaction can be sustained; hence this fusion reaction is inertially confined. Most of these fusion concepts for space-based propulsion are spin-offs of fusion reactor technology from the US Department of Energy (DOE) terrestrial fusion research programs. However, the figures-of-merit for a propulsion system are so different (e.g. specific impulse, specific mass, etc.) from those of a terrestrial power station generating low-cost electricity, a fusion reactor type (technology) selected for space propulsion and power applications will likely be very different than one selected for terrestrial electric power production [75].

Frisbee [75] has summarized two ICF space propulsion systems for missions that may be 50 or more years in the future. The first concept called VISTA (Vehicle for Interplanetary Space Transport Applications) has been studied by a team of DOE/NASA centers to allow a fast (i.e. 60 day round trip) manned-Mars mission with a 100 t payload in a vehicle sized at 5800 t, of which 4100 t is hydrogen expellant and 40 t is D–T fuel. The VISTA ICF engines would produce a jet power of 30 000 t at 30 Hz operation rate (i.e. 30 D–T pellets are ignited per second in the magnetic thrust chamber) and a specific impulse of $166\,600\,\text{m s}^{-1}$. This concept design is based on very optimistic assumptions regarding the success of present inertial confinement fusion research efforts and on the spacecraft technology expected to be available by the year 2020. The British Interplanetary Society conducted a design study in the 1970s to evaluate the feasibility of ICF propulsion for interstellar travel [75]. The vehicle was called Daedalus and was designed for an interstellar flyby with a total velocity of 0.1 c (i.e. the speed of light). Daedalus was engineered as a two-stage vehicle with a total mass at ignition of 53 500 t. The first stage carries 46 000 t of propellant and would produce a thrust of 7.5×10^6 N. The burn time is estimated to be about 2 years. The second stage carries 4000 t of propellant and would produce a thrust of 6.6×10^5 N during its burn time of also about 2 years. The final net payload would be 830 t. The specific impulse for each stage is approximately 10^7 m s^{-1} or 0.03 c. The D–^{3}He propellant mass would be 30 000 t of ^{3}He and 20 000 t of D. In order to obtain this enormous amount of the very rare ^{3}He isotope, the atmosphere of Jupiter would be "mined" (via liquefaction and isotope separation techniques) using floating factories suspended from hydrogen balloons. Nuclear-fission rockets would ferry the propellant up to the Daedalus vehicle [75]. While this scheme would certainly involve hydrogen isotopes in extremely novel roles, the major advances in primary and secondary technology are almost certainly generations away from the present.

8.4.7
Summary and Conclusions

For over forty years hydrogen has contributed to development and implementation of space technology. Its unique combination of chemical and physical properties led to very diverse applications from being the most efficient fuel for chemical propulsion systems to providing cryogenic temperatures for specialized scientific instruments. Not only does hydrogen contribute to spacecraft power generation and management via nickel–hydrogen batteries and fuel cells, it serves as a source of potable water to the human occupants during space flights. The remarkable reversible chemistry of metal hydrides gives various devices ranging from sorption cryocoolers, to batteries, to thermal switches, to ultrapure gas sources. It is expected that some of the new roles for hydrogen described in Section 8.4.6 will become more prominent during the next few decades. However, very strong incentives currently exist to address the fundamental issue of storing the maximum quantities of hydrogen in the smallest possible volumes – whether to enhance its propulsion capabilities or extend the size and duration of space missions.

Acknowledgments

This paper was prepared at the Jet Propulsion Laboratory, California Institute of Technology, under a contract with the National Aeronautics and Space Administration. We also appreciate the information and assistance provided by D. H. DaCosta (Ergenics Hydride Solutions), J. E. Elvander (Boeing), R. C. Longsworth (SHI-APD Cryogenics, Inc.) and J. W. Reiter (Swales Aerospace).

References

1 Kordesch, K., Simander, G. (1996) *Fuel Cells and Their Applications*, Wiley-VCH, Weinheim, Germany.
2 EG&G Services and National Energy Technology Laboratory (2000) *Fuel Cell Handbook*, 5th edn, report for US Department of Energy, October.
3 Carrette, L., Friedrich, K.A., Stimming, U. (2000) *ChemPhysChem*, **1**, 162–193.
4 Schönbein, C.F. (1839) *Philos. Mag (III)*, **14**, 43.
5 Grove, W.R. (1839) *Philos. Mag (III)*, **14**, 127.
6 Grove, W.R. (1842) *Philos. Mag (III)*, **21**, 417.
7 Appleby, A.J., Foulkes, F.R. (1989) *Fuel Cell Handbook*, Van Nostrand Reinhold, New York.
8 George, R.A., Bessette, N.F. (1998) *J. Power Sources*, **71**, 131–137; Bessette, N.F., Borglum, B.P., Schichl, H., Schmidt, D.S. (2001) *Siemens Power J.*, **1**, 10–13.
9 Raak, H. (2001) *Sulzer Tech. Rev.*, **3**, 4–7.
10 Spedding, P.L. (1973) *J. Electrochem. Soc.*, **120**, 1049–1052.
11 Horvath, A.L. (1985) *Handbook of Aqueous Electrolyte Solutions*, Ellis Horwood Ltd., Chichester.
12 Brandon, N.P., Skinner, S., Steele, B.C.H. (2003) *Annu. Rev. Mater. Res.*, **33**, 182–213.
13 Dimitrova, P., Friedrich, K.A., Stimming, U., Vogt, B. (2002) *Solid State Ionics*, **150**, 115–122.
14 Ralph, T.R., Hards, G.A., Keating, J.E., Campbell, S.A., Wilkinson, D.P., Davis, M., St Pierre, J., Johnson, M.C. (1997) *J. Electrochem. Soc.*, **144**, 3845–3857.
15 Stonehart, P. (1990) *Ber. Bunsen-Ges. Phys. Chem.*, **94**, 913–921.
16 Ruge, M., Büchi, F.N., Brooman, E.W., Doyle, C.M., Cominellis, C., Winnick, J. (eds.) (2001) *Proceedings of the Symposium "Energy and Electrochemical Processes for a Cleaner Environment"*, PV 2001–23, 165–173 Electrochemical Society, Pennington, NJ.
17 Chang, H., Kim, J.R., Cho, J.H., Kim, H.K., Choi, K.H. (2002) *Solid State Ionics*, **148**, 601–606.
18 Schmitz, A., Tranitz, M., Wagner, S., Hahn, R., Hebling, C. (2003) *J. Power Sources*, **118**, 162–171.
19 Maynard, H.L., Meyers, J.P. (2002) *J. Vac. Sci. Technol. B*, **20**, 1287–1297.
20 Heinzel, A., Hebling, C., Muller, M., Zedda, M., Muller, C. (2002) *J. Power Sources*, **105**, 250–255.
21 International Energy Agency (2003) *Key World Energy Statistics 2003*, Paris.
22 Friedlmeier, G., Friedrich, J., Panik, F. (2001) *Fuel Cells*, **1**, 92–96.
23 Schmid, J.E.H.P. (2003) *Handbook of Fuel Cells Fundamentals, Technology and Applications*, Vol. 4 (eds. W. Vielstich, H.A. Gasteiger, A. Lamm), John Wiley & Sons, Ltd, Chichester, pp. 1167–1171.
24 Rodriques, A., Fronk, M., McCormick, B. (2003) *Handbook of Fuel Cells Fundamentals, Technology an Applications*, Vol. 4 (eds. W. Vielstich, H.A. Gasteiger, A. Lamm), John Wiley & Sons, Ltd, Chichester, pp. 1172–1179.
25 Matsuo, S. (2003) *Handbook of Fuel Cells Fundamentals, Technology and Applications*, Vol. 4 (eds. W. Vielstich, H.A. Gasteiger, A. Lamm),

John Wiley & Sons, Ltd, Chichester, pp. 1180–1183.
26. Rodatz, P., Garcia, O., Guzzella, L., Büchi, F.N., Bärtschi, M., Tsukada, A., Dietrich, P., Kötz, R., Scherer, G.G., Wokaun, A. (2003) *Proceedings of the Session "Fuel Cell Power for Transportation", SAE 2003 World Congress.*, Detroit MI, USA, March 3–6, 2003, vol. **1741**, pp. 77–88.
27. von Helmolt, R. (2002) Fuel Cells in the Automobile, *Proceedings of the f-cell*, Stuttgart, October 14–15.
28. Schuler, A., Schild, J., Batawi, E., Rüegge, A., Tamas, M., Doerk, T., Raak, H., Dogwiler, B. (2002) *Proceedings of the 5th European Solid Oxide Fuel Cell Forum, Lucerne*, July 1–5, 2002, pp. 446–452.
29. Wagner, H.A. (2003) *Handbook of Fuel Cells Fundamentals, Technology an Applications*, Vol. **4** (eds. W. Vielstich, H.A. Gasteiger, A. Lamm), John Wiley & Sons, Ltd, Chichester, pp. 1224–9.
30. Veyo, S.E., Fukuda, S., Shockling, L.A., Lundberg, W.L. (2003) *Handbook of Fuel Cells Fundamentals, Technology an Applications*, Vol. **4** (eds W. Vielstich, H.A. Gasteiger, A. Lamm), John Wiley & Sons, Ltd, Chichester, pp. 1276–1289.
31. Strasser, K. (2003) *Handbook of Fuel Cells Fundamentals, Technology an Applications*, Vol. **4** (eds W. Vielstich, H.A. Gasteiger, A. Lamm), John Wiley & Sons, Ltd, Chichester, pp. 1201–1214.
32. Kickulies, M. (2002) *Proceedings of the Fuel Cell World Conference*, Lucerne July 1–5, 2002, pp. 278–285.
33. Arthur. D. Little Inc. (2001) *Cost Analysis of Fuel Cell Systems for Transportation*, report for US Department of Energy.
34. Binder, M.J., Taylor, W.R., Holcomb, F.H. (2000) *Proceedings of the Fuel Cell 2000 Conference*, Lucerne July 10–14, 2000, pp. 109–15.
35. Jung, M., Kroeger, H.H. (1970) US Patent 3,511,710.
36. Lee, J.-Y. (1997) US Patent 5,599,640.
37. Amendola, S. (1998) US Patent 5,804,329.
38. Amendola, S., Onnerud, P., Kelly, M., Petillo, P., Binder, M. (1999) *Proc. Electrochem. Soc.*, **98-15** (selected Battery Topics), 47, 194th Meeting of the Electrochemical Society, November 1–6, 1998, Boston, MA.
39. Amendola, S., Onnerud, P., Kelly, M., Petillo, P., Sharp-Goldman, S., Binder, M. (1999) *J. Power Sources*, **84** (1), 130.
40. Amendola, S., Petillo, P., Kelly, M., Sharp-Goldman, S., Binder, M. (1999) *195th Meeting of the Electrochemical Society*, May 2–6, Seattle, WA.
41. Suda, S. (2002) US Patent 6,358, 488.
42. Pecsok, R.L. (1953) *J. Am. Chem. Soc.*, **76**, 2862.
43. Stockmayer, W.H., Rice, D.W., Stephenson, C.C. (1955) *J. Am. Chem. Soc.*, **77**, 1980.
44. Jasinski, R. (1965) *Electrochem. Tech.*, **3**, 40.
45. Indig, M.E., Snyder, R.N. (1962) *J. Electrochem. Soc.*, **109**, 1104.
46. Li, Z.P., Liu, B.H., Arai, K., Suda, S. (2003) *J. Electrochem. Soc.*, **150** (7), A868.
47. Li, Z.P., Liu, B.H., Arai, K., Morigasaki, N., Suda, S. (2003) *J. Alloys Compd.* **356–357**, 469.
48. Strasser, K. (1990) *J. Power Sources*, **29**, 149.
49. Lee, S.-M., Kim, J.-H., Lee, H.-H., Lee, P.S., Lee, J.-Y. (2002) *J. Electrochem. Soc.*, **149** (5), A603.
50. Liu, B.H., Li, Z.P., Suda, S. (2003) *J. Electrochem. Soc.*, **150** (3), A398.
51. Kubokawa, M., Yamashita, M., Abe, K. (1968) *Denki Kagaku oyobi Kogyo Butsuri Kagaku*, **36**, 778.
52. Liu, B.H., Li, Z.P., Suda, S. (2004) *Electrochim. Acta*, **49**, 3097.
53. Westerkamp, H. (1939) *ATZ*, **S.** 523 (in German).
54. Oehmichen, M. (1942) *Hydrogen as Engine Fuel, Deutsche Kraftfahrtforschung, Heft 68*, VDI-Verlag, Berlin (in German).
55. Pehr, K., Burckhardt, S., Koppi, J., Korn, T., Patsch, P. (2002) *ATZ* (in German).
56. Athenstaedt, U. (1993) *MTZ*.

57 Schüers, A., Abel, A., Fickel, H., Preis, M., Artmann, R. (2002) *MTZ* (in German).

58 MAN Nutzfahrzeuge AG (1996) *MAN – Hydrogen Powertrain for City Buses*, Technical Information of MAN Nutzfahrzeuge AG, Nürnberg (in German).

59 Furuhama, S., Kobayashi, Y. (1982) Hydrogen Cars with LH_2-Tank, LH_2-Pump and Cold GH_2-Injection Two-Stroke Engine, SAE Paper 820349.

60 Mazda Motor Corp. (2003) *Company Website, History of Rotary since 1967*, http://www.mazda.com/history/rotary/index.html, 24.11.2003.

61 Lambe, H., Watson, J.S. (1992) *Int. J. Hydrogen Energy*, **17** (7), 513–25.

62 Prechtl, P., Dorer, F. (1999) *MTZ Int.* (12).

63 Rottengruber, H., Wiebicke, U., Zeilinger, K., Woschni, G. (2000) *MTZ Int.* (2).

64 Berckmüller, M., Rottengruber, H., Eder, A., Brehm, N., Elsässer, G., Müller-Alander, G., Schwarz, C. (2003) Potentials of a Charged SI-Hydrogen Engine, SAE Paper 2003-01-3210.

65 Rottengruber, H. (1999) *Nitrogen Oxide Formation in the Hydrogen Diesel Engine*, TU, München, ISBN 3-89791-047-0 (in German).

66 Witt, J.L. (1999) *Analysis of Thermodynamical Losses of a SI-Engine under Conditions of Variable Cam-Phasing*, Technische Universität, Graz (in German).

67 Göschel, B. (2003) The Hydrogen Combustion Engine as the Drive System for the BMW of the Future, 24th International Engine Symposium, Vienna, 16 May 2003.

68 Kiesgen, G., Klüting, M., Bock, C., Fischer, H. *The new 12-Cylinder Hydrogen Engine in the 7 Series: The H_2 ICE Age has begun*, SAE Paper 2006-01-0431.

69 Enke, Graber, Hecht, Starr, *The Bivalent V12-Engine of the BMW Hydrogen 7*, MTZ International 06/2007, Germany.

70 Peschka, W. (1992) *Liquid Hydrogen – Fuel of the Future*, Springer-Verlag, Wein, Austria.

71 Kinard, G.E. (1998) The commercial use of liquid hydrogen over the last 40 years, in *Proceedings of International Cryogenic Engineering Conference*, vol. **ICEC-17** (eds D. Dew-Hughes, R.G. Scurlock, and J.H.P. Watson). Inst. Phys. Publ., Bristol, UK, pp. 39–44.

72 *Space Launch Report*, http://www76.pair.com/tjohnson/slr.html.

73 McDowell, J. (1997) *Jonathan's Space Home Page*, http://www.planet4589.org/space/ [Harvard University, 1997 – present].

74 Wade, M. (ed.) *Encyclopedia Astronautica*, http://www.astronautix.com/.

75 Frisbee, R.H. (ed.) (2005) *Advanced Space Propulsion Concepts [Online Database]*, http://eis.jpl.nasa.gov/sec353/apc/. Pasadena California: NASA Jet Propulsion Laboratory. Last Updated March 2002. Accessed on February 25, 2004.

76 Collaudin, B., Passvogel, T. (1999) *Cryogenics*, **39**, 157–165.

77 Donabedian, M. (ed.) (2003) *Spacecraft Thermal Control, Handbook – Volume II: Cryogenics*, The Aerospace Press, El Segundo, CA.

78 Collaudin, B., Rando, N. (2000) *Cryogenics*, **40**, 797–819.

79 Holmes, W., Cho, H., Hahn, I., Larson, M., Schweickart, R., Volz, S. (2002) *Cryogenics*, **41**, 865–870.

80 Hughes, J.L., Herr, K.C. (1973) *Cryogenics*, **13**, 513–519.

81 Bowman, R.C., Jr. (2003) *J. Alloys Compds.*, **356–357**, 789–793.

82 Bowman, R.C., Jr., Kiehl, B., Marquardt, E. (2003) "Closed-Cycle Joule–Thomson Cryocoolers," in *Spacecraft Thermal Control Handbook, Volume II: Cryogenics* (ed. M. Donabedian), The Aerospace Press, El Segundo, CA, pp. 187–216.

83 Bard, S., Wu, J., Karlmann, P., Cowgill, P., Mirate, C., Rodriguez, J. (1995) "Ground Testing of a 10 K Sorption Cryocooler Flight Experiment (BETSCE)," in *Cryocoolers 8*, (ed.

R.G. Ross, Jr.). Plenum, New York, pp. 609–621.

84 Bard, S., Karlmann, P., Rodriguez, J., Wu, J., Wade, L., Cowgill, P., Russ, K.M. (1997) "Flight Demonstration of a 10 K Sorption Cryocooler," in *Cryocoolers 9* (ed. R.G. Ross, Jr.). Plenum, New York, pp. 567–576.

85 Bowman, R.C., Jr., Karlmann, P.B., Bard, S. (1998) "Post-Flight Analysis of a 10 K Sorption Cryocooler," in *Advances in Cryogenic Engineering*, vol. **43** (ed. P. Kittel), Plenum, New York, pp. 1017–24.

86 Wade, L.A., Bhandari, P., Bowman, R.C., Jr., Paine, C., Morgante, G., Lindensmith, C.A., Crumb, D., Prina, M., Sugimura, R., Rapp, D. (2000) "Hydrogen Sorption Cryocoolers for the Planck Mission," in *Advances in Cryogenic Engineering*, vol. **45** (eds Q.-S. Shu, *et al.*), Kluwer Academic/Plenum, New York, pp. 499–506.

87 Collaudin, B., Linder, M., Rando, N. (2003) Herschel-planck and the next steps in space cryogenics, in *Proceedings of 19th International Cryogenic Engineering Conference (ICEC-19)* (eds. G. Gistau-Baguer, P. Seyfert), Narosa Publishing, New Delhi, pp. 493–502.

88 Bhandari, P., Prina, M., Bowman, R.C., Jr., Paine, C., Pearson, D., Nash, A. (2004) *Cryogenics*, **44**, 395–401.

89 Donabedian, M. (2003) "Stored Solid Cryogen Systems," in *Spacecraft Thermal Control, Handbook – Volume II: Cryogenics* (ed. M. Donabedian). The Aerospace Press, El Segundo, CA, pp. 31–9.

90 Bartschi, B.Y., Morse, D.E., Woolston, T.L. (1996) *John Hopkins APL Technical Digest*, **17**, 215–25.

91 *The James Webb Space Telescope homepage*, http://ngst.gsfc.nasa.gov/.

92 Dunlop, J., Giner, J., van Ommering, G., Stockel, J. (1975) Nickel-Hydrogen Cell, U.S. Patent 3,867,199.

93 Giner, J., Dunlop, J. (1975) *J. Electrochem. Soc.*, **122**, 4–11.

94 Dunlop, J., Rao, G.M., Yi, T.Y. (1993) *NASA Handbook for Nickel-Hydrogen Batteries*, NASA Reference Publication, p. 1314.

95 Thaller, L.H., Zimmerman, A.H. (2003) *Nickel-Hydrogen Life Cycle Test: Review and Analysis*, The Aerospace Press, El Segundo, CA.

96 Dunlop, J. (1984) "Nickel-Hydrogen Batteries," in *Handbook of Batteries & Fuel Cells*, Ch. 4 (ed. D. Linder). McGraw-Hill, New York.

97 Rogers, H.H., Krause, S.J., Levy, E., Jr. (1978) Design of long life nickel-hydrogen cells, in *Proceedings of 28th Power Sources Conference*, June 1978.

98 Hass, R.J., Chawathe, A.K., van Ommering, G. (1988) Space station battery system design and development, in *Proceedings of 23rd Intersociety Energy Conversion Engineering Conference*, vol. 3, 577, August 1988.

99 Ruetschi, P., Meli, F., Desilvestro, J. (1995) *J. Power Sources*, **57**, 85–91.

100 Fetcenko, M.A., Ovshinsky, S.R., Young, K., Reichman, B., Fierro, C., Koch, J., Martin, F., Mays, W., Ouchi, T., Sommers, B., Zallen, A. (2002) *J. Alloys Compd.*, **330–332**, 752–759.

101 DaCosta, D.H., Golben, M., Tragna, D.C. (2002) Metal hydride systems for the hydrogen planet, in *Proceedings of 14th World Hydrogen Energy Conference* (eds. T.K. Bose and R.D. Venter), CD-Publication, Montreal, Canada, paper A-108; Golben, M., Nechev, K., DaCosta, D.H. (1997) A low pressure bipolar Nickel-Hydrogen battery, in *Proceedings of 12th Annual Battery Conference* (eds H.A. Frank, E.T. Seo), IEEE97TH8226, Long Beach, CA, January, 1997, pp. 307–12.

102 Low, G.M., "The spaceships", *Apollo Expeditions to the Moon*, Ch. 4, Section 3, http://www.hq.nasa.gov/office/pao/History/SP-350/ch-4-3.html.

103 Sauer, R.L., Calley, D.J., "Potable Water Supply," *Biomedical Results of Apollo*, Ch. 4, Section 6, http://lsda.jsc.nasa.gov/books/apollo/S6CH4.htm.

104 Command and Service Module Performance, *Apollo 15 Mission Report*, Chapter 6, http://www.hq.nasa.gov/office/pao/History/alsj/a15/a15mr-6.htm.

105 Electrical Power Systems, *NSTS 1988 News Reference Manual – Space Shuttle Orbiter Systems,* http://science.ksc.nasa.gov/shuttle/technology/sts-newsref/sts-eps.html

106 *Cassini-Huygens Mission to Saturn & Titan Homepage,* http://saturn.jpl.nasa.gov/index.cfm, NASA Jet Propulsion Laboratory, Pasadena California.

107 DaCosta, D. (March 2004) HERA-Ergenics, Personal Communication.

108 Prina, M., Kulleck, J.G., Bowman, R.C., Jr. (2002) *J. Alloys Compd.,* **330–332**, 886–891.

109 Johnson, D.L., Wu, J.J. (1997) "Feasibility Demonstration of a Thermal Switch for Dual Temperature IR Focal Plane Cooling," in *Cryocoolers 9* (ed. R.G. Ross, Jr.). Plenum, New York, pp. 795–805.

110 Bugby, D., Marland, B. (2003) "Cryogenic Thermal Switches," in *Spacecraft Thermal Control Handbook, Volume II: Cryogenics* (ed. M. Donabedian). The Aerospace Press, El Segundo, CA, pp. 347–370.

111 Rosso, M.J., Golben, P.M. (1987) *J. Less-Common Met.,* **131**, 283–292.

112 Lynch, F.E. (1991) *J. Less-Common Met.,* **172–174**, 943–958.

113 Mueller, P.J., Batty, J.C., Zubrin, R.M. (1996) *Cryogenics,* **36**, 815–822.

114 Salerno, L.J., Kittel, P. (1999) *Cryogenics,* **39**, 381–388.

115 Hastings, L.J., Plachta, D.W., Salerno, L., Kittel, P. (2002) *Cryogenics,* **41**, 833–839.

116 Orton, G.F., Scuderi, L.F. (1998) A hypersonic cruiser concept for the 21st century, Paper No. 985525, *1998 World Aviation Conference (SAE/AIAA),* Anaheim, CA, September, 1998.

117 Cockrell, C.E., Jr., Auslender, A.H., Guy, R.W., McClinton, C.R., Welch, S.S. (2002) Technology roadmap for dual-mode scramjet propulsion to support space-access vision vehicle development, Paper No. AIAA-2002–5137, *11th AIAA/AAAF Inter. Space Planes & Hypersonic Systems and Technologies Conference,* Orleans, France, September, 2002.

118 Powell, J., Maise, G., Paniagua, J. (2004) Is NTP the key to exploring space? *Aerospace America,* **42** (1), 36–42.

Index

A
absorbers, switchable   324
absorption
– bulk   107–108
– coating effects   105–107
– hydrogen   94
absorption curves   221
activation energy
– adsorption   117
– decomposition of complex hydrides   223
active components, fuel cells   346–348
active layer, gas diffusion electrodes   347
adsorbate, distance to substrate   176
adsorbed hydrogen atoms, chemical potential   118
adsorption
– activation energy   117
– coating effects   105–107
– hydrogen   94, 173–188
adsorption pressure   219
AF *see* antiferromagnetic coupling
air supply, fuel cells   351
air/fuel ratio   374–375
airships   14–16
alkali metal alanates   234–236
alkaline earth metal alanates   234–236
alkaline electrolysis   158–159
– manufacturers   158
alkaline fuel cell (AFC)   344, 363–364, 369–371
alkaline solutions, aqueous   245–246
alloys
– binary   100
– enthalpy/entropy of formation   320–321
– mischmetal nickel   137
Allred-Rochow electronegativity   189
all-solid-state switchable mirror   323
aluminum hydrides, Ti-doped sodium aluminum hydrides   213–233
ammonia, hydrogen storage   242–244

anion structure, alanates   236
antibonding orbital   110–111
antiferromagnetic (AF) coupling   269, 272–273
aqueous alkaline solutions   245–246
Ariane   385
Aristotle   7
Arrhenius analysis   107–108
Arrhenius equation   220
Arrhenius plot   117, 222
asymmetry ratios, magnetic   274
atomic data, hydrogen isotopes   72
atomic magnetic moment   273
automotive applications
– fuel cells   352, 355–358
– hydrogen-ICE   379–381
autothermal reforming, biomass   41–42

B
back seeding   120
bacteria, hydrogen storage   238–240
band structure   307
– calculated   292
– local model   193
batteries   337
– metal hydride-based   326
– nickel–hydrogen *see* nickel–hydrogen batteries
– Ni–MH *see* NiMH batteries
BETSCE sorption cryocooler   391
binary alloys, surface segregation   100
binary hydrides   188–189
biomass
– autothermal reforming   41–42
– combustion   41–42
– energy   39–40
– hydrogasification   41–42
– hydrogen production   40–42, 51
– methanation   41–42

Hydrogen as a Future Energy Carrier. Edited by A. Züttel, A. Borgschulte, and L. Schlapbach
Copyright © 2008 WILEY-VCH Verlag GmbH & Co. KGaA, Weinheim
ISBN: 978-3-527-30817-0

biomass (*continued*)
– renewable carbon resource 42–43
– shift reaction 41–42
– thermal conversion 41–42
biomass gasification 155
bionics 240
biosphere cycles 47–48
bipolar stacking, fuel cells 349
bipropellant liquid propulsion systems 385
black state 300–302, 309
boil-off rate, hydrogen 172
bond dissociation energies 88
bonding, complex transition metal hydrides 200–201
bonding orbital 110–111
borohydride complex ion 245–246
borohydride fuel cell *see* direct borohydride fuel cell
borohydride solutions, crystallization 248
Boudouard reaction 150–151
Boyle, Robert 7
brilliant eyes ten-kelvin sorption cryocooler experiment *see* BETSCE
buffer layers, catalytic 316–318
bulk absorption, relevance of surface reactions 107–108
Bunsen, Robert 11
Buran space shuttle 385

C

calculated band structures 292
calorific value of mixture 374
Calvin–Benson cycle 239
cap layers 278
– catalytic 316–318
capacity, storage *see* storage capacity
carbon budget, global 44
carbon cycle 37–39
– replacement by hydrogen cycle 43–45
carbon flows 39
carbon hydrates, hydrogen storage 240–241
carbon hydrides 240–242
carbon materials, hydrogen storage 178–184
carbon nanotubes 179–180
carbon reservoirs 2
carbon resources, renewable 42–43
Carnot efficiency 340
catalysis
– borohydride fuel cells 369–370
– deabsorption 220–223
– hydrogen dissociation and recombination 108–125
– hydrogen generation 130

– hydrolysis 249
– reabsorption 220–223
– Sabatier reaction 406
catalytic cap layers/buffer layers 316–318
cation environments 195
Cavendish, Henry 8, 13
cell polarization, direct borohydride fuel cells 368
cells
– fuel cells *see* fuel cells
– galvanic 337–339
Challenger disaster 17–20, 400
chemical elements *see* elements
chemical loading 280
chemical potential 99–100
– adsorbed hydrogen atoms 118
chemical reactions *see* reactions
chemisorption 98–99, 175–177
– heat of 99
– metal oxides 102–104
– metal surfaces 101–102
– potential energy 96
chloralkaline electrolysis 161
climate 34–35
CO_2 emissions 34–35
– reduction 35–37
coal
– gasification 49
– hydrogen production 149–155
– reserves 30–31
coating, thin film 105–107
coexistence of Mg and MgH_2 300–302
combinatorial research 316–321
combustion
– air/fuel ratio 374–375
– biomass 41–42
– hydrogen 9, 91
– internal combustion engines 371–373
– methane, propane and gasoline 91
common pressure vessel (CPV) 397, 400–401
complex formation, "interstitial" hydrides 199–200
complex hydrides 211–237
– encapsulation 231–232
– non-transition metal 203–213
– safety aspects 231–232
– thin films 283
– transition metal *see* transition metal hydrides
complexes
– metal ammine 243–244
– metal–hydrogen 195
– mononuclear 196–198
– polynuclear 198–199

components, fuel cells 346–348
composition, fossil fuels 27–28
compounds
– ammonia-based 242–244
– intermetallic 190
compressed hydrogen gas, volumetric/gravimetric density 169
compressibility 80
compression ratio 378
computational methods
– force field calculations 229
– Monte Carlo simulations 175
– $NaAlH_4$ systems 225–231
concentration dependence
– dielectric function 286
– switchable mirror properties 281, 285
conductivity
– specific 347
– temperature dependence 287
conversion, energy see energy
cooling
– efficiency 389
– fuel cells 351
copper, hydrogen adsorption 115–116, 119–121
cost, fuel cells 361–363
coupled systems, exchange coupled 267–270
coupling
– (anti-)ferromagnetic 269
– interlayer exchange see interlayer exchange coupling
covalent bonds, hydrogen 88
coverage, atomic hydrogen on Cu(100) 120
CPV see common pressure vessel
cryocooler
– BETSCE 391
– Joule–Thomson 389
– liquid hydrogen sorption 390
cryogenic tanks 404
cryogenic temperatures, hydrogen adsorption 181–183
cryogenics applications, hydrogen 388–394
cryostat, solid hydrogen 392
crystal structure
crystallization, borohydride solutions 248
cubic YH_3 298–300
Curie temperature, ferromagnetic film 267
current–voltage characteristics, fuel cells 342
cycles
– biosphere 47–48
– carbon 37–39
– geochemical 39
– global water cycle 46
– hydrogen see hydrogen cycle
100 cycles test 224
cyclic stability, Ti-doped sodium aluminum hydrides 223–224

D

d-band 111
DBFC see direct borohydride fuel cell
deabsorption, catalyzed 220–223
decomposition
– activation energy 223
– Ti-doped sodium aluminum hydrides 223
decomposition temperature, tetrahydroborates 207
defect, point 103
defect-free surface 104
dehydrogenation
– reactions 129
– reversible 236
density
– compressed hydrogen gas 169, 173
– energy see energy density
– gravimetric 169
– hydrogen in metallic hydrides 195
– liquid hydrogen 173
– p-hydrogen 76
– power 368
– solid hydrogen 173
– volumetric 169
deposition, RE hydride thin films 279
deposition techniques, magnetic films 267
desorption, hydrogen 114–115
desorption enthalpy, complex transition metal hydrides 203
desorption pressure 219
detonability limits, hydrogen–air–water vapor 92
detonation 90–94
deuterium 12
diatomic hydrogen see molecular hydrogen
dielectric function
– concentration dependence 286
– magnesium hydride 297
Diesel-cycle gaseous-fuel engines 373
diffusion
– gas diffusion electrodes 347–348
– hydrogen 85–90
– opaque materials 314–316
– switchable mirrors 312–314
diffusion coefficients, hydrogen 85
dihydride state 281–282, 285
dihydrogen see molecular hydrogen

direct borohydride fuel cell (DBFC) 125, 130, 134–135, 364–371
– catalysts 369–370
– electricity generation 367
– electrochemistry 365
– electrode kinetics 365–366
– thermodynamics 365–366
direct methanol fuel cell (DMFC) 241, 344, 364–365, 369
disasters
– Challenger 17–20, 400
– Hindenburg 14–17, 66
dissociation
– catalyzed 108–125
– hydrogen molecule 105–107
– metal hydrides 218
dissociation energies, covalent hydrogen bonds 88
dissociation reaction, hydrogen molecule 89
distribution, hydrogen 56
DMFC *see* Direct methanol fuel cells
Döbereiner platinum lighter 10
domain switching, optical 293
doped $NaAlH_4$ systems, reversible 225–231
dosing 122–123
dual mode nuclear propulsion system 407
dual-fuel vehicle 381
dynamic hydriding/dehydriding process 253

E

economic factors, hydrogen production 54
effective medium approach 106
efficiency
– Carnot 340
– cooling 389
– electrochemical 341
– fuel cells 340–343, 353–354
– thermodynamic 378–379
electric power train, hybrid 357–359
electrical conductivity *see* conductivity
electrical properties, switchable mirrors 286–288, 305–306
electricity 155–163, 162
– generation 367
electrochemical efficiency 341
electrochemical loading 266
electrochemistry, borohydride fuel cells 365
electrochromical devices 321–323
electrode kinetics, borohydride fuel cells 365–366

electrode potential, equilibrium 339
electrode stack designs 398
electrodes
– gas diffusion 347–348
electrolysis 5
– alkaline 158–159
– chloralkaline 161
– hydrogen production 50–51, 155–163
– solid-polymer 160
– using renewable energy 161–163
– water 155–157
electrolytes
– fuel cells 344–346
– specific conductivity 347
electrolytic loading 280
electromigration, switchable mirrors 312–314
electron correlation models 291
electronegativity 87
– Allred-Rochow 189
electronic band structure 307
elements
– binary hydrides 189
– hydrides 87
empirical models, metal hydrides 192–195
encapsulation, complex hydrides 231–232
endothermic reactions, equilibrium constants 152
energy
– activation *see* activation energy
– conversion 41, 59–63
– dissociation 88
– Gibbs *see* Gibbs energy
– potential *see* potential energy
– production 49–63
– renewable *see* renewable energy
energy consumption, primary 25–26
energy density, batteries and fuel cells 337
engine torque 376
enthalpy
– desorption 203
– hydrogen 78, 81–83
– hydrogen production 150–151
enthalpy diagram 209
enthalpy of formation
– alloys 320–321
– hydrides 191–194
enthalpy of segregation 100
entropy
– hydrogen 77–78
– n-hydrogen 83
entropy of formation
– alloys 320–321
– hydrides 191–194

environmental effects
– fossil fuels 33–35
– hydrogen economy 65
– hydrogen production 54
environment, global 47–48
epitaxial switchable mirrors 294–295
epitaxial thin films, stress 282–283
equation of state (EoS), hydrogen 77–81
equations
– first rocket equation 382
– ideal gas law 119
– Langmuir isotherm 174, 176
– Lennard–Jones potential 175
– Stirling approximation 119
– Van der Waals 79, 167
– Van't Hoff equation 192
equilibrium constants, exothermic and endothermic reactions 152
equilibrium coverage, atomic hydrogen on Cu(100) 120
equilibrium electrode potential
– fuel cells 339
ESA 392–394
ex situ hydrogen loading 279–281
exchange coupled systems 267–270
excrescence, vermicular 230
exhaust water 244–256
exothermic reactions, equilibrium constants 152
explosion properties, hydrogen, methane, propane and gasoline 91
explosive hazards 93

F

F *see* ferromagnetic coupling
Faraday efficiency 156
Fe/V superlattices 270–271
– ideal interfaces 275
– magnetization 273
ferromagnetic (F) coupling 269
ferromagnetic film, Curie temperature 267
fiber optic hydrogen sensors 325–326
film deposition techniques 267
films, thin *see* thin films
fire hazards 92–93
first rocket equation 382
Fischer–Tropsch process 155
flammability limits 91–92
fluorinated surfaces 137–138
force field calculations 229
formation, fossil fuels 27–28
fossil fuels 3
– advantages and uses 25–27
– consumption growth 23–25

– environmental impact 33–35
– formation and composition 27–28
– future trends 35–37
– global reserves and production 29–33
– hydrogen production 149–154
– proven reserves 30–31
– replacement by hydrogen cycle 43–45
free electron metals 121
fuel cell power train 357–359
fuel cell stacks 348–349
fuel cell systems 348–354
– low power 350
fuel cells 14, 60–63
– active components 346–348
– AFC *see* alkaline fuel cell
– air supply 351
– alkaline
– automotive applications 352, 355–358
– bipolar stacking 349
– cooling 351
– cost 361–363
– current–voltage characteristics 342
– DBFC *see* direct borohydride fuel cell
– DMFC *see* direct methanol fuel cell
– efficiency 340–343, 353–354
– electrolytes 344–346
– equilibrium electrode potential 339
– Gibbs energy 338–340
– history 336–337
– hydrogen as fuel 364–371
– hydrogen/oxygen 339
– lifetime 363
– losses 342, 353
– MCFC *see* molten carbonate fuel cell
– micro-fuel cells 369
– PAFC *see* phosphoric acid fuel cell
– PEMFC *see* PEMFC
– portable applications 354–355
– power and heating 62–63
– proton exchange membrane (PEM) 61
– SOFC *see* solid oxide fuel cell
– space applications 361
– temperature range and reactants 343
– thermodynamics 337
fuels
– air/fuel ratio 374–375
– biomass energy 39–40
– gasoline 91
– fossil *see* fossil fuels
– gaseous 373
– hydrogen 16–20, 23–67, 127, 371–372
– rocket fuel 382–388
– space shuttle 16–20
fugacity 81

functionalized materials
– magnetic heterostructures 265–275
– switchable mirrors 275–327
fusion, inertial confinement 408

G

Gaia hypothesis 43
galvanic cells 337–339
gas chain 336
gas cylinders, high pressure 167–170
gas diffusion electrodes 347–348
gas loading/unloading 266, 279–280
gas reserves 30–31
gas turbines 59–60
gaseous hydrogen, physical properties 78
gaseous-fuel engines 373
gasification
– biomass 41–42
– coal 49,
gasoline, combustion and explosion properties 91
gas–solid surface interactions 132
geochemical cycle 39
Gibbs energy 77–78, 81
– fuel cells 338–340
– n-hydrogen 83
– water electrolysis 155–156
global carbon budget 44
global environment, biosphere cycles 47–48
global reserves, fossil fuels 29–33
global water cycle 46
graphene sheets 180
graphitic nanofibers 179–180
gravimetric density, compressed hydrogen gas 169
greenhouse effect 3
grid independent power 356
Grove, Sir William Robert 10–11, 336
GW approximation (GWA) 291, 293, 295

H

H see hydrogen
Haber, Fritz 11
halogens cycle 48
hazards
– explosive 93
– fire 92–93
– hydrogen economy 65–66
– preventive measures 93
heat
– specific heat ratio 378
heat engines, efficiency 341
heat of chemisorption 99

heat of solution 99
heat release rate 377
heating, fuel cells 62–63
Helmont, Jan Baptista van 7
heterostructures, magnetic 265–275
high pressure gas cylinders 167–170
high temperature, hydrogen adsorption 183–184
Hindenburg disaster 14–20, 66
history
– fuel cells 336–337
– hydrogen 7–20
homogeneous ternary hydride phase 229
homogenous water gas reaction 150
hybrid electric power train 357–359
hydride ion 129–131
hydride thin films 277–279
hydride-forming intermetallic compounds 190
hydrides 87
– binary 188–189
– carbon see carbon hydrides
– complex see complex hydrides
– formation see metal hydride formation
– homogeneous ternary hydride phase 229
– intermetallic 188–189
– metal see metal hydrides
– non-transition metal complex see non-transition metal complex hydrides
– organic see organic hydrides
– surface engineering 132–138
– transition metal complex see transition metal hydrides
hydrocarbons, hydrogen production 149–155
hydrogasification, biomass 41–42
hydrogen
– adsorption and absorption 94
– applications 127
– as a fuel 16–20, 23–67, 127, 371–372
– chemical properties 85–90
– combustion 9
– combustion and explosion properties 91
– conversion to energy 59–63
– cryocoolers 388–394
– density in metallic hydrides 195
– diffusion 85–90, 314–316
– diffusion coefficients 85
– dissociation 108–125, 134
– equation of state 77–81
– flammability limits 91–92
– four states 125–132
– fuel cells 364–371

- hazards 92–93
- history 7–20
- ignition and detonation 90–94
- interaction with solid surfaces 94–108
- o-hydrogen 73–74
- p-hydrogen 73–74
- present scenario 48–49
- primitive phase diagram 76, 166
- rocket fuel 382–388
- safety concepts 93
- space applications 381–410
- state transitions 126
- toxicology 93

hydrogen adsorption 173–188
- cryogenic temperatures 181–183
- high temperature 183–184
- measuring techniques 177–178
- metal hydrides 189–192

hydrogen as a fuel, internal combustion engines (ICE) 371–372
hydrogen atoms, chemical potential 118
hydrogen bomb 12
hydrogen concentration dependence 281, 285
hydrogen cryostat 392
hydrogen cycle
- energy production 49–63
- implementation 63–67
- key elements 45–46
- replacement of fossil fuel 43–45
- water 46–47

hydrogen desorption
- mass spectra 124
- TPD spectra 115

hydrogen distribution 56
hydrogen functionalized materials *see* functionalized materials
hydrogen gas 71–94
- compressed 169

hydrogen generation
- hydrolysis 248–249
- protide 130
- systems and devices 250–252

hydrogen in transportation 13–14
hydrogen ion 128
hydrogen isotopes 71–72
hydrogen liquefaction 171
hydrogen loading 266–267, 279–281
hydrogen molecule 72–75
- dissociation 89, 105–107
- physical properties 75–77,

hydrogen power train 380
hydrogen powered space launch vehicles 384

hydrogen production
- biomass 40–43, 51
- coal gasification 49
- cost 5
- electrolysis 50–51, 155–163
- environmental, economic and scaling factors 54
- from carbon and water 154
- from coal and hydrocarbons 149–155
- natural gas reformation 49–50
- non-renewable methods 49–55
- nuclear power 49
- photo-biological 53
- photo-electrochemical 52–53
- reactions 149–154
- renewable methods 50–55
- thermochemical 51–52
- thermophysical 52

hydrogen recombination, catalysis 108–125
hydrogen sensors, fiber optic 325–326
hydrogen sorption cryocooler 390
hydrogen sources, water 246
hydrogen spark ignition engine 373–379
hydrogen storage 56–59, 95
- alanates 234–236
- ammonia and ammonia-based compounds 242–244
- basic methods 168
- carbon hydrides 240–242
- carbon materials 178–184
- complex transition metal hydrides 195–203
- high pressure gas cylinders 167–170
- in molecular form 165–172
- indirect *see* indirect hydrogen storage
- liquid hydrogen 170–172
- lithium nitride and imide 236
- metal hydrides 133, 188–195
- metal–organic frameworks 186–187
- non-transition metal complex hydrides 203–213
- organic hydrides 237–244
- reversible *see* reversible hydrogen storage
- saturation capacity 183
- silicate structures 184–186
- space shuttle 387, 404–405
- zeolites 184–186

hydrogen storage materials
- lightweight 318–320
- reversible 212

hydrogen transmission 55–56
hydrogen/oxygen fuel cell, overall reaction 339

hydrogenation
- reactions   129
- reversible   236
hydrogen-ICE   372–373, 379–381
hydrogenography   320–321
hydrogen-to-carbon ratio   201
hydrolysis   248–250
hydroxonium ion   128
hysteresis, optical properties   288–290

I

ICE *see* internal combustion engines
ideal gas law   119
ideal interfaces, Fe/V superlattices   275
IEC *see* interlayer exchange coupling
ignition, hydrogen   90–94
image processing   356
imide, lithium   236
implementation, hydrogen cycle   63–67
in situ deposition, RE hydride thin films   279
indicator layers   311–316
indirect hydrogen storage   244–256
inertial confinement fusion   408
integrated thermal desorption spectra   208
Intensity map, reflection   310
interaction potential energy, hydrogen molecule   73
interfaces
- ideal   275
- solid–liquid   249–250
interlayer exchange coupling (IEC)   268–270, 272
intermetallic compounds, hydride-forming   190
intermetallic hydrides   188–189
internal combustion engines (ICE)   371–373
International Energy Agency (IEA)   35
International Space Station   399
"interstitial" hydrides, complex formation   199–200
inversion curve   84–85
ionization potential   89
Ising model   272
isoreticular MOFs   187
isotherm
- Langmuir   174, 176
- pressure–composition   216–217, 299–300, 304

J

Joule, James   81
Joule–Thomson coefficient   84
Joule–Thomson cryocooler   389
Joule–Thomson cycle   172
Joule–Thomson effect   81–84
Justi, Eduard   13

K

Kelvin, William Thomson, Lord of   81
kinetics, tetrahydroborates   208–211
Kirchhoff, Gustav   11

L

Langmuir isotherm   174, 176
laser deposition, pulsed   279
lattice defects, STM images   103
lattice planes   98
Lavoisier, Antoine Laurent   9
layering, self-organized   309
LDA *see* Local-density approximation
Lennard-Jones potential   96, 175, 190
lifetime
- fuel cells   363
- nickel–hydrogen batteries   403
ligands   196
lightweight hydrogen storage materials   318–320
Linde cycle   172
liquefaction, hydrogen   171
liquid electrolyte Gd–Mg switchable mirror   296
liquid hydrogen   170–172
- density   173
- physical properties   78
- storage vessels   171–172
liquid hydrogen sorption cryocooler   390
liquid propulsion systems, bipropellant   385
liquid–solid interface   249–250
liquid–solid surface interactions   133
lithium nitride and imide   236
loading, *ex situ*   279–281
loading technique   266–267
local band-structure model   193
Local-density approximation (LDA)   291–294
Long March vehicle   386
losses, fuel cells   342, 353
low power fuel cell systems   350
low temperature properties, p-hydrogen   76
Lurgi process   12

M

magnesium hydride, optical properties   295–296
magnesium–rare earth films, switchable mirrors   295–303

magnesium–transition metal films, switchable mirrors 303–311
magnetic asymmetry ratio 274
magnetic heterostructures 265–275
magnetic saturation field 272
magnetization, Fe/V superlattices 273
magneto-optic Kerr effect 272
manufacturers
– alkaline electrolysis 158
– solid-polymer electrolysis 160
mass spectrometry 124
MCFC *see* molten carbonate fuel cell
melting point 247
metal ammine complexes 243–244
metal hydrides
– batteries 326
– dissociation 218
– empirical models 192–195
– formation 109, 121–124
– hydrogen adsorption process 189–192
– hydrogen density 195
– hydrogen storage 133
– hydrogen storage 188–195
– NiMH batteries *see* NiMH batteries
– non-transition metals 203–213
– optical properties 275–327
– space suit cooling 405
– thin films 283
– transition metals *see* transition metal hydrides
metal oxides, chemisorption 102–104
metal surfaces 101–102, 113
metal–hydrogen complexes 195
metal–insulator transition 286–288, 305–306
metallic RE films 277
metal–organic frameworks (MOFs), hydrogen storage 186–187
methane
– biomass methanation 41–42
– combustion and explosion properties 91
methanol, hydrogen storage 240–241
MI transition *see* metal–insulator transition
micro-fuel cells 369
microscopic shutter effect 296–298
military applications, portable fuel cells 356
mirrors, switchable 275–327
mischmetal nickel alloys 137
mobile communication 356
MOFs *see* metal–organic frameworks
molecular beam deposition 279
molecular hydrogen 126–128
– storage 165–172,

molten carbonate fuel cell (MCFC) 345, 349, 359
monatomic hydrogen 128–129
mononuclear complexes 196–198
Monte Carlo simulations 175
multilayers, tailoring of optical properties 302–303

N

$NaAlH_4$ *see* sodium aluminum hydrides
$NaBH_4$ *see* sodium borohydride
nanofibers, graphitic 179–180
nanosized surface structures 136–137
nanotubes 179–180
NASA 16–20, 361, 383, 401
natural gas 3, 27–28
– reformation 49–50
NECAR 359
nickel alloys, mischmetal 137
nickel–hydrogen batteries
– International Space Station 399
– lifetime 403
– pressure vessels 397, 400–401
– reactions 395–396
– space application 394–404
– space performance characteristics 402
NiMH (nickel metal hydride) batteries, secondary 134
nitride, lithium 236
nitrogen cycle 48
NO_x-emissions, hydrogen spark ignition engine 373–375
non-renewable hydrogen production 49–55
non-transition metal complex hydrides 203–213
normal-hydrogen (n-hydrogen), triple point 77
nuclear power, hydrogen production 49
nuclear propulsion system, dual mode 407

O

oil reserves 30–31
optical domain switching 293
optical properties
– hysteresis 288–290
– magnesium hydride 295–296
– metal hydrides 275–327
– switchable mirrors 284, 306–308
– tailoring 302–303
orbitals, overlapping 110–111
organic hydrides 237–244
Organization of Petroleum Exporting Countries (OPEC) 24
ortho-hydrogen (o-hydrogen) 73–74

Otto-cycle engine 376–378
oxides, metal 102–104

P

PAFC *see* phosphoric acid fuel cell
Paracelsus 7
para-hydrogen (p-hydrogen) 73–74
– low temperature properties 76
– triple point 77
paramagnetic layers 268, 270
partition functions, H_2 and H 118
PEFC *see* PEMFC
Peierls distortion models 290–291
PEM (proton exchange membrane) fuel cell 61
PEMFC 127–128, 344, 364, 366
– automobile applications 249
– hydrogen dissociation 134
– hydrogen storage device 250–251
– thermodynamics 365
periodic table of the elements 87, 189
phase diagram, hydrogen 76, 166
phosphoric acid fuel cell (PAFC) 344–345, 364
photo-biological hydrogen production 53
photo-electrochemical hydrogen production 52–53
photosynthesis 2–3, 237–238
– hydrogen storage 238–240
– overall chemical reaction 238
physisorption 98–99, 173–175
pixel-by-pixel switching 294
Planck spacecraft 393
p-metal hydrides 203–204
point defects, STM images 103
polarization, cell 368
polycrystalline thin films, stress 282–283
polymer electrolyte fuel cell (PEFC) *see* PEMFC
polymer electrolyte membrane fuel cell *see* PEMFC
polymeric hydrides 203–204
polynuclear complexes 198–199
portable applications, fuel cells 354–355
potential
– chemical *see* chemical potential
– fuel cell electrodes 339
– ionization 89
– Lennard-Jones *see* Lennard-Jones potential
potential energy
– chemisorption 96
– hydrogen approaching metal surfaces 113

– hydrogen molecule on metal surface 109
– interaction 73
powder patterns 225, 227
power and heating, fuel cells 62–63
power density 368, 372
power trains 357–359
– hydrogen 380
precoverage, surface 105–107
pressure, ad-/desorption 219
pressure vessels, nickel–hydrogen batteries 397, 400–401
pressure–composition isotherm 216–217, 299–300, 304
primary energy consumption 25–26
production
– fossil fuels 29–33
– hydrogen 40–43
projection, H_2 states 112
propane, combustion and explosion properties 91
propulsion system
– dual mode nuclear 407
– spacecrafts 385
protide 129–131, 245
– $NaBH_4$ synthesis 133–134, 254
protium 128–129
proton 128
proton exchange membrane (PEM) fuel cell 61
proton–proton separation 73
pulsed laser deposition 279

Q

quaternary 3d transition metal hydrides 198

R

Raney-nickel 158
rare-earth (RE) hydride thin films 277–279
– *in situ* deposition 279
– switchable mirrors 283
reabsorption, catalyzed 220–223
reactants, fuel cells 343
reactions
– Boudouard reaction 150–151
– Calvin–Benson cycle 239
– dissociation of hydrogen molecule 89
– equilibrium constants 152
– Fischer–Tropsch process 155
– homogenous water gas reaction 150
– hydrogen production 149–154
– hydrogen/oxygen fuel cell 339
– hydrogenation and dehydrogenation 129
– nickel–hydrogen batteries 396

- photosynthesis   238
- Sabatier   406
- solid-state   196
- soot reactions   150–151
- yield   153

reactions, surface   107–108
reactivity, surfaces   136
recombination, catalyzed   108–125
reflection
- hydrogen concentration dependence   285
- Intensity map   310

reflection spectra, switchable mirrors   303, 308
reformation, natural gas   49–50
renewable carbon resources   42–43
renewable energy
- electricity   162
- electrolysis   161–163

renewable hydrogen production   4, 50–55
reprocessing, $NaBH_4$ 253–255
reserves, fossil fuels   30–31
resources, renewable carbon   42–43
reversible doped $NaAlH_4$ systems   225–231
reversible hydrogen storage
- carbon hydrides   241–242
- materials   212

reversible hydrogenation/dehydrogenation   236
Rochow see Allred-Rochow electronegativity
rocket fuel   382–388

S

Sabatier reaction   406
safety
- complex hydrides   231–232
- fire and explosive hazards   93
- hydrogen economy   65–66

saturation capacity, hydrogen storage   183
saturation field, magnetic   272
scaling factors, hydrogen production   54
scanning tunnel microscopy (STM), point defect images   103
seasonal hydrogen system   242
secondary batteries, Ni–MH   134
segregation
- enthalpy   100
- surface   98–101

self-organized layering   309
sensors   357
- hydrogen   325–326

shift reaction, biomass   41–42
shutter effect, microscopic   296–298
Sievert's method   177

silicate structures, hydrogen storage   184–186
single pressure vessel (SPV)   397, 400–401
smart windows   324
sodium aluminum hydrides   213–233, 225–231
sodium borohydride   133–134, 253–255
SOFC see solid oxide fuel cell
solid hydrogen, density   173
solid hydrogen cryostats   392
solid oxide fuel cell (SOFC)   345–346, 349, 359
solid polymer electrolyte fuel cell (SPEFC) see PEMFC
solid solutions   191
solid surfaces, interaction with hydrogen   94–108
solid–liquid interface   249–250
solid-polymer electrolysis (SPE)   160
solid-state reactions   196
solubility, borohydrides   246–247
solutions
- alkaline   245–246
- crystallization   248
- heat of   99

soot reactions   150–151
sorption cryocoolers   390–391
sorption mechanism, tetrahydroborates   208–211
space applications
- fuel cells   361
- hydrogen   381–410
- nickel–hydrogen batteries   394–404

space launch vehicles, hydrogen powered   384
space shuttle   16–20, 386
- Buran   385
- hydrogen and oxygen storage systems   387
- hydrogen storage   404–405
- main engine   388

space suit cooling, metal hydrides   405
spark ignition engine   373–379
sp-band   110–111
SPE see solid-polymer electrolysis
specific conductivity, fuel cell electrolytes   347
specific heat ratio   378
specific surface area   183
spectra
- reflection   303, 308
- thermal desorption   208
- TPD   115, 119, 123
- transmission   303, 308

SPEFC see PEMFC
sputtering   279, 319
SPV see single pressure vessel , 398
stacks, fuel cells   348–349
stationary applications, fuel cells   358–359
sticking coefficient   116–117
– molecular hydrogen on Cu(110)   120
sticking probability   116–117
Stirling approximation   119
STM see scanning tunnel microscopy
storage, hydrogen see hydrogen storage
storage capacity   183, 214
– Ti-doped sodium aluminum hydrides   223–224
storage vessels, liquid hydrogen   171–172
stress, in thin films   282–283
submarines, fuel cell propulsion system   360
substrate–adsorbate distance   176
sulfur cycle   48
superlattices, Fe/V see Fe/V superlattices
surface area, specific   183
surface engineering, hydrides   132–138
surface interactions   132–133
surface precoverage   105–107
surface reactions, relevance for bulk absorption   107–108
surface reactivity   136
surface segregation   98–101
– binary alloys   100
surface structure   97–98
– nanosized   136–137
surfaces
– defect-free   104
– fluorinated   137–138
– interaction with hydrogen   94–108
– metal   101–102
switchable absorbers   324
switchable mirrors   275–327
– all-solid-state   323
– applications   311–321
– diffusion and electromigration   312–314
– electrical properties   286–288, 305–306
– epitaxial   294–295
– first-generation   283
– liquid electrolyte   296
– optical properties   306–308
– reflection and transmission spectra   303, 308
– second generation   295–303
– theoretical models   290–294
– thin film   277–283
– third generation   303–311
– transmission   281
syngas   154

T

T metal see transition metal . . .
tailoring
– optical properties   302–303
– thermodynamic   232–233
TDS see thermal desorption spectroscopy
temperature dependence, conductivity   287
temperature programmed desorption (TPD)
– hydrogen desorption   114–115
– spectra   114–115, 119, 123
temperature range, fuel cells   343
temperatures
– cryogenic see cryogenic temperatures
– decomposition   207
– high see high temperatures
– low see low temperatures
terminal hydrogen ligands   196
ternary hydride phase, homogeneous   229
ternary 3d transition metal hydrides   198
tetrahydroborates
– decomposition temperature   207
– sorption mechanism and kinetics   208–211
– stability   206–208
– structure   205–206
thermal conversion, biomass   41–42
thermal desorption spectroscopy (TDS)   178
thermochemical hydrogen production   51–52
thermochromic devices   323
thermodynamic efficiency   378–379
thermodynamic tailoring   232–233
thermodynamics
– adsorbed hydrogen   98–101
– borohydride fuel cells   365–366
– fuel cells   338–339
– hydrogen   77–81
– PEMFC   365
– Ti-doped sodium aluminum hydrides   215–220
thermogravimetry   178
thermophysical hydrogen production   52
thin film coating   105–107
thin films
– magnesium–rare earth   295–303
– magnesium–transition metal   303–311
– rare-earth hydride see rare-earth hydride thin films

- stress 282–283
- switchable mirrors 277–283
thin films, complex metal hydrides 283
Thomson, William *see* Kelvin
titanium-doped sodium aluminum hydrides 213–233
- decomposition 223
- thermodynamic properties 215–220
torque, engine 376
toxicology, hydrogen 93
TPD *see* temperature programmed desorption
transition metal hydrides
- bonding 200–201
- complex 195–203, 237
- hydrogen storage properties 202
- ternary and quaternary 3d 198
transmission
- hydrogen 55–56
- hydrogen concentration dependence 281, 285
- hysteresis 289
transmission spectra, switchable mirrors 303, 308
trihydride state 281–282, 285
triple point, p- and n-hydrogen 77
tritium 12

U
unloading, hydrogen 266

V
Van der Waals equation 79, 167
- coefficients 80
Van der Waals force 79
Van't Hoff diagram, dissociation 218
Van't Hoff equation 192
Van't Hoff plot 194
vapor pressure, p-hydrogen 76
vermicular excrescence 230
vessels
- liquid hydrogen 171–172
- pressure 397, 400–401
viscosity, borohydrides 247
visualization, hydrogen diffusion 314–316
volumetric density, compressed hydrogen gas 169
volumetry 177

W
water
- electrolysis 155–157
- exhaust 244–256
- global cycle 46–47
- hydrogen source 246
water gas reaction, homogenous 150
wavefunction, hydrogen molecule 72, 74
windows, smart 324

X
yield 153

Y
yttria-stabilized zirconia (YSZ) 345–346

Z
zeolites, hydrogen storage 184–186
zeppelin 14–16
zero emission vehicles (ZEV) 61–62, 359
- historic 14
z-scheme, photosynthesis 239